Methods of Environmental Impact Assessment

Second Edition

The Natural and Built Environment Series

Editors: Professor John Glasson and Professor Mick Bruton

Introduction to Environmental Impact Assessment: 2nd Edition
John Glasson, Riki Therivel and Andrew Chadwick

Public Transport: 3rd Edition
Peter White

Urban Planning and Real Estate Development
John Ratcliffe and Michael Stubbs

Landscape Planning and Environmental Impact Design: 2nd Edition
Tom Turner

Controlling Development
Philip Booth

Partnership Agencies in British Urban Policy
Nicholas Bailey, Alison Barker and Kelvin MacDonald

Planning, the Market and Private House-Building
Glen Bramley, Will Bartlett and Christine Lambert

British Planning Policy in Transition
Mark Tewdwr-Jones

Development Control
Keith Thomas

Forthcoming:

Urban Regeneration: A critical perspective
Sue Brownill and Neil McInroy

Methods of Environmental Impact Assessment

Second Edition

Edited by Peter Morris and Riki Therivel

Spon Press
Taylor & Francis Group

LONDON AND NEW YORK

First published 1995 by UCL Press
Reprinted 2000

Second edition published 2001
by Spon Press
11 New Fetter Lane, London EC4P 4EE

Simultaneously published in the USA and Canada
by Spon Press
29 West 35th Street, New York, NY 10001

Reprinted 2003

Spon Press is an imprint of the Taylor & Francis Group

Typeset in Goudy by Graphicraft Limited, Hong Kong

Printed and bound in Great Britain by St Edmundsbury Press, Bury St Edmunds, Suffolk

British Library Cataloguing in Publication Data
A catalogue record for this book is available from the British Library

Library of Congress Cataloging in Publication Data
Methods of environmental impact assessment / edited by Peter Morris and Riki
Therivel. – 2nd ed.
    p. cm. – (The natural and built environment series)
  Includes bibliographical references and index.
   1. Environmental impact analysis – Great Britain.   2. Environmental impact
analysis – European Union countries.   I. Title: Environmental impact assessment.
II. Morris, Peter, 1934–   III. Therivel, Riki, 1960–   IV. Series.

TD194.68.G7 M48   2001
333.7′14 – dc21   00-068766

ISBN 0-415-23958-3 (hb)
ISBN 0-415-23959-1 (pb)

Contents

Appendices

List of contributors

Jeremy Biggs, Gill Fox, Pascale Nicolet, Mericia Whitfield and Penny Williams work for the Ponds Conservation Trust Policy and Research (PCTPR) division based at Oxford Brookes University.

Rosemary Braithwaite works in the Archaeology Team of Hampshire County Council's Planning Department.

Mike Breslin is Principal at ANV, an acoustic and noise consultancy based in Milton Keynes.

Andrew Brookes is the Options Appraisal Manager at the National Centre for Risk Analysis and Options Appraisal, The Environment Agency.

Greg Callaghan is a Transport Planner at Peter Brett Associates.

Andrew Chadwick is a Research Associate with the Impacts Assessment Unit in the School of Planning, Oxford Brookes University.

Kelly P. Davis is a GIS/RS Professor in the Postgraduate GIS Program, Niagara College, Ontario, Canada.

Derek M. Elsom is a Professor in the Geography Department, Oxford Brookes University, and a member of the university's Impacts Assessment Unit.

Roy Emberton is a Technical Director of WSP Environmental Limited.

John Glasson is Professor of Planning and Head of the School of Planning at Oxford Brookes University, and the Research Director of the university's Impacts Assessment Unit. He is also Pro Vice Chancellor of Research.

Philip Grover is a Senior Lecturer in Historic Conservation in the School of Planning, Oxford Brookes University.

Martin J. Hodson is a Principal Lecturer in Environmental Biology in the School of Biological and Molecular Sciences, Oxford Brookes University.

David Hopkins works in the Archaeology Team of Hampshire County Council's Planning Department.

John Lee is a Database Manager/Research Assistant in the School of Biological and Molecular Sciences, Oxford Brookes University.

Peter Morris was a Principal Lecturer (now retired) in Ecology in the School of Biological and Molecular Sciences, Oxford Brookes University.

Jeremy Richardson is an Environmental Consultant at Scott Wilson (London Office).

Agustin Rodriguez-Bachiller is a Senior Lecturer in Quantitative Methods in the School of Planning, Oxford Brookes University.

Chris Stapleton is Managing Director of the Bell Cornwell Partnership Environmental Consultants Ltd.

Riki Therivel is a Director at CAG Consultants. She is also a Visiting Professor in Environmental Impact Assessment in the School of Planning, Oxford Brookes University.

Stewart Thompson is a Senior Lecturer in Ecology in the School of Biological and Molecular Sciences, Oxford Brookes University.

David Thurling is a Principal Lecturer in Environmental Biology in the School of Biological and Molecular Sciences, Oxford Brookes University.

Graham Wood is a Senior Lecturer in Environmental Impact Assessment and Quantitative Methods in the School of Planning, Oxford Brookes University.

Preface and acknowledgements

The idea of a book on methods of environmental impact assessment first arose during the writing of the first edition of *Introduction to Environmental Impact Assessment* (Glasson et al. 1994). We realised that very few books existed on how EIA should be carried out for specific environmental components such as air, flora and fauna, or socio-economics, and that none was written for the UK/EU context. Since then, *Introduction* has gone through a second edition, and the first edition of *Methods of Environmental Impact Assessment* has become more dated than we would like. Together with *Introduction*, this new edition aims to provide a comprehensive coverage of the theory and practice of EIA in the UK and EU in the early 2000s.

The book is aimed at people who organise, review, and make decisions about EIA; at environmental planners and managers; at students taking first degrees in planning, ecology, geography and related subjects with an EIA content; and at postgraduate students taking courses in EIA or environmental management. It explains what the major concerns of the EIA component specialists are, how data on each environmental component are collected, what standards and regulations apply, how impacts are predicted, what mitigation measures can be used to minimise or eliminate impacts, what some of the limitations of these methods are, and where further information can be obtained. It does not aim to make specialists out of its readers; to do so would require at least one book per environmental component. Instead it aims to foster better communication between experts, a better understanding of how EIAs are carried out, and hopefully better EIA-related decisions.

Like its sister volume, this book emphasises best practice – what ideally should happen – as well as minimal regulatory requirements. EIA is a constantly evolving and improving process. If the trends of the last two decades continue, today's EIA best practice will be tomorrow's minimal regulatory requirement.

The basis of this book is a unit on Oxford Brookes University's MSc course in Environmental Assessment and Management, entitled Methods of Environmental Impact Assessment. The unit is taught by a range of university staff and outside specialists who have practical expertise in EIA. Most of the chapters in the first edition were written by the people who originally taught on the course. In this edition, the original chapters have been updated or rewritten by (often a combination of) the original authors, staff now teaching the unit, and outside practitioners. There is also a new second part, consisting of chapters on "cross-cutting" methods that can be applied to, and can often facilitate integration between, many of the environmental components discussed in the first part.

We are very grateful to all the authors, original and new, for their excellent contributions. We no longer run either the MSc course or the Methods unit: Peter is in retirement and Riki works for CAG Consultants and so teaches less. However, the course and unit have been taken over very effectively by Elizabeth Wilson, Stewart Thompson and Graham Wood. We are also grateful to John Glasson for his continuing support. Finally, we are grateful for the help of Roger Barrowcliffe (Environmental Resources Management, London) who also provided Figures 8.1 and 8.2; ESRI UK for permission to use Figure 16.1; and Derek Whitely and Rob Woodward (both of Oxford Brookes University) for the line drawings.

Notes on text format

Key words or phrases are highlighted in **bold** at appropriate points in the text, e.g. in paragraphs in which they are explained.

Terms highlighted in ***bold italics***, at least the first time they appear in a chapter, are defined in the glossary.

Reference is necessarily made to numerous acronyms, e.g. of relevant organisations. Where the full names of acronyms are not given in the text, they can be found in Appendix A (together with the meanings of chemical symbols and of quantitative units and symbols).

Part I

Methods for environmental components

1 Introduction

Riki Therivel and Peter Morris

1.1 EIA and the aims of the book

This book aims to improve practice of environmental impact assessment (EIA) by providing information about how EIAs are, and should be, carried out. Although it focuses on the UK context in its discussion of policies and standards, the techniques it discusses apply universally. This introductory chapter (a) summarises the current status of EIA, and the legislative background in the UK and EU, (b) explains the book's structure, and (c) considers some trends in EIA methods.

Formal EIA can be defined as "a process by which information about the environmental effects of a project is collected, both by the developer and from other sources, and taken into account by the relevant decision making body before a decision is given on whether the development should go ahead" (DoE 1995). It can also be defined more simply as "an assessment of the impacts of a planned activity on the environment" (UNECE 1991). In addition to the decision on whether a project should proceed, an EIA will consider aspects such as ***project alternatives*** and mitigation measures that should be implemented if the development is allowed. The findings of an EIA are presented in a document called an Environmental Statement *or* (as in this book) ***Environmental Impact Statement*** (EIS). The overall EIA process is explained and discussed in this book's 'sister volume' *Introduction to environmental impact assessment* (Glasson *et al.* 1999).

EIAs involve individual assessments of aspects of the environment (e.g. population, landscape, heritage, air, climate, soil, water, fauna, flora) likely to be significantly affected by a proposed project. This book focuses on assessment methods (practical techniques) used in the part of the EIA process concerned with analysing a development's impacts on these ***environmental components***.

1.2 The EIA process

1.2.1 Introduction

Figure 1.1 summarises the main EIA procedures that will be followed in the assessment of any environmental component. The figure assumes that the developer has conducted feasibility studies, and that *screening* has already been carried out – and these assumptions are made in the chapters. Screening is discussed in Glasson *et al.* (1999).

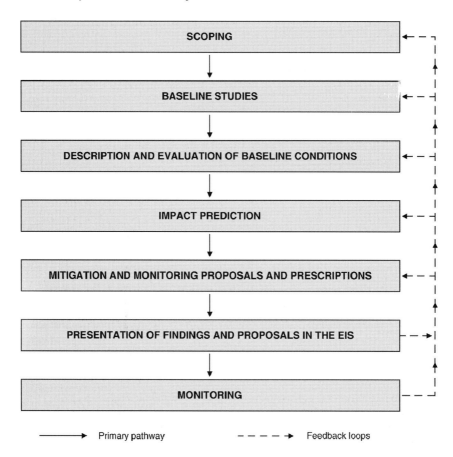

Primary pathway   – – – – ➔ Feedback loops

The model illustrates the stepwise nature of EIA, but also the requirement for continuous reappraisal and adjustment (as indicated by the feedback loops).

Figure 1.1 Procedures in the assessment of an environmental component for an EIA.

1.2.2 Scoping and baseline studies

Scoping is an essential first step in the assessment of a component. The main aims are:

- to identify at an early stage (when the project design is relatively amenable to modification) what key *receptors*, impacts and project alternatives to consider, what methodologies to use, and who to consult. UK Government policy also advocates an appraisal-led design process, and various documents (e.g. MAFF/ WO 1996) provide guidance on identifying the **preferred option** from an environmental perspective;
- to ensure that resources and time are focused on important impacts and receptors;
- to establish early communication between the developer, consultants, statutory consultees and other interest groups who can provide advice and information;
- to warn the developer of any constraints that may pose problems if not discovered until later in the EIA process.

The scoping exercise should provide a ground plan for subsequent steps by making a preliminary assessment of:

- the project's **potential impacts** on component receptors, estimated from the project description (including its size, construction requirements, operational features and secondary developments such as access roads) and the nature of components and receptors;
- the **impact area/zone** within which impacts are likely to be effective, estimated from the impact types and the nature of the surrounding area and environmental components, e.g. impacts on air or water may be effective at considerable distances from the project site;
- possible **mitigation measures**;
- the need and potential for **monitoring**;
- the **methods and levels of study** needed to obtain reliable baseline information that can be used to evaluate the baseline conditions, make accurate impact predictions, and formulate adequate mitigation measures and monitoring procedures. The selection of methods should involve consideration of:

    * the impact and receptor variables on which the studies will focus, and the accuracy and precision needed for each;
    * the most appropriate methods for collecting, analysing and presenting information;
    * the resource requirements and timing considerations, especially for field surveys;
    * constraints such as the time and resources available.

Scoping checklists (see Table 1.1, p. 7) are a useful scoping tool, particularly for tasks such as identifying key impacts and receptors, and selecting appropriate consultees and interest groups. The findings of the scoping exercise should be documented in a **scoping report** that is made available to the developer, participating consultants, and consultees. However, lack of detailed information at the scoping stage means that scoping estimates and decisions should be reassessed in the light of baseline information gained as the EIA progresses.

Baseline studies form the backbone of component assessments. It is only when they provide sound information on the socio-economic or environmental systems in the impact area that valid impact predictions can be made, and effective mitigation and monitoring programmes formulated.

The distinction between baseline studies and scoping is not clear cut because (a) consultation should be ongoing, and (b) scoping includes gathering information, much of which is effectively baseline material that can at least form the starting point for more detailed studies. In both stages, it is usually possible to compile some of the required information, by means of a **desk study**. A thorough search should be made because (a) gathering existing information is generally less expensive and time-consuming than obtaining new data, and (b) it is pointless to undertake new work that merely duplicates information that already exists. However:

- Scoping will usually require brief site visits (e.g. for reconnaissance or to confirm features identified on maps) – perhaps including walkover surveys. Such initial visits are best undertaken by several members of the EIA and design team, so that relationships between components can be identified.

- In most cases, existing baseline data will be inadequate or out of date, and it will be necessary to obtain new information by some form of **field survey**.

The **description and evaluation of baseline conditions** should include:

- a clear presentation of methods and results;
- indications of limitations and uncertainties, e.g. in relation to data accuracy and completeness;
- an assessment of the value of key receptors and their sensitivity to impacts.

1.2.3 Impact prediction

Impact prediction is fundamental to EIA, and the likely impacts of a project should be considered for all environmental components. In order to predict the impacts of a development it is also necessary to consider changes in the baseline conditions that may occur in its absence (a) prior to its initiation, which can be several years after production of the EIS, and (b) during its projected lifetime. These can be assessed in relation to the current baseline conditions and information on past, present and predicted conditions and trends. Most of the relevant information will have to be sought through the desk study although comparison of field survey data with previous data can help to elucidate recent trends. Appendix C gives sources of historical information.

According to the EIA legislation (§1.4) impact prediction should include assessment of:

- **Direct/primary impacts** – that are a direct result of a development.
- **Indirect/secondary impacts** – that may be 'knock-on' effects of (and in the same location as) direct impacts, but are often produced in other locations and/or as a result of a complex pathway.
- **Cumulative impacts** – that accrue over time and space from a number of developments or activities, and to which a new project may contribute.

An additional possibility is **impact interactions** – between different impacts of a project, or between these and impacts of other projects – that result in one or more additional impacts, e.g. $(A + B) \rightarrow C$.

All impacts may be **positive** (beneficial) or **negative** (adverse), **short, medium,** or **long term, reversible** or **irreversible,** and **permanent** or **temporary.**

Ideally, impact prediction requires:

- a good understanding of the nature of the proposed project, including project design, construction activities and timing;
- knowledge of the outcomes of similar projects and EIAs, including the effectiveness of mitigation measures;
- knowledge of past, existing or approved projects which may cause interactive or cumulative impacts with the project being assessed;
- predictions of the project's impacts on other environmental components that may interact with that under study;
- adequate information about the relevant receptors, and knowledge of how these may respond to environmental changes/disturbances.

Table 1.1 Commonly used aids to impact identification and prediction

Method	Features
Checklists	Useful for identifying key impacts and ensuring that they are not overlooked, especially in scoping. Can include information such as data requirements, study options, questions to be answered, and statutory thresholds – but not generally suitable for detailed analysis.
Matrices	Mainly used for impact identification, but provide the facility to show cause–effect links between impact sources (plotted along one axis) and impacts (plotted along the other axis). They can also indicate features of impacts such as their predicted magnitudes and whether they are likely to be localised or extensive, short or long term, etc.
Flowcharts and networks	Can be useful for identifying cause–effect relationships/links/pathways: between impact sources; between sources and impacts; and between primary and secondary impacts.
Mathematical/ statistical models	Based on mathematical or statistical functions which are applied to calculate deterministic or probabilistic quantitative values from numerical input data. They range from simple forms, that can be employed using a calculator or computer spreadsheet, to sophisticated computer models that incorporate many variables. They need adequate/reliable data, can be expensive, may not be suitable for 'off the peg' use. The results usually require validation.
Maps and GIS	Maps can indicate feature such as impact areas, and locations and extents of receptor sites. Overlay maps can combine and integrate two or three 'layers', e.g. for different impacts and/or environmental components or receptors. GIS can analyse a number of layers, and has facilities for the input and manipulation of quantitative data, including modelling (§16.5).

Methods of impact prediction vary both between and within EIA components. For example, the assessment of impact **magnitude** (severity) may be qualitative or quantitative. Qualitative assessments usually employ ratings such as *neutral, slight, moderate, large* – applied to both negative and positive impacts. Quantitative assessments involve the measurement or calculation of numerical values, e.g. of the level of a **pollutant** in relation to a statutory threshold value.

There are several **standard techniques** that can be used to aid impact prediction in assessments of most environmental components. These are reviewed in Glasson *et al.* (1999) and briefly summarised in Table 1.1. In addition, increasing use is being made of environmental risk assessment (ERA). This also employs statistical modelling, and techniques such as **event tree analysis** (§14.4.1) which is a form of flowchart analysis. ERA is particularly relevant to the prediction of impacts from accidents, and is embodied, for example, in the *Control of Major Accident Hazards (COMAH) Regulations* (SI 1999/743) which implement the EU COMAH Directive 96/82/EC on the control of major accident hazards involving dangerous substances (see HSE & EA/SEPA 1999).

An additional task in impact prediction is the assessment of **impact significance**, which is the 'product' of an impact's characteristics (magnitude and extent in space and time) and the value, sensitivity/fragility and recoverability of the relevant receptor(s). It therefore requires an evaluation of these receptor attributes – which should have been carried out in the baseline evaluation.

Impact prediction is often poorly addressed, perhaps because it is the most difficult step in EIA. Direct impacts are usually relatively easy to identify, but accurate prediction of indirect and cumulative impacts can be much more problematic. Three useful guides on assessment of cumulative impacts are USCEQ (1997), CEAA (1999) and EC (1999), all of which are available online from the relevant websites, which also provide other EIA information and links.

Whatever methods are employed, impact prediction is not an exact science. There are bound to be uncertainties (that can sometimes be expressed as ranges) which should be clearly stated in the EIS.

1.2.4 Mitigation

Mitigation measures aim to avoid, minimise, remedy or compensate for the predicted adverse impacts of the project. They can include:

- selection of alternative production techniques, and/or locations or alignments (of linear projects);
- modification of the methods and timing of construction;
- modification of design features, including site boundaries and features, e.g. landscaping;
- minimisation of operational impacts, e.g. pollution and waste;
- specific measures, perhaps outside the development site, to minimise particular impacts;
- measures to compensate for losses, e.g. of amenity or habitat features.

Much of the environment damage caused by developments occurs during the **construction phase**, and a problem is that construction is usually contracted to a construction company who will not have participated in the EIA process, and over whom the developer may have little control (Wathern 1999). Consequently, there is a need to provide **construction phase management plans**, ideally as part of overall project environmental management plans (see §1.6). In addition, because project specifications frequently change between publication of the EIS and the start or completion of construction (often for unforeseeable reasons) developers sometimes employ site environmental managers to ensure (a) that such modifications take account of environmental considerations, and (b) that construction phase mitigation measures are carried out.

Different mitigation measures will be needed in relation to specific impacts on different environmental components and receptors. The EIS should provide detailed prescriptions for the proposed measures (that clearly relate to specific impacts), indicate how they would actually be put in place, and propose how they might be modified if unforeseen post-project impacts arise. A primary consideration is the likely significance of post-mitigation **residual impacts**, and care is needed to ensure that a mitigation measure does not generate new impacts, perhaps on receptors in other environmental components.

Best practice dictates that the **precautionary principle** (advocated in EU and UK environmental policy) should be applied, i.e. that mitigation should be based on the possibility of a significant impact before there is conclusive evidence that it will occur. Similarly, on the basis of the EU principles that preventive action is preferable to remedial measures, and that environmental damage should be rectified at source (see §1.4), the best mitigation measures should involve modifications to the project rather than containment or repair at receptor sites, or compensatory measures such as habitat creation – which should normally be considered only as a last resort (see §11.6.3).

In addition to mitigation, government guidelines suggest that opportunities for **environmental enhancement** (improvement of current environmental conditions and features) should be sought in EIA. For instance, this is one of the duties of the Environment Agency (EA), especially in relation to coastal and flood defences (MAFF 1999).

1.2.5 Presentation of findings and proposals in the EIS

The information presented in the EIS must be clear and, at least in the non-technical summary, should be in a form that can be understood by 'non-experts' without compromising its integrity. It should also be 'transparent', e.g. in relation to limitations and uncertainties. Presentation methods vary between components, but can include the use of maps, graphs/charts, tables and photographs.

The EIS must be an integrated document, and this will necessitate assessing the component in relation to others, e.g. to evaluate its relative importance, and ensure that potential conflicts of interest have been addressed (see §1.6).

1.2.6 Monitoring

Monitoring can be defined as the continuous assessment of environmental or socio-economic variables by the systematic collection of specific data in space and time. It can be strictly continuous, e.g. using recording instruments, but more commonly involves periodic repeat data collection, usually by the same or similar methods as in baseline surveys. Monitoring in EIA can include:

- **Baseline monitoring** – which may be carried out over seasons or years to quantify ranges of natural variation and/or directions and rates of change, that are relevant to impact prediction and mitigation. This can avoid the frequent criticism that baseline studies are only 'snapshots' in time. However, time constraints in EIA usually preclude lengthy survey programmes, and assessments of long-term trends normally have to rely on existing data.
- **Compliance monitoring** – which aims to check that specific conditions and standards are met, e.g. in relation to emissions of pollutants.
- **Impact and mitigation monitoring** – which aims to compare predicted and actual (residual) impacts, and hence to determine the effectiveness of mitigation measures.

Unless otherwise specified, 'monitoring' in EIA normally refers to impact and mitigation monitoring, which is also sometimes called auditing. There is often considerable uncertainty associated with impacts and mitigation measures, and it is responsible best practice to undertake monitoring during both the construction and post-development phases of a project. Monitoring is essential to learn from both successes and failures. For example:

- It is the only mechanism for comparing predicted and actual impacts, and hence of checking whether mitigation measures have been put in place, testing their effectiveness, and evaluating the efficiency of the project management programme;
- If mitigation measures are amenable to modification, it should still be possible to reduce residual impacts identified during monitoring (feedback loop in Fig. 1.1);
- It can provide information about responses of particular receptors to impacts;
- It is the only means of EIA/EIS evaluation and of identifying mistakes that may be rectified in future EIAs. For example, it will provide information that can be used to assess the adequacy of survey and predictive methods, and how they may be improved. Thus, a principal aim of monitoring should be to contribute to a cumulative database that can facilitate the improvement of future EIAs (Clark 1996).

Monitoring is not strictly part of the EIA process, is not statutory in the UK, and can be expensive. Consequently, in spite of government guidance that it should be undertaken (e.g. MAFF/WO 1996), lack of monitoring is a serious deficiency in current EIA practice.

1.3 The current status of EIA

Since the first EIA system was established in the USA in 1970, EIA systems have been set up worldwide and have become a powerful environmental safeguard in the project planning process. In Europe, EU Directive 85/337/EEC and amending Directive 97/11/EC on EIA (EC 1985, 1997) set the legal basis for individual member states' EIA regulations. More than 300 EISs are currently prepared annually in the UK alone, and the new amendments will undoubtedly further increase this number.

Fifteen years after Directive 85/337 became operational, most of the parties involved in the EIA process in Europe are becoming experienced: many environmental consultancies have prepared some form of EIS; most competent authorities have received several, and environmental groups and statutory consultants are becoming increasingly adept at using EIA as a tool for environmental protection, although shortages of resources often restrict their input to the EIA process. EIA quality is improving, but is often only just satisfactory (Glasson *et al.* 1999). Less positively, most members of the public have never seen an EIS, and it is unclear to what extent EIAs are used in decision-making.

In recent years, new tools, techniques and approaches have been developed which complement and support the EIA process. For example:

- Mapping software and geographical information systems (GIS) (Chapter 16) now allow much more effective analysis and presentation of information than in the past;
- There is a rapid expansion in the range and availability of information databases, including remote sensed data (Chapter 15) and other digital data suitable for GIS;

- The Internet now provides ready access to a wealth of information, including legislation documents and other publications, databases and software;
- In ecology and landscape analysis, although legislation and government guidelines still focus on protecting designated areas, there is a shift from "save the best and leave the rest" to consideration of the 'wider countryside' and characterisation of areas, with the aim of promoting their uniqueness and joint diversity (CC & EN 1998);
- More emphasis is being placed on environmental enhancement, not just mitigation of negative impacts;
- Although monitoring is still not mandatory, it is being encouraged in government guidelines;
- Evolving approaches to public participation – for instance, 'visioning' conferences and community mapping exercises – allow local residents' perceived impacts to be better understood and taken into consideration in EIA;
- Strategic environmental assessment (SEA) – which is EIA of the local, regional and national policies, plans and programmes that set the context of project-based EIAs – is becoming more common.

1.4 EIA legislation

Several important internationally accepted principles underlie the recent rapid growth in EIA and SEA. The World Commission on Environment and Development espoused the principle of **sustainable development** in its report of 1987 (WCED 1987), and this was further elucidated at the UN Conference on Environment and Development (UNCED) 1992 – the 'Rio Earth Summit' (Quarrie 1992). The UK Government endorsed this and has promoted sustainable development, e.g. DETR (1999a). The European Commission promotes sustainability in its *Fifth Action Programme on the Environment* (EC 1993), defining "features of sustainability" as being:

- to maintain the overall quality of life;
- to maintain continuing access to natural resources;
- to avoid lasting environmental damage;
- to consider as sustainable a development that meets the needs of the present without compromising the ability of future generations to meet their needs.

In particular, the Programme stresses that:

- preventive action is preferable to remedial measures;
- environmental damage should be rectified at the source;
- the polluter should pay the costs of measures taken to protect the environment;
- environmental policies should form a component of other European policies.

The new *sixth environment action programme* (EC 2001) also promotes a strategic approach to environmental protection, and EIA is an example of this approach.

EU Directives 85/337/EEC and 97/11/EC require that, for a specified list of project types (Annex I of Directive 97/11), EIA *must* be carried out. EIA *may* be carried out for projects in another list (Annex II), depending on the characteristics and location of the project, and the characteristics of the potential impacts (Annex III). The required contents of the EIS are given in Annex IV. These are:

1. Description of the project, including in particular:

    • a description of the physical characteristics of the whole project and the land-use requirements during the construction and operational phases;
    • a description of the main characteristics of the production processes, for instance, nature and quantity of the materials used;
    • an estimate, by type and quantity, of expected residues and emissions resulting from the operation of the proposed project.

2. An outline of the main alternatives studied by the developer and an indication of the main reasons for this choice, taking into account the environmental effects.
3. A description of the aspects of the environment likely to be significantly affected by the proposed project, including, in particular, population, fauna and flora, soil, water, air and climate, material assets, including the architectural and archaeological heritage, landscape and the inter-relationship between the above factors.
4. A description of the likely significant effects of the proposed project on the environment resulting from:

    • the existence of the project,
    • the use of natural resources,
    • the emission of pollutants, the creation of nuisances and the elimination of waste, and
    • the description by the developer of the forecasting methods used to assess the effects on the environment.

5. A description of the measures envisaged to prevent, reduce and where possible offset any significant adverse effects on the environment.
6. A non-technical summary of the information provided under the above headings.
7. An indication of any difficulties (technical deficiencies or lack of know-how) encountered by the developer in compiling the required information.

Directive 97/11/EC (EC 1997), which became operational on 14 March 1999, expanded the requirements of Directive 85/337/EEC by:

• requiring EIA for a wider range of projects, and upgrading of some Annex II projects to Annex I status;
• giving criteria (including the concept of **"sensitive environments"** and a list of specified types of sensitive environments) for choosing which Annex II projects require EIA;
• strengthening the procedural requirements concerning transboundary impacts (where pollution from one country affects another country);
• requiring developers to include an outline of the main alternatives that they studied and explain the reasons for the final choice between alternatives;
• allowing developers to request an opinion from the *competent authority* on the scope of an EIA;
• requiring competent authorities to make public the main reasons on which project decisions are based and the main mitigation measures required.

Clearly, these amendments will affect, and improve, EIA practice in the EU.

In the UK, EIA Directives are implemented by about 40 regulations – mainly Statutory Instruments (SIs). The core regulations are the *Town and Country Planning (Environmental Impact Assessment) (England and Wales) Regulations 1999* (HMSO 1999) and the equivalent regulations in Scotland (SE 1999a). The requirements of each regulation differ slightly, but all are essentially variants of the core regulations. Schedules 3 and 4 of the regulations (which are equivalent to Annexes III and IV of the EC Directive) are particularly relevant to this book.

Government guidance on the EIA procedures is given in DETR (1999b, 2000) and SE (1999b), and additional guidance on the preparation of EISs is given in DoE (1995). Guidance at the EU level is available from the EC-EDG website (Appendix A). EIA procedures are further discussed in Glasson *et al.* (1999), which also presents a wide range of further literature on the topic.

Increasingly, EIA is also being carried out informally in situations where it is not mandatory, but where developers feel that its structured approach would help in project management or in speeding up the planning process (Hughes & Wood 1996). Moreover, authorities such as the Environment Agency (EA) frequently produce or require informal **environmental appraisals** for projects not requiring statutory EIA. The principles and procedures described in this book also apply to such informal assessments.

In addition to the specific EIA legislation, there is a wide range of legislation that affects individual EIA components, key examples of which are referred to in the relevant chapters.

1.5 Book structure

1.5.1 Overall structure

The book is divided into two main parts. The first part, as in the book's first edition, discusses EIA methods for a range of environmental components. Table 1.2 shows how the chapters correspond to the components itemised for particular attention in the EU and UK legislation. The book includes some components not specifically listed in the regulations but often discussed in practice, namely noise, transport, geology and geomorphology. Chapters 2 and 3 deal with socio-economic impacts. Chapters 4–7 deal with impacts that are partly socio-economic and partly physical: noise, landscape, transport, and archaeology and other material and cultural assets. Chapters 8–10 cover the physical environment in terms of air and climate, soils/geology/geomorphology, and water. Chapters 11–13 cover 'flora and fauna' in terms of their ecology in terrestrial, freshwater and coastal environments. Because of its particular importance in the coastal zone, geomorphology is now included in Chapter 13. Impacts on agriculture are primarily covered in Chapter 9 on soils, but are further addressed in the chapters on socio-economics.

The second part of the book considers 'cross-cutting' EIA methods: risk assessment and management in Chapter 14; remote sensing in Chapter 15; GIS in Chapter 16; and the new approach to environmental capital in Chapter 17. These chapters have been added since the first edition because the techniques can be applied to, and can often facilitate integration between, many of the environmental components discussed in the first part; and because they are likely to be increasingly used in EIA.

Table 1.2 The book's coverage of the environmental components listed in Annex IV of Directive 97/11/EC and Schedule 4 of the UK regulations

Environmental component	*Chapter number and title*
Population	2. Economic impacts 3. Social impacts 4. Noise 5. Transport
Landscape	6. Landscape
Material assets, including the architectural and archaeological heritage	2. Economic impacts 3. Social impacts 7. Archaeological and other material and cultural assets
Air, climatic factors	8. Air quality and climate
Soil	9. Soils, geology and geomorphology
Water	10. Water
Fauna and flora	11. Ecology – overview and terrestrial ecology 12. Freshwater ecology 13. Coastal ecology and geomorphology

There are seven Appendices: Appendix A is a list of useful addresses, acronyms, chemical symbols, and quantitative units and symbols used in the text; Appendix B lists key UK Government environmental authorities and agencies; Appendix C gives sources of historical information; and Appendices D–G give information on aspects relating mainly to the chapters on ecology.

1.5.2 Chapter structure

The chapters in the first part of the book are all similar in structure; each includes the main EIA steps for the assessment of an environmental component (outlined in §1.2). The main chapter sections are:

- introduction
- definitions and concepts
- legislative background and interest groups
- scoping and baseline studies
- impact prediction
- mitigation
- monitoring

The subjects covered cannot all be discussed in depth in a book of this size. Each chapter aims to provide an overview of the subject, but some aspects are more pertinent to some components than others, and are therefore discussed in greater depth in those chapters. Two problems are: (a) each component covers a large, complex subject and only brief mention can be made of many aspects including specific methods; (b) the wide range of subjects covered by the different chapters means that a reader is likely to be familiar with some but not others. These problems are addressed in three ways:

- each chapter's 'definitions and concepts' section provides some background information on the subject;
- the glossary provides definitions of terms, each of which is highlighted in bold italics when it first appears in a chapter (see Notes on Text Format, p. xv);
- the chapters aim to act as springboards for further reading by making frequent reference to other texts that contain extensive bibliographies and/or details of specific techniques.

The chapters in the second part of the book are necessarily somewhat different and individual in structure, although some retain features of the environmental component chapters.

1.6 Integration of component assessments

Although the chapters in this book are presented as separate entities, in practice the individual environmental component assessments would be integrated together, and be part of the wider process of project planning. Clearly, an EIA must involve a **team of experts** on the various components, and in many cases on different aspects of a given component. Close co-ordination is needed to avoid duplication of effort, while ensuring that important aspects are not omitted. This is particularly important for interrelated components such as soils, geology, air, water, and ecology. In addition, the EIS must be an integrated document in which relationships between components are clearly explained. The use of GIS can facilitate the integration and comparison of data on different components.

It follows that there must be an **EIA co-ordinator** who will ensure that (a) cross-component consultation is carried out throughout the EIA process, and (b) appraisals are conducted to consider aspects such as components' relative importance, the relative significance of different impacts, interactions between impacts, possible conflicts of interest, and distributional effects. For example:

- One sector of the community, or part of the impact area, may be particularly affected by multiple developments, or by the concentration of a project's impacts; lower socio-economic groups, for instance, are more likely to suffer from traffic accidents, air pollution and noise (Lucas & Simpson 2000). Identification of the groups/areas most strongly affected can be facilitated by use of GIS or simply by a table listing receptors (e.g. particular socio-economic groups, sensitive sites) on one axis, and the main impacts of a project on the other axis. A more equitable distribution of impacts may then be sought, or strongly affected groups may be compensated in some way.

- It is important to ensure that mitigation measures proposed for different environmental components are consistent with those for other components, and do not themselves cause negative impacts. For instance, tree plantings which reduce visual impacts could have beneficial side-effects for noise, but could intrude on archaeological remains, and might have positive or negative ecological impacts.

The appraisals can include the use of scenarios and **sensitivity analysis** – of the effects (on an appraisal) of varying the projected values of important variables. Another useful tool is the use of an *audit trail*, which can be particularly beneficial if further EIA analysis is needed because the project changes substantially between the time when it is approved and when it is built. Ideally, final assessment should result in the preparation of a list of proposed planning conditions/obligations and an **Environmental Management Plan** (EMP) for the proposed development, to be included in the EIS or presented in a separate document (Brew & Lee 1996).

1.7 The broader context and the future of EIA methods

Projects are not planned, built, operated and decommissioned in isolation, but within regional, national and international processes of change that include other projects, programmes, plans and policies. The aim of assessing cumulative impacts (§1.2.3) is to take these into account as far as possible in relation to a single development project. However, some projects are so inextricably related to other projects, or their impacts are so clearly linked, that a joint EIA of these projects should be carried out. For instance, if a gas-fired power station requires the construction of a new pipeline and gas reception/processing facility to receive the gas, and transmission lines to carry the resulting electricity, these projects should be considered together in an EIA, despite the fact that each requires EIA under different regulations.

Other projects are 'growth-inducing', i.e. necessary precursors to other projects. For instance, a new motorway may induce the construction of motorway service stations, hypermarkets or new towns; or the infrastructure provided for one project may make a site more attractive, or may present economies of scale, for further development. Although it is probably not feasible to consider induced impacts in detail in an EIA, the EIA should at least acknowledge the possibility of these further developments.

The broadening of EIA's remit to encompass other projects may allow trade-offs to be made between impacts and between projects. For instance, an environmentally beneficial "**shadow project**" may be proposed to neutralise the negative impacts of a development project. An example of this is the 'creation' of a new waterfowl feeding ground on coastal grassland as compensation for the loss of tidal mudflat feeding grounds caused by the Cardiff Bay Barrage. However, shadow projects need to be treated with caution. For instance, it can be argued that the provision of a coastal grassland area does not effectively compensate for the loss of tidal mudflats because it is a different habitat supporting different wildlife communities.

In the future, project EIAs may also be set in the context of SEAs of sectoral or regional policies, plans and programmes (Kleinschmidt & Wagner 1998, Partidario & Clark 2000, Therivel & Partidario 1996). SEAs can reduce the time and cost of EIA, and even eliminate the need for certain types of EIA (Bass 1998). SEA is not yet a legal requirement in the EU, but a draft Directive is expected to be agreed in

2001 (see EC-EDG website). In the UK, local authorities (LAs) have reviewed the environmental impacts of their development plans since 1992 as a result of PPG 12 (DoE 1992), and SEA is becoming more widespread at the national and regional government levels. The Government has published a range of guidance on how to carry out SEAs (DoE 1993, DETR 1998, 1999c). Remote sensing and GIS are likely to be particularly useful in SEA because of their ability to cover large areas and handle large amounts of data.

EIA and SEA should be, and are increasingly being, linked to other related techniques. For example:

- Project design is increasingly being influenced by environmental concerns. There is increasing awareness of the need to minimise resource use in building construction and use, and greater application of techniques such as passive solar heating, photovoltaics and greywater recycling, and of innovative construction methods such as straw bale and earth-sheltered housing and self-build schemes;
- There is increasing use of environmental risk analysis and risk management (Chapter 14).
- The environmental capital approach proposed by CC *et al.* (2001) (Chapter 17) can be used to develop management plans for areas of various sizes, based on an analysis of the benefits and disbenefits that they provide: it is likely to provide a particularly useful early input to the project design process.
- Integrated pollution prevention and control (IPPC) legislation and techniques bring together analyses of the impacts of new developments on air, water and soils.
- Village/community mapping exercises can help to identify features that are particularly valued by local residents.
- Life-cycle analyses can help to identify the impact of buildings from the production of the materials used to build them through to their ultimate dismantling and disposal.
- Sustainability checklists can be used by development control officers to ensure that all developments – not just those for which EIA is required – minimise their environmental impacts.

At the global level, environmental policy is experiencing a general move away from a narrow emphasis on the protection of current environmental resources, and towards a broader promotion of sustainability. The EC model of sustainable development (§1.4), which suggests that economic, social and environmental objectives are mutually reinforcing, is not without its critics. For instance, Levett (1998) notes that environmental life-support systems are a precondition for economic and social development, and that they thus require unconditional protection. He also suggests that the economy itself is a social construct, and that quality of life can be independent of economic growth. Others (e.g. CC *et al.* 2001, LGA 1997, Therivel *et al.* 1992) suggest that development should be more clearly constrained by environmental/sustainability targets, and that environmental assessment should be a vehicle for ensuring that such thresholds are not exceeded.

Finally, concern about wider distributional impacts – for instance, about whether some countries are 'importing' sustainability at the cost of making environmental conditions in other countries unsustainable – is likely to lead to more evolved forms of public participation and political negotiations, but ultimately to a more equitable approach to development and the environment.

References

Inclusion of an Internet address indicates that the publication can be accessed at the site.

Bass R 1998. *Quantifying the environmental impacts of land use plans.* Paper presented at the International Association of Impact Assessment annual conference. Christchurch, New Zealand.

Brew D & N Lee 1996. The role of environmental management plans in the EIA process. *EIA Newsletter (12)*, Manchester University EIA centre.

CC (Countryside Commission), English Heritage, English Nature and Environment Agency 2001. *Quality of life capital*, prepared by CAG Consultants and Land Use Consultants. Cheltenham: CC. (http: www.qualityoflifecapital.org)

CC & EN (Countryside Commission and English Nature) 1998. *The character of England: landscape, wildlife and natural features.* Cheltenham: CC.

CEAA (Canadian Environmental Assessment Agency) 1999. *Cumulative effects assessment practitioners' guide.* Hull, Quebec: CEAA. (http://www.ceaa.gc.ca/publications_e.htm)

Clark BD 1996. *Monitoring and auditing in environmental assessment – improving the process.* London: IAE & EARA Joint Annual Conference.

DETR 1998. *Policy appraisal and the environment: policy guidance.* London: DETR. (http://www.environment.detr.gov.uk/appraisal/index.htm)

DETR 1999a. *A better quality of life: a strategy for sustainable development for the United Kingdom* CM4345. London: DETR. (http://www.environment.detr.gov.uk/sustainable/quality/index.htm)

DETR 1999b. *Circular 2/99, Environmental impact assessment.* London: TSO.

DETR 1999c. *Proposals for a good practice guide on sustainability appraisal of regional planning guidance.* London: DETR. (http://www.planning.detr.gov.uk/rpg/sustain/goodprac/index.htm)

DETR 2000. *Environmental Impact Assessment: A guide to the procedures.* London: HMSO. (http://www.planning.detr.gov.uk/eia/guide/index.htm)

DoE 1992. *Planning Policy Guidance Note 12: Development plans and regional planning guidance.* London: HMSO. (http://www.planning.detr.gov.uk/ppg12/index.htm)

DoE 1993. *Environmental appraisal of development plans: a good practice guide.* London: HMSO.

DoE 1995. *Preparation of environmental statements for planning projects that require environmental assessment: a good practice guide.* London: HMSO.

EC (European Commission) 1985. Council Directive on the assessment of the effects of certain private and public projects on the environment (85/337/EEC). *Official Journal of the European Communities* L 175/40. Brussels: European Commission.

EC 1993. *Towards sustainability: fifth action programme on the environment.* Brussels: European Commission. (http://europa.eu.int/comm/environment/)

EC 1997. Council Directive 97/11/EC amending Directive 85/337/EEC on the assessment of the effects of certain public and private projects on the environment. Brussels: *Official Journal of the European Commission* L073/5-21. (http://europa.eu.int/comm/environment/eia)

EC 1999. *Guidelines for the assessment of indirect and cumulative impacts as well as impact interactions.* Brussels: EC DG XI. (http://europa.eu.int/comm/environment/)

EC 2001. *Environment 2010: Our future, our choice. The sixth environment action programme of the European Community 2001–2010.* Brussels: European Commission. (http://europa.eu.int/comm/environment/)

Glasson J, R Therivel & A Chadwick 1999. *Introduction to environmental impact assessment,* 2nd edn. London: UCL Press.

HMSO 1999. *The Town and Country Planning (Environmental Impact Assessment) (England and Wales) Regulations 1999, SI 1999 No 293.* London: HMSO. (http://www.hmso.gov.uk/si/si19990293.htm or via http://www.planning.detr.gov.uk/eia/assess/index.htm)

HSE (Health and Safety Executive) & EA/SEPA 1999. *Guidance on the environmental risk assessment aspects of COMAH safety reports.* (http://www.environment-agency.gov.uk)

Hughes J & C Wood 1996. Formal and informal environmental assessment reports: their role in UK planning decisions. *Land Use Policy* **13**(2), 101–113.

Kleinschmidt V & D Wagner (eds) 1998. *Strategic environmental assessment in Europe: Fourth European workshop on environmental impact assessment.* Dordrecht: Kluwer Academic.

Levett R 1998. Urban housing capacity and the sustainable city. 4: Monitoring, measuring and target setting for urban capacity. In *Urban housing capacity: what can be done?* London: Town and Country Planning Association and Joseph Rowntree Foundation.

LGA (Local Government Association) 1997. *Sustainability in development control.* Research Report, prepared by CAG Consultants. London: LGA.

Lucas K & R Simpson 2000. *Transport and accessibility: the perspectives of disadvantaged communities.* Research paper for the Joseph Rowntree Foundation. London: Transport Studies Unit, University of Westminster.

MAFF 1999. *High level targets for flood and coastal defence and elaboration of the Environment Agency's flood defence supervisory duty.* London: MAFF.

MAFF/WO 1996. *Code of practice on environmental procedures for flood defence operating authorities,* PB 2906. London: MAFF.

Partidario MR & R Clark 2000. *Perspectives on strategic environmental assessment.* London: Earthscan.

Quarrie J (ed.) 1992. *Earth Summit '92 – The United Nations Conference on Environment and Development, Rio de Janeiro 1992.* London: Regency Press.

SE (Scottish Executive) 1999a. *The Environmental Impact Assessment (Scotland) Regulations 1999, Circular 15/1999.* Edinburgh: SE. (http://www.scotland.gov.uk)

SE 1999b. *Planning Advice Note (PAN) 58 – Environmental impact assessment.* Edinburgh: SE. (http://www.scotland.gov.uk)

Therivel R & MR Partidario (eds) 1996. *The practice of strategic environmental assessment.* London: Earthscan.

Therivel R, E Wilson, S Thompson, D Heaney & D Pritchard 1992. *Strategic environmental assessment.* London: Earthscan.

UNECE (United Nations Economic Commission for Europe) 1991. *Policies and systems of environmental impact assessment.* Geneva: UNECE.

USCEQ (US Council on Environmental Quality) 1997. *Considering cumulative effects under the National Environmental Policy Act.* Washington DC: Executive Office of the President. (http://www.whitehouse.gov/CEH/ *or* ceq.eh.doe.gov/)

Wathern P 1999. Ecological impact assessment. In *Handbook of environmental impact assessment,* Vol. 1, J Petts (ed.), 327–46. Oxford: Blackwell Science.

WCED (World Commission on Environment and Development) 1987. *Our common future.* Oxford: Oxford University Press.

2 Socio-economic impacts 1: overview and economic impacts

John Glasson

2.1 Introduction

Major projects have a wide range of impacts on a locality – including bio-physical and socio-economic – and the trade-off between such impacts is often crucial in decision-making. Major projects may offer a tempting solution to an area's, especially a rural area's, economic problems, which, however, may have to be offset against more negative impacts such as pressure on local services and social upheaval, in addition to possible damage to the physical environment. Socio-economic impacts can be very significant for particular projects and the analyst ignores them at his/her peril. Nevertheless they have often had a low profile in EIA.

This chapter begins with an initial overview of socio-economic impacts of projects/developments, which explains the nature of such impacts. Economic impacts, including the direct employment impacts and the wider, indirect impacts on a local and regional economy are then discussed in more detail. The chapter dovetails with Chapter 3 which focuses on related impacts such as changes in population levels and associated effects on the social infrastructure, including accommodation and services. Several of the methods discussed straddle the two chapters and will be cross-referenced to minimise duplication. Chapters 2 and 3 draw in particular on the work of the Impacts Assessment Unit (IAU) in the School of Planning at Oxford Brookes University, which has undertaken many research and consultancy studies on the socio-economic impacts of major projects.

2.2 Definitions and concepts: socio-economic impacts

2.2.1 Origins and definitions

Socio-economic impact assessment (SIA) developed in the 1970s and 1980s mainly in relation to the assessment of the impacts of major resource development projects, such as nuclear power stations in the US, hydro-electric schemes in Canada and the UK's North Sea oil and gas related developments. The growing interest in socio-economic impacts, partly stimulated by the introduction of the US *National Environmental Policy Act* of 1969 and subsequent amendments of 1977, generated some important studies and publications, including the works of Wolf (1974), Lang & Armour (1981), Finsterbusch (1980, 1985), and Carley & Bustelo (1984). It also led to considerable debate on the nature and role of SIA. Some authors refer to social impact assessment; others refer to socio-economic impact assessment. Some see SIA as an integral part of EIA, providing the essential "human elements" complement to the often narrow bio-physical focus of many EISs "from the perspective of the social impact agenda, this meant: valuing people 'as much as fish' . . ." (Bronfman 1991). Others see SIA as a separate field of study, a separate process, and some authors raise the legitimate concern that SIA as an integral part of EIA runs the risk of marginalisation and superficial treatment. Chapters 2 and 3 of this text focus on the wider definition of socio-economic impacts, *within* the EIA process.

Wolf (1974), one of the pioneers of SIA, adopted the wide-ranging definition of SIA as "the estimating and appraising of the conditions of a society organised and changed by the large scale application of high technology". Bowles (1981) has a similarly broad definition: "the systematic advanced appraisal of the impacts on the day to day quality of life of people and communities when the environment is affected by development or policy change". A more lighthearted, but often relevant approach to definition can be typified as the "grab bag" (Carley & Bustelo 1984) or 'Heineken' approach – with SIA including all those vitally important, but often intangible impacts which other methods cannot reach.

More recently a major study by the Interorganisational Committee on Guidelines and Principles for Social Assessment (1994) defined social impacts as "the consequences to human populations of any public or private actions that alter the ways in which people live, work, play, relate to one another, organise to meet their needs, and generally cope as members of society." Social impacts are the "people impacts" of development actions. Social impact assessments focus on the human dimension of environments, and seek to identify the impacts on people and who benefits and who loses. SIA can help to ensure that the needs and voices of diverse groups and people in a community are taken into account.

2.2.2 Socio-economic impacts in practice: the poor relation?

The early recognition, by some analysts, of the importance of socio-economic impacts in the EIA process and in the resultant EISs has been partly reflected in legislation. The definition of the environment, as included in the 1979 US CEQ regulations, addresses bio-physical components and socio-economic factors and characteristics. The EU Directive 85/337/EEC (EC 1985), outlined in §1.4, requires a description of possible impacts on human beings. Furthermore, the UK Government has produced guidance which suggests that "certain aspects of a project including numbers employed and where they will come from should be considered within an environmental statement" (DoE 1989). Yet despite some legislative impetus, the consideration of social and economic impacts has continued to be the poor relation in EIA and in EISs (Glasson & Heaney 1993). There may be several reasons for this which can be summed up by the general perceptions that socio-economic impacts seldom occur, are invariably negative, and cannot easily be measured. Such perceptions are, of course, a gross distortion. Socio-economic impacts invariably follow from a development, they are often positive, they can be measured and they are important. Indeed the key trade-offs in the decisions on projects often revolve around the balancing of socio-economic benefits (usually employment) against bio-physical costs. Socio-economic impacts are important because the economic fortunes and lifestyles and values of people are important.

In a review of the coverage of socio-economic impacts in EISs produced in the UK between 1988 and 1992, Glasson and Heaney showed that from a sample of 110

EISs, only 43% had considered socio-economic impacts at all. Coverage was better than (a low) average for power station, mixed development and mineral extraction projects. Within those EISs which included socio-economic impacts, there was more emphasis on economic impacts (particularly direct employment impacts) than social impacts. Both operational and construction stages of projects were considered, although with more emphasis on the former. The geographical level of analysis was primarily local, with only very limited coverage of the wider regional scale and no consideration of impacts at the national level. There was very limited use of techniques; where they were included they were primarily economic or employment *multipliers*. Quality was also generally unsatisfactory; only 36% of EISs that considered socio-economic impacts were considered to deal with the economic impacts adequately or better. For social impacts, the figure was only 15%.

Socio-economic impacts merit a higher profile. A United Nations study of EIA practice in a range of countries advocated a number of changes in the EIA process and in the EIS documentation (UNECE 1991). These included giving greater emphasis to socio-economic impacts in EIA. Box 2.1 highlights the important links

---

Box 2.1 Importance of social impacts in EIA

To quote UNEP (1996):

"There is often a direct link between social and subsequent biophysical impacts. For example, a project in a rural area can result in the in-migration of a large labour force, often with families, into an area with low population density. This increase in population can result in adverse biophysical impacts, unless the required supporting social and physical infrastructure is provided at the correct time and place.

Additionally, direct environmental impacts can cause social changes, which, in turn, can result in significant environmental impacts. For example, clearing of vegetation from a riverbank in Kenya, to assist construction and operation of a dam, eliminated local tsetse fly habitats. This meant that local people and their livestock could move into the area and settle in new villages. The people exploited the newly available resources in an unsustainable way, by significantly reducing wildlife populations and the numbers of trees and other wood species which were used as fuel wood. A purely 'environmental' EIA might have missed this consequence because the social impacts of actions associated with dam construction would not have been investigated.

The close relationship between social and environmental systems makes it imperative that social impacts are identified, predicted and evaluated in conjunction with biophysical impacts. It is best if social scientists with experience of assessing social impacts are employed as team members under the overall direction of a team or study leader who has an understanding of the links between social and biophysical impacts."

And the World Bank (1991):

"Social analysis in EA is not expected to be a complete sociological study nor a cost-benefit analysis of the project. Of the many social impacts that might occur, EA is concerned primarily with those relating to environmental resources and the informed participation of affected groups.

Social assessment for EA purposes focus on how various groups of people affected by a project allocate, regulate and defend access to the environmental

resources upon which they depend for their livelihood. In projects involving indigenous people or people dependent on fragile ecosystems, social assessment is
particularly important because of the close relationship between the way of life of
a group of people and the resources they exploit. Projects with involuntary resettlement, new land settlement and induced development also introduce changes in
the relationships between local people and their use of environmental resources."

between social and bio-physical impacts with particular reference to developing
countries. It also cautions against over-ambitious SIA. In a different context, in a
survey of academics on the effectiveness of the US *National Environmental Policy
Act*, Canter & Clark (1997) draw out five priorities for the future, one of which
is the need for better integration of bio-physical and socio-economic factors and
characteristics. A starting point in raising the SIA profile is to clarify the various
dimensions of socio-economic impacts.

2.2.3 The scope of socio-economic impacts

A consideration of socio-economic impacts needs to clarify the type, duration, spatial extent and distribution of impacts; that is, the analyst needs to ask the questions
what to include, over what period of time, over what area and impacting whom?

An overview of **what to include** is outlined in Table 2.1. There is usually a functional relationship between impacts. Direct economic impacts have wider indirect
economic impacts. Thus direct employment on a project will generate expenditure
on local services (e.g. for petrol, food and drink). The ratio of local to non-local labour
on a project is often a key determinant of many subsequent impacts. A project
with a high proportion of in-migrant labour will have greater implications for the
demography of the locality. There will be an increase in population, which may also
include an influx of dependants of the additional employees. The demographic
changes will work through into the housing market and will impact on other local
services and infrastructure (e.g. on health and education services), with implications
for both the public and private sector (see Fig. 2.1).

In some cases, population changes themselves may be initiators of the causal
chain of impacts; new small settlements (often primarily for commuters) would fit
into this category. Development actions may also have socio-cultural impacts. A new
settlement of 15,000 people may have implications for the lifestyles in a rural, small
village based environment. The introduction of a major project, with a construction
stage involving the employment of several thousand people over several years, may
be viewed as a serious threat to the quality of life of a locality. Social problems may
be associated with such development, which may generate considerable community
stress and conflict. In practice, such socio-cultural impacts are usually poorly covered
in EISs, being regarded as more intangible and difficult to assess.

The question of **what period of time** to consider in SIA raises in particular the
often substantial differences between impacts in the construction and operational
stages of a project. Major utilities (such as power stations and reservoirs) and other
infrastructure projects, such as roads, may have high levels of construction employment but much lower levels of operational employment. In contrast, manufacturing
and service industry projects often have shorter construction periods with lower levels
of employment, but with considerable employment levels over projects which may

Table 2.1 What to include? – types of socio-economic impacts

1. **Direct economic**
   - local and non-local employment
   - characteristics of employment (e.g. skill group)
   - labour supply and training
   - wage levels
2. **Indirect/wider economic/expenditure**
   - employees' retail expenditure
   - linked suppliers to main development
   - labour market pressures
   - wider multiplier effects
3. **Demographic**
   - changes in population size (temporary and permanent)
   - changes in other population characteristics (e.g. family size, income levels, socio-economic groups)
   - settlement patterns
4. **Housing**
   - various housing tenure types
   - public and private
   - house prices
   - homelessness and other housing problems
5. **Other local services**
   - public and private sector
   - educational services
   - health services; social support
   - others (e.g. police, fire, recreation, transport)
   - local finances
6. **Socio-cultural**
   - lifestyles/quality of life
   - gender issues; family structures
   - social problems (e.g. crime, illness, divorce)
   - community stress and conflict; integration, cohesion and alienation

extend for several decades. The closure of a project may also have significant socio-economic impacts; unfortunately these are rarely covered in the initial assessment. Socio-economic impacts should be considered for all stages of the life of a development. Interestingly, nuclear reactor decommissioning did become a project requiring mandatory environmental assessment under Directive 97/11/EC (EC 1997). Even within stages, it may be necessary to identify sub-stages, e.g. peak construction employment, to highlight the extremes of impacts that may flow from a project. Only through monitoring can predictions be updated over the life of the project under consideration.

What area to cover in SIA raises the often contentious issue of where to draw the boundaries around impacts. Boundaries may be determined by several factors. They may be influenced by estimates of the impact zone. Thus, for the construction stage of a major project, a sub-regional or regional boundary may be taken, reflecting the fact that construction workers are willing to travel long distances daily for short-term, well-paid employment. On the other hand, permanent employees of an operational development are likely to locate much nearer to their work. Other determinants of the geographical area of study may include the availability of data (e.g. for counties and districts in the UK), and policy issues (e.g. providing spatial impact data related to the areas of responsibility of the key decision-makers involved in a project). Different socio-economic impacts will often necessitate the use of different

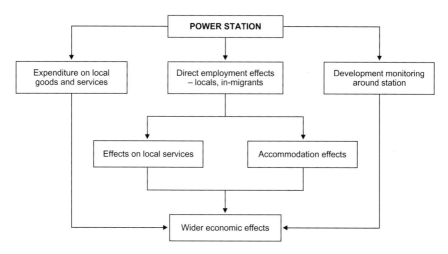

Figure 2.1 Example of linkages between socio-economic impacts for a power station project.

geographical areas, reflecting some of the determinants already discussed. As noted earlier, EISs in practice have focused on local areas. This may provide a very partial picture; economic impacts often have wider regional, and occasionally national and international implications.

The question of **who will be affected** is of crucial importance in EIA, but is very rarely addressed in EISs. The differential effects of development impacts do not fall evenly on communities; there are usually winners and losers. For example, a new tourism development in a historic city in the UK may benefit visitors to the city and tourism entrepreneurs, but may generate considerable pressures on a variety of services used by the local population. Distributional impacts can be analysed by reference to geographical areas and/or to groups involved (e.g. local and non-local; age groups; socio-economic groups; employment groups).

SIA should also pay particular attention to vulnerable sections of the population being studied – the elderly, the poor, and minority or ethnically distinctive groups – and to areas which may have particular value to certain groups in terms of cultural or religious beliefs. In this context, an interesting development in the USA, after long campaigning by black and other ethnic groups, is the Clinton "*Executive Order on Federal Actions to Address Environmental Justice in Minority Populations and Low Income Populations*" (White House 1994). Under this Order, each federal agency must analyse the environmental effects, including human health, economic and social effects, of federal actions, including effects on minority and low-income communities, when such analysis is required under NEPA. Similarly, but from the wider perspective of the World Bank (1991), Box 2.2 provides some examples of the key social differences that may be environmentally significant.

There are of course many other dimensions to impacts besides the areas discussed here, including adverse and beneficial, reversible and irreversible, quantitative and qualitative, and actual and perceived impacts (see Glasson *et al.* 1999). All are relevant in SIA. The distinction between actual and perceived impacts raises the distinction between more 'objective' and more 'subjective' assessments of impacts. The impacts of a development perceived by residents of a locality may be significant

---

Box 2.2 Examples of social differences which may be environmentally significant

Communities are composed of diverse groups of people, including, but not restricted to the intended beneficiaries of a development project. Organised social groups hold territory, divide labour and distribute resources. Social assessment in EA disaggregates the affected population into social groups which may be affected in different ways, to different degrees and in different locations. Important social differences which may be environmentally significant include ethnic or tribal affiliation, occupation, socio-economic status, age and gender.

Ethnic/Tribal groups. A project area may include a range of different ethnic or tribal groups whose competition for environmental resources can become a source of conflict. Ethnicity can have important environmental implications. For example, a resettlement authority may inadvertently create competition for scarce resources if it grants land to new settlers while ignoring customary rights to that land by indigenous tribal groups.

Occupational groups. A project area may also include people with a wide array of occupations who may have diverse and perhaps competing interests in using environmental resources. Farmers require fertile land and water, herders require grazing lands, and artisans may require forest products such as wood to produce goods. A project may provide benefits to one group while negatively affecting another. For example, while construction of dams and reservoirs for irrigation and power clearly benefits farmers with irrigation, it may adversely affect rural populations engaged in other activities living downstream of the dam.

Socio-economic stratification. The population in the project area will also vary according to the land and capital that they control. Some will be landless poor, others will be wealthy landowners, tenant farmers or middlemen entrepreneurs. Disaggregating the population by economic status is important because access to capital and land can result in different responses to project benefits. For example, tree crop development may benefit wealthy farmers, but displace the livestock of poor farmers to more marginal areas.

Age and gender. A social assessment should include identification of project impacts on different individuals within households. Old people may be more adversely affected by resettlement than young people. Men, women and children play different economic roles, have different access to resources, and projects may have different impacts on them as a result. For example, a project that changes access to resources in fragile ecosystems may have unanticipated impacts on local women who use those resources for income or domestic purposes.

(World Bank 1991)

---

in determining local responses to a project. They can constitute an important source of information to be considered alongside more 'objective' predictions of impacts.

2.3 Baseline studies: direct and indirect economic impacts

2.3.1 Understanding the project/development action

Socio-economic impacts are the outcome of the interaction between the characteristics of the project/development action and the characteristics of the 'host' environment. As a starting point, the analyst must assemble baseline information on both sets of characteristics.

The assembling of relevant information on the characteristics of the project would appear to be one of the more straightforward steps in the process. However, projects have many characteristics and for some, relevant data may be limited. The drafting of a **direct employment labour curve** is the key initial source of information (see Fig. 2.2). This shows the anticipated employment requirements of the project. To be of maximum use it should include a number of dimensions, including in particular the duration and categories of employment. The labour curve should indicate the anticipated labour requirements for each stage in the project life cycle.

For the purposes of prediction and further analysis, there may be a focus on certain key points in the life cycle. For example, an SIA of peak construction employment could reveal the maximum impact on a community; an analysis of impacts at full operational employment would provide a guide to many continuing and long-term impacts. The labour curve should also indicate requirements by employment or skill category. These may be subdivided in various ways according to the nature of employment in the project concerned, but often involve a distinction between managerial and technical staff, clerical and administrative staff and project operatives. For a construction project, there may be a further significant distinction in the operatives category between civil works operatives. A finer disaggregation still would focus on the particular trades or skills involved, including levels of skills (e.g. skilled/semi-skilled/unskilled) and types of skills (e.g. steel erector, carpenter, electrician).

Projects also have **associated employment policies** that may influence the labour requirements in a variety of ways. For example, the use/type of shift working and the approach to training of labour may be very significant in determining the scope for local employment. An indication of likely wage levels could be helpful in determining wider economic impacts into the local retail economy. An indication of the main developer's attitude/policy to subcontracting can also be helpful in determining the wider economic impacts for the local and regional manufacturing and producer services industries.

Hopefully, the initial brief from the developer will provide a good starting point on labour requirements and associated policies. But this is not always the case, particularly where the project is a 'one-off' and the developer cannot draw on comparative experience from within the firm involved. In such cases the analyst may be able to draw on EISs of comparative studies. However, many major projects are at the forefront of technology and there may be few national, or even international, comparators available. There may be genuine uncertainty on the relative merits of different designs for a project, and this may necessitate the assessment of the socio-economic impacts of various possibilities. For example, an assessment by the IAU at Oxford in 1987 for the Hinkley Point C power station proposal, considered the socio-economic impacts for both pressurised water reactor and advanced gas-cooled reactor designs (Glasson *et al.* 1987).

Projects also have a tendency to change their characteristics through the planning and development process and these may have significant socio-economic implications. For example, the discovery during the early stages of project construction of major foundation problems may necessitate a much greater input of civil works operatives. Major projects also tend to have a substantial number of contractors, and it may be difficult to forecast accurately without knowledge of such subcontractors, and indeed of the main contractor. Such uncertainties reinforce the necessity of regular monitoring of project characteristics throughout project planning and development.

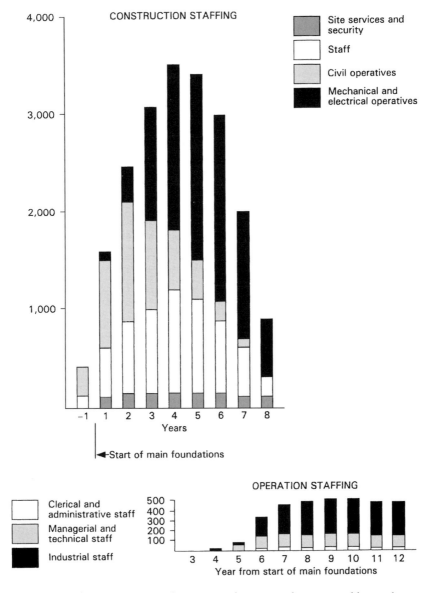

Figure 2.2 Labour requirements for a project disaggregated in time and by employment category.

2.3.2 *Establishing the economic environment baseline*

Defining the **'host' economic environment area** depends to some extent on the nature of the project. Some projects may have significant national or even international employment implications. The construction of the Channel Tunnel had wide-ranging inter-regional economic impacts in the UK, bringing considerable benefits to areas well beyond Kent and the South-East region of England, for example to the West Midlands (Vickerman 1987). Many projects have regional or sub-regional economic impacts, and almost all have local economic impacts. As noted in §2.2, it can be useful to make a distinction between the anticipated construction and operational daily commuting zones for a project. The former is invariably much larger in geographical area than the latter, possibly extending up to 90 minutes one-way daily commuting time from the project. For these areas, and for the wider region and nation as appropriate, it is necessary to assemble data on current and anticipated labour market characteristics, including size of labour force, employment structure, unemployment and vacancies, skills and training provision.

The **size of the labour force** provides a first guide to the ability of a locality to service a development. Information is needed on the economically active workforce (i.e. those males and females in the 16 to retirement age bands). This then needs disaggregation into industrial and/or occupational groups to provide a guide to the economic activities and employment types in the study area(s). An industrial disaggregation would identify, for example, those in agriculture, types of manufacturing and services. In the UK, the *Standard Industrial Classification* (SIC) provides a template of categories (Table 2.2). An occupational disaggregation indicates particular skill groups (Table 2.3). Data on unemployment and vacancies provides indicators of the pressure in the labour market and the availability of various labour groups. It should be disaggregated by length of unemployment, as well as by skill category and location. Data should also be collected on the provision of training facilities in an area. Such facilities may be employed to enhance the quality of labour supply.

In the UK, the provision of labour market data comes from various, and changing, sources. The national Department of Employment is a primary source, and a guide to available data is provided in Table 2.4. The National Online Manpower Information Service (NOMIS) computerised database is a particularly useful source of employment and unemployment data at various geographical levels. Department

Table 2.2 UK broad Standard Industrial Classification (SIC) (since 1992)

Division of industry	SIC
A	Agriculture, hunting and forestry
B	Fishing
C	Mining and quarrying
D	Manufacturing
E	Electricity, gas and water supply
F	Construction
G	Wholesale and retail trade
H	Hotels and restaurants
I	Transport, storage and communication
J	Financial intermediaries
K	Real estate, renting and business activity
L	Public administration and defence
M	Education
N	Health and social work
O	Other community, social and personal service activity
P	Private households with employed persons
Q	Extra-territorial organisations and bodies

Source: Office for National Statistics (ONS).

Table 2.3 UK broad occupational categories

i	Professional and managerial occupations
ii	Intermediate occupations (clerical and related)
iii (N)	Skilled occupations – non-manual
iii (M)	Skilled occupations – manual (craft or similar)
iv	Partly skilled occupations (general labourers)
v	Unskilled occupations
vi	Armed forces and inadequately described

Source: Office for National Statistics.

Table 2.4 Major UK employment/earnings data sources

Labour Market Trends (incorporating Employment Gazette) – published monthly. This is the major source on employment. At the regional level there is monthly information on employment, redundancies, vacancies, unemployment, and annual information on number of employees (age/sex/SIC), activity rates, seasonal unemployment and new employment data. Breakdowns by Local Authority Areas and Parliamentary constituencies are also available. There are also occasional labour force projections (male/female/total) by region.

New Earnings Survey – produced annually since 1971. It relates to earnings of employees by industry, occupation, region, etc. at April each year. Part E of the six parts includes detailed analysis of earnings (weekly, hourly) by occupation and industry for regions.

Skills and Enterprise Network – introduced in 1991 and produced quarterly. It provides an important data source on skills, employment and training.

Labour Market Quarterly Report – provides a commentary, including tables and charts, on current labour market trends and the implications for training, employment and unemployment, and includes special features on particular labour market topics. It includes some regional data.

National Online Manpower Information System (NOMIS) (http://www.dur.ac.uk/) – a pay-as-you-use nationally networked information system offering rapid access and integrated analysis for data on employment, unemployment, job vacancies, population and migration, at various geographical levels.

Source: Updated from Glasson (1992).

of Employment regions may also provide useful annual and more frequent reviews of the employment situation in their region. A basic geographical area for the Department of Employment data is the Travel to Work Area (TTWA). Another important UK source of data is the Census of Population. The results of the 1991 census include information on the economic activity, workplace and transport to work of the population. The statutory local and structure plans for the area under consideration also provide valuable employment data; this may be complemented by data in advisory regional strategies and reviews/studies which are well developed in some regions (e.g. South-East via SERPLAN).

In some areas, the sources noted may be enhanced by various one-off studies, including, for example, skills audits which seek to establish the current and latent skills provision of an area. In the UK, a network of Training and Enterprise Councils or Local Enterprise Companies exist in each region, and provide a useful contact, particularly on training information. Predictably, the various data sources do not use the same geographical bases; in particular the discrepancy between TTWAs and local authority areas can cause problems for the analyst. The latter should also be aware of the influence of 'softer' data – for example, information on possible developments in other major projects in a locality which may have labour market implications for the project under consideration. Data on other 'host' area economic characteristics – such as wage levels, characteristics of the retail economy, and detailed characteristics of local manufacturers – may be more limited, although some local authorities do produce very useful industrial directories.

Local economic impacts may also be influenced by the policy stance(s) of the host area. For many localities the possibility of employment and local trade gains from a project may be the only perceived benefits. There may be a desire to maximise such gains and to limit the **leakage** of multiplier benefits (see §2.5). This may result in an authority taking a policy stance on the percentage of 'local' labour to be employed on a project. For example, in an extreme case, Gwynedd County Council negotiated, through the use of an Act of Parliament, a very high percentage of local labour for the construction of the Wylfa nuclear power station on Anglesey. A local position may also be taken on the provision of training facilities. There may be concern about the possible local employment 'boom–bust' scenario associated with some major projects, which may of course bring caution into the setting of high local employment ratios.

2.3.3 Clarifying the issues

Consideration of **project and 'host' environment characteristics** can help to clarify key issues. Denzin (1970) and Grady *et al.* (1987) remind us that issue specification should be rooted in several sources, and they advocate the use of the philosophy of "**triangulation**" for data (the use of a variety of data sources), for investigators (the use of different sets of researchers), for theory (the use of multiple perspectives to interpret a single set of data) and for methods (the use of multiple methods). Thus, the use of quantitative published and semi-published data, as outlined, should be complemented by the use of key informant interviews, working groups (e.g. of developer, local planning officers, councillors, and representatives of interest groups) and possibly public meetings.

Whilst many direct and indirect employment impacts will be specific to the case in hand, the following **key questions** (Murdock *et al.* 1986) tend to be raised in most cases:

- "What proportions of project construction and operation jobs are likely to be filled by local workers, as compared to in-migrants, and what are the likely origins of the in-migrant workers?
- What is likely to be the magnitude of the secondary (indirect and induced) employment resulting from project development? What proportions of these jobs will be filled by local workers?

- How will local businesses be affected by rapid growth resulting from a major project? For example, will development provide opportunities for expansion or will local firms experience difficulty competing with new chain stores and in attracting and retaining quality workers?"

2.4 Impact prediction: direct employment impacts

2.4.1 *The nature of prediction*

Prediction of socio-economic impacts is an inexact exercise. Ideally the prediction of the direct employment impacts on an area would be based on information relating to the recruitment policies of the companies involved in the development, and on individuals' decisions in response to the new employment opportunities. In the absence of firm data on these and related factors, predictions need to be based on a series of **assumptions** related to the characteristics of the development and of the locality. These could, for example, include the following:

- the labour requirement curves for construction and operation will be as provided by the client;
- local recruitment will be encouraged by the developer with a target of 50%;
- employment on the new project will be attractive to the local workforce by virtue of the comparatively high wages offered.

Predictive approaches may use **extrapolative methods**, drawing on trends in past and present data. In this respect, use can be made of comparative situations and the study of the direct employment impacts of similar projects. Unfortunately the limited monitoring of impacts of project outcomes reduces the value of this source, and primary surveys may be needed to obtain such information. Predictive approaches may also use **normative methods**. Such methods work backwards from desired outcomes to assess whether the project, in its environmental context, is adequate to achieve them. For example, the desired direct employment outcome from the construction stage of a major project may be X% local employment.

Underpinning all prediction methods should be some clarification of the **cause–effect relationships** between the variables involved. Figure 2.3 provides a simplified flow diagram for the local socio-economic impacts of a power station development. Prediction of the local (and regional as appropriate) labour recruitment ratios is the key step in the process. Non-local workers are, by definition, not based in the study area. Their in-migration for the duration of a project will have a wider range of secondary demographic, accommodation, services and socio-cultural impacts (as discussed in Chapter 3). The wider economic impacts on, for example, local retail activity, will be discussed further in this chapter. The key determinants of the local recruitment ratios are the labour requirements of the project, the conditions in the local economy, and relevant local authority and developer policies on topics such as training, local recruitment and travel allowances. It is possible to quantify some of the cause–effect relationships, and various **economic impact models**, derived from the multiplier concept, can be used for predictive purposes. These will be discussed further in §2.5.

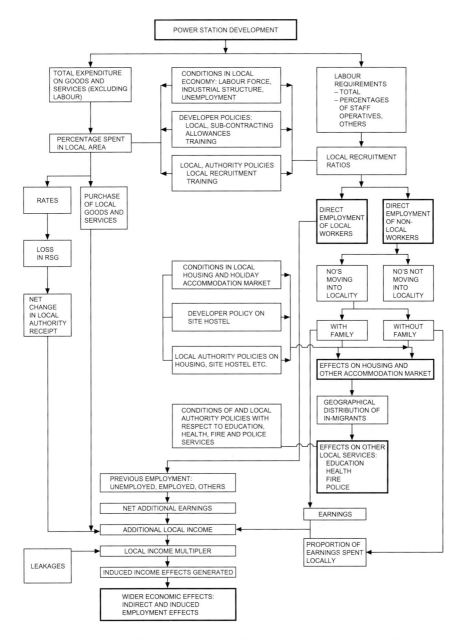

Figure 2.3 A cause–effect diagram for the local socio-economic impacts of a power station proposal.

Source: Glasson *et al.* (1987).

Note: RSG = Rate Support Grant

Whatever prediction method is used, there will be a degree of **uncertainty** attached to the predicted impacts. Such uncertainty can be partly handled by the application of probability factors to predictions, by sensitivity analysis, and by the inclusion of ranges in the predictions (see Glasson *et al.* 1999, Chapters 5 and 9).

2.4.2 Predicting local (and regional) direct employment impacts

Disaggregation into project stages, geographical areas and employment categories is the key to improving the accuracy of predictions. For example, the construction stage of major projects will usually involve an amalgam of professional/managerial staff, administrative/secretarial staff, local services staff (e.g. catering, security) and a wide range of operatives in a variety of skill categories. Most projects will involve civil-works operatives (e.g. plant operators, drivers), and most will also include some mechanical and electrical activity (e.g. electricians, engineers). For each employment category there is a labour market, with relevant supply and demand characteristics. Guidance on the mix of local/non-local employment for each category can be obtained from comparative studies and from the best estimates of the participants in the process (e.g. from the developer, from the local employment office). Hopefully, but in practice not often, guidance will be informed by the monitoring of direct employment impacts in practice.

As a normal rule, the more specialist the staff, the longer the training needed to achieve the expertise, and the more likely that the employee will not come from the immediate locality of the project. Specialist professional staff and managerial staff are likely to be brought in from outside the study area; they may be transferred from other sites, seconded from headquarters or recruited on the national or international market. Only a small percentage may be recruited from the local market, which may simply just not have the expertise available in the numbers necessary. On the other hand, local services staff (e.g. security, cleaning, catering), and to a slightly lesser extent secretarial and administrative staff, may be much more plentiful in most local labour markets, and the local percentage employed on the project may be quite high, and in some cases very high. Other skill categories will vary in terms of local potential according to the degree of skill and training needed. There may be an abundance of general labourers, but a considerable shortage of coded welders.

Comparative analysis of the disaggregated employment categories is likely to produce broad bands for the level of local recruitment. These can then be refined with reference to the conditions applicable to the particular project and locality under consideration. For example, high levels of unemployment in particular skill categories in the locality may boost local recruitment in those categories. Normative methods may also come into play. The developer may introduce training programmes to boost the supply of local skills. Table 2.5 provides an example of the sort of estimates which may be derived. Whilst the predictions may still use ranges, a prediction from the disaggregated analysis is much more robust than taking employment as an homogenous category.

A further level of **micro-analysis** would be to predict the employment impacts for particular localities within the study area, and for particular groups, such as the unemployed. A further level of **macro-analysis**, used in some EISs, would include an estimate of the total person days of employment per year generated by the project (e.g. 10,000 employment days in 2001).

Table 2.5 Example of predicted employment of local and non-local labour for the construction stage of a major project

	Total labour requirements	Local labour		Non-local labour	
		%	range	%	range
Site services, security and clerical staff	300	90	250–290	10	10–50
Professional, supervisory and managerial staff	430	15	50–80	85	350–380
Civil operatives	500	55	250–300	45	200–250
Mechanical and electrical operatives	1520	40	550–670	60	850–970
Total	2750	44	1100–1340	56	1410–1650

Local labour: Employees already in residence in the Construction Daily Commuting Zone before being recruited on site. **Non-local labour**: All other employees.

2.5 Impact prediction: wider economic impacts

2.5.1 The range of wider economic impacts

In addition to the direct local (and/or regional) employment effects, major projects have a range of **secondary** or **indirect impacts**. The workforce, which may be very substantial (and well paid) in some stages of a project, can generate considerable retail expenditure in a locality, on a whole range of goods and services. This may be a considerable boost for the local retail economy; for example, IAU studies of the impact of power station developments suggest that retail turnover in adjacent medium and small towns may be boosted by at least 10% (Glasson *et al.* 1982). The projects themselves require supplies ranging from components from local engineering firms, to provisions for the canteen. These can also boost the local economy.

Such demands create employment, or sustain employment, additional to that directly created by the project. As will be discussed in Chapter 3, the additional workforce may demand other services locally (e.g. health, education), and housing, which may generate additional construction. These demands will create additional employment. Training programmes associated with a project may bring other economic benefits in terms of a general upgrading of skills. Overall, the net effect may be considerably larger than the original direct injection of jobs and income into a locality, and such wider economic impacts are invariably regarded as beneficial.

However, there can be **wider economic costs**. Existing firms may fear the competition for labour which may result from a new project. They may lose skilled labour to high-wage projects. There may be inflationary pressures on the housing market and on other local services. Major projects may be a catalyst for other development in an area. A road or bridge can improve accessibility and increase the economic potential of areas. But major projects may also cast a shadow over an area in terms of alternative developments. For example, large military projects, nuclear power stations, mineral extraction projects and others may have a deterrent impact on other activities, such as tourist – although the construction stage and the operation of many projects can be tourist attractions in themselves, especially when aided by good interpretation and visitor centre facilities.

2.5.2 Measuring wider economic impacts: the multiplier approach

The analysis of the wider economic effects of introducing a major new source of income and employment into a local economy can be carried out using a number of different techniques (Brownrigg 1971, Glasson 1992, Lewis 1988, McNicholl 1981). The three methods most frequently used are (a) the economic base multiplier model, (b) the input–output model, and (c) the Keynesian multiplier, although it should be added that the percentage of EISs including such studies is still small.

The **economic base multiplier** is founded on a division of local (and/or regional) economies into basic and non-basic activities. Basic activities (local/regional supportive activities) are seen as the 'motors' of the economy; they are primarily oriented to markets external to the area. Non-basic activities (regional dependent activities) support the population associated with the basic activities, and are primarily locally oriented services (e.g. retail services). The ratio of basic to non-basic activities, usually measured in employment terms, is used for prediction purposes. Thus an X increase in basic employment may generate a Y increase in non-basic employment. The model has the advantages, and disadvantages, of simplicity (Glasson 1992).

Input–output models provide a much more sophisticated approach. An input–output table is a balancing matrix of financial transactions between industries or sectors. Adapted from national input–output tables, regional or local tables can provide a detailed and disaggregated guide to the wider economic impacts resulting from changes in one industry or sector. However, unless an up-to-date table exists for the area under study, the start-up costs are normally too great for most EIA exercises. Batey *et al.* (1993) provide an interesting example of the use of input–output analysis to assess the socio-economic impacts of an airport development.

For several reasons – primarily related to the availability of appropriate data at a local level – the **Keynesian multiplier approach** has been used in several studies and is discussed in further detail here. The basic theory underlying the Keynesian multiplier is simple: "a money injection into an economic system, whether national or regional, will cause an increase in the level of income in that system by some multiple of the original injection" (Brownrigg 1974). Mathematically this can be represented at its most simple as:

$$Y_r = K_r J \tag{1}$$

where: Y_r is the change in the level of income in region r
         J is the initial income injection (or multiplicand)
         K_r is the regional income multiplier

If the initial injection of money is passed on intact at each round, the multiplier effect would be infinite. The £X million initial injection would provide £X million extra income to workers, which in turn would generate an extra income of £X million for local suppliers, who would then spend it, and so on ad infinitum. But the multiplier is not infinite because there are a number of obvious leakages at each stage of the multiplier process. Five important leakages are:

s   the proportion of additional income saved (and therefore not spent locally)

t_d the proportion of additional income paid in direct taxation and National Insurance contributions

m   the proportion of additional income spent on imported goods and services

u   the marginal transfer benefit/income ratio (representing the relative change in transfer payments, such as unemployment benefits, which result from the rise in local income and employment)

t_i the proportion of additional consumption expenditure on local goods which goes on indirect taxation (e.g. VAT).

The multiplier can therefore be formulated as follows:

$$K_r = \frac{1}{1-(1-s)(1-t_d-u)(1-m)(1-t_i)} \qquad (2)$$

Substituting (2) into (1) then gives:

$$Y_r = \frac{1}{1-(1-s)(1-t_d-u)(1-m)(1-t_i)}J \qquad (3)$$

Thus, when applied to the multiplicand J, the multiplier K_r gives the accumulated wider economic impacts for the area under consideration, as in equation (3). The Keynesian multiplier can be calculated in income or employment terms. The various leakages normally reduce the value of local and regional multipliers in practice to between 1.1 and 1.8; in other words, for each £1 brought in directly by the project, an extra £0.10–0.80 are produced indirectly. The size of the import leakage is a major determinant, since the bigger the leakage, the smaller the multiplier. Leakages increase as the size of the study area declines, and decreases as the study area becomes more isolated. Thus, of the UK regions, Scotland has the highest regional multiplier (Steele 1969). Local (county and district level multipliers) normally vary between 1.1 and approximately 1.4.

Keynesian multiplier studies have been used particularly extensively in **tourism impact studies** (Eadington & Redman 1991, Fletcher & Archer 1991), and more recently in the assessment of the impact of higher education on local and regional economies. Universities can have very significant local economic impacts. The direct employment associated with them is the most obvious of these impacts, and universities are often amongst the largest single employers in their local labour markets. A CVCP study (1994) lists some 20 published university local economic impact studies. Such work has been undertaken by universities, reflecting a desire to demonstrate their local economic significance (Lincoln et al. 1993). A recent study for Oxford Brookes University (Chadwick & Glasson 1998) showed that this medium-sized university generated local expenditure of approximately £100 million per year, and over 2000 (fte) local jobs, making it one of the major employers in the city.

In practice, EIA studies will probably limit such analyses to gross estimates of the wider economic impacts at perhaps the peak construction and full operation stages. But it is possible to disaggregate also with reference to the various employee groups. A study of the predicted local socio-economic impacts of the construction and operation of the proposed Hinkley Point C nuclear power station illustrates the variations, with higher multipliers associated with in-migrants with families (1.3–1.5) than with unaccompanied in-migrants (1.05–1.11) (Glasson *et al.* 1988). The Keynesian multiplier model, with modifications as appropriate, is well suited to the assessment of the wider economic impacts of projects. But it can only be as good as the information sources on which it is based to construct both the multiplicand and the multiplier. Predictive studies of proposed developments are more problematic in this respect than studies of existing developments, although knowledge of the latter can inform prediction.

2.5.3   Assessing significance

Socio-economic impacts, including the direct employment and wider economic impacts, do not have recognised standards. There are no easily applicable "state of local society" standards against which the predicted impacts of a development can be assessed. Whilst a reduction in local unemployment may be regarded as positive, and an increase in local crime as negative, there are no absolute standards. Views on the significance of economic impacts, such as the proportion and types of local employment on a project, are often political and arbitrary. Nevertheless it is sometimes possible to identify what might be termed **threshold or step changes** in the socio-economic profile of an area. For example, it may be possible to identify predicted impacts which threaten to swamp the local labour market, and which may produce a 'boom–bust' scenario. It may also be possible to identify likely high levels of leakage of anticipated benefits out of a locality, which may be equally unacceptable.

In the assessment of significance, the analyst should be aware of the philosophy of **'triangulation'** noted earlier. Multiple perspectives on significance can be gleaned from many sources, including the local press, which can be very powerful as an opinion former, other key local opinion formers (including local councillors and officials), surveys of the population in the host locality, and public meetings. All can help to assess the significance, perceived and actual, of various socio-economic impacts. A very simple analysis might measure the column-cm of local newspaper coverage of certain issues in the planning stage of the project; a survey of local people might seek to calculate simple measures of agreement (MoA) with certain statements relating to economic impacts. MoA is defined as the number of respondents who agree with the statement, minus the number who disagree, divided by the total numbers of respondents. Thus, an MoA of 1 denotes full agreement; −1 denotes complete disagreement.

2.6   Mitigation and enhancement

Most predicted economic impacts are normally encouraged by the local decision-makers. However, there may be concern about some of the issues already noted, such as the poaching of labour from local firms, the swamping of the local labour market, or the shadow effect on other potential development. In such cases, there may be attempts to build in formal and/or informal controls, such as 'no-poaching agreements'. The fear of the 'boom–bust' scenario may lead to requirements for a compensatory 'assisted area' package for other employment with the demise of employment associated with the project in hand. A number of studies of post-redundancy employment experiences have been undertaken in recent years in the UK. Some relate to traditional industries such as coal-mining, shipbuilding and steel (Hinde 1994, Turner & Gregory 1995). A number of studies have been associated with the restructuring of the defence and aerospace sectors (Bishop & Gripaios 1993). There have also been studies of the end of construction programmes (Glasson & Chadwick 1997, Armstrong *et al.* 1998).

However, in general the focus for economic impacts is more on measures to **enhance benefits**. When positive impacts are identified there should be a concern to ensure that they do happen and do not become diluted. The potential local employment benefits of a project can be encouraged through appropriate skills training programmes for local people. Targets for the proportion of local recruitment may be set. Various measures, such as project open days for potential local suppliers and a register of local suppliers, may help to encourage local links and to reduce the leakage of wider economic impacts outside the locality. The use of good management practices, including a local liaison committee which brings together the operator and community representatives, and a responsive complaints procedure, can help mitigation and enhancement. A commitment to monitoring, and the publication of monitoring data, can also make a major contribution to effective mitigation.

2.7   Monitoring

Previous stages in the EIA process should be designed with monitoring in mind. **Key indicators** for monitoring direct employment impacts include: levels and types of employment, by local and non-local sources and by previous employment status; trends in local and regional unemployment rates; and the output of training programmes. All these indicators should be disaggregated to allow analysis by employment/skill category. Relevant data sources include developer/contractor returns, monthly unemployment statistics, and training programme data; these can be supplemented by direct survey information. Key indicators of the wider economic impacts include: trends in retail turnover, the fortunes of local companies and development trends in the locality. Some guidance on such indicators may be gleaned from published data. The project developer may also provide information on the distribution of subcontracts, but surveys of, for example, workforce expenditure, and the linkages of local firms with a project, may be necessary to gain the necessary information for useful monitoring.

Monitoring is currently not mandatory for EIA in the UK. There are few comprehensive studies to draw on. The work of the IAU at Oxford on monitoring the local socio-economic impacts of the construction of Sizewell B (Glasson & Chadwick 1988–97) provides one of the few examples of a longitudinal study of socio-economic impacts in practice. It shows the significance of direct employment and wider economic impacts for the local economy. At peak over 2000 local jobs were provided, but with a clear emphasis on the less skilled jobs. Local skills have been upgraded through a major training programme, and whilst some local companies have experienced recruitment difficulties as a result of Sizewell B, the impact did not appear to be too significant. A group of about 30 to 40 mainly small local companies have benefited substantially from contracts with the project. Although the actual level of project employment was higher than predicted, many of the predictions made at the time of the public inquiry have stood the test of time, and the key socio-economic condition of encouraging the use of local labour has been fulfilled.

2.8 Conclusions

Socio-economic impacts are important in the EIA process. They have traditionally been limited to no more than one EIS chapter, and often a small late chapter, if they have been included at all. Our placing of such impacts early in this text, and in two chapters, emphasises our concern to indicate their importance in a comprehensive EIA. The discussion has outlined the broad characteristics of such impacts and discussed economic impacts in more detail, with a particular focus on approaches to establishing the information baseline and to prediction. Some predictive methods can become complex. This may be appropriate for major studies; for smaller studies, some of the simpler methods may be more appropriate. The non-local: local employment ratio associated with a project has been identified as a key determinant of many subsequent socio-economic effects.

References

Armstrong HW, M Ingham & D Riley 1988. *The effect of the Heysham 2 power station on the Lancaster and Morecambe economy*. Lancaster: Department of Economics, Lancaster University.

Batey P, M Madden & G Scholefield 1993. Socio-economic impact assessment of large-scale projects using input–output analysis: a case study of an airport. *Regional Studies* **27**(3), 179–192.

Bishop P & P Gripaios 1993. Defence in a peripheral region: the case of Devon and Cornwall. *Local Economy* **8**, 43–56.

Bowles RT 1981. *Social impact assessment in small communities*. Toronto: Butterworth.

Bronfman LM 1991. Setting the social impact agenda: an organisational perspective. *Environmental Impact Assessment Review* **11**, 69–79.

Brownrigg M 1971. The regional income multiplier: an attempt to complete the model. *Scottish Journal of Political Economy* **18**.

Brownrigg M 1974. *A study of economic impact: the Stirling University*. Edinburgh: Scottish Academic Press.

Canter L & R Clark 1997. NEPA effectiveness – a survey of academics. *Environmental Impact Assessment Review* **71**, 313–328.

Carley MJ & ES Bustelo 1984. *Social impact assessment and monitoring: a guide to the literature*. Boulder, Colorado: Westview Press.

Chadwick A & J Glasson 1998. *Oxford Brookes University – local economic impacts*. Working Paper 174. Oxford: Oxford Brookes University.

CVCP (Committee of Vice-Chancellors and Principals) 1994. *Universities and communities*. A report by CURDS, University of Newcastle for CVCP: London.

Denzin NK 1970. *Sociological methods: a source book*. Chicago: Aldine Publishing.

DoE (Department of the Environment) 1989. *Environmental assessment: a guide to the procedures*. London: HMSO.

Eadington W & M Redman 1991. Economics and tourism. *Annals of Tourism Research*, **18**(1), 41–56.

EC (European Commission) 1985. Council Directive on the assessment of the effects of certain private and public projects on the environment (85/337/EEC). *Official Journal of the European Communities* **L 175/40**, 5 July 1985, Brussels: European Commission.

EC 1997. Council Directive 97/11/EC amending Directive 85/337/EEC on the assessment of the effects of certain public and private projects on the environment. Brussels: *Official Journal of the European Commission* L073/5-21. (http://europa.eu.int/comm/environment/eia)

Finsterbusch K 1980. *Understanding social impacts: assessing the effects of public projects*. Beverly Hills, California: Sage Publications.

Finsterbusch K 1985. State of the art in social impact assessment. *Environment and Behaviour* **17**(2), 193–221.

Fletcher J & B Archer 1991. The development and application of multiplier analysis. In *Progress in Tourism*, Vol. 3. C Cooper (ed.), 28–47. London: Belhaven.

Glasson J 1992. *An introduction to regional planning*. London: UCL Press.

Glasson J & A Chadwick 1988–1997. *The local socio-economic impacts of the Sizewell 'B' PWR construction project*. Impacts Assessment Unit, Oxford Brookes University.

Glasson J & A Chadwick 1997. Life after Sizewell B: post-redundancy experiences of locally recruited construction employees. *Town Planning Review* **68**(3), 325–345.

Glasson J & D Heaney 1993. Socio-economic impacts: the poor relations in British EISs. *Journal of Environmental Planning and Management* **36**(3), 335–343.

Glasson J, R Leavers, J Porter & S Squires 1982. *A comparison of the social and economic effects of power stations on their localities*. Oxford Polytechnic: Power Station Impacts Team.

Glasson J, MJ Elson, D van der Wee, B Barrett 1987. *The socio-economic impact of the proposed Hinkley Point 'C' power station*. Oxford Polytechnic: Power Station Impacts Team.

Glasson J, D van der Wee, B Barrett 1988. A local income and employment multiplier analysis of a proposed nuclear power station development at Hinkley Point in Somerset. *Urban Studies* **25**, 248–261.

Glasson J, R Therivel & A Chadwick 1999. *Introduction to Environmental Impact Assessment*, 2nd edn. London: UCL Press.

Grady S, R Braid, J Bradbury, C Kerley 1987. Socio-economic assessment of plant closure: three case studies of large manufacturing facilities. *Environmental Impact Assessment and Review* **26**, 151–165.

Hinde K 1994. Labour market experiences following plant closure: the case of Sunderland's shipyard workers. *Regional Studies* **28**, 713–724.

Interorganisational Committee on Guidelines and Principles for Social Assessment 1994. Guidelines and Principles for Social Impact Assessment. *Impact Assessment* **12**(summer), 107–152.

Lang R & A Armour 1981. *The assessment and review of social impacts*. Ottawa, Canada: Federal Environmental Assessment and Review Office.

Lewis JA 1988. Economic impact analysis: a UK literature survey and bibliography. *Progress in Planning* **30**(3), 161–209.

Lincoln I, D Stone & A Walker 1993. *The impact of higher education institutes on their local economy: a review of studies and assessment methods*. Report by NERU, University of Northumbria for CVCP: London.

McNicholl IH 1981. Estimating regional industry multipliers: alternative techniques. *Town Planning Review* **55**(1), 80–88.

Murdock SH, FL Leistritz & RR Hamm 1986. The state of socio-economic impact analysis in the USA: limitations and opportunities for alternative futures. *Journal of Environmental Management* **23**, 99–117.

Steele DB 1969. *Regional multipliers in Britain*. Oxford Economic Papers No. 19.

Turner R & M Gregory 1995. Life after the pit: the post-redundancy experiences of mineworkers. *Local Economy* **10**, 149–162.

UNECE (United Nations Economic Commission for Europe) 1991. *Policies and systems of environmental impact assessment*. Geneva: UNECE.

UNEP (United Nations Environment Programme) 1996. *Environmental Impact Assessment: Issues, Trends and Practice*. Stevenage: SMI Distribution.

Vickerman R 1987. Channel Tunnel: consequences for regional growth and development. *Regional Studies* **21**(3), 187–197.

White House 1994. *Memorandum from President Clinton to all heads of all departments and agencies on an executive order on federal actions to address environmental injustice in minority populations and low income populations*. Washington DC: White House.

Wolf CP (ed.) 1974. *Social impact assessment*. Washington: Environmental Design Research Association.

World Bank 1991. *Environmental assessment sourcebook*, Vol. 1, Chapter 3. *World Bank Technical Paper No. 139*. Washington DC: World Bank.

3 Socio-economic impacts 2: social impacts

Andrew Chadwick

3.1 Introduction

Chapter 2 discussed how the workforce involved in the construction and operation of any major project is likely to be drawn partly from local sources (within daily commuting distance of the project site) and partly from further afield. Those employees recruited from beyond daily commuting distance can be expected to move into the locality, either temporarily during construction or permanently during operation. Some of these employees will bring families into the area. In-migrant employees and their families will exert a number of impacts on their host localities:

- They will result in an increase in the **population** of the area and possibly in changes to the age and sex structure of the local population.
- They will require **accommodation** within reasonable commuting distance of the project site.
- They will place additional demands on a range of **local services**, including schools, health and recreational facilities, police and emergency services.
- They may have **financial implications** for the local authorities in the area, with additional costs of service provision set against an increase in revenues.
- They may have **other social impacts**, such as changes in the local crime rate or in the social mix of the area's population.

3.2 Definitions

The geographical extent of social impacts, i.e. the **impact area**, will depend largely on the residential location of in-migrant workers and their families. In-migrant employees can be expected to move into accommodation within reasonable commuting distance of the project site, although the definition of what constitutes a reasonable distance will depend on the project stage (construction or operation), as well as local settlement patterns and the local transport network. Monitoring data from similar projects elsewhere should indicate the likely extent of daily commuting and thus the likely boundaries of the impact area. These boundaries can be defined in various ways, for example in terms of a fixed distance or radius from the project site or, more usually, in terms of administrative or political areas such as local authority districts (LADs), health authority areas or school catchment areas.

3.3 Baseline studies

3.3.1 Demography – establishing the existing baseline

The demographic impact of any development will depend on the project-related changes in population in relation to the existing population size and structure in the impact area. It is therefore necessary to establish the existing population baseline in the impact area (i.e. size and age/sex structure). The most useful source of population data in Great Britain, particularly for small geographical areas, is the *Census of Population*. This is carried out once every ten years, most recently in April 1991. Since all households in Great Britain are included in the census, reliable information is available at all geographical levels, from census enumeration districts (covering 150–200 households) upwards. Census data are available in a wide range of formats, both published and unpublished:

- The series of *County Monitors* and more detailed *County Reports*, published by the Office for National Statistics (ONS), provide data for counties and LADs in England and Wales. Similar reports are produced for health authority areas, and data for local authorities in Scotland are published by the General Register Office (GRO).
- For areas smaller than a LAD, or for user-defined areas, the census *Small Area Statistics* or more detailed *Local Base Statistics* should be used. These are available direct from ONS/GRO, from commercial census agencies and via on-line computerised databases such as NOMIS.

Further details on the various types of census data available can be found in Dale & Marsh (1993), Denham (1992), Leventhal *et al.* (1993), LGMB (1992), OPCS (1992) and Openshaw (1995).

The great strengths of the census are its comprehensiveness and the availability of data for small or user-defined areas. Its main weakness is that it is only undertaken once every ten years. Given the delay in the processing and publication of results, the latest data are sometimes more than a decade out of date (Openshaw 1995). Between censuses, it is therefore necessary to consult other sources to obtain an up-to-date picture of population size and structure in the impact area. The most often used of these sources are the official *Mid-year population estimates (Series PP1)*, published annually by ONS. These are published in *Key Population and Vital Statistics, local and health authority areas* and *Population Estimates, Scotland*. In addition, most local authorities produce their own population estimates, both for the authority as a whole and for its constituent parts (i.e. wards or parishes). These estimates may be based either on primary data collection or the use of various proxy measures of population change since the latest census, such as changes in the electoral roll or doctors' registrations (see England *et al.* 1985, Healey 1991). A number of commercial market analysis companies also produce census-based population estimates for small or user-defined areas. Details of these companies are contained in Leventhal *et al.* (1993).

3.3.2 Projecting the demographic baseline forward

The data sources outlined above allow the existing population baseline in the impact area to be established. But it may also be desirable to project this baseline forward, ideally to the expected times of peak construction and full operational activity for the proposed development. A number of data sources are available to guide this process. Sub-national population projections are published by ONS/GRO (*Series PP3*). Long-term projections (up to 25 years ahead) are produced about every three to five years, and short-term projections are published in intervening years. The projections are available for local and health authority areas, but not for individual LADs (other than London boroughs, metropolitan districts and unitary authorities). Population projections and forecasts are often also produced by local authorities themselves. These are used by authorities as inputs to their land use planning work (e.g. structure plan preparation) and to estimates of future service requirements (e.g. school places). Projections are usually available for LADs and in some cases are disaggregated to ward or parish level (see Congdon & Batey 1989, England *et al.* 1985, Woods & Rees 1986). Some commercial market analysis companies also produce population projections for small areas.

These various sources have **limitations** as means of projecting forward the population baseline for relatively small geographical areas. Projections for smaller areas (e.g. LADs) tend to be less reliable than those for larger areas (e.g. counties or regions). This is because net migration is usually a more important determinant of population change for smaller areas; and migration flows are much more difficult to predict than the number of births and deaths. The sources also differ in the extent to which they simply project forward past trends in an unmodified way. For example, ONS stresses that its population projections are not 'forecasts', in that they take no account of the potential effects of changes in local planning policies. These are often designed to counteract past trends, for example, to slow down the rate of population and housing growth in an area. Local authority forecasts are much more likely to incorporate such anticipated policy effects and may therefore be preferable, although of course the intended policy effects may not materialise in practice.

3.3.3 Accommodation – establishing the existing baseline

The 1991 census, as well as providing population data, is also the most useful source of data on the **housing stock** in small geographical areas. The census provides two alternative measures of the housing stock in an area – the number of dwellings and the number of household spaces. These two measures differ only slightly, with the former always being somewhat lower than the latter. The census provides a breakdown of both household spaces and dwellings, according to their tenure (i.e. whether they are owner occupied, privately rented, rented with a job or business, or rented from a housing association or local authority). The amount of vacant accommodation is identified, as is accommodation which is not used as a main residence – this includes second homes and some holiday accommodation (e.g. self-catering cottages).

All of this information, although providing a very detailed picture of the available housing stock, relates to the position at the time of the latest census and will therefore need to be updated. This can be achieved by using a number of sources. The DETR, NAW and SE publish data on the number of new houses completed (by the private and public sector), and existing homes renovated and demolished, for each LAD and unitary authority. These data are published each quarter in *Local Housing Statistics – England*, *Quarterly Welsh Housing Statistics* and *Housing Trends in*

Scotland. This information, perhaps supplemented by more detailed development control data from local authorities themselves, should allow any significant changes in the overall size of the housing stock since the latest census to be estimated. Local estate agents are a useful source of data on current house price and rent levels. House price data are also published by some of the larger building societies, providing data for local authority areas and selected postal towns and cities (see, for example, the *Halifax Quarterly House Price Index*).

During the construction stage of any development, some **in-migrant employees** are likely to move into bed and breakfast establishments, hotels, caravans or other types of **tourist accommodation**. It is therefore necessary to establish how much of such accommodation is available in the impact area, and to determine typical occupancy levels. Any unoccupied accommodation (e.g. outside the peak tourist season) could be used by in-migrant employees without affecting the availability of accommodation for other existing users. Regional tourist boards, local authorities and tourist information centres all maintain databases or lists of accommodation establishments within their areas of jurisdiction. Details of each individual establishment are often available, including the location, number of rooms and charges/tariffs. A detailed picture of the existing stock of accommodation can therefore be obtained. When combining lists prepared by different organisations for the same geographical area, care should be taken to avoid the double-counting of establishments.

Information on existing occupancy levels in tourist accommodation is published regularly by the English, Welsh and Scottish Tourist Boards. Each of these bodies carries out a monthly survey of occupancy levels in a sample of hotels (and also in self-catering, caravan and camping accommodation in the Scottish survey). Results show the monthly bed and room occupancy rates in the establishments surveyed, and are available for Tourist Board regions and selected sub-regional areas (typically either individual counties or groupings of two or three adjacent counties). Results for each of the English Tourist Board (ETB) Regions are published annually in *Regional Tourism Facts* by the ETB. Data for smaller geographical areas may be available from the numerous area-specific tourism studies carried out by the regional tourist boards, local authorities, academics and consultants.

3.3.4 Projecting the accommodation baseline forward

Non-project related changes in the local housing stock can be estimated most easily by using simple **trend projection methods**. These are typically based on the assumption that existing rates of housebuilding (net of demolitions and other losses, such as changes of use) will continue for the foreseeable future. Data on current and recent housebuilding rates are published on a regular basis by the DETR, NAW and SE. Such methods, although easily applied, are rather crude, in that they take no account of possible changes in the state of the national economy or in local rates of population and household growth; they also fail to allow for the influence of local planning policies on the scale and location of new housebuilding.

An alternative approach would be to use estimates of future population and household growth in the area to predict the likely demand for new houses. *Local authority population and household forecasts* are likely to be particularly relevant. High and low estimates of household growth are usually made by local authorities, using different assumptions about net migration, employment and household formation (see England *et al.* 1985, Field & MacGregor 1987, King 1987, CPRE 1994). Of course, the anticipated increase in the number of households in an area may not be met by an equivalent increase in the housing stock. This is because local planning policies may be intended to meet only part of the projected increase in households (Bramley & Watkins 1995, Bramley *et al.* 1995). The extent, phasing and location of new housebuilding envisaged by local planning authorities is indicated by the *housing allocations in approved structure plans and adopted local plans.*

Likely changes in the stock of tourist and other temporary accommodation are difficult to predict, although regional tourist boards and local authorities may be able to indicate the scale of any significant additional provision, either already under construction or with outstanding planning permission.

3.3.5 Local services

In-migrant employees and their families will place demands on a wide range of services provided by local authorities, health authorities and other public bodies. In the space available, it is not possible to discuss each of these service areas in detail. The bulk of this section therefore examines one service area – **local education services** – as an example of how the existing service baseline might be established and projected forward. Other service areas are briefly discussed at the end of the section.

The **number and type of schools and further education colleges** within the impact area can be obtained directly from local education authorities (LEAs) (for LEA-maintained schools and colleges) and the Department for Education and Employment (DfEE), NAW and SE (for grant-maintained and independent – i.e. private sector – schools and colleges). LEAs will be able to indicate the existing number of pupils on school rolls and the total available capacity (in permanent and temporary accommodation), both for the LEA area as a whole and for each individual school. An age breakdown of existing pupils should also be available. This information can be used to determine the extent to which the available capacity in LEA schools is currently being utilised, across the authority as a whole and for individual schools, high school catchment areas or age groups. Current pupil/teacher ratios can be obtained direct from the LEA or from the DfEE's annual publication *Statistics of Education: Schools in England* (and similar NAW and SE publications).

Information on significant **planned changes in school capacity** due to the closure, amalgamation or enlargement of existing schools and the opening of new schools should be obtained from the LEA concerned. All LEAs also produce forecasts of future pupil numbers, both for the authority as a whole and for individual schools. These are derived in some cases from the authority's own population and household projections, and should incorporate the effects of anticipated non-project in-migration (see Jenkins & Walker 1985). These data sources will allow any significant anticipated changes in pupil numbers and the utilisation of capacity within the impact area to be identified.

Information on **other public services**, such as recreation, police, fire and social services, should be obtained directly from the relevant local authority department. For **health services**, Family Health Service Authorities and Regional Health Authorities will be able to provide a wide range of data on existing medical, dental and pharmacy services, as well as hospital facilities in the impact area.

3.4 Impact prediction

3.4.1 *Population changes*

Changes in population caused by a major project can include both direct and in-direct increases. The **direct increase** will consist of in-migrant employees and any other family members brought into the locality. A number of separate estimates are therefore required to determine the population changes directly due to the project: (a) the total number of employees moving into the impact area, during both the construction and operational stages of the development; (b) the proportion of these in-migrant employees bringing their family; and (c) the characteristics of these families (i.e. their size and age structure).

The total number of employees moving into the impact area

Chapter 2 has outlined the methods available for predicting the mix of local and in-migrant employees associated with the construction and operation of major projects. *During the construction stage*, the build-up in the number of in-migrant workers will reflect the build-up of the construction workforce and changes in the local labour percentage. At the end of the construction stage, most in-migrant workers will move out of the impact area and return to their original address or another construction project elsewhere. However, a small proportion may establish local ties, especially during a lengthy construction project, and may decide to remain in the area. A construction project spanning several years may therefore result in a small permanent increase in the local population. *During operation*, the main flow of in-migrant employees will usually occur at a relatively early stage, with subsequent in-migration limited to that caused by the normal turnover of employees.

The proportion of in-migrant employees bringing their family

During the construction stage, only a minority of in-migrant employees – mainly those on long-term contracts – are likely to bring their family into the area. The precise proportion will depend on various factors:

- the length of the construction programme (for projects lasting only a few months, it is likely to be negligible; for projects spanning several years, the proportion may reach at least 10–20%);
- the location and accessibility of the project site, which will determine the relative merits of weekly commuting and family relocation;
- conditions in the national and local housing markets (a depressed national housing market or sharp inter-regional house price differentials may discourage house and family relocation);
- the availability of suitable family accommodation, schools and other amenities in the locality.

During the operational stage, the vast majority of in-migrant employees will relocate permanently to the area, although there may be some initial delay whilst suitable accommodation is found and existing properties are sold. Those employees with partners or children can be expected to bring them into the area (with the exception of a small number of weekly commuters). The precise proportion of employees with families will depend on the age and sex profile of the in-migrant workforce. For example, a younger workforce might be expected to contain a higher proportion of single, unattached employees who will not bring families into the area.

The characteristics of in-migrant families

Once the likely number of in-migrant families has been determined, it is necessary to estimate the average size and broad age structure of these families. The usual approach to estimating the size of in-migrant families is to use detailed census data on household headship. The census shows the average size of households of different types, classified according to the age, sex and marital status of the head of household. Therefore, if it was considered likely that most in-migrant families would contain a married, male head of household, aged 20–59 years, the average size of this type of household – either nationally or in the impact area – could be calculated. For projects with a younger anticipated workforce, the average size of households with married male heads aged, say, 20–44 years could be calculated instead. This method assumes that the household characteristics at the time of the 1991 census will remain largely unchanged; it also requires some knowledge (or guesswork) about the age and sex profile of the in-migrant workforce.

Let us assume that the method outlined above suggests that each in-migrant family will contain an average of 3.2 persons. It could then be assumed that each of these families would consist of two adults of working age (the in-migrant employee and partner) and an average of 1.2 other family members – mainly dependent children up to 18 years old, but also including a small proportion of 'adult' children (over 18 years old) still living with their parents and perhaps some elderly relatives. The precise proportion of adult children and elderly relatives should ideally be derived from monitoring data, but in the absence of such information, a rough guestimate may be required. Information on the age structure of the 0–18-year-old population is available from a number of sources, and this can be used as the basis of predictions of the ages of dependent children brought into the area. The current age breakdown of 0–18 year olds is provided by the 1991 census, the mid-year population estimates and local authority population estimates. The projected future age breakdown of this group can be obtained from the various population projections and forecasts outlined in §3.3.2. The census also provides an age breakdown of children (and others) moving into particular LADs or counties during the 12 months prior to the census date.

The precise age distribution of dependent children will of course depend on the age profile of their parents. For example, a younger workforce will tend to have a higher proportion of pre-school children than might be suggested by the data sources above, whereas an older workforce may have a higher proportion of secondary school children. Some fine-tuning of the age distribution revealed by the data sources above may therefore be required, to take account of the expected age profile of the project workforce. The age breakdown of the workforce should ideally be estimated by obtaining information on the age of employees on similar projects elsewhere. Such information should be readily available to the project developer (for operation) or its contractors (for construction).

As well as the direct population increase due to the arrival of in-migrant project employees and families, the development may give rise to **indirect population impacts**.

These impacts can arise in two main ways. First, some locally recruited project employees will leave local employers to take up jobs on the project. This will result in local job vacancies, some of which may be filled by in-migrants. Indirect employment may also be created in local industries supplying or servicing the project, or in the provision of project-related infrastructure. Again, some of these jobs may be taken by in-migrant employees. The scale of the resulting additional in-migration is very difficult to estimate, but its possible existence should at least be acknowledged (see Clark *et al.* 1981 for some possible estimation methods). A second source of indirect impacts arises from the fact that some locally recruited project employees might have migrated out of the impact area if the project had not gone ahead, especially if alternative job opportunities locally were limited. The project may therefore lead to a reduction in out-migration from the area. Again, the extent of any such reduction is difficult to predict. It is likely to be significant only in areas experiencing a static or declining population, net out-migration and limited or declining employment opportunities.

3.4.2 The significance of population changes

The significance of project-related population changes will depend on three main factors: (a) the existing population size and structure in the impact area (i.e. the population baseline); (b) the geographical distribution of the in-migrant population; and (c) the timing of the population changes. Put simply, if in-migrants are few relative to the existing population and have a similar age and sex structure, are distributed over a wide area and do not all arrive at once, then the impacts are unlikely to be significant. The first step in assessing significance is therefore to **express the estimated project-related population increase as a percentage of the baseline population** in the impact area. The predicted age structure of in-migrants should be compared with the baseline age structure, and any significant differences outlined (DoE 1995).

The next step is to estimate the likely **geographical distribution of in-migrants**. Population changes may be quite localised, rather than being evenly distributed throughout the impact area. However, in the absence of information from monitoring studies, the precise distribution of in-migrants is difficult to predict. The simplest approach would be to assume that the number of employees moving into a particular settlement would be a positive function of that settlement's size and a negative function of its distance from the project site. In practice, the predictions derived from this type of model would need to be modified to allow for the characteristics of the particular locality. These could include: the expected location of future housebuilding in the impact area; differences in the availability and price of various types of housing; and the attractiveness of each settlement in terms of school and other facilities and general environment. **The timing of the arrival of in-migrant employees** and the associated population changes will largely follow the expected build-up in the project workforce. However, during the construction stage, most in-migrant families are likely to arrive in the early stages, given that families will tend to be brought by those employees on long-term contracts for the duration of the project.

The nature and significance of population impacts will change as the project progresses through the various stages of its life cycle. In-migrant employees and their families will become older. In addition, during the operational stage – which may span several decades – there may be some natural increase from the original in-migrant population. These changes can be estimated by using a simple 'cohort

survival' method, applying age-specific birth and death rates to the original population (see Field & MacGregor 1987). Some allowance may also need to be made for the turnover of employees on the project. As older employees retire, they will tend to be replaced by younger employees, with younger families. This process will counteract, but not completely reverse, the tendency for the in-migrant population to become older.

3.4.3 Accommodation requirements

The total amount of accommodation required will be determined by the size of the **in-migrant workforce** and the extent to which accommodation is shared. Methods to estimate the total number of in-migrant employees were outlined in Chapter 2. Sharing of accommodation is likely to be minimal amongst the permanent **operational workforce**, since most in-migrant employees will be accompanied by their families. However, there may be a limited amount of sharing amongst younger, single employees, especially in rented accommodation. During the **construction stage**, sharing may be much more significant, especially amongst those employees using rented, caravan and perhaps B&B accommodation. Estimates of the likely extent of sharing should be incorporated into any predictions of the demand for accommodation by the construction workforce. Otherwise, the amount of accommodation required is likely to be over-estimated, perhaps significantly. Published monitoring studies of recent construction projects, although limited in number in the UK, may provide an indication of the likely extent of sharing (e.g. see Glasson & Chadwick 1995).

The type and location of accommodation required will also differ in the operational and construction phases. The vast majority of in-migrant **operational employees** are likely to relocate permanently to the impact area. Most will wish to purchase a property in the area, although a small proportion may prefer private rented accommodation. This latter group will include younger, single employees and a small number of weekly commuters not relocating their family. There may also be some demand for social rented accommodation, from local authorities and housing associations. The likely mix between owner-occupied, private and social rented accommodation requirements can be roughly estimated by using census data – the 1991 census provides information on the tenure of all households moving address during the 12 months to April 1991. Separate tenure patterns can be identified for different types of move (e.g. moves within the same LAD, inter-county or inter-regional moves). This information is also available for different age groups, according to the age of the head of household. These data could perhaps be combined with the expected age profile of the operational workforce, to produce estimates of the likely tenure patterns of in-migrant households.

Predicting the likely **mix of accommodation used by in-migrant construction workers** is a more complicated exercise. A wider range of accommodation is likely to be suitable, including B&B, caravan and other types of tourist accommodation. A further complication is that, for larger construction projects, the developer may decide to provide accommodation specifically for the workforce. The extent of such provision will have important implications for the take up of other types of accommodation. Because the local supply of different types of accommodation and the extent of developer provision will vary from one locality and project to another, the precise mix of accommodation used can vary considerably from project to project.

Monitoring data, even if they are available, may therefore provide only a rough indication of the likely take up of each type of accommodation.

In the absence of developer provision, the vast majority of in-migrant construction workers are likely to use private rented, B&B/lodgings or caravan accommodation. The use of each type of accommodation can be roughly estimated by drawing on the available monitoring data from other construction projects, adjusted to allow for the particular supply characteristics in the impact area, i.e. the amount of each type of accommodation available, its location, cost and existing occupancy levels (see §3.3.3). For example, if the local supply of tourist accommodation is very limited, concentrated in highly priced hotels at some distance from the project site and is usually fully occupied, the proportion of employees using such accommodation is likely to be relatively low.

Some construction workers may wish to purchase properties in the locality. The number is likely to be minimal during construction projects lasting only a few months, but may be more significant (at least 10%) in cases where construction activity spans several years. The proportion of in-migrant employees buying properties will be closely linked to the proportion bringing families into the impact area. However, since some families will prefer to use rented accommodation, the number of owner-occupied properties required is likely to be lower than the total number of in-migrant families.

In certain cases, the project developer may decide to make specific accommodation provision for the construction workforce. This may involve negotiations with the local planning authority over the provision of additional caravan sites or the expansion of existing sites. In other cases, the developer may wish to provide purpose-built hostel accommodation, located on or adjacent to the construction site. This typically consists of single bedrooms and associated catering, recreational and other facilities. To the extent that such provision is made, the proportion of in-migrant employees using other types of accommodation will be lower than would otherwise have been the case.

It may be helpful to provide estimates of the demand for different types of accommodation in various alternative scenarios, e.g. without any hostel or additional caravan provision, with a small hostel or with a larger hostel. Such estimates will themselves help to clarify the need for such developer provision. The precise geographical distribution of the accommodation taken up by in-migrant employees is difficult to predict: §3.4.2 outlined a possible approach.

3.4.4 The significance of accommodation requirements

The project-related demand for local accommodation is likely to result in a net change in the amount of accommodation available in the impact area. On the one hand, the availability of accommodation will be reduced by the take up of local accommodation by project employees and their families. This accommodation would otherwise have been available to local residents and non-project in-migrants. On the other hand, to the extent that project-related demands are met by the release of unoccupied or under-occupied accommodation and/or the bringing forward of speculative house building development, the amount of accommodation available locally will be higher than would otherwise have been the case. The balance between these two types of change will represent the net change due to the project. This should then be expressed as a percentage of the existing (or projected) stock of accommodation in the impact area. Similar calculations can be made for each separate type of accommodation and for particular settlements or areas within the impact area.

In extreme cases, the net decline in the availability of accommodation due to the project may be such that the project-related and non-project demands for accommodation may outstrip the available local supply. Assessment of such pressures requires projections of the following:

- the likely project-related demand for accommodation (as outlined earlier in the section);
- the likely non-project demand for accommodation by local residents and non-project in-migrants (derived from the projected growth in population and households, as outlined in §3.3.2 and §3.3.4);
- likely changes in the local supply of accommodation, including project-induced changes, such as the release of unoccupied and under-occupied accommodation and the bringing forward of speculative development.

Cases in which the project results in a shortfall in the local supply of accommodation are likely to require the consideration of mitigation measures (DoE 1995). However, in practice, pressure on one locality is likely to be relieved by the diversion of demand (both project and non-project) into adjacent localities. Unless seen as undesirable, this may eliminate the need for mitigation measures.

3.4.5 The demand for local services

In-migrant employees and their families will place demands on a wide range of services provided by local authorities and other public bodies. The demand for these services will largely reflect the age and sex distribution of the in-migrant population (see §3.4.1). For example, in the case of **health and personal social services**, the number of young children and elderly people will be a critical determinant of demand. In such cases, rough estimates of likely demand can be obtained by combining the predicted age and sex structure of the in-migrant population with age and sex-specific data on visiting rates to or by doctors, health visitors or social workers. The latter can be obtained from local and health authorities.

In the case of **education services**, demand also clearly depends on the age structure of the in-migrant population, since provision must be made for all children between the ages of 5 and 16. However, there are complications, given that this provision can be made either by the state or by the independent sector and that some children below and above compulsory school age may also require school or college places. The remainder of this section provides an example of the calculations involved in estimating the number of additional school places likely to be required locally in response to an influx of project employees.

Predicting the demand for additional local school places requires separate estimates of:

- *The total number of children aged 0–18 years* brought into the impact area by in-migrant employees (see §3.4.1).
- *The number of these children below compulsory school age (0–4 years), aged 5–16 and above school leaving age* (see §3.4.1).

- *The proportion of those children below compulsory school age likely to require nursery education in the impact area.* Information on the proportion of this age group currently attending nursery schools, both nationally and in individual LEAs, can be obtained from the Department for Education and Employment (DfEE) *Statistical Bulletin.* The proportion attending nursery schools in the relevant LEA area could then be assumed to apply to the in-migrant children associated with the project. This assumes that there will be no changes in LEA policies on the provision of nursery education before the project gets underway.
- *The proportion of children aged 5–16 attending primary and secondary schools in the impact area.* All children of school age can be assumed to attend either LEA, grant-maintained or independent schools in the impact area.
- *The proportion of children above school leaving age (16–18 years) likely to remain in full- or part-time education in the impact area.* Data on the proportion of this age group attending secondary schools or further education colleges are published in the *Statistical Bulletin* and are updated annually. This information is available for individual LEAs, as well as nationally. Again, these proportions could be assumed to apply to the children brought into the area by in-migrant employees. In practice, the likelihood of this age group remaining in education depends very much on the availability of alternatives, in the form of employment and training opportunities (Bradford 1993). It must therefore be assumed that there will be no changes in the relative attractiveness of these various alternatives.

Some adjustments may be needed to the resulting estimates to allow for the fact that a proportion of the school places taken up locally will be in the **independent sector**. The DfEE *Statistical Bulletin* shows the proportion of pupils of different ages attending independent schools, in England as a whole, regions and some sub-regions. For example, in 1990, the proportion of pupils in England attending independent schools was 5% for 5–10 year olds, 9% for 11–15 year olds and 19% for those aged 16 or over. These national proportions could be assumed to apply to the children brought into the area, again assuming no changes in the relative import-ance of the state and independent sectors before the project gets underway. The estimated number of pupils attending independent schools could then be subtracted from the total school place requirement to show the number of places required in LEA and grant-maintained schools and colleges.

The demand for additional school places is unlikely to be evenly distributed throughout the impact area. The extent to which demand is geographically concen-trated or dispersed will determine the total number of schools affected and the likelihood of strains on educational provision in individual schools. The distribution of school place requirements will largely reflect the place of residence of in-migrant families. Unfortunately, the latter is difficult to predict in the absence of relevant monitoring data: §3.4.2 outlined a possible approach to prediction, but it may be helpful to present a series of estimates based on different assumptions about the concentration or dispersal of in-migrant families.

3.4.6 The significance of demands on local services

An important indicator of the significance of local service impacts is the extent to which **capacity thresholds** are exceeded as a result of the demands arising from the in-migrant population. Let us consider the example of the demand for local school places. If the current accommodation capacity in a school is expected to be almost fully utilised in the absence of the project, and pupil/teacher ratios are already high, then even a small project-induced increase in pupil numbers may create a need for additional classrooms and/or extra teaching staff. In the absence of such additional provision, the result may be overcrowding and an unacceptable increase in class sizes. By contrast, a large increase in pupil numbers in a school with a considerable amount of under-utilised capacity and low pupil/teacher ratios may be much less significant. Increases in pupil numbers in such schools may still be important, even if they do not put the available capacity under pressure. Class sizes will be larger than would otherwise have been the case, and additional staff time may need to be devoted to individual assessments of incoming pupils. Assessment of significance therefore requires information not only on the likely project-related increase in demand, but also on the existing (and projected) utilisation of service capacity.

In certain circumstances, additional service demands may be seen as beneficial. For example, an influx of pupils into a small rural primary school with declining rolls may safeguard the future of the school, either in the short term (during construction) or in the medium to long term (during operation). The nature and significance of local service impacts will change as the project progresses through its various stages. The in-migrant population, including children, will tend to become older, with the result that the type of services demanded will tend to change over time. For example, there will tend to be a shift away from nursery and primary school demand towards secondary school demand. This tendency will be counterbalanced to some extent by the turnover of employees (bringing new, younger, families into the area) and by births in the original in-migrant families.

3.4.7 Local authority finances

Local authority finances can be affected by changes in both their revenue and their expenditure.

Implications for revenue

Major projects can affect the revenues received by their host local authorities in two main ways. First, in-migrant employees buying properties in the impact area will become liable to pay council tax in the local authority area into which they have moved. The likely **increase in council tax receipts** can be roughly estimated by multiplying the predicted number of in-migrant employees purchasing properties locally by the existing average council tax payment in the LAD concerned. Methods to estimate the proportion of in-migrant employees buying properties were outlined in §3.4.3. Information on existing council tax levels is published annually by the DETR (*Council Tax Levels, England*), NAW and SE (*Welsh and Scottish Local Government Financial Statistics*). If project employees purchase houses mainly in higher than average price bands, this simple method will under-estimate the actual increase in receipts.

The second way in which the project will affect local authority revenues is through the population changes brought about by the arrival of in-migrant employees and families. These changes will affect the **standard spending assessment** of the local authority concerned. The standard spending assessment (SSA) is central government's assessment of how much it would cost the local authority to provide a typical or standard level of service. SSAs are a key determinant of the distribution of *revenue support grant* from central government to individual authorities. The SSA consists of a basic amount per head of population, with various weightings to reflect particular local circumstances such as the number of primary and secondary school pupils, daily commuting flows into and out of the area and the extent of social deprivation. The population data used in the calculation of SSAs are the official mid-year estimates published by ONS/GRO. To the extent that in-migrant employees and their families are picked up in these official estimates, the SSA for the authority concerned should be adjusted upwards. However, these adjustments will not take place immediately. There is usually about a two year time lag between an actual increase in population and the resulting increase in revenues.

In principle, if a project results in a 5% increase in a local authority's population (with no changes in the structure of that population), then this should be reflected – after a time lag – in a 5% increase in the authority's SSA and – all other things remaining equal – a similar increase in revenue support grant. It should therefore be possible to estimate the likely increase in SSA and revenues associated with the project-induced increase in population. In practice, things are rather more complicated. First, not all in-migrant employees or families will be picked up by the mid-year population estimates, especially during the construction stage. Construction employees not bringing families into the area are unlikely to appear on the electoral register in the impact area, re-register with a local doctor or appear on local property registers for council tax purposes. They are therefore unlikely to be picked up by any of the data sources used to arrive at the mid-year estimates. As a result, any increase in the authority's SSA is unlikely to fully match the actual percentage increase in population. Any estimates of increased SSAs due to the project must incorporate some allowance for this under-recording of the actual population increase, at least during construction.

A second problem is that the increase in the SSA due to the project will reflect not only the size of the project-induced population increase, but its precise structure (e.g. the number and ages of children); the latter is more difficult to estimate accurately. A final problem is that an increase in the SSA for an authority does not necessarily produce an equivalent percentage increase in revenue support grant. For example, if the project results in an increase in the number of council tax payers (as outlined above), the additional revenue received will be taken into account in determining the amount of grant distributed to the authority.

Contrary to popular opinion, local authorities in the UK do not benefit directly from the payment of **non-domestic rates** by the project developer during construction or operation. Receipts from non-domestic rates are pooled nationally and then redistributed to individual authorities on a per capita basis. However, to the extent that the project-related population increase is recorded in the official mid-year estimates, the authority should receive increased receipts from the non-domestic rate pool. The likely increase can be calculated roughly by multiplying the existing per capita receipts from the pool by the expected increase in local population. However, for the reasons given above, the recorded increase in population in official estimates may not fully reflect the actual increase (especially during construction) and this should be allowed for in any estimates. Information on existing levels of revenue support grant, SSAs and receipts from the non-domestic rate pool are published annually by the Chartered Institute of Public Finance and

Accountancy's (CIPFA's) Statistical Information Service, in *Finance and General Rating Statistics* (for English and Welsh local authorities) and *Rating Review* (for Scottish authorities).

Implications for expenditure

In-migrant employees and their families will place demands on a range of services provided by local authorities, health authorities and other public bodies. These service demands will entail additional expenditure for the authorities concerned. Let us take the example of the arrival of pupils into LEA schools within the impact area. Methods to estimate the likely number of such pupils were outlined in §3.4.5. But how can the additional expenditure necessitated by the arrival of these pupils be estimated?

The simplest approach would be to multiply the expected number of pupils by the existing annual average cost per pupil in the LEA concerned. Data on average expenditure per pupil in each individual LEA are published annually by CIPFA in *Education Statistics, Estimates* (for the current year) and *Actuals* (for the previous year). This average cost method has two main weaknesses. First, costs per pupil vary according to the age group involved – they are invariably higher for secondary school pupils than for primary school pupils. A more sophisticated approach would therefore involve combining estimates of the expected numbers of in-migrant pupils in particular age groups with the average cost per pupil for each of these age groups. The CIPFA publications noted above provide data on costs per pupil for each LEA, broken down into single year age bands.

A second and more fundamental weakness of the average cost approach is that it fails to distinguish between fixed and variable costs in service provision. **Fixed (overhead) costs** do not vary in response to changes in the number of pupils in individual schools. Examples include most of the costs associated with school buildings, maintenance, heating, cleaning, rates and central support and management functions. **Variable costs** are those which change in response to changes in pupil numbers. Examples include capitation allowances (which are based on the number and ages of children on the roll at the beginning of each year) and teachers' salaries (if the increase in pupil numbers results in additional staff being taken on). Existing average costs per pupil include both fixed and variable costs, and are therefore unlikely to be a reliable guide to the actual costs incurred as a result of a marginal increase in the number of pupils. In schools with considerable surplus capacity, in which the arrival of pupils does not create a need for additional staff or accommodation, the additional cost per pupil is likely to be considerably lower than existing average costs in the LEA as a whole. Estimates of additional expenditure must therefore be carefully justified, with a clear distinction being drawn between the fixed and variable cost elements.

Similar estimates can also be made of the additional expenditure incurred in **other service areas**, such as police, fire, recreation and personal social services. Expenditure in some of these service areas may be rather unresponsive to small changes in population, unless critical capacity thresholds are likely to be approached. Information on existing local authority expenditure per head of population in these service areas is again available from CIPFA's Statistical Information Service. The proposed development may also necessitate the provision of improved infrastructure by the local authority (or authorities) concerned. This will typically include the construction of new roads or the improvement of existing ones. The local authority will normally require the developer to fund the full capital cost of such provision.

3.4.8 The significance of changes in local authority finances

Predictions should be made of the future stream of project-induced revenues and expenditures, ideally for each year of the construction and operational stages of the project life. Although these two streams may balance over the lifetime of the project, there are likely to be periods during which there are shortfalls or surpluses. Any significant shortfalls in revenues, and their likely timing and duration, should be noted. For many projects, the build-up of revenues is likely to lag behind the need for additional expenditure. For example, additional population will create immediate demands on local services, but will be reflected in increased revenue support grant only after a time lag. The construction stage may also see little increase in revenues, with most in-migrant employees not buying properties locally and not being recorded in official population estimates.

3.4.9 Other social impacts

Other social impacts can be wide-ranging and may include:

- increased crime levels locally, particularly during the construction stage, associated with an influx of young male itinerant employees into the impact area;
- changes in the occupational and socio-economic mix of the population; and
- linked to the above, problems in the integration of incoming employees and families into the local community and community activities. There may be a clash of lifestyles or expectations between incomers and the existing host community.

An extensive literature concerned with the assessment of such social and cultural impacts is available, much of it written from a North American perspective. Further details are provided in §3.7. Prediction of such impacts is difficult, but is likely to require at least a comparison of the predicted age, sex and occupational profile of in-migrants with that of the existing population in the impact area. The latter can be determined largely by reference to census data, as outlined in §3.3.1. Monitoring studies may be helpful in indicating the likely scale of certain impacts (e.g. see Glasson & Chadwick 1995 for an assessment of the impact of a major construction project on local crime levels).

3.5 Mitigation

A number of approaches to the mitigation of **demographic impacts** are available. The most basic would be to encourage the maximum recruitment of labour from within daily commuting distance of the project site, thereby reducing the number of employees and families moving into the impact area. Possible methods to encourage the use of local labour by developers and contractors were discussed in Chapter 2. In addition, during the construction stage, developer policies on travel, accommodation and relocation allowances might be used to influence the relative attractiveness of daily and weekly commuting versus relocation. Such policies might lead to some reduction in the proportion of in-migrant employees relocating and bringing families into the area.

The mitigation of local **accommodation impacts** is likely to involve attempts either to provide additional accommodation for the workforce or to encourage the use of unoccupied or under-occupied accommodation in the impact area. Encouragement of the sharing of accommodation would also be a useful mitigation measure, but it is uncertain how this could be carried out in practice. The provision of accommodation specifically for the workforce, in the form of purpose-built hostel or additional caravan accommodation, has already been discussed in §3.4.3. The success of such provision as a mitigation measure will depend on its attractiveness in relation to the alternatives available locally, in terms of location, facilities and cost. The release of unoccupied accommodation is rather more difficult to influence. During construction, one approach might involve the placing of advertisements in the local press requesting those willing to provide workforce accommodation to contact the developer. This may alert potential providers of accommodation to the opportunities presented by the project. In some circumstances, it may be considered desirable to encourage the use of local B&B and other tourist accommodation (e.g. to boost occupancy levels outside a short tourist season). This could be achieved by the compilation of a directory of local accommodation establishments by the developer, and its use by contractors and individuals seeking accommodation in the area.

Impacts on **local services and local authority finance** can be partially mitigated by the direct provision of certain facilities by the developer. Examples might include a medical centre and fire-fighting equipment and staff located on the project site, as well as recreational facilities for the workforce. Developer funding of additional local authority provision necessitated by the project is also likely to be requested. Funding of local community projects may also be offered as partial compensation for the adverse impacts of the project.

3.6 Monitoring

Existing monitoring of demographic and social impacts is limited, other than for large-scale energy and resource development projects (Chadwick & Glasson 1999). Ideally, such monitoring should consist of three key elements. The first of these is the establishment of **administrative systems to ensure a regular flow of information on key parameters**, including at the very least the total numbers directly employed on the project and the mix of local and in-migrant employees. During most construction projects, the developer is likely to request this type of information from the contractors on site as a routine part of project management, for example to monitor earnings levels, bonuses and allowances across the construction site. The provision of such information can be made a contractual requirement. Existing monitoring systems can therefore often be used with only minimal modifications. For most projects, information on the operational workforce should be directly available to the developer via its own personnel records. However, this will not be the case for certain developments, such as business parks or retail projects, where several employers occupy the floorspace provided by the developer. In such cases, the developer (or perhaps the local authority) may wish to establish data collection systems covering all occupants, with the submission of information being requested on a regular basis.

The systems described above will, at best, only indicate the total number of employees moving into the impact area. Information on the number of these employees bringing families, the characteristics of these families, the type and location of accommodation taken up and the use of local services can only be obtained directly from the workforce itself. The second component of any monitoring system must therefore be a **periodic survey of the project workforce**. This is likely to involve interviewing a sample of the workforce, with care taken to ensure a representative coverage of all types of employees. Such surveys can also be used to obtain information on other issues, such as workforce expenditure and journey to work patterns. Survey work of this type might be repeated on an annual basis, at least during the initial stages of the development.

The final element in any monitoring system should be the **monitoring of various social and economic trends within the impact area**. These can include regular monitoring of house prices or rent levels, the amount of housebuilding, occupancy levels in local B&B and other accommodation, school rolls, doctors' list sizes or crime levels. Such trends should be compared with those in suitable control areas, including the wider region or sub-region; comparison with national trends may also be appropriate. In addition, periodic surveys of local service providers (e.g. headteachers or doctors) may provide a useful source of monitoring data.

3.7 Further reading

Useful data sources in the assessment of economic and social impacts include **census data** and a range of other **official statistics** published by government. In the UK, a number of guides to the use of census data have been published, mainly in response to the release of 1991 census data. These include Dale & Marsh (1993), Denham (1992), Leventhal *et al.* (1993), LGMB (1992), OPCS (1992) and Openshaw (1995). Useful guides to other UK official statistics can be found in Healey (1991), Mort (1992) and ONS (1996). Recent data are also available from the ONS website.

Government guidance on the assessment of socio-economic impacts is rather limited at present, although a number of examples can be found in North America, Australia and New Zealand, as well as in international aid agencies. Examples include ADB (1991, 1994), CEPA (1994), Lang & Armour (1981), ODA (1995), SIAWG (1995), and USAID (1993). Other useful guidance can be found in ICGPS (1995) and Shell International Exploration & Production (1996).

A number of **general texts** on EIA include some discussion of socio-economic impacts and their assessment. Examples include Barrow (1997), Canter (1995), Clark *et al.* (1981), Colombo (1992), DoE (1995), Erickson (1994), Petts & Eduljee (1994), and Vanclay & Bronstein (1995). The incorporation of socio-economic impacts into EIA is also discussed in Bond (1995), Dale & Lane (1995), Dale *et al.* (1997), Glasson & Heaney (1993), Kirkpatrick & Lee (1997), Kolhoff (1996), Newton (1995) and Pellizzoni (1992).

Specialist texts on socio-economic and social impact assessment, mainly written from a North American perspective, include Branch *et al.* (1984), Burdge (1994a,b), Canter *et al.* (1985), Finterbusch *et al.* (1983, 1990), Halstead *et al.* (1984), Lang & Armour (1981), Leistritz & Murdoch (1981), Maurice & Fleischman (1983), Taylor *et al.* (1995), and Wildman & Baxter (1985). Other useful references include Becker (1995), Burdge & Vanclay (1995), Leistritz (1994), and Leistritz *et al.* (1994).

Specific impact or development types, or aspects of socio-economic assessment have also generated a considerable literature. For example, the socio-economic impacts of **major projects**, mainly in relation to large-scale energy and resource development projects, are discussed in Buchan & Rivers (1990), Cocklin & Kelly (1992), Denver Research Institute (1982), Gilmore *et al.* (1980), Glasson & Chadwick (1995), Hill *et al.* (1998), and Leistritz & Maki (1981). In a related area, the social impacts of rapid 'boomtown' development, largely in a North American context, are discussed in England & Albrecht (1984), Freudenburg (1984), and Thompson & Bryant (1992). The social impact of **tourism development** is another area highlighted in the literature. Examples include Beekhuis (1981), and Shera & Matsuoka (1992).

The **monitoring** of socio-economic impacts is examined in Bisset & Tomlinson (1988), Chadwick & Glasson (1999), Denver Research Institute (1982), Gilmore *et al.* (1980), and Glasson (1994). More general reviews of the field of socio-economic and social impact assessment can be found in Burdge (1987), Burdge & Vanclay (1996), Finterbusch (1995), Freudenburg (1986), Lane (1997), McDonald (1990), Murdoch *et al.* (1986), Rickson *et al.* (1990), and Wildman (1990).

A number of publications provide an overview of experience with socio-economic impact assessment in **specific countries**. UK and European experience is discussed in Glasson & Heaney (1993), Juslen (1995), Newton (1995), Pellizzoni (1992) and Pinhero & Pires (1991). US and Canadian practice is reviewed in Denq & Altenhofel (1997), Finterbusch (1995), Gagnon (1995), Haque (1996), Lang & Armour (1981), Maurice & Fleischman (1983), and Murdoch *et al.* (1986). The development of socio-economic impact assessment in Australia and New Zealand is reflected in an extensive literature. Examples include Beckwith (1994), Buchan & Rivers (1990), CEPA (1994), Cocklin & Kelly (1992), Dale & Lane (1995), Dale *et al.* (1997), Howitt (1989), Lane (1997), Rivers & Buchan (1995), Seebohm (1997) and SIAWG (1995).

Social impact assessment in **developing countries**, and for projects financed by international aid agencies, is discussed in ADB (1991, 1994), Burdge (1990), Derman & Whiteford (1985), Finterbusch *et al.* (1990), Fu-Keung Ip (1990), Henry (1990), Jiggins (1995), ODA (1995), Ramanathan & Geetha (1998), Rickson *et al.* (1990), Suprapto (1990), and USAID (1993).

References

ADB (Asian Development Bank) 1991. *Guidelines for social analysis of development projects*. Manila: ADB.

ADB 1994. *Handbook for incorporation of social dimensions in projects*. Manila: Social Development Unit, ADB.

Barrow CJ 1997. *Environmental and social impact assessment: an introduction*. London: Arnold.

Becker HA 1995. Demographic impact assessment. In *Environmental and social impact assessment*, F Vanclay & DA Bronstein (eds), 141–151. Chichester: Wiley.

Beckwith JA 1994. Social impact assessment in Western Australia at a crossroads. *Impact Assessment* **12**(2), 199–213.

Beekhuis JV 1981. Tourism in the Caribbean: impacts on the economic, social and natural environment. *Ambio* **X**(6), 325–331.

Bisset R & P Tomlinson 1988. Monitoring and auditing of impacts. In *Environmental impact assessment: theory and practice*, P Wathern (ed.), 117–128. London: Unwin Hyman.

Bond AJ 1995. Integrating socio-economic impact assessment into EIA. *Environmental Assessment* **3**(4), 125–127.

Bradford M 1993. Population change and education: school rolls and rationalisation before and after the 1988 Education Reform Act. In *Population matters – the local dimension*, AG Champion (ed.), 64–82. London: Chapman.

Bramley G, W Bartlett & C Lambert 1995. *Planning, the market and private housebuilding.* London: UCL Press.

Bramley G & C Watkins 1995. *Circular projections: household growth, housing development and the household projections.* London: CPRE.

Branch K, DA Hooper, J Thompson & JC Creighton 1984. *Guide to social impact assessment: a framework for assessing social change.* Boulder, Colorado: Westview.

Buchan D & MJ Rivers 1990. Social impact assessment: development and application in New Zealand. *Impact Assessment Bulletin* 8(4), 97–105.

Burdge RJ 1987. Social impact assessment and the planning process. *Environmental Impact Assessment Review* 7(2), 141–150.

Burdge RJ 1990. The benefits of social impact assessment in third world development. *Environmental Impact Assessment Review* 10(1–2), 123–134.

Burdge RJ 1994a. *A community guide to social impact assessment.* Middleton, Wisconsin: Social Ecology Press.

Burdge RJ 1994b. *A conceptual approach to social impact assessment.* Middleton, Wisconsin: Social Ecology Press.

Burdge RJ & F Vanclay 1995. Social impact assessment. In *Environmental and social impact assessment*, F Vanclay & DA Bronstein (eds), 31–65. Chichester: Wiley.

Burdge RJ & F Vanclay 1996. Social impact assessment: a contribution to the state-of-the-art series. *Impact Assessment* 14(1), 59–86.

Canter LW (ed.) 1995. *Environmental impact assessment*, 2nd edn. New York: McGraw-Hill.

Canter LW, B Atkinson & FL Leistritz 1985. *Impact of growth: a guide for socio-economic impact assessment and planning.* Chelsea, Michigan: Lewis.

CEPA (Commonwealth Environmental Protection Agency) 1994. *Social impact assessment.* Barton, ACT, Australia: CEPA.

Chadwick A & J Glasson 1999. Auditing the socio-economic impacts of a major construction project: the case of Sizewell B nuclear power station. *Journal of Environmental Planning and Management* 42(6), 811–836.

Clark BD, K Chapman, R Bisset, P Wathern & M Barrett 1981. *A manual for the assessment of major development proposals.* London: HMSO.

Cocklin C & B Kelly 1992. Large-scale energy projects in New Zealand: whither social impact assessment? *Geoforum* 23(1), 41–60.

Colombo AG (ed.) 1992. *Environmental impact assessment.* Dordrecht, The Netherlands: Kluwer Academic.

Congdon P & P Batey (eds) 1989. *Advances in regional demography: information, forecasts and models.* London: Belhaven Press.

CPRE (Council for the Protection of Rural England) 1994. *The housing numbers game.* London: CPRE.

Dale A & C Marsh (eds) 1993. *The 1991 census users' guide.* London: HMSO.

Dale AP & MB Lane 1995. Queensland's Social Impact Assessment Unit: its origins and prospects. *Queensland Planner* 35(3), 5–10.

Dale AP, P Chapman & ML McDonald 1997. Social impact assessment in Queensland: why practice lags behind legislative opportunity. *Impact Assessment* 15(2), 159–179.

Denham C 1992. *An introduction to the 1991 census.* Market Research Society Census Briefing.

Denq F & J Altenhofel 1997. Social impact assessments conducted by federal agencies. *Impact Assessment* 15(3), 209–231.

Denver Research Institute 1982. *Socio-economic impacts of power plants.* EPRI EA-2228. Palo Alto, California: Electric Power Research Institute.

DoE (Department of the Environment) 1995. *Preparation of environmental statements for planning projects that require environmental assessment: a good practice guide.* London: HMSO.

Derman W & S Whiteford (eds) 1985. *Social impact analysis and development planning in the third world.* Boulder, Colorado: Westview.

England JL & SL Albrecht 1984. Boomtowns and social disruption. *Rural Sociology* **49**, 230–246.

England JR, KI Hudson, RJ Masters, KS Powell & JD Shortridge (eds) 1985. *Information systems for policy planning in local government.* Harlow: Longman.

Erickson PA 1994. *A practical guide to environmental impact assessment.* London: Academic Press.

Field B & B MacGregor 1987. *Forecasting techniques for urban and regional planning.* London: UCL Press.

Finterbusch K 1995. In praise of SIA – a personal review of the field of social impact assessment: feasibility, justification, history, methods, issues. *Impact Assessment* **13**(3), 229–252.

Finterbusch K, LG Llewellyn & CP Wolf (eds) 1983. *Social impact assessment methods.* Beverley Hills, California: Sage.

Finterbusch K, LJ Ingersol & LG Llewellyn (eds) 1990. *Methods for social impact analysis in developing countries.* Boulder, Colorado: Westview.

Freudenburg WR 1984. Differential impacts of rapid community growth. *American Sociological Review* **49**, 697–705.

Freudenburg WR 1986. Social impact assessment. *Annual Review of Sociology* **12**, 451–478.

Fu-Keung Ip D 1990. Difficulties in implementing social impact assessment in China: methodological considerations. *Environmental Impact Assessment Review* **10**(1–2), 113–122.

Gagnon C 1995. Social impact assessment in Quebec: issues and perspectives for sustainable community development. *Impact Assessment* **13**(3), 272–288.

Gilmore JS, DM Hammond, JM Uhlmann, KD Moore, DC Coddington 1980. The impacts of power plant construction: a retrospective analysis. *Environmental Impact Assessment Review* **1**, 417–420.

Glasson J 1994. Life after the decision: the importance of monitoring in EIA. *Built Environment* **20**(4), 309–320.

Glasson J & A Chadwick 1995. *The local socio-economic impacts of the Sizewell B PWR power station construction project, 1987–1995: summary report.* Report to Nuclear Electric plc. Oxford: School of Planning, Oxford Brookes University.

Glasson J & D Heaney 1993. Socio-economic impacts: the poor relations in British environmental impact statements. *Journal of Environmental Planning and Management* **36**(3), 335–343.

Halstead JN, RA Chase, SH Murdoch & FL Leistritz 1984. *Socioeconomic impact management.* Boulder, Colorado: Westview.

Haque EE 1996. The integration of regional economic impact assessment with social impact assessment: the case of water improvement service projects in rural Manitoba, Canada. *Impact Assessment* **14**(4), 343–369.

Healey MJ (ed.) 1991. *Economic activity and land use: the changing information base for local and regional studies.* Harlow: Longman.

Henry R 1990. Implementing social impact assessment in developing countries: a comparative approach to the structural problem. *Environmental Impact Assessment Review* **10**(1–2), 91–101.

Hill AE, CL Seyfrit & MJE Danner 1998. Oil development and social change in the Shetland Islands 1971–1991. *Impact Assessment and Project Appraisal* **16**(1), 15–25.

Howitt R 1989. Social impact assessment and resource development: issues from the Australian experience. *Australian Geographer* **20**(2), 153–166.

ICGPS (Interorganizational Committee on Guidelines and Principles for Social Impact Assessment) 1995. Guidelines and principles for social impact assessment. *Environmental Impact Assessment Review* **15**(1), 11–43.

Jenkins J & JR Walker 1985. School roll forecasting. In *Information systems for policy planning in local government*, JR England et al. (eds), 96–112. Harlow: Longman.

Jiggins J 1995. Development impact assessment: impact assessment of aid projects in non-western countries. In *Environmental and social impact assessment*, F Vanclay & DA Bronstein (eds), 265–281. Chichester: Wiley.

Juslen J 1995. Social impact assessment: a look at Finnish experiences. *Project Appraisal* **10**(3), 163–170.

King D 1987. *The Chelmer population and housing model.* Paper presented at Regional Studies Association workshop on Regional Demography. London: London School of Economics.

Kirkpatrick C & N Lee (eds) 1997. *Sustainable development in a developing world: integrating socio-economic appraisal and environmental assessment.* Cheltenham: Edward Elgar.

Kolhoff AJ 1996. Integrating gender assessment study into environmental impact assessment. *Project Appraisal* **11**(4), 261–266.

Lane M 1997. Social impact assessment: strategies for improving practice. *Australian Planner* **34**(2), 100–102.

Lang R & A Armour 1981. *The assessment and review of social impacts.* Ottawa: Federal Environmental Assessment and Review Office.

Leistritz FL 1994. Economic and fiscal impact assessment. *Impact Assessment* **12**(3), 305–318.

Leistritz FL & KC Maki 1981. *Socio-economic effects of large-scale resource development projects in rural areas: the case of McLean County, North Dakota.* Fargo, North Dakota: Department of Agricultural Economics, North Dakota State University.

Leistritz FL & H Murdoch 1981. *The socio-economic impact of resource development: methods of assessment.* Boulder, Colorado: Westview.

Leistritz FL, RC Coon & RR Hamm 1994. A microcomputer model for assessing socioeconomic impacts of development projects. *Impact Assessment* **12**(4), 373–384.

Leventhal B, C Moy & J Griffin (eds) 1993. *An introductory guide to the 1991 census.* Henley on Thames: NTC Publications.

LGMB (Local Government Management Board) 1992. *Working with the census.* London: LGMB.

Maurice EV & WA Fleischman (eds) 1983. *Sociology and social impact analysis in federal resource management agencies.* Washington: US Department of Agriculture, Forest Service.

McDonald GT 1990. Regional economic and social impact assessment. *Environmental Impact Assessment Review* **10**(1–2), 25–36.

Mort D 1992. *UK statistics: a guide for business users.* Aldershot: Ashgate.

Murdoch SH, FL Leistritz & RR Hamm 1986. The state of socioeconomic impact analysis in the United States of America: limitations and opportunities for alternative futures. *Journal of Environmental Management* **23**, 99–117.

Newton JA 1995. *The integration of socio-economic impacts in environmental impact assessment and project appraisal.* MSc dissertation, University of Manchester (UMIST) (mimeo).

ODA (Overseas Development Administration) 1995. *A guide to social analysis for projects in developing countries.* London: HMSO.

ONS (Office for National Statistics) 1996. *Guide to official statistics – 1996 edition.* London: HMSO.

OPCS (Office of Population Censuses and Surveys) 1992. *An introduction to the 1991 census and its output.* London: OPCS.

Openshaw S (ed.) 1995. *Census users' handbook.* Cambridge: GeoInformation International.

Pellizzoni L 1992. Sociological aspects of EIA. In *Environmental impact assessment,* AG Colombo (ed.), 313–334. Dordrecht, The Netherlands: Kluwer Academic.

Petts J & G Eduljee 1994. *Environmental impact assessment for waste treatment and disposal facilities.* Chichester: Wiley.

Pinhero P & AR Pires 1991. Social impact analysis in environmental impact assessment: a Portuguese case study. *Project Appraisal* **6**(1), 2 only.

Ramanathan R & S Geetha 1998. Socio-economic impact assessment of industrial projects in India. *Impact Assessment and Project Appraisal* **16**(1), 27–31.

Rickson RE, T Hundloe, GT McDonald & RJ Burdge (eds) 1990. Social impact of development: putting theory and methods into practice. *Environmental Impact Assessment Review* **10**(1–2) (special issue).

Rivers MJ & D Buchan 1995. Social assessment and consultation: New Zealand cases. *Project Appraisal* **10**(3), 181–188.

Seebohm K 1997. Guiding principles for the practice of social assessment in the Australian water industry. *Impact Assessment* **15**(3), 233–251.

Shell International Exploration and Production 1996. *Social impact assessment – HSE manual.* The Hague: Shell.

Shera W & J Matsuoka 1992. Evaluating the impact of resort development on an Hawaiian island: implications for social impact assessment policy and procedures. *Environmental Impact Assessment Review* **12**(4), 349–362.

SIAWG (Social Impact Assessment Working Group) 1995. *Social impact assessment in New Zealand: a practical approach.* Wellington: Town and Country Planning Directorate, Ministry of Works.

Suprapto RA 1990. Social impact assessment and environmental planning: the Indonesian experience. *Impact Assessment Bulletin* **8**(1–2), 25–28.

Taylor CN, BC Hobson & CG Goodrich 1995. *Social assessment: theory, process and techniques.* Centre for Resource Management, Lincoln University, New Zealand.

Thompson JG & D Bryant 1992. Fiscal impact in a western boomtown: unmet expectations. *Impact Assessment* **10**(3).

USAID (US Agency for International Development) 1993. *Handbook No. 3: project assistance (Appendix 3F – social soundness analysis).* Washington, DC: USAID.

Vanclay F & DA Bronstein (eds) 1995. *Environmental and social impact assessment.* New York: Wiley.

Wildman P 1990. Methodological and social policy issues in SIA. *Environmental Impact Assessment Review* **10**(1–2), 69–79.

Wildman PH & GB Baxter 1985. *The social assessment handbook: how to assess and evaluate the social impact of resource development on local communities.* Sydney: Social Impact.

Woods R & P Rees (eds) 1986. *Population structures and models.* London: Allen & Unwin.

4 Noise

Riki Therivel and Mike Breslin

4.1 Introduction

Virtually all development projects have noise impacts. Noise during construction may be due to such activities as land clearance, piling, and the transport of materials to and from the site. During operation noise levels may decrease for some forms of developments such as science parks or new towns, but may remain high or even increase for developments such as new roads or industrial processes. Demolition is a further cause of noise. As a result, despite the fact that EU Directives 85/337/EEC and 97/11/EU (§1.4) do not require noise to be analysed, the EIAs for most projects do consider noise.

Noise is a major and growing form of **pollution**. It can interfere with communication, increase stress and annoyance, cause anger at the intrusion of privacy, and disturb sleep, leading to lack of concentration, irritability, and reduced efficiency. It can contribute to stress-related health problems such as high blood pressure. Prolonged exposure to high noise levels can cause deafness or partial hearing loss. Noise can also affect property values and community atmosphere. A noise attitude survey carried out by the Building Research Establishment in 1991 found that more than half of the homes in England and Wales were exposed to noise levels over the standards recommended by the World Health Organisation: 47% of respondents were affected by traffic noise, 41% by aircraft noise, 13% by train noise, and more than 4% by construction noise. In just the three years between 1992 and 1995, noise complaints received by environmental health officers rose by almost 50% (DETR 1997). The Royal Commission on Environmental Pollution (1994) estimates that noise from traffic alone costs £1.2–5.4 billion each year in the form of productivity losses, decreased house values, and cost of abatement measures.

Although most EIAs – and this chapter – are limited to the impact of noise on people, noise may also affect animals and in certain (highly unusual) cases EIAs will need to include specialist studies on these impacts. Bregman & Mackenthun (1992) summarise previous studies on animals' reactions to noise, and impacts of disturbance (including noise) are discussed in §11.5.5. Although noise is linked to vibration, this chapter deals only with noise; most EIAs do not cover vibration. It should be noted, however, that for some studies (particularly major railway projects and/or projects involving substantial demolition or piling) vibration effects can be significant and a full vibration assessment must be carried out. In the UK the principal vibration standards to be considered are British Standards 6472 and 7385 (BSI 1992, 1993).

4.2 Definitions and concepts

4.2.1 Definitions

Noise is unwanted sound. This definition holds within it one of the core aspects of noise impact assessment: namely it deals with peoples' subjective responses ('unwanted') to an objective reality ('sound'). The physical level of noise does not directly correspond to the level of annoyance it causes (think about your favourite CD and your neighbours'/parents' reaction to it), yet it is the annoyance caused by noise that is important in EIA. Noise impact assessment revolves around the concept of quantifying and 'objectifying' peoples' personal responses. The following definitions and concepts all relate to this issue.

   Sound consists of pressure variations detectable by the human ear. These pressure variations have two characteristics, frequency and amplitude. Sound **frequency** refers to how quickly the air vibrates, or how close the sound waves are to each other (in cycles per second, or Hertz (Hz)). For example, the sound from a transformer has a wavelength of about 3.5 m, and hums at a frequency of 100 Hz; a television line emits waves of about 0.03 m, and whistles at about 10,000 Hz or 10 kHz. Frequency is subjectively felt as the **pitch** of the sound. Broadly, the lowest frequency audible to humans is 18 Hz, and the highest is 18,000 Hz. For convenience of analysis, the audible frequency spectrum is often divided into standard octave bands of 32, 63, 125, 250, 500, 1 k, 2 k, 4 k and 8 kHz.

   Sound **amplitude** refers to the amount of pressure exerted by the air, which is often pictured as the height of the sound waves. Amplitude is described in units of pressure per unit area, micropascals (µPa). The amplitude is sometimes converted to sound **power**, in picowatts (10^{-12} watts), or sound intensity (in 10^{-12} watts/m^2). Sound intensity is subjectively felt as the **loudness** of sound. However, none of these measures are easy to use because of the vast range which they cover (see Table 4.1). As a result, a logarithmic scale of **decibels** (dB) is used. A sound level in decibels is given by

$$L = 10 \log_{10}(P/p)^2 \text{ dB},$$

where P is the amplitude of pressure fluctuations, and p is 20 µPa, which is considered to be the lowest audible sound. The sound level can also be described as

$$L = 10 \log_{10}(I/i) \text{ dB},$$

where I is the sound intensity and i is 10^{-12} watts/m^2, or by

$$L = 10 \log_{10}(W/w) \text{ dB},$$

where W is the sound power, and w is 10^{-12} watts. The range of audible sound is generally from 0 dB to 140 dB, as is shown in Table 4.1.

   Because of the logarithmic nature of the decibel scale, a doubling of the power or intensity of a sound, for instance, adding up two identical sounds, generally leads to an increase of 3 dB, not a doubling of the decibel rating. For example, two lorries, each at 75 dB, together produce 78 dB. Multiplying the sound power by ten (e.g. ten lorries)

Table 4.1 Sound pressure, intensity and level

Sound pressure (µPa)	Sound power (10^{-12} watt) or intensity level (10^{-12} watt/m^2)	Sound level (dB)	Example
200,000,000	100,000,000,000,000	140	threshold of pain
	10,000,000,000,000	130	riveting on steel plate
20,000,000	1,000,000,000,000	120	pneumatic drill
	100,000,000,000	110	loud car horn at 1 m
2,000,000	10,000,000,000	100	alarm clock at 1 m
	1,000,000,000	90	inside underground train
200,000	100,000,000	80	inside bus
	10,000,000	70	street-corner traffic
20,000	1,000,000	60	conversational speech
	100,000	50	business office
2,000	10,000	40	living room
	1,000	30	bedroom at night
200	100	20	broadcasting studio
	10	10	normal breathing
20	1	0	threshold of hearing

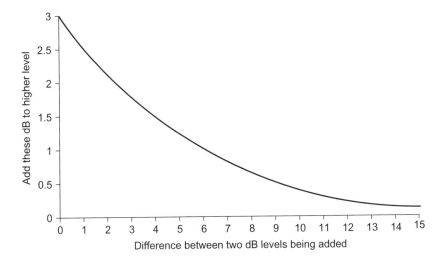

Figure 4.1 Adding two sources of sound.

leads to an increase of 10 dB. Figure 4.1 shows how the dB increase can be calculated if one noise source is added to another. Box 4.1 shows two examples of these principles.

Subjectively, a change of 3 dB is generally held to be barely detectable by the human ear under normal listening circumstances, providing that the change in sound pressure level is not accompanied by some change in the character of the sound.[1]

---

1 This fundamental principle, however, is currently the subject of debate. For instance, the *Design Guide for Roads and Bridges* (DoT 1993) asserts that abrupt changes as small as 1 dB in, say, road traffic noise can bring appreciable benefits or disbenefits. However, long-term significant effects are unlikely from changes of less than 3 dB (DETR 1997).

Box 4.1 *Adding sound levels: examples*

Adding sources with different levels
Assume three sources with sound levels of 59 dB, 55 dB and 61 dB. Start with two of these, e.g. 59 and 55 dB. Take the higher: 59. Calculate the difference between the two levels being added: $59 - 55 = 4$. Figure 4.1 shows that about 1.4 dB needs to be added to the higher level: $59 + 1.4 = 60.4$. To add the third level, repeat the process using 60.4 (i.e. 55 + 59) and 61. The total of all three is about 63.7 dB.

The same procedure could be carried out with a different combination of the three levels. For instance, start with 61 and 59. The difference is 2. Figure 4.1 shows that about 2 dB needs to be added to the higher figure: $61 + 2 = 63$. Repeating the process with 63 and 55 gives about 63.7 dB.

Adding ten equal levels
Assume that all of ten sound levels are at 50 dB. Remember that two equal sound levels added together equal one level plus 3 dB (as in the far left of Fig. 4.1). Start from top left:

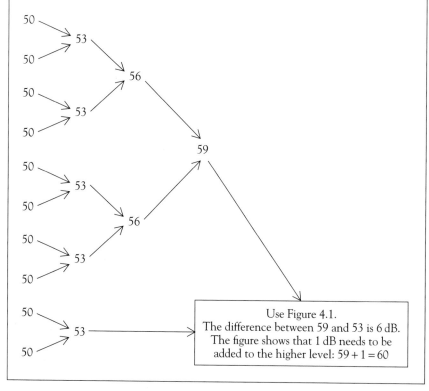

A change of 10 dB is broadly perceived as a doubling/halving of loudness. Consequently, the logarithmic decibel scale, in addition to simplifying the necessary manipulation of a very large range of sound pressures/intensities, is conveniently related to the human perception of loudness.

The human ear is more sensitive to some frequencies than to others (think of fingernails on a blackboard). It is most sensitive to the 1 kHz, 2 kHz, and 4 kHz

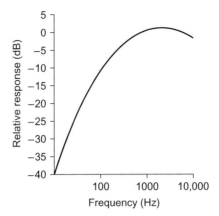

Figure 4.2 A-weighting curve.

octaves, and much less sensitive at the lower audible frequencies. For instance, tests of human perception of noise have shown that a 70 dB sound at 4 kHz sounds as loud as a 1 kHz sound of about 75 dB, and a 70 dB sound at 63 Hz sounds as loud as a 1 kHz sound of about 45 dB. Since most sound analyses, including those in EIA, are concerned with the loudness experienced by people rather than the actual physical magnitude of the sound, an **A-weighting curve** is used to give a single figure index which takes account of the varying sensitivity of the human ear; this is shown in Figure 4.2. Most sound-measuring instruments incorporate circuits that carry out this weighting automatically, and all EIA results should be A-weighted (dB(A)). Other weightings exist, but are rarely used.

Noise levels are rarely steady: they rise and fall with the types of activity taking place in the area. Time-varying noise levels can be described in a number of ways. The principal measurement index for environmental noise is the **equivalent continuous noise level**, LA_{eq} (DoE 1995). The LA_{eq} is a notional steady noise level which, over a given time, would provide the same energy as the time-varying noise: it is calculated by averaging all of the sound pressure/power/intensity measurements, and converting that average into the dB scale. Most environmental noise meters read this index directly. LA_{eq} has the dual advantages that it (a) takes into account both the energy and duration of noise events, and (b) is a reasonable indicator of likely subjective response to noise from a wide range of different noise sources.

In the UK, in addition to LA_{eq}, statistical indices are used as the basis of some types of noise assessment. LA_{90}, the dB(A) level which is exceeded for 90% of the time, is used to indicate the noise levels during quieter periods, or the **background noise**. Industrial noise, or noise from stationary plant, is often assessed against the background noise level (BS 4142). LA_{10}, the dB(A) level which is exceeded for 10% of the time and which is representative of the noisier sounds, is used as the basis of road traffic noise assessment in the UK.[2] Note that, in all cases, $L_{10} \geq L_{eq} \geq L_{90}$.

---

2 LA_{10} is not necessarily a better indicator of subjective response to road traffic noise than LA_{eq}. It owes its continued use in the UK to its appearance in legislation and because the Department of Transport's guidance document *Calculation of Road Traffic Noise* (DoT 1988) is formulated in terms of LA_{10} noise level.

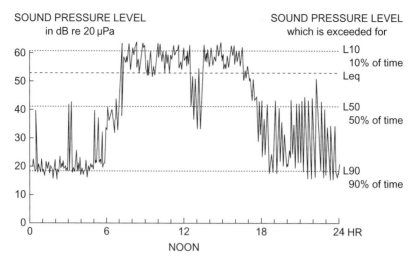

Figure 4.3 Sound levels exceeded for stated percentage of the measurement period.

In addition to LA_{eq} and the statistical indices it can be useful to consider the maximum noise level, the LA_{max}. The LA_{max} can be particularly important when night-time noise and the potential for sleep disturbance is considered.

Many noise standards specify the length of time over which noise should be measured. For instance, the Noise Insulation Regulations 1975 are based on measures of $dBLA_{10}$ (18 h); the average of the L_{10} levels, in dB(A), measured in each hour between 6am and midnight. Mineral Planning Guidance Note 11 refers to $dBLA_{eq}$ (1 h), the equivalent continuous noise level, in dB(A), during one hour of a weekday. When considering noise criteria which are expressed in terms of LA_{eq}, the measurement period can be particularly important. The slow passage of an HGV at a distance of 10 m, for instance, may give rise to a 12-sec LA_{eq} of 75 dB(A), a 5-min LA_{eq} of 61 dB(A) and a 1-h LA_{eq} of 50 dB(A).

4.2.2 Factors influencing noise impacts

The principal physical factors that influence how much effect a sound will have upon a potentially affected receptor are the **level of the sound** being assessed and the **level of other sounds** which also affect the receptor. In turn these are determined by several factors.

First, as one gets further away from a source of sound in the environment, the level of noise from the source decreases. The principal factor contributing to this is probably **geometric dispersion of energy**. As one gets further away from a sound source, the sound power from the source is spread over a larger and larger area (think of the way that ripples diminish from a stone thrown into a pond). The rate at which this happens is between 3 dB per doubling of distance for very big sources (such as major roads) and 6 dB per doubling of distance for comparatively small sources (for instance, an individual small piece of machinery). It is because of this principle that noise fades rapidly near a noise source, but slowly far from it (it is why, for instance, motorways can be heard over such long distances).

The next most important factor in governing noise levels at a distance from a source is **whether the propagation path from the noise source to the receiver is obstructed**. If there is a large building, a substantial wall or fence, or a topographic feature that obscures the line of sight, this can reduce noise levels by, typically, a further 5–15 dB(A). The amount of attenuation (reduction) depends upon the geometry of the situation and the frequency characteristics of the noise source. Trees, unfortunately, do not generally act as effective barriers.

If the sound is travelling over a reasonable distance (generally hundreds rather than tens of metres), the **type of ground** over which it is passing can have a substantial influence on the noise level at the receiver. If the sound is passing at a reasonably low physical level over soft ground (grassland, crops, trees, etc.) there will be an additional attenuation to that due to geometric dispersion. It should be noted, however, that only soft ground attenuation or barrier attenuation (i.e. not both) should generally be included in calculations.

Beyond these simplest physical characteristics it may be necessary to consider other physical characteristics of the sound being assessed. In particular it may be important to consider whether the sound is **impulsive** (it contains distinct clatters and thumps), **tonal** (whine, scream, hum) or whether it contains **information content** (such as speech or music). Other physical effects that may have to be considered, if detailed noise calculations are to be carried out, could include reflection and meteorological effects.

Probably the most important aspect of reflection that needs to be considered is whether the propagation model being used calculates **free-field** (at least 3.5 m from reflective surfaces other than the ground) or **facade** (1 m from the facade of the potentially affected receptor). PPG24 suggests a facade value is 3 dB higher than the free-field level determined for the same location, and the DoT's (1988) *Calculation of Road Traffic Noise* suggests a 2.5 dB differential. In reality, facade effects vary from source to source and depending on whether the soundfield is directional or diffuse. Whether calculation or measurement results are free-field or facade is critical, however, as the differentials that have to be assumed are considerable. Other reflection effects occur where hard surfaces act as acoustic mirrors, increasing the sound pressure level or intensity (not the power) of a source. This may need to be considered where detailed calculations are being carried out.

Meteorological effects generally only need to be considered where calculations are being made over large distances (upwards of 100 m or so). Wind speed and direction can affect noise levels. A gentle positive wind (the wind blowing from the noise source to the receptor) slightly increases noise levels compared with calm conditions, but a negative wind has a larger effect (i.e. it reduces noise levels more than a positive wind increases them). Some propagation models have a positive wind component allowance built into them, others allow the modelling of noise levels under different meteorological conditions. Clearly, as distances increase from a noise source, the degree of certainty to which noise levels can be estimated rapidly diminishes. Where large distances are involved, and noise level estimates are critical (as they can be for power stations or large petrochemical plants for instance) it is essential that the conditions for which any noise predictions are expected to hold are clearly defined.

4.3 Legislative background and interest groups

Noise is controlled in three ways: by controlling overall noise levels, setting limits on the emission of noise sources, and keeping people and noise apart. The local authority environmental health officer's view will be sought by the planning authority when an application is received. He/she will be able to identify issues of particular concern and advise on the most appropriate regulations and guidance for appraising a given development project, so the developer should discuss plans with him/her prior to submission.

The overarching regulations and guidance that apply to most developments are the *Control of Pollution Act 1974*, the *Environmental Protection Act 1990*, and *Planning Policy Guidance Note 24: Planning and Noise* (PPG24). Under the Control of Pollution Act a local authority can control noise from construction sites and designate noise abatement zones in which specified types of development may not exceed specified noise levels. The Environmental Protection Act makes statutory nuisances, including noise from a premises which is prejudicial to health or a nuisance, subject to control by the local authority. PPG24 gives guidance to local authorities in England and Wales on how to minimise noise impacts. It discusses issues to be considered when applications for noisy and noise-sensitive developments are made, advises on the use of planning conditions to minimise noise, and proposes noise exposure categories for new residential development (see Table 4.2). The local planning authority may also require a Section 106 obligation concerning noise to be agreed before granting planning permission.

Further legislation and guidance applies to specific types of developments: the key ones are reviewed in Table 4.3. A longer discussion can be found in e.g. Garbutt (1992), Hughes (1992), Smith *et al.* (1996) or Williams (1997).

Table 4.2 Noise exposure categories from Planning Policy Guidance Note 24

Noise source		A	B	C	D
road traffic	07:00–23:00	< 55 dB(A)	55–63	63–72	> 72
	23:00–07:00	< 45 dB(A)	45–57	57–66	> 66
rail traffic	07:00–23:00	< 55 dB(A)	55–66	66–74	> 74
	23:00–07:00	< 45 dB(A)	45–59	59–66	> 66
air traffic	07:00–23:00	< 57 dB(A)	57–66	66–72	> 72
	23:00–07:00	< 48 dB(A)	48–57	57–66	> 66
mixed sources	07:00–23:00	< 55 dB(A)	55–63	63–72	> 72
	23:00–07:00	< 45 dB(A)	45–57	57–66	> 66

A – Noise need not be considered as determining factor in planning application
B – Noise should be taken into account when determining planning applications and, where appropriate, conditions imposed to ensure an adequate degree of protection against noise
C – Planning permission should not normally be granted. If it is, conditions should be imposed to ensure a commensurate degree of protection against noise
D – Planning permission should normally be refused

Table 4.3 Noise regulations, standards and guidelines

Type of project	*Key regulations, standards and guidelines*	*Comments*
Road	Land Compensation Act 1973	Allows people whose enjoyment of their property has been reduced by public works to be compensated, and allows regulations to be enacted to determine when compensation is due. To date only the Noise Insulation Regulations 1975 have been introduced, which apply to new highways.
	Noise Insulation Regulations 1975 (SI 1975/1763) Noise Insulation (Amendment Regulations) 1988 Memorandum on the Noise Insulation (Scotland) Regulations TRL Supplementary Report 425 *Rural Traffic Noise Predictions – An Approximation*	The regulations and amendments require highway authorities to provide noise insulation for residential properties if they are (a) within 300 m from a new or altered highway, (b) not subject to compulsory purchase, demolition or clearance, (c) not already receiving a grant for noise insulation works, (d) subject to 18-h L_{10} noise levels over 67.5 dB(A), (e) subject to an increase of at least 1 dB(A) over the existing noise level, and (f) on a new road which contributes at least 1 dB(A) to the final noise level. The memorandum is the Scottish equivalent of this report. The supplementary report gives procedures for how changes in noise levels can be approximated for quiet rural locations.
	Calculation of Road Traffic Noise (DoT/WO 1988)	Gives procedures for predicting noise in areas where noise is dominated by traffic noise; this can be extrapolated to distances of up to 300 m from the road. Calculations incorporate information about traffic volume, vehicle speeds, the percentage of HGVs, the road gradient, road surface, and distance from source to receiver. This procedure must be used for the Noise Insulation Regulations.
	Design manual for roads and bridges (DMRB) Vol. 11 (DoT 1993)	Gives procedures for assessing the impact of road schemes where traffic increases or decreases of 25% of more (about 1 dB(A)) are expected in the year the scheme opens.
	New Approach to Appraisal (NATA) (DETR 1998)	Provides a framework for assessing the impacts of road scheme options, using the DMRB methods (see §5.6.2).
Airport	BS5727: 1979	Provides methods for measuring, analysing and describing aircraft noise.

Table 4.3 (*continued*)

Type of project	Key regulations, standards and guidelines	Comments
Railway	*Railway Noise and the Insulation of Buildings* (DoT 1991)	Recommends noise insulation standards for new railway lines, and reviews noise indices and planning conditions related to existing rail transport.
Industrial	BS4142 *Rating Industrial Noise Affecting Mixed Residential and Industrial Areas*: 1990	Provides methods for determining the increase in noise levels from new buildings and plant, and the likelihood of this increase causing complaints (based on background and predicted noise levels).
Mineral workings, construction and other open sites	*Mineral Planning Guidance Note 11, The Control of Noise at Surface Mineral Workings* (DoE 1993)	Gives guidance for determining background noise at proposed surface minerals workings, predicting and assessing their noise impacts, and ensuring that these impacts are kept within acceptable limits.
	BS5228 *Noise Control on Construction and Open Sites: 1984/1992*	Presents indices for noise from opencast coal extraction, piling operations and similar works, and gives guidance on how such noise can be measured, assessed and controlled.

Other relevant legislation includes the *Public Health Act 1961, Health and Safety at Work, etc. Act 1974, Motor Vehicles (Construction and Use) Regulations 1978, Road Traffic Regulation Act 1984, Civil Aviation Act 1982, Local Government (Miscellaneous Provisions) Act 1982, Town and Country Planning Act 1990, Town and Country Planning (Scotland) Act 1972*, BS8233 on sound insulation and noise reduction for buildings, local authority byelaws, and building regulations which require houses and flats to be built to prescribed noise insulation standards. Various EC Directives control noise from vehicles, aircraft and construction plant. Individuals may resort to common law if they suffer annoyance from noise; this generally involves proving the existence of a private nuisance, namely an unlawful interference with their land, their use and enjoyment of their land, or some right enjoyed by them over the land or connected with it.

4.4 Scoping and baseline studies

The EIA scoping stage identifies relevant potential noise sources, identifies the people and resources likely to be affected by the proposed development's noise (the receivers), and determines noise monitoring locations. The baseline studies involve identifying existing information on noise levels, carrying out additional noise measurements at appropriate locations where necessary, and considering future changes in baseline conditions. These stages – which are interlinked and do not necessarily happen consecutively – are discussed below.

The project details should be analysed and **each potential source of noise impact identified**. Both on-site and off-site sources should be considered and (where appropriate) both the construction and operational stages. Each source of impact should be considered and a judgement made with regard to (a) carrying out further detailed assessment; (b) carrying out further but less detailed assessment; or (c) discarding the source of impact from the main EIA stage on the grounds that any resultant effects are highly unlikely to be significant. The reasoning for the ranking of sources of impact should be made explicit. This process enables the EIA proper to concentrate on assessing noise from the sources of impact most likely to give rise to significant effects.

Ultimately the effects of noise are dictated by the characteristics of the **potentially affected receptors**. Various maps can help to identify noise receptors in the area, but this should be confirmed by a site survey. The people affected by a development are not only local residents but also users of public places such as parks and footpaths, and of other outdoor areas such as private playing fields and fishing lakes. EIAs should identify any potentially particularly noise-sensitive receivers such as schools, hospitals, and recording studios.

Sites for monitoring are normally determined in consultation with the environmental health officer, and possibly also with the local community. Where there are only a limited number of receivers, monitoring will normally be carried out on all of them. However, where there are many receivers, for instance, along a proposed road or rail line, representative receivers will need to be identified. Particularly noise-sensitive receivers are normally all monitored. A systematic approach is required, splitting potentially affected receptors and resources into residential, non-residential and noise-sensitive, and non-residential and not noise-sensitive. Clearly the latter class of resources (perhaps factories and other industrial premises for instance) can be scoped out. Noise-sensitive non-residential resources may need a further degree of sub-classification (a major broadcast studio may be potentially more sensitive than a shopping centre for instance). It is advisable, however, to treat residential receptors uniformly. Although individual sensitivities to noise vary enormously, the aim of the assessment should be to evaluate the likely response of 'normal' communities.

Because noise is primarily a local impact, only limited existing information can be obtained from desktop studies, and virtually all EIAs rely on noise measurements carried out at the site. Information about the wider area may be gleaned from the CPRE/ASH Consultants' maps of 'tranquil areas', which combine information about quiet areas (determined by distance from major roads, rail lines, airports, and built-up areas) and areas with little visual intrusion (e.g. by pylons). Local authority environmental audits may include noise data, but are unlikely to be site-specific.

Measurement of ambient noise is normally achieved by carrying out measurements at the potentially most affected noise-sensitive receptors. Every effort should be made to carry out measurements at the times when the new source will be operating and with typical ambient conditions (normal prevailing wind, no rain, dry roads and during normal weekdays and weekends as appropriate). If under particular conditions (e.g. a specific wind direction) higher background levels commonly occur, these are also recorded. For some projects (wind farms for instance) it may be appropriate to carry out assessments for a range of climatic conditions; care should be taken, however, to exclude the effects of atypical climatic conditions, such as temperature inversions. The noise survey may also record the quietest conditions that typically occur in an area (e.g. on a quiet Sunday morning). This is because the biggest increase in noise caused by a proposed development will be in comparison with these quiet conditions.

Sound measuring equipment is portable and battery-powered, and usually consists of (a) a microphone, which converts changes in ambient pressure into an electrical quantity (usually voltage), (b) a sound level meter which amplifies the voltage signals, averages them, and converts them to dB, (c) an analyser which records noise descriptors (e.g. L_{eq}, L_{10}) over a period of time, and (d) a reference sound source against which to calibrate the equipment. Several of these will normally be incorporated into the same piece of machinery. The sound level meter will have different types of settings, corresponding to different ways of averaging voltage; slow (over 1 sec), fast (0.125 sec), and sometimes peak and impulse. A windshield should always be used for environmental noise measurements.

The precise procedures for measuring sound – for instance, the length of time of measurement, location of equipment, and measurement levels – are sometimes specified in the relevant regulations or guidelines (see §4.3). It is generally advisable to agree the noise monitoring regime with the relevant environmental health officer, who will have a good understanding of local conditions and any particular 'hot spots'. A typical survey strategy may include a limited number of long-term unattended measurement positions (24 h or more) and several sample positions where a number of (at least) 15 min attended sample measurements are carried out.

Broadly, noise measurements involve:

- taking note of the equipment used, including manufacturer and type;
- taking note of the date, weather conditions, wind speed, and wind direction;
- calibrating the sound meter and microphone;
- setting up the microphone at the appropriate site (check relevant guidelines/ legislation for details);
- noting the precise location where measurements are taken (e.g. on a map or using grid references);
- taking measurements using the criteria from the relevant guidelines (e.g. continuous for 24 h, or for 1 h; using fast weightings for traffic or slow for construction noise);
- noting start and finish times, identifying the principal influences on the noise environment (particularly the major influences on the LA_{eq}, LA_{90} and LA_{max}) during the measurement period, and any other factors (e.g. whether the equipment was attended or not) that could affect the measurements; and
- checking the calibrations.

Table 4.4 gives an example of baseline noise data. Generally an EIA includes such data, a description of how they were collected, and a map showing the location of the measurement points. Where noise monitoring is carried out during construction and operation, the same measurement points will generally be used.

A final stage of scoping and baseline studies is to consider whether baseline noise levels are likely to change in the future in the absence of the proposed development. For instance, if a development is proposed near an industrial complex that is currently under construction, then the future baseline is likely to change. In some cases the future baseline may be established through calculations, particularly intensification of a route corridor where the level of noise from the existing traffic can be readily calculated.

Table 4.4 Example of baseline sound data

Date	Start of period	Sound levels, in dB(A)					Comments
		L_{90}	L_{50}	L_{10}	LA_{max}	L_{eq}	
1 April	1500	56	57	60	62	58	mostly traffic noise
	2200	46	49	53	55	50	traffic, dog barking
2 April	0720	55	57	59	61	57	traffic, birdsong

The most important things to be noted are generally:
- principal influence on LA_{eq}
- principal influence on LA_{90}
- and whether the samples can be considered representative.

4.5 Impact prediction

The aim of noise prediction in EIA is to identify the changes in noise levels which may occur, both in the short and long terms, as a result of the development; and the significance of these factors.

Predicting noise levels is a complex process which incorporates a wide range of variables, including:

- existing and likely future baseline noise levels,
- the type of equipment, both mobile and fixed, used at the site (see BS 5228 for indicative sound levels from mobile plant; Table 4.5 gives examples of typical sound levels from construction equipment);
- the duration of various stages of construction and operation,
- the time of day when the equipment is used,
- the actions of the site operator,
- the location of the receivers and their sensitivity to noise,
- the topography of the area, including the main forms of land use and any natural sound barriers,
- meteorological conditions in the area.

These will affect the amount and type of sound coming from the site (e.g. type of equipment, duration of workings), how that sound travels (e.g. distance between

Table 4.5 Examples of typical sound levels from construction equipment (BS 5228)

Type of equipment	Sound level, in dB(A), at 7 m
unsilenced pile-driver	110
unsilenced truck scraper, grader	94
unsilenced pneumatic drill	90
unsilenced compressor	85
concrete breaker	85
crane	85
unsilenced generator	82
sound reduced compressor	70

Table 4.6 Example of noise predictions

receiver no.	noise source	distance (m)	sound power level at source (dB(A))	distance correction (dB(A))	screening attenuation (dB(A))*	soft ground attenuation (dB(A))*	predicted L_{eq} at receiver (dB(A))	ambient noise levels (LA_{eq})	increase in noise (dB(A)L_{eq})
1	loading operations	470	110	−61.4	−5	0	43.6	52.4	0.6
2		335	110	−58.5	0	−8.2	43.3	42.9	3.2
3		135	110	−50.6	0	0	59.4	60.1	2.6

* Either screening/barrier or soft ground attenuation is valid for a given site, not both.

source and receptor, topography, meteorology), and the response of the receptors (e.g. timing of workings, sensitivity to noise).

Essentially, noise level prediction involves predicting the sound power level at the source; predicting the sound level at each monitoring site (which represents certain receivers) using corrections for factors such as distance, screening and ground attenuation; and adding the new sound levels to the ambient levels. Table 4.6 shows an example. Where a development project has multiple sound sources that are close together, they will normally be considered together as one source (by adding their levels using Fig. 4.1, p. 67). Where multiple sound sources are not close together, each source's sound level at each receiver is calculated, and these sound levels are then added together (again using Fig. 4.1) for each receiver: Box 4.2 gives a very basic example to illustrate these principles.

Detailed procedures for predicting sound levels from different types of development, and different stages of development (construction, operation, decommissioning) are specified in many of the regulations listed in §4.3. The procedures are too cumbersome and diverse to discuss in detail here; they are often set up as computer models. The reader is referred to the relevant regulations and standards for further information. It may be necessary to carry out noise monitoring at a similar existing activity or development in order to predict the effects of a proposal.

The **significance** of changes in noise levels generally depends on the number of people affected, and how badly they are affected. The latter is the difference between the current ambient sound levels at the receivers, and the predicted future sound levels (i.e. ambient plus additional new sound). Considerable, but not unchallenged,

---

Box 4.2 Noise predictions for dispersed multiple sound sources

Assume that a receiver will be affected by sound from three dispersed sources:

- Source A emits 95 dB at 1 m, and is 64 m from the receiver
- Source B emits 97 dB at 1 m, and is 128 m from the receiver
- Source C emits at 109 dB at 1 m, and is 256 m from the receiver

Take the basic principle from § 4.2.2 that a doubling of distance reduces sound by 6 dB: −6 dB at 2 m, −12 dB at 4 m . . . −36 dB at 64 m, −42 dB at 128 m, −48 dB at 256 m (*Note* – in practice this reduction will depend on many other factors, so the principle should be used as a broad rule of thumb only). The additional sound at the receiver will thus be 59 dB from source A (95 dB to start with, minus 36 dB because it is 64 m away), 55 dB from source B, and 61 dB from source C. The total additional sound at the receiver will be 59 + 55 + 61 dB: Box 4.1 shows that this is about 63.7 dB.

Table 4.7 Example of noise significance criteria (adapted from Arup Environmental 1993)

Criterion	*Construction noise*	*Traffic noise*
Severe adverse	Noise above traffic noise insulation thresholds for > 8 wk; insulation or permanent rehousing required	> 15 dB increase
Major adverse	Noise above traffic noise insulation thresholds for < 8 wk; insulation or temporary rehousing required	10–15 dB increase
Moderate adverse	Noise above ambient levels for > 8 wk, but below traffic noise insulation thresholds	5–10 dB increase
Minor adverse	Noise above ambient levels for < 8 wk, but below traffic noise insulation thresholds	3–5 dB increase
None	Noise at or below ambient levels	< 3 dB increase

consensus exists about the significance of noise impacts. A change of 3 dB is barely detectable whereas a change of 10 dB corresponds subjectively to a doubling or halving of loudness; Table 4.7 suggests possible significance criteria. The World Health Organisation suggests that daytime outdoor noise levels should be below 50 dB LA_{eq} to prevent significant community annoyance, but in cases where there are other reasons to be in an area, like good schools, people may tolerate up to 55 dB LA_{eq} (WHO 1988). PPG24 implies that 55 dB LA_{eq} may be considered a general environmental health goal (see Table 4.3).

Within this overall framework, however, variations exist. An increase in noise in an area already subjected to high noise levels may be more significant than a similar increase in an area with lower noise levels. The same level of noise at a noise-sensitive location will be more significant than that at a less sensitive location. If the new source is a road and the area is already dominated by road traffic noise, then it is unlikely that the subjective response will be dramatically greater than any calculated change in noise levels would suggest. A new industrial source, however, could be tonal or impulsive or a new specialist commercial source (say, perhaps, a cinema complex or a night-club) may give rise to appreciable levels of low-frequency noise. In these instances, a description of the impact in terms of change or absolute levels of A-weighted sound pressure levels may not be an adequate indicator to allow potential effects to be assessed, and more detailed descriptions will be necessary.

4.6 Mitigation

Mitigation will be necessary if the noise from the proposed development is likely to exceed the levels recommended in the relevant standards (see §4.4). However, it may be useful to implement noise mitigation measures even if standards are met, to prevent annoyance and complaints and as part of best practice procedures, provided that an appreciable community benefit is likely to result. The best noise mitigation is that which is integrated into the project design: the siting of machinery and buildings, choice of equipment, and landscaping to reduce noise are all easiest, cheapest and most effective if they are designed in rather than pasted on near the end.

For a new potentially 'noisy' project, mitigation of noise is best carried out at the source, before the noise has escaped. Failing this, barriers and the siting of buildings can be used to obscure the line of sight from noise sources to potentially affected noise-sensitive locations. As a last resort noise can be controlled at the receiver's end through the provision of, say, secondary glazing or other noise insulation measures.

Control of noise at the source can take a number of forms. First, the equipment used or the modes of operation can be changed to produce less noise. For instance, rotating or impacting machines can be based on anti-vibration mountings. Internal combustion engines must be fitted with silencers. Airplanes can be throttled back after a certain point at take-off, to reduce their noise. Traffic can be managed to produce a smooth flow instead of a noisier stop-and-start flow, and use of quieter road surfacing materials can reduce tyre noise. Well-maintained equipment is generally quieter than poorly maintained equipment.

Second, the source can be sensitively located. It can be located (further) away from the receivers, so that noise is reduced over distance. A buffer zone of undeveloped land can be left between a new road and a residential area. The development can be designed so that its noisier components are shielded by quieter components; for instance, housing can be shielded from a factory's noise by retail units. Natural or artificially-constructed topography or landscaping can be used to screen the source.

The source can be enclosed to insulate or absorb the sound. Sound insulation reflects sound back inside an enclosure or barrier, so that sound outside the enclosure is reduced. However, merely enclosing the source is not the optimum solution, since the noise reverberates within the enclosure, and effectively increases the strength of the enclosed sound. Providing sound absorption within the enclosure avoids this happening. Sound absorption occurs where the enclosure or barrier absorbs the sound, converting it into heat. Most enclosures are constructed of both insulating and absorbing materials.

Details of requirements for noise enclosures and their effectiveness are very complex and require specialist knowledge. The reader is referred to the relevant standards and to textbooks on noise control (see e.g. Smith *et al.* 1996, or SRL 1991). However, some general points can be made here. Methods of measuring sound insulation usually distinguish between airborne sound (noise) and structural sound (vibration), and any reference to insulation should distinguish between them. Broadly, the ability of a panel to resist the transmission of energy from one side of the panel to the other, or its transmission loss, will depend on (a) the mass of the panel (more mass = more transmission loss), (b) whether it is layered or not, and with or without discontinuities between the layers, (c) whether it includes sound absorbing material, and (d) whether it has any holes or apertures.

Acoustic fencing or other screens, either at the source or at the receiver, can also reduce noise by up to 15 dB. The effectiveness of screens depend on their height and width (larger is better), their location with respect to the source or receiver (closer is better), their form (wrapped around the source or receiver is better), their transmission loss, their position with respect to other reflecting surfaces, the area's reflectivity, and whether they have any holes or apertures.

Noise screens can consist of topographical features or tree plantings as well as artificial materials. For instance, earth mounds (bunds) are often built alongside roads to absorb and reflect traffic noise away from nearby buildings. Thick areas (≥ 30 m) of dense trees and underbrush may reduce noise by up to 3–4 dB at low frequencies and 10–12 dB at high frequencies; although thinner tree belts have little actual effect on noise, the visual barrier they form can make people think that noise levels have been reduced. A mixture of deciduous and coniferous trees will give maximum noise reduction in the summer, and some reduction in the winter when the leaves of the deciduous trees have fallen. It must be remembered that saplings take time to mature, and are unlikely to reduce noise for several years after planting.

Control of noise at the receiver's end is often similar to that at the source. Good site planning can minimise the impact of noise; for instance, in a house by a busy road the more noise-sensitive rooms (e.g. bedroom, living room) can be shielded from the road noise by the less noise-sensitive rooms (e.g. kitchen, bathroom). A screen can be erected to reflect sound away from the receiver, for instance an acoustical screen between a highway and house. The equivalent of a noise enclosure can be achieved by soundproofing a house using double-glazed windows. The *Land Compensation Act 1973* requires highway authorities to insulate houses affected by noise over a certain level.

4.7 Monitoring

Any conditions imposed as part of a project's planning permission are enforceable, including conditions related to noise. These can apply not only to noise levels (e.g. during construction, operation; during the day, night), but also to noise monitoring to be conducted by the developer (e.g. distance from the site boundary, frequency). If no planning conditions are set, local environmental health officers can still monitor noise from a site, for instance, in response to complaints by local residents to determine whether it is a statutory nuisance.

There are presently no requirements to compare any noise-monitoring data with the noise predictions made in EIAs. A best-practice EIA could propose not only noise-related planning conditions, but also a noise-monitoring programme, and relate its findings to the EIA to improve future noise-prediction methodologies. The sites and noise measurement techniques used in carrying out baseline noise surveys should be such that comparable monitoring data can later be collected. However, given the current lack of legislative requirements for monitoring, this is unlikely to occur.

4.8 Conclusion

This has been only a brief introduction to a very technically complex topic. Noise prediction requires expert input, and probably computer models. Readers are strongly urged to familiarise themselves with the relevant regulations and standards (see §4.3) as well as standard texts on acoustics and noise control.

References

Arup Environmental 1993. *Redhill Aerodrome Environmental Statement.* London: Ove Arup & Partners.

Bregman JI & KM Mackenthun 1992. *Environmental impact statements.* Chelsea, Michigan: Lewis.

BSI (British Standards Institute) 1992. *British Standard guide to evaluation of exposure to vibration in buildings (1Hz–80Hz).* BS 6472. London: BSI.

BSI 1993. *Evaluation and measurement for vibration in buildings. Part 2. Guide to damage levels from groundborne vibration.* BS 7385: Part 2. London: BSI.

DETR (Department of Environment, Transport and the Regions) 1997. *Digest of environmental statistics 1997.* London: The Stationery Office.

DETR 1998. *Guidance on the new approach to appraisal.* London: HMSO. (http://www.detr.gov.uk/itwp/appraisal/guidance/index.htm)

DoE (Department of the Environment) 1993. *Mineral Planning Guidance Note 11: The control of noise at surface mineral workings.* London: HMSO.

DoE 1995. *Preparation of environmental statements for planning projects that require environmental assessment. A good practice guide.* Environmental Planning Research Programme. London: HMSO.

DoT/WO (Department of Transport, Welsh Office) 1988. *Calculation of road traffic noise.* London: HMSO.

DoT 1991. *Road accidents Great Britain 1990.* London: HMSO.

DoT 1993. *Design manual for roads and bridges.* Vol. 11: *Environmental assessment.* London: HMSO.

Garbutt J 1992. *Environmental law – a practical handbook.* London: Chancery.

Hughes D 1992. *Environmental law,* 2nd edn. London: Butterworths.

Royal Commission on Environmental Pollution 1994. *Eighteenth report: Transport and the environment.* London: The Stationery Office.

Smith BJ, RJ Peters & S Owen 1996. *Acoustics and noise control,* London: Longman.

SRL (Sound Research Laboratories Ltd) 1991. *Noise control in industry,* London: Chapman & Hall.

WHO (World Health Organisation) 1988. *Environmental criteria 12: Noise.* Geneva: WHO.

Williams M 1997. *Environmental noise and vibration measurement and standards.* Harrow: Brüel & Kjær.

5 Transport

Jeremy Richardson and Greg Callaghan

5.1 Introduction

Increasingly, transport is seen as a key factor in the approval, design and likely success of prospective new developments. Developments require good access for both their employees and customers as well as the need for optimum servicing arrangements, all of which will affect the surrounding transport and highway network, which in turn will impact upon the delivery of sustainable planning policies. In short, there has been a fundamental shift in thinking over the last five years regarding transports role and importance within the development agenda.

Recent government policy seeks to promote sustainable developments where the aim is to lead development proposals away from a reliance on private car access (i.e. 'out of town' development sites) to brownfield sites in urban areas with established public transportation corridors. These policies have been developed in order to reduce the need for travel by:

- reducing the growth in the length and number of motorised journeys;
- encouraging alternative means of travel which have less environmental impact;
- reducing reliance on the private car.

The government is therefore committed to a **sustainable development** strategy seeking to reduce the need to travel through the use of land use planning and transport planning. Indeed, a framework of locational policies and guidance, such as *Planning Policy Guidance Notes* 6 and 13 (§5.3.2) will help to achieve the above aims. Development proposals will therefore need to be supplemented by complementary transport objectives, such as: car parking restrictions; increased provision for pedestrians and cyclists; traffic management measures; public transport improvements; and park-and-ride proposals.

Two methodologies that can be employed by transportation planners are of particular interest to those seeking information on transport network impacts for an EIA. They are *Transport Assessments* and the *New Approach to Appraisal* (NATA).

A Transport Assessment evaluates the impacts of a proposed development on the transport network. Transport impacts may take account of the numbers of cars, buses, bicycles, trains; the frequency and reliability of service; and the origins and destination of travellers. However, as well as assessing the activity on the network, any proposed physical changes to the infrastructure, such as bus priority measures, a road widening scheme or the addition of a high-occupancy vehicle lane, needs to be assessed against impacts on the environment and their contribution towards developing a sustainable transport system. Consequently, the government has published new guidance on appraising road investment proposals – the **New Approach to Appraisal**, which DETR (1998a) formulated in consultation with the relevant environment agencies, namely the Countryside Agency (CA), English Heritage (EH), English Nature (EN), and the Environment Agency (EA) (see Appendix B). NATA is a new and developing tool but DETR plans to adopt it as the standard assessment method for any new development (requiring public money) that necessitates significant changes to the transport infrastructure.

A wide range of potential transport and environmental impacts are associated with a new development, including noise and vibration, air quality, biodiversity, community severance, visual intrusion, traffic generation, and economic regeneration. The significance of these impacts is dependent on the measures proposed to improve access to the development. Access is a function of the nature, location and size of the development and more importantly the choice of transport mode. The nature of the development, and any proposed transport provision, will determine the nature of trips to and from the site, as well as the potential for achieving a modal shift through increased public transport provision.

In recent years the Government's Roads Programme has been dramatically cut, in response to the fact that new roads attract new/induced traffic and as such lead to an increased rate in traffic growth. In the future there will be a reduced amount of large-scale trunk road schemes built compared to the early 1990s. This chapter reviews the policy and planning context and the techniques to evaluate the transport impacts of a land use development. However, the techniques are broadly applicable to both the assessment of the transport impacts of a project, and the assessment of the environmental impacts of a transport scheme such as a road bypass.

5.2 Definitions and concepts

There are several modes of transport, including vehicular traffic, heavy and light rail, cycling and walking, which are of significance to new developments. Vehicular traffic can be further subdivided into private cars and taxis, vans, goods vehicles, buses, motorcycles and pedal cycles. The exact nature of the impact of traffic on the network depends on a number of factors, including the relative composition of the categories outlined above, the nature and make up of the road network and the surrounding land use.

Hughes (1994) states it is possible to describe a stream of traffic on a length of road at a particular time with reference to:

- Highway link capacity
- Junction capacity
- Driver delay/queuing time
- Speed
- Number of accidents or accident rate
- Proportion of heavy goods vehicles
- Number of bus movements
- Pedestrian cycle flows crossing the road
- Frequency of access
- Turning movements

- Location and type of on-street car parking
- The nature of frontage land uses.

On the rail network the pertinent factors are:

- Line capacity (single or dual)
- Station capacity (stairwells, platform width, etc.)
- Platform length
- Rolling stock passenger capacity
- Frequency of service and station wait time
- Junction capacity and signalling
- Lay over capacity
- Proportion of freight trains
- Proportion of stopping and non-stopping services
- Speed.

5.3 Legislative background

5.3.1 *Environmental assessment*

The European EIA Directive 85/337 and the *Amendment Directive 97/11/EC* set out the criteria and procedures for undertaking an EIA of certain projects. Although neither the Directive nor the key relevant statutory instrument[1] specifically mention the need to assess traffic and transport impacts, it is clear that in order to assess the impacts on the environment properly the traffic and transport impacts must be included. Government good practice guidance on the preparation of an environmental statement (DETR 1999) states, "Traffic associated with a new development can have a wide range of often adverse environmental impacts".

The Design manual for roads and bridges (DMRB) Volume 11 (DoT 1993) sets out the procedures and methodologies for assessing the environmental impacts of proposed new roads. It can also be used to assess the secondary impacts of increased levels of traffic due to a new development. However, it is an extremely detailed tool designed to assess a single mode of transport (large-scale trunk road schemes) and consequently its applicability for assessing the transport impacts of non-road schemes is limited.

The Transport White Paper, Transport: A new deal, better for everyone (DETR 1998b) has set a new policy framework within which transport planning must be set. The emphasis in on planning integrated transport systems which provide travellers with a viable alternative to the private car. Integrated transport seeks to promote modes such as the bus and train, walking and cycling, and to direct development to locations which reduce the need to travel. This White Paper has been developed to tackle the problems associated with increased congestion and pollution. At the local level county councils and unitary authorities are required to produce Local Transport Plans (LTPs). The LTP is seen as the most important delivery mechanism of

1 The Town and Country Planning (Environmental Impact Assessment) (England and Wales) Regulations 1999 (SI No 293).

the integrated transport strategy. All LTP policies must be assessed against objectives concerning sustainable transport, the environment and the economy. Targets and indicators are used to monitor the success of these policies.

The Road Traffic Reduction Act (DETR 1997) places responsibility on county councils and unitary authorities in England, Scotland and Wales to produce a report on the assessment of the levels of road traffic in the area and a forecast of the growth in the levels. Furthermore, the report must set targets for a reduction in the levels of traffic or in traffic growth in the area.

5.3.2 Planning policy

Development in the United Kingdom is controlled through the *Town and Country Planning Act 1990* as amended by the Planning and Compensation Act 1991 [Town and Country Planning (Scotland) Act 1972] and the Local Government Act 1972 [Local Government (Scotland) Act 1973]. The legislation states that the planning authority must consult the corresponding highway authority if the proposed development involves access to a highway or is likely to increase traffic movements on the local highway network. For motorways and other trunk roads the Highway Authority is the Secretary of State for Transport. For other classified roads the local authority, which may be the county council or a unitary authority, is regarded as the local highway authority.

The requirement to carry out a **Transport Assessment** (TA) is non-statutory but is contained in government guidance (PPG13) (see below). Furthermore, Section 106 of the *Town and Country Planning Act 1990*, together with Section 278, can be used by the local planning authority to secure developer contributions to ameliorate the adverse impact of the development. Increasingly, Section 106 is used to secure funding for transport infrastructure and accessibility improvements. Local authorities see the TA as a key instrument in negotiating transport improvements through Section 106 agreements. For example, a TA may identify a development requirement to improve the access to the site, through the building of a new access road, the improvement of a junction, or by introducing new or expanded public transport services. The Section 106 negotiations have moved away from developers traditionally providing for highway improvements and are now more actively used for providing public transport and softer modes infrastructure improvements.

Planning Policy Guidance Note 6 (PPG6) (DoE 1996) on Town Centres and Retail Development states that, for retail developments over 2500 square metres, the planning application should be supported by evidence as to the impact of the development including:

- the accessibility by a choice of means of transport, assessing the proportion of employees and visitors/customers likely to arrive by different modes;
- the likely changes in travel patterns over the catchment area; and, where appropriate,
- environmental impacts.

Importantly, PPG6 requires that a **'sequential' test** be used to determine the acceptability of the site location allocated for development. This test requires both the local authority and the developer to direct development to brownfield sites that are within the urban area and are well served by existing transport corridors. This test is designed to prevent urban sprawl and reduce the need to travel. As a result, a proposal for an 'out of town' development will only be considered if the developer can provide evidence that no alternative town centre or edge of centre sites are available. PPG6 also provides guidance on the accessibility of development types in relation to public transport, in terms of acceptable walking distances to public transport facilities.

Planning Policy Guidance Note (PPG13) (DETR 2001) on Transport and the Environment aims to integrate transport and land use planning so that there is a reduced need to travel. It states that "A system of Transport Assessments to be submitted alongside applications for major developments are to replace the existing Traffic Impact Assessments". These assessments are to outline the potential modal split of trips to and from the proposed site, and measures to improve access by public transport, walking and cycling and reduce the number and impacts of motorised journeys associated with the proposal. PPG13 states that guidance will be issued by the DETR on the contents and preparation of TAs.

5.3.3 Other guidance

Guidance on traffic impact assessment has been produced by the Institution of Highways and Transportation (IHT 1994). The Institute of Environmental Assessment has developed *Guidelines for the environmental assessment of road traffic* (IEA 1993) which sets out the methodology for assessing the environmental impacts of traffic generated by non-road developments. This set of guidance is already deemed by some to be outdated, but it does form the current basis for the assessment of development schemes.

The *New Approach to Appraisal* (NATA) (§5.1) was originally developed as part of the review of the roads programme that was undertaken by the Labour Government. Only road schemes that were able to perform well against NATA's five overarching objectives were permitted, namely those that:

- protect and enhance the built and natural **environment**;
- improve **safety** for all travellers;
- contribute to an efficient **economy**, and support sustainable economic growth in appropriate locations;
- promote **accessibility** to everyday facilities for all, especially those without a car;
- promote the **integration** of all forms of transport and land use planning, leading to a better, more efficient transport system.

Subsequently, the government has issued guidance on the use of NATA as a general appraisal tool for investment in transport schemes. At present the government is still in the process of further developing NATA so that it can appraise strategic multi-modal transport corridor studies.

NATA seeks to bring together all the information from transport impact assessment, cost–benefit assessment, and EIA (if one is required; if not, methods are suggested for appraising the impacts on the environment), into one **Appraisal Summary Table** (AST) (see Table 5.1, p. 96). The AST aims to provide the decision-maker with both qualitative and quantitative information pertaining to all the impacts of the scheme in an unbiased format. Importantly the NATA guidance states that, although NATA will usually be used within the detailed design phase of a transport scheme or Phase Three of a trunk road scheme, it is recommended for all stages in the development, e.g. during the feasibility and initial design. Consequently, it is possible that a draft NATA will be available when the EIA is undertaken.

5.4 Interest groups and sources of information

County councils and unitary authorities are responsible for developing Local Transport Plans that set out a five-year transport planning framework. Yearly progress reports have to be submitted to the DETR on the implementation and monitoring of the plan. County councils and unitary authorities also are responsible for meeting the targets set in their road traffic reduction reports.

Local authorities are also responsible for developing Air Quality Strategies designed to meet the government's national objectives on air quality (see Chapter 8). Where these objectives are likely to be exceeded, an action plan must be drawn up detailing how poor air quality will be improved.

Local authorities' social exclusion strategies designed to include vulnerable groups such as the long-term unemployed, will depend upon an efficient public transport system in order to be successful. These obligations ensure the relevant authorities have responsibility towards the transport, environment and social impacts of a development.

Non-statutory interest groups which one may find useful to consult include the Council's Local Agenda 21 groups, non-governmental organisations such as Transport 2000, the Council for the Protection of Rural England, Friends of the Earth and local community groups.

Information regarding traffic flows are generally held by highway authorities. They may collect their own data as part of the process of preparing their local transport plan or their road traffic reduction report. They may also possess information on pedestrian and cycling flows, and bus and rail services in their area. Some authorities will possess traffic or multi-modal transport models which simulate the transport network in the area and allow one to forecast future scenarios relating to traffic growth or future developments. Local authorities are also likely to hold information on the DETR's national traffic census, which includes surveys of traffic flows on major roads (both trunk and principal roads) undertaken every three years. Furthermore, the DETR holds additional information from:

- national traffic surveys;
- trunk road network management;
- appraising infrastructure movements;
- research and monitoring studies.

5.5 Scoping and baseline studies

5.5.1 Transport Assessment: impact of development on transport network

The *Guidelines on Traffic Impact Assessment* (IHT 1994) state that if the development is likely to result in an increase in traffic by 10%, or by 5% in congested or sensitive areas, a Traffic Impact Assessment (TIA) should be produced. TIAs have now been superseded by Transport Assessments (TAs).

The Institute of Environmental Assessment guidelines state that environmental impacts from traffic are likely to be significant if there is either a predicted increase in traffic flow, or numbers of heavy goods vehicles, of 30% or more, or within a sensitive area where traffic flows will increase by 10%.

Prior to a TA a **scoping study** should be undertaken. The IHT Guidance on TIAs lists the key points to be considered within a TIA, which will still form the basis of the new TAs:

- the proposal, its size and the existing land use;
- existing traffic surveys;
- whether the development involves the relocation of an existing use;
- potential modal split;
- potential traffic generation from the site;
- the critical time period;
- whether the local highway network will need modification;
- whether adjacent links or junctions will become overloaded;
- areas of the impact of the proposal;
- when the site is expected to become fully operational;
- whether there are significant phases to the development;
- the assessment years;
- the level of car parking provision.

In addition to the above, it is now critical to assess the **current public transport provision** in the area and whether improvements to the public transport accessibility of the site will need to be made.

In order to develop the scoping study into a full TA, a certain amount of baseline data will be required. This information can be summarised as follows:

- vehicle flows;
- information on pedestrian, cycle and other road user movements;
- information on public transport accessibility;
- information on current public transport usage;
- road safety problems;
- site development constraints;
- secondary measurements such as noise and air quality measurements.

Baseline **traffic flow** data are required for links and turning movements at junctions over either a peak hour, 12-h or 24-h period. These baseline surveys will need to take account of variations in traffic flows, and will be used to establish existing peak flows and where these will correspond with the predicted peak hour flow of the development proposal. There may also be seasonal variations due to school holidays or tourism. From this information an assessment of the remaining capacity of links and junctions can be made.

Public transport capacity will also need to be assessed in the area of development sites. This will relate to service frequency, reliability, boarding and alighting information, origin and destination of customers, location of routes and number of seats on buses and trains, as well as information on journey reliability. (This assessment may result in the need for infrastructure improvements, such as bus lanes and bus priority measures in order to make services more reliable and more attractive.)

Current **cycle facilities** will need to be assessed in the vicinity of the development site, together with current flows and potential for mode shift towards cycling. A number of improvements may then be incorporated into developments, including:

- cycle paths in order to segregate cyclists from other road users, thus increasing safety;
- secure and sheltered locking facilities at interchanges and other end destinations;
- traffic management measures to create a safe feeling for cyclists by reducing traffic speeds and improving crossing facilities.

The modal shift towards cycling will need to be encouraged by the developer through the promotion of sensitive design and integration with the surrounding transport network. This should also be supported through the development of Green Travel Plans (discussed later in this chapter).

Pedestrian flows need to be analysed in the context of the location of major generators. These flows will relate to major attractors for pedestrians such as shops, schools and public transport facilities, which in turn relate to the location of pedestrian crossing facilities. Within the context of development proposals better facilities for pedestrians need to be considered. Development proposals should encourage more people to walk and there are a number of measures available to achieve this:

- environmental and public realm improvements;
- traffic management and calming schemes;
- pedestrianisation schemes;
- improved security and safety, including lighting and CCTV;
- more direct and 'desirable' routes;
- improved pavement widths;
- improved and desirable links and crossing facilities.

(IHT 1994)

Finally, the issue of **road safety** is important in determining the location and number of accidents on the surrounding network. Historical accident data will establish trends and groupings of accidents and will lead to the potential development of traffic management measures aimed at mitigating these accidents. This can be developed in association with measures to improve facilities for cyclists and pedestrians.

5.5.2 The New Approach to Appraisal (NATA)

Guidance on NATA (DETR 1998a) states that "all major proposals should be subject to full new approach of appraisal. For low cost proposals certain specific proposals may be set aside". The five overarching objectives of NATA give rise to the following sub-objectives or criteria:

- **Environment**
    * noise
    * local air quality
    * landscape

- * biodiversity
- * heritage
- * water

- **Safety**
- **Economy**

  - * journey times and vehicle operating costs
  - * cost
  - * journey time reliability
  - * regeneration

- **Accessibility**

  - * pedestrians and others
  - * access to public transport
  - * community severance

- **Integration**

All transport schemes must be assessed on their impact on these objectives. As indicated in Box 5.1, methods of collecting baseline information vary with subject area, and in many cases are described in other chapters. However, the baseline data required will relate quite specifically to the NATA assessment methods – which are described in §5.6.2.

5.6 Impact prediction and evaluation

5.6.1 Transport Assessment: impact of development on the transport network

Predicting the transport impact of a development requires background information on the proposed development. This information relates to the number of trips attracted to the site, the modal split of those trips, the distribution of the trips, the assignment of those trips, and the impact of these changes on the public transport and local road network.

Proposed development

The type of development proposed will determine the nature of its transport impacts. Within any TA the description of the development must include the following information:

- the size and mix of development types on the site;
- numbers of employees, visitors, customers or residents;
- number of proposed parking spaces;
- existing or former use;
- existing planning consents on the site;
- proposed access arrangements;
- hours of operation;

Box 5.1 NATA baseline survey methods

Environment
- **Noise** and **air quality** – data are obtained using the methods recommended in DMRB (DoT 1993). Further information on baseline surveys is given in Chapter 4 (noise) and Chapter 8 (air quality)
- **Landscape** – methods are described in Chapter 6
- **Biodiversity** – methods are described in Chapter 11
- **Heritage** – methods are described in Chapter 7
- **Water** – methods are described in Chapter 10

Safety
Baseline information is collected as part of the transport assessment described in §5.5.1.

Economy
Baseline information is needed on journey times and vehicle operating costs as well as the cost of construction and maintenance delay charges. The present value of the scheme must be identified. Reliability of journey time requires information on existing traffic levels, and the link and junction capacity of the network. Existing regeneration objectives for the location need to be made explicit.

Accessibility
Information on existing access to public transport, community severance, and pedestrians and other user information can be obtained from the transport assessment. Information is required on the number of pedestrians experiencing new severance and the severity to which they are subjected, and the number of pedestrians experiencing relief.

Integration
Information regarding the compatibility of land use, transport and other policies and plans needs to be appraised. These policies and plans need to cover Central Government policies and guidance, Regional Planning Guidance, *Development Plans* and local transport plans. Integration between the different modes of transport should also be assessed.

- servicing arrangements;
- phasing of development.

Number of trips and modal choice

There are a number of methods which predict the trip attraction of a site:

- first principles;
- comparisons with similar existing developments in similar areas;
- databases (i.e. TRICS);
- complex traffic models (i.e. SATURN or TRIPS).

The **first principles method** is based on a number of assumptions. Some of them may be based on survey data but where this does not exist best judgement can be used. Assumptions may include average car occupancy or % of long-distance travellers versus local ones. First principles as a method for predicting trips and modal choice is not commonly accepted by local authorities; it may be used for developments that do not have a high trip attraction or are unique. This method can be difficult to quantify.

Comparisons with similar developments are often undertaken when the database method does not provide sufficient information relating to specific use classes, or locations or modal shifts. This is the simplest method of trip assessment and it may be prudent to survey more than one similar site in order to assess a range of data. This method can be used to survey in more detail specifics relating to modal shift, car occupancy, cycle flows and pedestrian activity, etc.

Databases are an accepted way of predicting the number of trips generated by a development proposal. The most widely accepted database is TRICS (Trip Rate Information Computer System), which provides trip information based on a range of developments contained within the database. Trip information relates to Gross Floor Area, location and Use Class, number of employees, etc. The information supplied by this database primarily relates to car-borne trips.

However, due to the increasing emphasis on modal shift away from the private car and the absence of a database relating to the number of public transport trips, the issue of the number of trips and modal choice is becoming more difficult. Several factors relate to the modal choice and number of trips to a site, including the public transport accessibility level of the site, the level of car parking, and complementary measures to encourage public transport (e.g. bus lanes or bus priority measures). The better the public transport provision, the more likely that modal shift will occur. It is also becoming more accepted that the levels of car parking provided on the development site will restrict the number of car-borne trips and encourage visitors to the site to use public transport facilities. The site-specific information contained in the TRICS database does provide public transport service information for some sites, but it is often better to obtain information based on similar sites in the same area as the development proposal.

It is also possible to assess the number of trips and modal split of a development using a **traffic model**. The most widely used traffic models are SATURN and TRIPS, which can be used to assess the traffic effects of major developments, as well as calculating trip distribution and assignment. However, these models are unlikely to be accurate for many land use types.

Trip distribution

Trip distribution – where they come from and go to – depends on numerous factors relating to the type of trips that are generated and the origins and destination of those trips. There are basically two types of trip: new and transferred. New trips are those that did not occur anywhere else on the transport network prior to the development. Transferred trips are trips that were previously made elsewhere, but subsequent to the opening of the development have transferred to it. For example, a housing development may be assumed to generate all new trips, whereas a shopping centre may mainly generate transferred trips.

A further subdivision of the above trips involves primary and non-primary trips. Primary trips are trips made for the sole purpose of visiting the site, whereas non-primary trips can be further subdivided into diverted and pass-by trips. Diverted trips deviate from their normal route to visit the development, whilst pass-by trips are made as part of another journey such as stopping off on the way home from work.

In order to assess the distribution of these trips a number of techniques can be employed, such as:

- prior knowledge of the catchment area, e.g. knowledge of the destinations of employees;
- distribution based on the current traffic patterns in the area;
- the use of travel time isochromes to assess the length of journeys;
- a Gravity Model based on population against travel time.

The distribution of public transport trips will be based on the existing public transport network, except for when areas of public transport deficiency have been highlighted and extra services are recommended as part of the development proposal.

Trip assignment

Once information on the number, type, and origin and destination of trips are known, they must be assigned to the network. One can use a range of techniques from crude guesswork, based on knowledge of the local highway network, to sophisticated computer modelling such as SATURN and TRIPS. These models provide the most robust way of assigning traffic based on a representation of the transport network connected to a database. The model assigns the trips in the database to the network according to a set of parameters describing the optimum route. These models also allow the assignment of future trips on the network, taking into account year on year growth in traffic and changes in the network.

Impact assessment

The above information, once calibrated, allows the TA to assess the impact of the development on both the highway and public transport networks. The TA should provide a framework for assessing how the development proposals will affect specific groups (IEA 1993):

- people at home
- people in work places
- sensitive groups including children, elderly and disabled
- sensitive locations, i.e. hospitals, places of worship, schools, historic buildings
- people walking
- people cycling
- open spaces, recreational sites, shopping areas
- sites of ecological/nature conservation value
- sites of tourist/visitor attraction.

Other affected parties (IHT 1994) are:

- road users (i.e. cars, buses, cyclists and pedestrians)
- local residents
- local community facilities
- public transport operators

- local authorities (highways, roads and planning)
- Department of the Environment, Transport and the Regions.

Issues to be covered in the impact assessment include traffic generation, junction delays, and the impact of delays on public transport. Junction capacities are tested using a variety of software, such as PICADY, ARCADY, OSCADY, LINSIG or TRANSYT, to assess priority junctions, roundabouts or signals, respectively.

The testing of the highway network will therefore lead to the identification of mitigation measures to relieve congestion. This may take the form of junction improvements, bus lanes, bus priority, cycling facilities or pedestrian crossing facilities.

5.6.2 NATA: *impact of road infrastructure on the environment*

Once the information on the traffic, cycling and pedestrian flows, distribution and routing is known it can be used within NATA to assess impacts on the environmental objectives (noise, air quality, landscape, biodiversity, heritage and water) together with safety, economy, accessibility and integration.

The **Appraisal Summary Table** (AST) contains three columns for evaluating the significance of the predicted impacts (Table 5.1). The first column is qualitative, and allows a textual description of the impacts. The next column is quantitative: it uses numbers to measure the scale of the impacts. The final column is the summary assessment, using: a **monetary scale**; **quantitative indicator**; or a **seven point textual scale** of the impacts (large, moderate, or small negative/adverse; neutral; and small, moderate, or large positive/beneficial).

To increase transparency and leave an audit trail of decision points, one or more worksheets accompany each objective. These set out the procedure for predicting and evaluating the impacts. A written comment is also recorded against each appraisal, from which the summary assessment textual ranking is determined.

The assessment methods for the environmental objectives are summarised in Box 5.2. The differences reflect both inherent differences in the subject areas and the different approaches of the statutory bodies that were asked to develop the criteria for landscape, biodiversity, heritage and water. For example, the Environment Agency uses a risk-based approach, in contrast to the objectives-led approaches by English Nature, the Countryside Agency and English Heritage. The same factors explain differences in two other key aspects of the system:

1. the environmental **features** (attributes) on which appraisal is focused (Box 5.3);
2. the **indicators** (criteria) that are used to evaluate the significance of impacts (Box 5.4).

Assessment of the remaining NATA components (safety, economy, accessibility, and integration) should be carried out as outlined below.

SAFETY

Information on safety required of the AST is obtained from the COBA evaluation, as outlined in DMRB Volume 11. Any associated special circumstances must be recorded, including the number of accidents and the number of personal injuries

Table 5.1 NATA Appraisal Summary Table

Proposal name	Option description
PROBLEMS	*Statement of problems*
OTHER OPTIONS	List of other options that have been, or could be, tested

OBJECTIVES		QUALITATIVE IMPACTS	QUANTITATIVE MEASURE	ASSESSMENT
ENVIRONMENT	Noise		No properties experiencing: – Increase in noise xxx	Net xxx properties experience *higher* noise levels
CO₂: *xxxx tonnes* added or removed	Local air quality		No. Properties experiencing: • better air quality xxx • worse air quality xxx	+/–xxx PM₁₀ +/–xxx NO₂
	Landscape		Not applicable	*Moderate Adverse*
	Biodiversity		Not applicable	*Neutral*
	Heritage		Not applicable	*Moderate beneficial*
	Water		Not applicable	*Large adverse*
SAFETY	–		Accidents xxx Deaths xxx Serious xxx Slight xxx	PVB £xxm xx% of PVC
ECONOMY	Journey times & veh. op. costs		Trunk road journey time savings: peak xxx mins; inter-peak xxx mins	PVB £xxm xxx% of PVC
	Cost		Not applicable	PVC £xxm
	Journey time reliability		Stress on key trunk road link: Before xxx%; After xx%	Yes No
	Regeneration		Serves regeneration priority area? Development depends on scheme?	Yes No
ACCESSIBILITY	Pedestrians and others		Not applicable	*Slight beneficial*
	Access to public transport		Not applicable	*Moderate beneficial*
	Community severance		Not applicable	*Large adverse*
INTEGRATION			Not applicable	*Positive*

Version of date	**Cost-benefit analysis:**	PVB £xxxm PVC £xxxm NPV £xxxm BCR xx

Box 5.2 NATA assessment methods for environmental objectives

Noise assessment is carried out as described in DMRB 11.3.7 (DoT 1993). The noise impact will depend on the time of day, flow, and type of traffic. The assessment considers the net number of properties in the assessment year that experience changes of greater than and less than 3 dB(A) between the 'do minimum' and 'with proposal' options. In order to calculate this the following information must be subtracted from one another:

- the number of residential properties for which the assessment year noise level with the proposal option is 3 dB(A) lower than in the do minimum option;
- the number of residential properties for which the assessment year noise level with the proposal option is 3 dB(A) higher.

For the AST only those properties experiencing a change in noise level from the base year of greater than 3 dB(A) should be taken into account.

Local air quality is assessed using the methodology for predicting air quality from traffic flow provided in DMRB 11.3.1. As indicators, NATA uses the objectives for NO_2 and PM_{10} of the *National Air Quality Strategy* (NAQS) (see §8.2.3). To assess compliance with the NAQS for the do minimum and the proposed option, the difference in roadside PM_{10} and NO_2 levels in 2005 is calculated using predicted traffic flows for each option. The objectives are:

- NO_2 – 21 parts per billion (ppb) expressed as an annual mean;
- PM_{10} – 50 micrograms per cubic metre ($\mu g/m^3$) expressed as 99% of a running 24 h mean.

The number of properties within each of the following bands is ascertained: 50–100 m from the roadside; 100–150 m from the roadside; and 150–200 m from the roadside.

A series of factors are than added to each band against which the number of properties is multiplied. This accounts for the diminishing impact of adverse air quality as one retreats from the roadside.

CO_2 levels are also predicted from the expected additional number of vehicle-km induced by each option. Calculating the change in level of CO_2 assesses the impact on global emissions.

Landscape [Countryside Agency] assessment uses the CA's countryside character and environmental capital approach (Chapter 17) to describe the baseline or character of the landscape and then evaluate the impact on it. The landscape is evaluated using the features listed in Box 5.3, and impacts are assessed using the indicators listed in Box 5.4. The appraisal also allows for Additional Mitigation, i.e. additional to normal mitigation measures that are expected to be incorporated in the project design.

Biodiversity [English Nature] is assessed from an evaluation of the nature conservation value of the features listed in Box 5.3 (primarily in sites) and an assessment of the ecological impact upon them using the indicators listed in Box 5.4. Further details of the appraisal are given in §11.5.8.

Heritage [English Heritage] is assessed using a four-part framework. All the stages are considered vital as the process of characterising and appraising the resource is as important as the final assessment score within the AST. **Part one** involves a description of the heritage component using the features in Box 5.3. This provides a picture of the historic built environment. **Part two** consists of "the appraisal of the character described under each attribute . . . to establish the significance of the site". This uses the indicators listed in Box 5.4, and seeks "to move away from a simple designation-based approach". **Part three** of the framework involves a description of the impact of the proposed project (see Box 5.4). **Part four** consists of the final Assessment score (on the seven-point text scale) together with a qualitative score based on a set of definitions.

Water [Environment Agency] is assessed for two features (water quality and land drainage/ flood risk) using a two-stage process. Stage 1 applies a risk-based approach for the nine indicators listed in Box 5.4. This provides assessments of negative impacts only. Stage 2 reappraises the impacts in relation to mitigation measures which may significantly reduce adverse effects or even produce positive impacts. Further details are given in §10.8.5.

Box 5.3 NATA environmental features (attributes)

Landscape
- **Pattern** – relationship between topography, elevation and degree of enclosure of landscape
- **Tranquillity** – degree of remoteness, isolation, lack of intrusion of built environment
- **Cultural** – distinctive local views, traditional field patterns, building styles, materials and archaeological remains
- **Landcover** – all types of land use in the area
- **Summary of character** – summarises and pulls together the relationships between the features

Biodiversity
- **Habitats**
- **Species and species groups**
- **Natural (geological) features** (including earth heritage sites)

Heritage
- **Form** – physical form of the site
- **Survival** – the extent to which the original fabric of the building remains
- **Condition** – includes the appearance and present management of the site
- **Complexity** – the diversity and the relationships of the elements that make up the site
- **Context** – the setting within the immediate surroundings

Water
- **Water quality**
- **Land drainage and flood defence**

(deaths, serious and slight injuries) over the 30-year life of a scheme. Finally, a monetary figure is assigned to the number of accidents saved.

ECONOMY

The economy section is divided into: journey times and vehicle operating costs; costs; reliability; and regeneration.

Journey times are usually predicted using a traffic model such as SATURN. The monetised costs and benefits are calculated using the COBA from the DMRB Volume 13. The costs and benefits of journey times during construction and maintenance are calculated using a QUADRO computer program (DMRB 14). For smaller proposals, appropriate techniques such as Form 502 should be used. The AST will show the journey times savings, and the benefits due to time savings and **vehicle operating costs**.

The **costs** category includes the present value of the land, property preparation, construction, maintenance and supervision.

At present there is not a fully developed methodology for appraising journey **reliability**. Research suggests that as the road becomes busier and nears its maximum capacity of cars that can use it over a given period of time, the reliability of journey times is reduced. The traffic flow is measured in Annual Average Daily Traffic

---

Box 5.4 NATA environmental indicators for evaluating the significance of impacts

Landscape

- **Description** – of the existing landscape, before the scheme is constructed
- **Scale it matters** – the policy level scale at which this feature matters, for example international, national, regional or local
- **Importance** – the reasons why this feature is important, such as reasons for a designation
- **Rarity** – the relative abundance of the feature or its trend in relation to a target feature
- **Substitutability** – whether the feature is replaceable within the given time period, e.g. 100 yr[1]
- **Impact** – the impact of the scheme on the feature, using the seven-point text scale.

Biodiversity

- **Site location**
- **Site designation** – statutory and non-statutory designation
- **Habitat type or species group** – e.g. dry heath, birds, invertebrates
- **Scale** (of importance) – international, national, regional or local
- **Importance** – e.g. reasons for designation
- **Rarity** – trend in relation to targets
- **Substitution possibilities** – e.g. potential for relocation or recreation[2]
- **Impact** – assessment of the impact of the scheme, using the seven-point text scale

Heritage

- **Scale it matters** – the policy level scale at which this feature matters, e.g. international, national, regional or local
- **Significance** – in terms of designations and other information, which can suggest levels of importance for the site
- **Rarity** – including aspects such as representativeness and fragility/vulnerability of other existing examples
- **Impact** – assessment of the impacts (physical, visual and cumulative) of the scheme, using the seven-point text scale.

Water

- **Water quality indicators** – general quality assessment (GQA) of the water chemistry, EU Freshwater Fish Directive, water abstraction points, groundwater vulnerability, location of wells/boreholes
- **Land drainage/flood defence indicators** – floodplain, watercourses, river corridors, flood risk
- **Impact** – risk-based negative impacts (five-point scale) which are reassessed in relation to mitigation measures

---

1 The concept of **substitutability** can be somewhat controversial as it allows valued landscape features/areas to be developed as long as "there is suitable land available locally to recreate the features being lost". Comments from relevant authorities, statutory bodies, organisations and local residents are also important. A preliminary judgement can be made using the following questions:
- does the development affect the locally distinctive pattern of landscape elements?
- how intrusive would the scheme be on the field of view and visual amenity?
- can the landscape accommodate further change?
2 Substitution of biodiversity features is also controversial because ecological systems are very difficult to recreate (see §11.6.3) – and it should normally be considered only as a last resort.

(AADT), and road capacity is expressed as Congestion Reference Flow. The ratio of flow to capacity is then used to describe the stress level of the road and is used as a proxy for journey time reliability.

There are no available techniques to monetise the benefits of **regeneration**, i.e. of helping to regenerate an area by improving the transport infrastructure. There are two categories associated with the regenerative impact of a scheme: (a) whether or not the scheme is a regenerative priority for the development, and (b) whether or not the scheme is necessary to ensure developments in the area are viable. In the latter case an infrastructure scheme associated with a development will score highly.

ACCESSIBILITY

Accessibility is divided into three sub-criteria: access to public transport; community severance; and impacts on pedestrians and others. Each is given a textual ranking.

For **public transport** access, the impact appraisal is based on access times to the route by non-motorised forms of transport, and the impact of the scheme on the reliability of the service on the route. The changes in access time combines the number of passengers affected by the scheme with changes in walking and/or wait time. Information on the impacts can be provided by specialised pedestrian models such as PEDROUTE and public transport Models such as TRIPS, which can model the impact on service and access time.

Community severance is appraised by calculating how non-motorised modes of transport, especially walking, are affected by the scheme. Both the extent of the severance in time and the numbers of people affected are used in assessing the significance of the impact.

The accessibility **impact on pedestrians and others** is calculated by assessing the impact of the scheme on journey times of all non-motorised forms of transport. Account is taken of delays and loss of amenity for non-motorised users, the extent of the impact on journey times, and the numbers of people involved.

INTEGRATION

This objective seeks to ensure that the transport scheme is integrated into and compatible with national, regional and local land use and transport policies and plans. This could involve the preparation of a compatibility matrix where all the policies considered relevant in the scoping stage are tested against each other to see whether they are compatible, neutral or incompatible. The NATA guidance suggests using positive and neutral as a scoring system.

5.7 Mitigation measures

Traditionally, the transport impacts of development were considered to be primarily traffic-related. Mitigation measures thus involved off-site highway works to reduce driver delay (e.g. junction improvements), pedestrian and cyclist delay (e.g. improved crossing facilities), community effects or accidents (e.g. traffic calming).

In the light of current government and local policies, the mitigation measures relating to developments have become more focused towards providing transportation alternatives and addressing environmental issues. Mitigation measures for the

Table 5.2 Mitigation measures: transport impacts of non-transport developments

Main factor	Mitigation issues	Further measures
Car parking	• Reduce car parking	• Green Travel Plan
Highway capacity	• Reduce trips • Traffic calming • Traffic management • Increase public transport	• Green Travel Plan
Pedestrians	• Traffic calming • Pedestrianisation • Improved safety • Signing • Environmental improvements • Wider pavements • Improved crossing facilities	• Improve lighting • CCTV • Reduce carriageway size
Cycling	• Improved safety • Shared ped routes • Improved crossing facilities • Restrictions on car parking • Traffic calming • Secure parking • Changing facilities	• Segregation of cyclists from road traffic
Traffic management	• Traffic calming	• Encourage walking • Improve local areas • Improve safety • Reduce traffic speeds • Improve public transport access
Public transport	• Bus priority • Real-time information • Upgrading of facilities • Diversion of bus routes • Introduction of new bus routes • Park-and-ride	• Bus lanes • Green Travel Plans • Guided bus

transport impacts of non-transport developments include the provision of public transport improvements, reductions in car parking, green travel plans, the promotion of pedestrian and cycling facilities, servicing, and traffic management measures together with compliance with more strategic transport measures, such as park-and-ride. Mitigation measures for the environmental impacts of transport infrastructure relate primarily to reducing noise, air and water pollution and visual intrusion; improving lighting; enhancing wildlife and ecology, amenity and recreation; and promoting the sustainable use of natural resources. These measures are summarised in Tables 5.2 and 5.3 (adapted from BRF 1999).

Table 5.3 Mitigation measures: environmental impacts of transport infrastructure

Main factor	Mitigation issues	Further measures
Noise pollution	• Noise barriers	• Reflective barriers • Absorbent barriers • Vegetative barriers
	• Road surfacing	• Porous asphalt • Whisper concrete • Thin surfacing
	• Traffic management	• Traffic calming
	• Engineering solutions	• Cuttings • Cut and cover • Optimum junction design
Air pollution	• Traffic management	• Reduced/improved traffic flow • Pedestrian priority • Speed restrictions
Water pollution	• Improved **runoff** • Interception of pollutants	
Visual intrusion	• Integration of development • Promotion/restriction of views • Promotion of gateways	
Improved Lighting		
Species and habitats	• Avoid, minimise, substitute or translocate	
Amenity & recreation	• Cyclist & pedestrian promotion	• Improve route • Create links • Provision of facilities • Covered walkways
	• Promote overall use of facilities	• Improved disabled facilities • Traffic signage • Enhance environment • Improved access • Bus lanes

5.8 Monitoring

Monitoring of the transport impacts of specific developments is useful but often neglected. Local authorities are required to monitor transport against targets and indicators set by the road traffic reduction reports and as part of the Local Transport Plan annual progress reports. Transport models need to be kept up to date, and require periodic surveys of bus service frequency and traffic levels to obtain realistic figures. Consequently relevant information may already be collected that could be used to monitor the impacts of the development.

It has now become standard practice with transport assessment to undertake a Green Travel Plan, normally at the request of the local authority as part of Local Agenda 21 policies. This plan will identify measures for encouraging the modal shift of employees or visitors from the private car to public transport, cycling and walking. These measures can range from the provision of showers and changing facilities to season tickets or bike loans. This, together with restrictions in car parking provision, has been successful at a number of development sites in achieving a mode shift away from the private car. Under PPG13, the production of a Green Travel Plan may be a planning requirement for a new development that is likely to generate significant traffic. Within this travel plan monitoring its success is an important requirement.

5.9 Conclusions

Transport planning exists in a rapidly changing policy arena. The sustainability agenda encourages land use planning and transport planning to co-ordinate so as to reduce the need to travel. Developments must seek to achieve a modal shift so that a greater proportion of trips are undertaken by public transport, walking and cycling. Methods to assess transport impacts reflect this changing agenda and seek to influence developments in a sustainable manner. These methods aim to maximise the use of more sustainable modes of transport and minimise the impact on the environment by bringing together the different modes and the different objectives of development and sustainability into one assessment tool. Both Transport Assessment and NATA are new and developing methods. Their effectiveness will depend on the way they are used, the intentions behind the proponents, and the stage in the development process at which they are employed. NATA is designed to be flexible: it can be used at the feasibility stage, before detailed design, using subjective judgement; more quantitative information can be completed at the later detailed design stage. This two-tier approach should enable environmental consideration to be incorporated into the decision-making process at an earlier stage and thus have a greater influence on the outcome.

References

BRF (British Roads Federation) 1999. *Old roads to green roads*. London: BRF.

DETR (Department of the Environment, Transport and the Regions) 1997. *The Road Traffic Reduction Act*. London: HMSO.

DETR 1998a. *Guidance on the new approach to appraisal*. London: HMSO. (http://www.detr.gov.uk/itwp/appraisal/guidance/index.htm)

DETR 1998b. *The Transport White Paper. Transport: A new deal, better for everyone.* London: HMSO.

DETR 1999. *Preparation of environmental statements for planning projects that require environmental assessment: a good practice guide.* London: HMSO.

DETR 2001. *Planning Policy Guidance Note 13. Transport.* London: HMSO. (http://www.planning.detr.gov.uk/ppg/ppg13/index.htm)

DoE (Department of the Environment) 1996. *Planning Policy Guidance Note 6. Town centres and retail developments.* London: HMSO.

DoE/DoT 1994. PPG13 *A guide to better practice.* London: HMSO.

DoT (Department of Transport) 1993. *Design manual for roads and bridges (DMRB),* Vol. 11: *Environmental assessment.* London: HMSO. (Sections updated periodically.)

Hughes A 1994. *Traffic.* In *Methods of environmental impact assessment,* P Morris & R Therivel (eds), 64–77. London: UCL Press.

IEA (Institute of Environmental Assessment) 1993. Guidance Note 1 *Guidelines for the environmental assessment of road traffic.* Lincoln: IEA.

IHT (Institution of Highways and Transportation) 1994. *Guidelines for traffic impact assessment.* London: IHT.

6   Landscape

Riki Therivel (based on Goodey 1995)

6.1 Introduction

> Landscape is an important national resource . . . an outstanding natural and cultural inheritance which is widely appreciated for its aesthetic beauty and its important contribution to regional identity and sense of place. Although it is subject to evolution and change, the landscape is recognised as a resource of value to future generations.
> (DoT 1993)

An attractive landscape can contribute to peoples' enjoyment of their built and natural environment; can attract investment and assist social and economic progress; and can promote biodiversity, reduce surface runoff, and provide carbon fixing (DETR 1998a).

The European EIA Directives and UK regulations require an EIA to identify, describe and assess the direct and indirect effects of a project on the landscape, and on the interaction of landscape and other impacts. Landscape impacts are probably the most subjective elements addressed by EIA.

6.2 Definitions and concepts

6.2.1 Landscape and its associations

In EIA, the term *landscape* commonly "refers to the appearance of the land, including its shape, texture and colours. It also reflects the way in which these various components combine to create specific patterns and pictures that are distinctive to particular localities" (LI/IEA 1995). The following factors contribute to the landscape:

- physical: geology, landform, climate and microclimate, drainage, soil, ecology
- human: archaeology, landscape history, land use, buildings and settlements
- aesthetic

  * visual, e.g. proportion, scale, enclosure, texture, colour, views
  * other senses, e.g. sounds, smells, tastes, touch

- associations:

  * historical, e.g. history of settlements, special events
  * cultural, e.g. well-known personalities, literature, painting, music (CC 1993).

Most EIAs distinguish between landscape impacts and visual impacts. **Landscape impacts** relate to the objective "changes in the fabric, character and quality of the landscape" (LI/IEA 1995): landscape impact assessments typically consider landscape character areas, individual landscape elements (e.g. church spire, prominent trees), and special interests such as designated landscapes. **Visual impacts** relate to the more subjective "changes in the available views of the landscape, and the effects of those changes on people" (LI/IEA 1995). They are essentially the subset of landscape impacts that deals with impacts on views, viewers and *visual amenity*.

As can be seen from the above list, the vast majority of landscape is **cultural**, rather than just natural heritage, and its assessment should take into account the values and associations of residents and visitors as well as professional interests. Most people seem to know what they like when it comes to viewing the landscape, and they frequently compose views for a glance, for contemplation, or for a photograph. Thus, many of the basic components of landscape are already stored in peoples' bank of experience. Contemporary popular meanings, media interpretations (Burgess 1990a) and perceptions of landscape (Goodey 1987, 1992) deserve attention in EIA. Research into "cultural landscapes" includes work on the cultural meanings implicit in landscape and its representation (Cosgrove 1984, 1985; Cosgrove & Daniels 1992; Rackham 1991), and Burgess's (1990b, 1993; Burgess *et al.* 1988) qualitative methodology for determining popular attitudes towards local landscapes and the recreation opportunities they provide. The (former) Countryside Commission (CC 1990, 1992, 1993) also aimed to reflect the breadth of public responses to valued landscapes.

Landscape is also linked to **ecology**. This has led the ASH consulting group (1993a) to develop the concept of ecological identity areas, which provides preliminary ecological assessment at the landscape scale.

Maps that link visual and sonic disturbance at a macroscale to illustrate **tranquil areas** – areas characterised by remoteness and sense of isolation – were similarly pioneered by ASH (1993b) and Rendel (1994). They define tranquil areas based on distance from highways, railways, pylons and aeroplane flight paths.

Although much of landscape and visual assessment is concerned with professional and public perceptions of the view – where a positive response to aesthetic attractiveness may be expected – users of the landscape are also concerned with **personal security**. Burgess's studies (1990b, 1993) of perceived opportunity and risk in urban fringe woodlands are especially significant in illustrating the fears expressed by wide sections of the community with regard to woodlands and open spaces (Painter 1992).

6.2.2 *Landscape quality and landscape character*

Over the years, various researchers have aimed to develop a hierarchy of landscape **quality** as a basis for landscape designations (see §6.3.1). Such approaches include classifications based on individual landscape elements, scenic quality and other measures of importance, sensitivity and capacity, and character and condition: they are summarised in Box 6.1.

More recently, there has been a shift away from the focus on landscape quality, and the implied management approach of 'preserve the best and leave the rest'. A new approach, pioneered by the Countryside Agency (CA 2000, CC *et al.* 1997), acknowledges the **character** of individual landscapes, the diversity of all the landscape

Box 6.1 Landscape quality evaluation methods, in rough chronological order (based on Hodge 1999)

Professional judgement involving intuitive methods, analytical methods, and measurement of landscape elements (Robinson *et al.* 1976). Most of these techniques fell into disrepute, as the choice of elements used for measuring and their weightings were found to be just as subjective as non-quantitative methods.

Landscape preferences of the public using indirect methods such as behaviour surveys and direct methods such as measuring preferences for photographs and sketches (Zube *et al.* 1975, Preece 1991).

Criteria-based analysis of importance used by the (then) Countryside Commission (1993) to evaluate landscapes for designation, but also applicable to any landscape:
- **Landscape as a resource**: the landscape should be a resource of at least national (regional, county, local) importance for reasons of rarity or representativeness.
- **Scenic quality**: it should be of high scenic quality, with pleasing patterns and combinations of landscape features, and important aesthetic or intangible factors.
- **Unspoilt character**: the landscape within the area generally should be unspoilt by large-scale, visually intrusive industry, mineral extraction or other inharmonious development.
- **Sense of place**: it should have a distinctive and common character, including topographic and visual unity and a clear sense of place.
- **Conservation interests**: in addition to its scenic qualities, it should include other notable conservation interests, such as features of historical, wildlife or architectural interest.
- **Consensus**: there should be a consensus of both professional and public opinion as to its importance, for example, as reflected through writings and paintings about the landscape.

Strength of character and condition. Strength of character denotes how closely the landscape matches the optimum profile of its particular landscape type or character area in terms of the typical patterns of characteristic features. Condition denotes how far the features of the landscape are removed from their optimum visual (physical condition) and functional state (ecological health of the remnants of semi-natural habitats). In areas of strong distinctive character, policies would aim to conserve and/or restore existing patterns. In areas where character is weak or in poor condition, policies would aim to create new landscapes and/or accommodate change (Warnock 1997).

Sensitivity and capacity. Landscape sensitivity relates to the potential visual impact of a development on landscape character or its quality, and is a function of landform and vegetation cover. The impact of new developments can be minimised if they are targeted at those landscapes which are least sensitive to change. New developments in sensitive landscapes should be limited and be designed in a way that reflects traditional patterns and styles. Landscape capacity relates to the potential of landscape to absorb development without adverse impacts. An assessment of landscape capacity also embraces landscape sensitivity (Warnock 1997, CA 1999).

types, and the benefits and services that landscapes provide. This approach, which is discussed at greater length at §6.3.2 and Chapter 17, changes the emphasis from landscape as 'scenery' to landscape as 'environment'.

There is still a debate amongst landscape professionals as to whether landscapes should be assessed on the basis of their quality or character, particularly because the boundaries of high-quality landscapes do not necessarily coincide with landscape character areas. There are arguments for both approaches (Hodge 1999). Professionals and the general public share a strong set of values about our finest landscapes, which are reflected in their designations: removal of these designations may well lead to their loss, which would be irreplaceable. On the other hand, all landscapes are important to those who live, work or visit them. Designated areas are often protected at the expense of the rest, so that non-designated areas often fail to be protected and are targeted by developers. The concept of landscape beauty is not timeless and is dependent on fashion and taste, so that today's judgement could be tomorrow's mistake (Hodge 1999).

6.2.3 Describing the landscape

Appleton (1975) developed a qualitative "**prospect and refuge**" approach to landscape aesthetics, which he later pursued into the arts. Tandy (1967) identified **areas of visual containment**, or landscape identity/character areas, as relatively self-contained zones whose landscape character changes little. In contrast, crossing a **visual barrier** such as a hilltop or a curve in a road, opens up a new vista. In terms of project impacts, **intrusion** – "the quality of an element or factor which appears to stand out to the detriment of a design; a serious visual problem or conflict" (Lucas 1991) – is likely to be one of the most significant issues raised in EIAs. Intrusion indices of various types have been used to qualify and graphically reflect the impact of development projects.

6.3 Legislative background and interest groups

6.3.1 Designations and regulations

When initiating a landscape appraisal, one should always identify the **landscape designations** applicable to, or adjacent to, the site under consideration. More than 20% of the area of England and Wales has been designated for landscape purposes as a result of such legislation as the *National Parks and Access to the Countryside Act 1949*, the *Countryside Act 1968*, and the *Wildlife and Countryside Act 1981*. The Countryside Agency (CA), Scottish Natural Heritage (SNH) and Countryside Commission for Wales (CCW) have the remit to protect and improve the landscape, and to provide new and improved opportunities for access to the countryside (Appendix B).

England's eight *National Parks*, covering almost 10,000 km², remain the most significant British statement of landscape quality, although their origins represented a predictable complexity of aesthetic, ecological and recreation interests. The New Forest is soon to become a National Park, and the South Downs are under consideration. Nationally designated *Areas of Outstanding Natural Beauty* (more than

20,000 km²), such as the Quantock Hills or the Forest of Bowland, again imply an aesthetic judgement as to quality – although the term 'natural' is controversial since even the most 'natural' areas are actually **semi-natural**. Other relevant non-statutory designations include more than 1000 km of *Heritage Coasts*, as well as *National Trails* such as the Pennine Way and the Ridgeway. The Midland New National Forest is a further national initiative.

Since 1987, approximately 65,000 ha have been managed by farmers under the *Environmentally Sensitive Areas* and *Countryside Stewardship* schemes to benefit wild-life and the landscape. *Heritage landscapes* may benefit from tax relief under the terms of the Inheritance Tax Act 1984.

The **cultural aspects** of the landscape mean that archaeological, building, town-scape and structural designations also need to be considered: *Parks and Gardens of Special Historic Interest, National Trust properties*, and areas subject to *Landscape Pavement Orders* in England, and *National Heritage Areas* and *National Scenic Areas* in Scotland, the latter two being subject to statutory designation. *Green Belts*, local areas of *Great Landscape Value*, or occasionally *Special Landscape Areas*, and *Country Parks* (England and Wales) or *Regional Parks* (Scotland) will also require consideration, the parks having statutory designation. In Scotland there is an additional array of local landscape and scenic designations. Local *Tree Preservation Orders* and *conservation area* designations will be significant, and some sites, such as the Ironbridge Gorge industrial archaeology complex in Shropshire, enjoy inter-national recognition by UNESCO as *World Heritage Sites*. It is also essential to compare any landscape designation with ecological/nature conservation areas (see Chapter 11).

6.3.2 *Landscape character assessments* (based on Hodge 1999)

As was suggested earlier, an emphasis on designations is increasingly seen as insuffi-cient in EIA, and landscape character areas are increasingly considered. Landscape character assessment was developed in the mid-1990s by SNH and the joint working of the Countryside Commission (now CA), English Nature and English Heritage, to encompass the commonplace and everyday landscapes as well as those that are designated. Their joint aims were to develop a national framework of geographical areas that described what the landscape was, how it had evolved, and how it could be changed positively, for example, by identifying opportunities for conservation, restoration or enhancement. Although separate national landscape characterisation programmes were carried out in England and Scotland, the strong parallels of their approaches led to their joint working in producing guidance on the preparation and use of landscape character assessments. Draft guidance was published in August 1999, and final guidance is expected in 2001.

England: Landscape characterisation in England began with a systematic analysis of the landscape based on altitude, land form, ecological characteristics, agricultural land capability, surface geology, farm types, settlement patterns, woodland cover, field and density patterns, visible archaeology, industrial history and designed park lands. A national dataset for each factor was prepared for every kilometre square in England. The data were analysed using TWINSPAN (multivariate cluster analysis) and GIS to produce a national typological map of national landscape character types. The National Joint Character of England map shows 159 terrestrial areas and

22 maritime character areas, defined in terms of their landscape, sense of place, wildlife and natural features. Interim Joint Summary Statements described the key natural and cultural characteristics of each character area. They have since been superseded by EN's **Natural Area** Profiles and the CA's eight volumes of Countryside Character descriptions based on the standard government administrative regions. The CA's descriptions are not prescriptive but make suggestions for the future of the landscape.

An increasing number of county and district authorities are undertaking more local, detailed landscape studies of their areas. Although much of this work is based on the CC's early (1987, 1993) guidance, it is now increasingly being based on characterisation.

Scotland: Unlike the CA's top-down approach to characterisation, SNH used a bottom-up approach in partnership with local planning authorities and other stakeholders. The methods and outputs were similar to those of the national Joint Character Mapping Exercise. The programme identified 360 landscape character types and produced 29 regional studies. Each regional study describes, maps and analyses the key landscape characteristics, and identifies forces of change that could alter/reduce the character and diversity of each landscape character type.

Wales: The CCW and the Wales Landscape Partnership have developed a GIS-based Landscape Assessment and the Decision Making Process (LANDMAP) for informing policies and decision-making. LANDMAP has four stages:

1. Orientation: A local steering group defines the information and output requirements.
2. Generation of aspect areas: Each area is subdivided geographically using agreed criteria, and is described by a team of specialists. Management recommendations are based on current land use and landscape values, e.g. designations, economic potential, opportunities for recreation and tourism. A public perception study is also carried out, and aspect information is fed into a GIS. Relevant information includes land use, settlement pattern, evaluated aspects, geology, vegetation and history.
3. Production of combined aspect areas: The information for each aspect is combined to produce a landscape assessment for the area, taking account of contextual issues (land use, settlement patterns) and additional information (e.g. planning designations, public perceptions). A GIS is used to test the scenarios of multiple landscape recommendations.
4. Monitoring and review: The local steering group decide whether there is a need for further study, how recommendations should be implemented, and the time period for monitoring and reviewing information (CCW 1998).

6.3.3 Guidance

Several manuals and websites provide guidance on landscape and visual impact assessment. Most are still based on the more traditional approaches to landscape quality and appraisal.

Landscape Assessment: A Countryside Commission Approach (CC 1987) was an early seminal guide which advised on the approach, practical methods and applications of landscape assessment. *Landscape Assessment Principles and Practice* (CCS 1991) established practical guidelines for landscape assessment, focusing particularly on assessing Scottish landscapes for designation. *Landscape Assessment Guidance* (CC 1993) updated the earlier documents, and became the basis for many EIA landscape assessments in the UK and abroad. It provides guidance on the approach, practical methods and applications of landscape assessment at various geographic scales; methods used in assessment (e.g. visual description, checklists, GIS approaches), each evaluated in terms of its effectiveness and appropriateness for various contexts; methods for consolidating the assessment and impact mitigation; and case studies.

The LI/IEA's (1995) *Guidelines for Landscape and Visual Impact Assessment* are specifically oriented to the needs of EIA practitioners, and provide good practice guidance for assessing the landscape and visual impacts of developments that require an EIA. They discuss good practice in landscape and visual impact assessment, present guidelines for such assessment, and address issues of consultation, review and implementation.

The (former) DoT's (1993) *Design manual for roads and bridges* explains how landscape appraisal should be carried out in the design of roads and bridges. The (former) DoE's (1995) guide on *Preparation of environmental statements for planning projects that require environmental assessment* gives concise advice on landscape and visual impact assessment. It suggests that landscape appraisal should include landscape description (landform, land cover, landscape elements), classification into areas of common character, and evaluation of how the landscape is perceived and its importance. Visual impact assessment should involve identifying a zone of visual influence for the proposal and evaluating how the development affects this zone based on representative viewpoints. Hankinson (1999) gives more detailed and up-to-date advice oriented to an international audience.

Looking back historically, Fabos (1979) provides an account of early literature in the field. Early landscape planning texts by Hackett (1971), Turner (1987) and Preece (1991) provide accessible introductory accounts of the subject, as do Lucas, (1991) and Bell's (1993) well-organised and illustrated reference manuals.

However, only the most recent guidance reflects the sea change in approaches to landscape appraisal discussed in §6.2.2. The DETR's (1998b) guidance on the *New Approach to Appraisal* (NATA) explains how to appraise the landscape impacts of new developments that are likely to have significant impact on the transport infrastructure (see Chapter 5). Its guidance document *By design: urban design in the planning system* (DETR 2000) also stresses principles of landscape characterisation, diversity and adaptability. As characterisation and environmental capital techniques become more commonly used, it is likely that the guidance will change as well.

6.4 Baseline studies

In EIA, baseline studies establish the parameters and structure for the following investigation. They need to be extensive and rigorous, establish a digestible account of the area and project concerned, and highlight specific details that will require later investigation. They should include a clear statement of purpose, initial consideration of the full range of landscape elements and meanings involved, application of a comprehensive and tested methodology, and clear communication in terms which can be understood and discussed by the wider community. This section suggests (a) steps in baseline landscape appraisal and description, and (b) criteria for evaluating landscapes.

6.4.1 Description of the baseline landscape

The boundaries of the landscape to be analysed – the project's *Zone of Theoretical Visibility* (ZTV) – need to be identified first. The 'project' being appraised may extend well beyond the boundaries of the project site, to encompass off-site construction and storage areas, associated workings such as pylons and pipelines, and possibly traffic to and from the site. Maps and site visits will later help to narrow the broad ZTV to a more defined *Zone of Visual Intrusion* (Hankinson 1999).

Desk studies provide the starting point for a landscape assessment. Useful sources of desktop information include:

- maps of land use, topography, geology, soil, hydrology
- historical maps and other data on archaeology, buildings, past uses of the site, etc. (Appendix C)
- local authority landscape characterisation exercises or assessments, guidelines, plans and GIS data that show landscape designations
- Countryside Agency information on character areas
- data on ecology and meteorology
- aerial photographs

These will provide initial information on:

- landform (topography, drainage)
- land cover, e.g. vegetation, how the land is managed (see Appendix F.6) and likely landscape processes and seasonal changes
- built-up areas and individual buildings
- other landscape features
- intervisibility, principal viewpoints, the likely visibility of the development
- designated areas
- cultural and historical associations

This information should be updated and fine-tuned during field surveys. All of these should be described in the EIA.

The site is then divided into areas that have internally consistent character: patterns of landform, land cover, scale, and degree of enclosure. These should be mapped and described, including a description of the character of each area, identification of key elements that contribute to that character, and an initial appraisal of the effect that development will have on those elements/character.

In terms of **visual assessment**, the baseline desk and field studies should fine-tune the *Zone of Visual Intrusion* by considering existing screening and topography, and conditions in winter as well as in summer, as leaf cover can provide substantial screening. Within this zone, those people potentially affected by the development should be identified. These include not only local residents but also people who work in the area, use the area for recreation, and travel through the area (on footpaths and bridleways as well as on roads). It is often also useful to identify, for later stages of analysis, representative viewers and particularly sensitive viewers. In some cases cultural associations may suggest additional principal viewpoints, for instance if a site has been painted from a particular point.

Much of the evidence derived from the landscape description can be synthesised into series of **visual representations** that illustrate landscape character areas, indicative and significant views, and evidence of special features. Methods for visual representation include:

- plans or maps of the site and its character areas, land uses, statutory designations, viewpoints, landscape quality, listed buildings, and/or tranquil areas
- photos from key viewpoints into and out of the proposed site; these should be clearly linked to a map that shows the viewpoints and other key locations
- diagrams of landscape features
- aerial photos
- videos

Reports by the CC (1993) and LI/IEA (1995) illustrate and evaluate these techniques, and discuss the value of overlay mapping and computer-aided landscape classification methods. Several aspects of visual impacts may be incorporated on the same map or figure.

6.4.2 Evaluation of the baseline landscape

In the **traditional approach** to landscape appraisal, it would be sufficient at this stage to identify the local, regional, national (and possibly international) designations and special interest groups affected; the management aspirations for the area by the CA and other bodies; and aspects of the cultural heritage that deserve special consideration. In terms of visual appraisal, the evaluation phase would identify how visible the site is for each relevant receiver/viewer: possible evaluation categories are full view, partial view, minimal view and no view.

The **newer approaches** to landscape appraisal would evaluate the landscape character areas by asking a series of questions:

- What benefits does the landscape in each character area provide? Examples include tranquillity, cultural heritage, sense of place, and land cover (e.g. agriculture, semi-natural habitats);
- To whom does the benefit matter, at what scale, and how important is it?
- How rare is the benefit? Is there enough of it?
- How could that benefit be substituted? (See Chapter 17.)

The results of such an evaluation should be a series of **management guidelines** that any development in that area should take into account. This exercise can provide a non-confrontational way of involving local residents, workers and visitors in project design and landscape impact mitigation. There is still much ambiguity about how best to carry out such an evaluation, for instance, in terms of terminology (e.g. feature v. benefit), how to judge importance, etc.: much of this will probably be cleared up in the next few years as more landscape evaluations are carried out.

6.5 Impact prediction

6.5.1 *Overview*

Effectively communicated predictions of the nature, likelihood and significance of changes that may occur as a result of the proposed development, over various sub-areas and periods and for various viewers, is at the heart of successful landscape/ visual impact assessment. Impact prediction begins with an understanding of the development, and the incorporation of good project design from the beginning. The landscape's likely evolution without the development is described in both written and graphic terms, and similar modes of description are then applied to the land-scape expected as a result of development. A structured comparison between these two descriptions will then highlight the magnitude and significance of the landscape and visual impact.

6.5.2 *Good project design*

Good project design and landscape/visual mitigation should be planned in at the start of the project. The DETR (2000) suggest seven objectives of urban design which are applicable to all developments:

- character: the development should have its own identity
- continuity and enclosure: it should clearly distinguish public and private spaces
- quality of the public realm: it should have attractive and successful outdoor areas
- ease of movement: it should be easy to get to and move through
- legibility: it should have a clear image, and be easy to understand
- adaptability: it should be able to change easily
- diversity: it should provide variety and choice

These objectives can be achieved by good project design, including use of land-scape issues as a criterion in the selection of the project site or process (e.g. landfill v. incineration); careful siting of major structures, access routes and parking, ma-terials storage, etc. in relation to visual receptors, ridgelines/valleys, etc.; sensitive choice of site levels; attention to the density, mix, height and massing of buildings; retention of special landscape features and provision of visual/ecological **buffer zones**; consideration of microclimates and the solar aspect of buildings; attention to ma-terials used and details such as openings and balconies; careful design of open spaces including plantings and fencing; and enhancement through new wildlife habitats, restoration of derelict land, and the provision of public open space and/or beautiful new landscapes (Barton *et al.* 1995, DETR 2000, Hankinson 1999).

Landscape design also has a part to play in other aspects of project design. For instance, it may need to be integrated with technical advice on water-holding facilities, acoustic fencing or bunding. Turner (1998) suggests other innovative examples of "environmental impact design", including provision of wild food, new footpaths, and conservation farming as part of an integrated approach to project/ landscape design. Circular 5/94, *Planning out crime*, explains how safety issues can be incorporated in project design. The management guidelines identified during the baseline evaluation (§6.4.2) should also be incorporated at this stage.

6.5.3 *Impact magnitude*

The project's location, dimensions (especially vertical), materials, colour, reflectivity, visible emissions, access routes, traffic volumes and construction programme will all need to be described in the EIA. The scale of the project may be identified and described in relation to existing landscape features (e.g. trees); balloons raised to the relevant height at the appropriate sites can also help to identify the height of buildings, although they can only be used on calm days.

The landscape with and without the project should then be described and compared to identify the degree of change that the project would bring about. The changes are likely to vary between different project stages (e.g. construction and operation, or phases of mineral extraction), and between seasons. Where the project involves substantial night-time lighting, this will also need to be considered. The predictions should discuss the duration and timing of impacts, and impacts with and without mitigation (§6.6).

Prediction of the visual impacts will involve predicting and illustrating the change in views from the key viewpoints identified earlier. For this, the EIA will need to discuss the number of people affected, the conditions under which they view the site (as residents, workers, driving by, etc.), the distance of their view, any screening of the view that they may experience (fencing, vegetation, etc.), and the duration and timing of the visual impacts.

No one technique for describing landscape and visual impacts fully captures the subject. Landscape change is usually registered by the observer in foot- or car-borne sequences of three-dimensional views, in subtle colour variation, and including visible movement as well as sound and textures: these are only partly replicable by even the most sophisticated 'virtual' techniques. Typical methods for describing landscape and visual impacts in EIA include:

- plans or maps showing the area over which the development can be seen (the zone of visual intrusion or influence) and the phasing of the development;
- sketches or artists' impressions of the development from previously identified viewpoints (particularly from representative viewers), with and without mitigation measures;
- cross-sections, perhaps with viewlines;
- photos from key viewpoints into and out of the proposed site;
- photographs with artists' impression overlays;
- photomontages of the proposed development, alternative options, and/or the site before, during and after the development;
- GIS methods;
- virtual environments.

Digital photography and computer programs are increasingly used in EIAs, not only to prepare maps, scale drawings, perspectives, cross-sections and photomontages, but as part of a process that incorporates initial surveys, impact analysis, presentation of alternative forms of mitigation, and preferred solutions (Branson *et al.* 1993). As RPS Clouston (1993) note, however, the graphic outputs will depend on the users involved:

> The critical point in photomontage is the addition of detail and the blending of new elements with old. It is on the colour matching and interpretation of detail that the final realistic effect will depend. The correct representation of tree

planting at, say, 15 years may be much more important to the landscape archi-
tect in the team than to the consulting engineer, who might well give a higher
priority to concrete finishes and road marking. The ultimate client may be just
as much concerned with limiting the cost of the exercise, and providing a
minimum accuracy level which will not give a misleading impression of the
proposals to the general public.

Steinitz (1990) illustrated the potential for user participation in the selection of
alternative landscape strategies using an integrated sequence of methods, including
GIS (Chapter 16). However, there remain at least two significant obstacles to the
uptake of these techniques: cost, which is likely to restrict the public availability of
methods and examples, and the scepticism of some professionals and the public,
who suspect that increasingly important value issues may be lost in technological
sophistication.

6.5.4 Impact significance

Impact significance is a combination of impact magnitude and the sensitivity of the
receiving landscape and viewers. For landscapes, designations will provide a starting point for assessing sensitivity, and the CC's (1993) criteria of importance (see
Box 6.1, p. 107) can be used for non-designated landscapes. The landscape/integrated
management objectives of the LANDMAP and Countryside Character descriptions
will also help to determine landscape sensitivity.

The sensitivity of viewers will depend on the amount of time that they see the view;
whether they live, work or play in the area; and their mental predisposition to the area.
One cannot assume a distance decay in interest or concern as the viewer recedes from
a site, nor that "people will get used to it". Major features such as power-station cooling
towers can be recognised positively as place markers, evidence of new technology, or
as attractive design by some, while remaining offensive intrusions to others.

Typical categories of sensitivity and magnitude are shown at Table 6.1. Impact
significance brings these together: for instance, high sensitivity with high magnitude
would be highly significant, whilst low sensitivity and high magnitude would be
moderately significant. Impact significance also involves judgements about whether
impacts are positive or negative. There are no quantitative criteria for assessing the
significance of landscape/visual impacts, and the appraisal is normally done through
professional judgement, taking into account any public comments from consultation. The results of the environmental capital exercise of §6.4.2 will allow a more
subtle analysis of impact significance based on the benefits that the landscape provides, as well as a more innovative approach to good project design and impact
mitigation. Methods for describing impact significance include maps showing the
degree of visual impact experienced by affected viewers, and tables that list representative viewers on one axis, and their number, distance of views, duration of
views, etc. on the other axis.

6.6 Mitigation and enhancement

As discussed earlier, good project design is a more effective, and often cheaper, way
to minimise negative and optimise positive landscape/visual impacts than *post hoc*

Table 6.1 Sensitivity and magnitude of landscape and visual impacts (LI/IEA 1995)

Landscape		
Sensitivity		Magnitude
For example, important components or landscape of particularly distinctive character susceptible to relatively small changes	high	Notable changes in landscape characteristics over an extensive area ranging to very intensive change over a more limited area
For example, a landscape of moderately valued characteristics reasonably tolerant of changes	medium	Moderate changes in localised area
For example, a relatively unimportant landscape, the nature of which is potentially tolerant of substantial change	low	Virtually imperceptible change in any components
Visual		
Sensitivity		Magnitude
For example, residential properties and public rights of way	high	For example, the majority of viewers affected/major change in view
For example, sporting and recreational facilities	medium	For example, many viewers affected/moderate change in view
For example, industry	low	For example, few viewers affected/minor change in view

'landscaping'. The surveyor and assessor should be part of a landscape design team so that the building(s) and landscaping are designed as creative environmental enhancements. However, a lack of regard for mitigation, and for complementary and comprehensive landscape design, has been one of the criticisms levelled at EIAs by local authorities (Fieldhouse 1993), and has led to authorities taking the initiative in proposing mitigation measures, often in the form of planning conditions.

The most common types of secondary/*post hoc* mitigation are (a) changes to building height, colour, shape and building materials; (b) mounding/bunding, which will in turn affect surface-water drainage and may itself cause visual intrusion; (c) planting both on and off the site, and retention of existing vegetation; and (d) other hard and soft landscaping (e.g. street furniture, design of access roads, lighting). Manipulation of views through the use of form or colour, or through the introduction of new landmark features may serve to deflect the eye, or imply cultural benefit.

Four characteristics are evident in landscape mitigation measures. First, the landscape consequent upon development is largely **the visible manifestation of all the physical, and some human, changes** achieved by the development. Although mitigation measures and environmental gain can aim to enhance views and improve habitat, their success is predicated on their integration with modifications to landform, drainage, surface and subsurface condition, local climate and access patterns. Landscape mitigation must never be regarded as the 'green icing' on a newly baked cake.

Ecological and visual suitability can be achieved only through an integrated team approach.

Second, **time** is the major factor in the success of landscape mitigation measures. Even a suburban retail park planted densely with semi-mature trees and decorative shrubs takes time to settle, both into the surrounding townscape, and in the eyes and minds of the users. Interested parties often propose that a development should be hidden from view, or at least 'integrated' with its surroundings in the shortest possible time. Rapid integration demands high investment in preparation, plant stock and management, and may not best serve longer-term ecological or land-use intentions for the area.

Third, there is a very strong tendency towards **heritage reference**. People's response to a new development is to search for historic landscape elements that can either be re-stated or contributed anew. Contemporary landscape values point strongly to this reinforcement of tradition, which has become enmeshed with a presumed 'ecological' approach to landscape design, with the consequent assumption that nothing that can be added to the landscape is as good or appropriate as what already exists. For example, the Essex Development Control Forum (1992) suggests that "There is a presumption in favour of conservation *in situ*, i.e. reinforcement of existing planting, management of areas, fencing to control grazing and overuse, etc. Only as a last resort should consideration be given to relocating seed beds or redistribution of wild flower seeds, etc.". This is the view taken by ecologists in relation to ecologically valuable habitats which are virtually impossible to recreate fully (see §11.6.3). However, if it is taken to mean that what is currently present is always better than anything that could be created, it effectively precludes innovative designs or the concept that new projects can help to improve the landscape or ecology of an area. There is no reason why a new development should not provide an essentially novel landscape experience, while still retaining valuable ecological systems within the local environment. New landscapes with new images and meanings may become the much-loved views of future generations.

The fourth characteristic of landscape mitigation is the importance of **process** in facilitating the desired landscape outcome. Many fundamental characteristics of the emerging landscape can be established through careful design of the development process itself, including the initial retention of local plant and surface materials, use of exposed substrata for visual effect, diversion of watercourses, and the pattern of vehicular use on-site during construction. Mitigation starts before the development begins, not after.

6.7 Monitoring

Monitoring and evaluation is gradually being recognised as an essential element in environmental management and EIA. The planning consent target, to which the EIA submission is linked in the British procedure, militates against client investment or professional interest following this short-term objective. Given that initial predictions of landscape impacts tend to be relatively uncertain, immediate change is easily recognised by the broad community, and mitigation requires time to be effective, it is important that landscape monitoring is undertaken.

A regular programme of specific, comparable, observations (and response measures where appropriate) would provide (a) an early-warning system for unexpected impacts of the development, permitting changes in the construction and/or mitigation procedures; (b) a learning experience which may feed directly into other projects; and (c) regular evidence as a basis for discussion with authorities or the public.

6.8 Concluding issues

A series of characteristics unique to landscape within the EIA process will shape the future development of methods in this area:

- Landscape planning is likely to become more important within the statutory planning process, encouraged by the scale of development and land-use issues, increasing conflicts over land use and appearance, and public interest in the look of the land.
- Given that landscape and visual impacts are inevitably qualitative, or quantitative summaries of qualitative observations, there must always be an admitted subjective element in the communication of baseline data, impact analysis and mitigation measures. The environmental capital approach could provide a useful framework for this, although its applicability will need to be further tested in practice.
- Continued development is to be expected in the area of visualisation, both for professional analysis and as a medium through which public views are sought. Care will be needed to ensure that the complex values evident in earlier debates on landscape/visual assessment are not obscured or forgotten in the drive to achieve technologically sophisticated solutions with attractive, if superficially convincing, outcomes.
- Although regional and national landscape values might seem to be established and enduring, their continual reappraisal and advancement through policy and investment will have a considerable impact on the landscape characteristics that we are encouraged to value. The current reappraisal of the purpose and presence of coastal flood defences is an example of this.

References

Appleton J 1975. Landscape evaluation: the theoretical vacuum. *Institute of British Geographers. Transactions* **66**, 120–123.

ASH Consulting Group 1993a. *Ecological identity areas: preliminary ecological assessment at the landscape scale.* Memorandum and plan sample. Didcot: ASH Consulting Group.

ASH Consulting Group 1993b. *South east tranquil areas.* London: Council for the Protection of Rural England/Countryside Commission.

Barton H, G Davis & R Guise 1995. *Sustainable settlements: a guide for planners, designers and developers.* Luton: University of the West of England and Local Government Management Board.

Bell S 1993. *The elements of visual design in the landscape.* London: E & FN Spon.

Branson B, S Foley & D Henderson 1993. EA case study: photomontage techniques. *Landscape Design* **224**, 40 only.

Burgess J 1990a. The influence of the media on people's relationships with, and attitudes to, nature and landscape. In *People, nature and landscape: a research review.* London: Landscape Research Group.

Burgess J 1990b. The production and consumption of environmental meanings in the mass media: a research agenda for the 1990s. *Institute of British Geographers. Transactions* **15**, 139–161.

Burgess J 1993. *Perceptions of risk in recreational woodlands in the urban fringe.* London: Department of Geography, University College London.

Burgess J, M Limb & CM Harrison 1988. Exploring environmental values through the medium of small groups. 1: Theory and practice. *Environment and Planning* **20**, 309–326.

CA (Countryside Agency) 1999. *Proceedings of Planning Workshop.* Cheltenham: Countryside Agency, 21 July 1999.

CA 2000. (http://www.countryside.gov.uk/)

CC (Countryside Commission) 1987. *Landscape assessment: a Countryside Commission approach*, CCP 18. Cheltenham: Countryside Commission.

CC 1990. *The Cambrian mountain landscape*, CCP293. Cheltenham: Countryside Commission.

CC 1992. *The Chilterns landscape.* Cheltenham: Countryside Commission.

CC 1993. *Landscape assessment guidance*, CCP423. Cheltenham: Countryside Commission.

CC et al. (Countryside Commission, English Heritage, English Nature, and Environment Agency) 1997. *What matters and why: environmental capital: a new approach.* Cheltenham: Countryside Commission.

CCS (Countryside Commission for Scotland) 1991. *Landscape assessment principles and practice.* Edinburgh: CCS.

CCW (Countryside Council for Wales) 1998. *LANDMAP Landscape assessment and the decision making process.* Draft handbook for consultation.

Cosgrove D 1984. *Social formation and symbolic landscapes.* Beckenham: Croom Helm.

Cosgrove D 1985. Prospect, perspective and the evolution of the landscape idea. *Institute of British Geographers. Transactions* **10**, 45–62.

Cosgrove D & S Daniels (eds) 1992. *The iconography of landscape.* Cambridge: Cambridge University Press.

DETR (Department of the Environment, Transport and the Regions) 1998a. *Planning for sustainable development: towards better practice.* (http://www.detr.gov.uk/)

DETR 1998b. *Guidance on the New Approach to Appraisal.* London: HMSO. (http://www.detr.gov.uk/itwp/appraisal/guidance/index.htm)

DETR 2000. *By design: urban design in the planning system – towards better practice.* (http://www.detr.gov.uk/)

DoE (Department of Environment) 1995. *Preparation of environmental statements for planning projects that require environmental assessment.* London: HMSO.

DoT (Department of Transport) 1993. *Design manual for roads and bridges*, Vol. 11: *Environmental assessment.* London: HMSO.

Essex Development Control Forum 1992. *The Essex guide to environmental assessment.* Chelmsford: The Essex Planning Officers' Association.

Fabos JG 1979. Planning and landscape evaluation. *Landscape Research* **4**(2), 4–9.

Fieldhouse K 1993. Question time. *Landscape Design* **218**, 15–18.

Goodey B 1987. Spotting, squatting, sitting or setting: some public images of landscape. In *Landscape meanings and values*, EC Penning-Rowsell & D Lowenthal (eds), 82–101. London: Allen & Unwin.

Goodey B 1992. Dix paysages Europeens dominants. *Paysage & Amenagement* **21** (October), 8–13.

Goodey B 1995. Landscape. In *Methods of environmental impact assessment*, P Morris & J Therivel (eds), 78–95. London: UCL Press.

Hackett B 1971. *Landscape planning: an introduction to theory and practice.* Newcastle upon Tyne: Oriel.

Hankinson M 1999. Landscape and visual impact assessment. In *Handbook of environmental impact assessment*, Vol. 1. J Petts (ed.), Ch. 16, 347–373. Oxford: Blackwell Science.

Hodge L 1999. *Landscape character assessment: a new force in structure plans*, MSc dissertation, School of Planning, Oxford Brookes University.

LI/IEA (Landscape Institute and Institute of Environmental Assessment) 1995. *Guidelines for landscape and visual impact assessment*. London: E & FN Spon.

Lucas OWR 1991. *The design of forest landscapes*. Oxford: Oxford University Press.

Painter K 1992. Different worlds: the spatial, temporal and social dimensions of female victimisation. In *Crime, policing and place: essays in environmental criminology*, DJ Evans, NR Fyfe & D Herberts (eds), 164–195. London: Routledge.

Preece R 1991. *Designs on the landscape*. London: Pinter (Belhaven).

Rackham O 1991. Landscape and the conservation of meaning. *Royal Society of Arts Journal* (January), 903–1013.

Rendel S 1994. Tranquillity – an essential commodity? *Town and Country Planning* **63**, 31–35.

Robinson DG, IC Laurie, JF Wagner & AL Traill 1976. *Landscape evaluation*. Centre for Urban & Regional Research, University of Manchester (for the Countryside Commission).

RPS Clouston 1993. EA case study: illustrative techniques. *Landscape Design* **224**, 29 only.

Steinitz C 1990. Toward a sustainable landscape with high visual preference and high ecological integrity: the loop road in Acadia National Park, USA. *Landscape and Urban Planning* **19**, 213–250.

Tandy C 1967. The isovist method of landscape survey. In *Methods of landscape analysis, symposium report*, 9–10, London: Landscape Research Group.

Turner T 1987. *Landscape planning*. London: Hutchinson.

Turner T 1998. *Landscape planning and environmental impact design*, 2nd edn. London: UCL Press.

Warnock S 1997. *Landscape character assessment and the town and country planning system*. Cheltenham: Countryside Commission.

Zube EH, RO Brush & JG Fabos 1975. *Landscape assessment: values, perceptions, and resources*. Stroudsburg, Pennsylvania: Dowden, Hutchinson & Ross.

7 Archaeological and other material and cultural assets

Rosemary Braithwaite, David Hopkins and Philip Grover (updated by Philip Grover)

7.1 Introduction

Europe has known some 500,000 years of human activity and settlement, from the earliest hunter gatherers to the present day. As a result, almost all sites on mainland Britain have had previous human occupation and are therefore of potential historical interest. The study of archaeological and other historical resources is important to (a) fulfil an innate curiosity about the past, since the origins and development, lifestyles, economy and industry of previous generations can be traced and understood through archaeological remains; (b) contribute to the sense of tradition and culture; and (c) promote a sense of national identity.

Archaeology is a vital component of recreation, since many people enjoy visiting archaeological sites and studying archaeological remains. It contributes to education; archaeological study is used as a basis for integrating the teaching of a number of other subjects, and can promote an understanding of the role of the past and its relevance to today's society. Britain's historic heritage is also important to the tourism industry. It attracts visitors from all over the world and, if well interpreted and presented it can be an important financial asset.

However, archaeological and other historical remains are a fragile and finite resource that needs to be carefully managed and conserved, and are therefore one of the many elements that need to be addressed in any EIA. On most sites these remains are not important enough to affect development, but a site's historical and cultural interest is always monitored by planning authorities, and EIAs should show that it has been considered.

7.2 Definitions and concepts

7.2.1 Overview

The DoE recommends that a site's "architectural and historic heritage, archaeological sites and features, and other material assets" should be described in an EIA, as well as the "effects of the development on buildings, the architectural and historic heritage, archaeological features, and other human artefacts" (DoE 1989). However, what precisely material and cultural assets are is open to interpretation. The general requirement to consider such aspects ensures that an EIA is comprehensive and that issues of strong local feeling or wider social and cultural heritage are considered. Some EIAs have interpreted material assets very widely, including, e.g., agriculture, forestry plantations, recreation and amenity, utilities and other services, communications, rights of way, and potential future resources. However, here material and cultural assets together are taken to be:

- archaeological remains, both above and below ground, i.e. buried remains and standing buildings;
- historic buildings and sites (including listed buildings, cemeteries and burial grounds, parks, gardens, village greens, bridges and canals);
- historic areas (including towns and villages in whole or in part – often designated as conservation areas);
- other structures of architectural or historic merit.

In practice, there is no precise distinction between archaeology and other aspects of the historic environment. For instance, English Heritage (EH), which is responsible for the major archaeological sites in England, is alternatively known as the Historic Buildings and Monuments Commission. Academic historians are concerned with the past on the basis of written evidence (in the UK, effectively from Roman times), whilst archaeologists use a much wider range of evidence and therefore may go back to the earliest human occupation (in the UK, perhaps 500,000 years ago to World War II). The legislation covering the historic environment is a patchwork of regulations and guidance which draws an arbitrary distinction between archaeology and ancient monuments on the one hand, and other aspects of the historic built environment such as listed buildings and conservation areas on the other. This chapter broadly follows the legislative distinction, so each section discusses first archaeology and then historic buildings and sites. However, it is recognised that this distinction is not always clear in practice and may not be applicable to all EIAs. Other structures of architectural or historic merit are addressed in Chapter 6, on landscape.

7.2.2 Archaeology

The range of archaeological evidence reflects the diversity of human experience: the need for water, food and shelter, the use of changing technologies, and the religious, cultural and political needs of society. The physical remains of human activity and endeavour are known as the archaeological resource. These remains range in size and complexity from individual objects used and discarded, to settlements. They include many details in the landscape, which itself is the product of human use and adaptation of the natural environment. The physical evidence may survive as earthworks such as burial mounds, hillforts, field banks and lynchets. They can also survive as structures such as buildings, canals, bridges and roads.

However, the majority of the archaeological resource is smaller and often hidden below ground, surviving as features such as pits, postholes, gullies and ditches cut into the subsoil. Very often the evidence is in the form of artefacts, like coins, pottery sherds, stone tools, and metal objects. Archaeological remains lie below many of the buildings and streets of British cities and towns. Over 600,000 archaeological sites are presently known in the UK, or about 200 per parish. The archaeological record is the sum of present archaeological knowledge, i.e. that part of the archaeological resource which has been identified to date. Table 7.1 summarises the principal archaeological periods and likely remains from these periods.

Table 7.1 Principal archaeological periods and likely remains (based on DoT 1993)

Period	Dates	Likely remains
Prehistoric	earliest Palaeolithic (~500,000 BC) to AD 43	from early rock shelters and stone artefacts to the circles, barrows, Celtic field patterns, farmsteads, villages and hillforts of the Late Iron Age
Roman	AD 43 to AD 410	native and immigrant farms, Roman towns and cities, military forts, roads
Medieval	5th–16th centuries	origins of most modern towns (e.g. postholes from wooden buildings, masonry), Norman castles, deserted villages, ridge and furrow agriculture
Post-medieval	late 16th to early 18th centuries	Civil War constructions, beginnings of industrial-scale extraction and manufacture, country houses and their parks and gardens
Industrial	mid-18th century onwards	buildings and infrastructure linked with industrialisation, industrial relics
Post-Industrial	World Wars	defences (e.g. pillboxes)

The rich pattern of archaeological remains that can be seen today is the result of the impact of successive generations on the remains left by previous generations. This process involves a degree of damage and destruction which is an inevitable part of the evolution of the archaeological record. However, the current threat to the archaeological resource is more significant than in the past due to the technological changes and the rapid increase in development that has occurred particularly since World War II. Today the archaeological record is more likely to be deleted than altered or added to. Land which has been marginal since prehistory has, with the use of modern machinery and chemicals, become viable for arable farming, with the resultant damage by ploughing and soil erosion. The increase in road building, housing and industrial developments and the need for materials for their construction continually depletes the archaeological resource. Already in some areas, post-war gravel extraction has been so extensive that the ability to understand the evolution of the landscape has been badly reduced. As a result of high land prices in towns, there is now a prevalence of deep basementing, below-ground carparks and substantial foundations to support high buildings. In some historic centres only a small proportion of the archaeological resource remains intact.

The significance of archaeological finds is derived both from the nature of the finds themselves in their contexts and from the interpretation archaeologists are able to put on them given contemporary understanding. Whilst the ability to learn about the past is based on the investigation and interpretation of archaeological remains, this investigation often results in the destruction of the archaeological resource being studied. Archaeological excavation aims to dismantle remains to their constituent parts in order to understand the processes by which they were formed (see §7.4.1). This work is closely documented, with all the elements drawn and photographed and the objects which are found removed and conserved. Although this enables future study and reinterpretation of the results, the site cannot actually be reconstructed. However archaeology is an evolving study and is constantly harnessing new technologies, techniques, procedures and theories. Preserving archaeological remains for future study is therefore important. Just as archaeologists today can learn substantially more from the archaeological resource than their counterparts of yesterday, so preserving a site *in situ* for future archaeologists will allow even more information to be gained. In addition the more visible sites that are used for tourism, recreation and education need to be preserved and conserved. Whilst the preservation of all remains would be impractical, and would lead to the stagnation of archaeology, the case for the preservation of the archaeological resource must always be carefully considered.

7.2.3 Historic buildings and sites

Historic buildings form the most visible and tangible of all aspects of the historic environment. They are a finite resource and cannot undergo change without cultural loss. The careful appraisal of their history and condition, together with their protection through effective policies and careful professional practice, can lead to improved decisions concerning their conservation. Three main sources of judgement apply to changes to the character of buildings, deriving from the disciplines of archaeology, architecture and architectural history.

In practice **listed buildings** should be seen as part of the wider historic environment which also includes archaeological remains. Unfortunately judgements on changes affecting listed buildings have often tended to focus purely on visual character rather than on a deeper appreciation of the intrinsic value of inherited or historically important building fabric. This emphasis on architectural character has in the past tended to give rise to facadism and imitative architectural styles, often of mediocre quality. A greater understanding of the impact of intervention by developers and a wider appreciation of the concept of stewardship on the part of building owners and local authorities needs to be encouraged if the special architectural and historic interest is to be properly safeguarded.

As well as individual buildings the **visible historic environment** can be defined in terms of areas. Important groups/ensembles of historic buildings, perhaps encompassing the core of a historic city or town, or indeed a whole settlement, are now recognised as important elements of the wider historic environment. Areas of special architectural or historic interest are frequently designated as 'conservation areas' and in many respects their management should be seen as analogous to that of historic buildings. These areas provide valuable points of reference in a rapidly changing world as well as representing the familiar and cherished local scene. Historic areas come in many forms but are typically characterised by important groups of historic buildings (not necessarily listed) based around a historic street pattern often with important urban squares or green spaces containing features such as mature trees. It has been estimated that designated conservation areas in the UK account for some 4% of the built environment. Together, listed buildings and conservation areas form a distinctive and finite part of the nation's cultural heritage. Development affecting this resource therefore needs careful management, and EIAs must include a full assessment of the particular value of the features in question.

7.3 Legislative background and interest groups

7.3.1 Archaeology

The principal legislation protecting the archaeological resource in England, Wales and Scotland is the *Ancient Monuments and Archaeological Areas Act 1979*. The equivalent legislation in Northern Ireland is the *Historic Monuments and Archaeological Object (NI) Order 1995*. In addition, the *Town and Country Planning Act 1990* and its Scottish equivalent (1997) affords protection to archaeological sites through the statutory planning process.

The *Ancient Monuments and Archaeological Areas Act 1979* provides legislative protection to a selection of archaeological sites or monuments which have been identified as being of national importance and included within a schedule maintained by the Secretary of State for Culture, Media and Sport. These are consequently referred to as **Scheduled Ancient Monuments**. Some ancient monuments of national importance are not yet scheduled and English Heritage (EH) is currently undertaking a review of the schedule, the *Monuments Protection Programme*, which is considerably increasing the number of scheduled sites. Any works to, or within, a Scheduled Ancient Monument likely to damage that monument require the prior consent of the Secretary of State; this consent is referred to as Scheduled Monuments Consent. Where consent is issued it is frequently subject to conditions to prevent damage or to limit damage to agreed levels and with appropriate archaeological recording. Unauthorised works which damage a Scheduled Ancient Monument are a criminal offence, and significant penalties exist. The act also protects the setting of such monuments. The Secretaries of State are advised on Scheduled Ancient Monuments and other archaeological, historical and heritage matters by the relevant heritage agencies, i.e. EH in England, Cadw in Wales, Historic Scotland (HS) in Scotland and The Environment and Heritage Service (EHS) in Northern Ireland (see Appendix B).

The 1979 Act also enables the designation of **Areas of Archaeological Importance** (AAIs). Five pilot AAIs were designated in 1984: York, Chester, Hereford, Exeter and Canterbury. Once an area is designated, developers are required to give six weeks 'operations notice' to the planning authority of any proposals to disturb the ground, tip on it or flood it. A designated 'investigating authority' then has the power to enter the site and, if necessary, undertake archaeological excavations for up to four months and two weeks. After that time the investigating authority must cease excavation but can continue to enter the site to record and inspect the works. This legislation did not address key areas such as preservation or funding, and has subsequently been overtaken by the procedures in PPG 16. No further AAIs have been designated.

The Town and Country Planning Acts enable local planning authorities (LPAs) to protect a wide range of archaeological remains through the planning process. Where development threatens to destroy remains, the authority can require appropriate investigation through a planning condition or legal agreement. In certain circumstances it can also secure the positive long-term management of sites. These provisions are usually expressed in the policies relating to archaeology within **development plans** (EH 1992a).

The impact of development on archaeology has been recognised as a material consideration within the planning system for some time. In 1990 the DoE issued *Planning Policy Guidance Note 16* on *Archaeology and planning* (DoE 1990), which describes how archaeological matters are to be dealt with in the English planning system. PPG16 is therefore an extremely useful and important reference document and should be carefully considered when preparing an EIA. Broadly PPG16 requires LPAs to acquire sufficient information to enable the full impact of a development to be considered. These powers had already been frequently used for the archaeological resource but were formalised by the EIA regulations. Accordingly, the manner in which archaeological considerations are already dealt with in the planning system is closely akin to the requirements of EIAs. PPG16's Welsh equivalent was published in 1991 (WO 1991) and its Scottish equivalent is *National Planning Policy Guidance 5* (NPPG5), *Archaeology and planning*. Both documents contain similar advice to that contained in PPG16.

7.3.2 Historic buildings and sites

The principal legislation governing the protection of historic buildings and sites in England and Wales is the *Planning (Listed Buildings and Conservation Areas) Act 1990*. For England, further central government guidance is to be found in *Planning Policy Guidance Note 15* (PPG15), which provides a full statement of government policies for the identification and protection of historic buildings, conservation areas and other elements of the historic environment. In Wales there is currently no direct equivalent to PPG15 although relevant policy and advice is contained in PPG1 (*General Principles*) and in Welsh Office circular 61/96 (*General Planning Guidance*). Parallel legislation exists for other parts of the UK in the form of the *Planning (Listed Buildings and Conservation Areas) (Scotland) Act 1997* and the *Planning (Northern Ireland) Order 1991*. In the case of Scotland the principal legislation is supplemented by *National Planning Policy Guidance 18* (NPPG18) and Historic Scotland's extremely comprehensive *Memorandum of Guidance on Listed Buildings and Conservation Areas 1998*.

Listed buildings

A listed building is one which has been included in a list compiled by central government as being of "special architectural or historic interest". A developer cannot demolish, alter or extend any listed building in a way that affects its architectural or historic character unless listed building consent has been obtained from the LPA, and listed buildings must be taken into account when LPAs undertake land use planning decisions. A small team of specialist investigators from EH identifies buildings to be listed: this method echoes that earlier employed for compiling schedules under the Ancient Monuments Acts 1882. In England in excess of 450,000 individual buildings are protected by listing, accounting for some 2% of the building stock. EH's proposals are closely scrutinised by the Secretary of State before confirmation, and similar scrutiny is applied to proposals by the other UK heritage agencies.

Central to architectural conservation – the conservation of whole buildings – is the definition of **building character**, for it is against this definition that judgements are made about the nature and extent of permitted changes. In England the criteria used to identify buildings of "special architectural or historic interest" are as follows:

- all buildings built before 1700 which survive in anything like their original condition;
- most buildings between 1700 and 1840, though selection is necessary;
- between 1840 and 1914 only buildings of definite quality and character, including the principal works of the principal architects;
- after 1914, selected buildings of high quality;
- less than 30 years old, only buildings of exceptional quality under threat;
- buildings less than 10 years old are not listed.

(Source: PPG15, Paragraph 6.11)

In choosing buildings for the list, the Secretary of State applies the following criteria derived from PPG15:

1. architectural interest by virtue of design, decoration, craftsmanship, building type and technique (e.g. displaying technological innovation or virtuosity), and significant plan forms;
2. historic interest (e.g. illustrating important aspects of the nation's social, economic, cultural or military history);
3. close historical association with nationally important people or events;
4. group value, especially where buildings comprise an important architectural or historic unity or a fine example of planning (e.g. squares, terraces or model villages).

Similar, but slightly differing principles have been established in other parts of the UK. Theoretically, listing applies to the whole of a property's curtilage, including objects and structures fixed to the building, although a detailed evaluation is needed to make a judgement about those features that are of worthwhile architectural or historic significance.

Listed buildings are graded to indicate their relative importance: Grade I buildings are of exceptional or outstanding interest, Grade II* are particularly important and of more than special interest, and Grade II are of special interest and warrant every effort being made to preserve them. Slightly different grades apply to Scotland and Northern Ireland.

LPAs are responsible for determining the majority of proposals affecting listed buildings. Decisions are made in accordance with national legislation, statutory local policy and in the context of central government guidance. However, there is considerable variation in the strength and quality of the protection afforded to listed buildings over the nation as a whole. Consequently, EIAs need to consider the policies of county and district councils as well as national legislation.

If an LPA considers a non-listed building to be of special architectural or historic interest and in danger of demolition or significant alteration, it can serve a *Building Preservation Notice*, which effectively lists the building for six months; this allows the Secretary of State to determine whether the building should be included in the statutory list or not. This is however an infrequently used power since compensation is payable in the event of the Notice not being upheld by the Secretary of State.

Conservation areas

According to the *Planning (Listed Buildings and Conservation Areas) Act 1990*, and parallel legislation outside England, conservation areas are sections of land or buildings designated by LPAs as being "of special architectural or historic interest, the character or appearance of which it is desirable to preserve or enhance". LPAs must have regard to conservation areas when exercising their planning functions, and conservation area consent must be obtained from the LPA before a building within a conservation area can be demolished.

Conservation areas have proved to be a popular and positive element of town planning since the passing of the original enabling legislation in 1967. There are now over 8000 conservation areas in England. It is the quality and interest of whole areas rather than individual buildings that is the prime concern of conservation areas. There is no standard specification for conservation areas. Whilst DoE Circular 8/87 has been superseded by PPG15, it did contain some useful guidance which is still of relevance in defining conservation areas. They

> ... will naturally be of many different kinds. They may be large or small, from whole town centres to squares, terraces and smaller groups of buildings. They will often be centred on listed buildings, but not always. Pleasant groups of other buildings, open spaces, trees, an historic street pattern, a village green or features of historic or archaeological interest may also contribute to the special character of an area.
> (DoE 1987)

The legislation and associated guidance encourages the involvement of local communities through conservation area advisory committees. The concept of conservation areas has found widespread support with the public as a whole in spite of early recalcitrance on the part of many councillors and continuing fears from the design disciplines.

Other legislation

There is no specific national legislation addressing the *World Heritage Sites* promoted by the UNESCO Convention for the Protection of the World Cultural and Natural Heritage; their protection lies in the importance given to them within the planning process and through policies relating to the development plans.

EH maintains a *Register of Parks and Gardens of Special Historic Interest in England*, namely sites that are regarded as an essential part of the nation's heritage. The register grades parks and gardens from Grade I of exceptional interest to Grade II of special interest. These sites are not afforded statutory protection, but are protected by recognition of their importance through the planning system, and policies relating to them in development plans. Historic Scotland and Scottish Natural Heritage compile a similar *Inventory of Gardens and Designed Landscapes in Scotland* (1998).

7.3.3 Interest groups and sources of information

In respect of sites of known or potential archaeological interest the county archaeologist, or the equivalent officer in unitary authorities, should be involved early in the EIA process. They advise on the care of archaeological sites, maintain the *Sites and Monuments Record* for their area (see §7.4.1), screen planning applications for archaeological impacts, and make recommendations to the planning committee. They will be able to make a rapid initial assessment (see §7.4.1) and suggest professional contacts (e.g. members of the Institute of Field Archaeologists with local knowledge and experience) if further specialist knowledge is required. In England each county has its own archaeologist, as do most unitary authorities and some district councils. In London the role of the county archaeologists is fulfilled by EH. Annex 2 of PPG16 contains addresses of county archaeologists. In many counties the Museum Service works closely with the county archaeologists.

In respect of listed buildings and conservation areas the local conservation officers should be involved early in the EIA process. They have specific detailed knowledge of historic buildings and conservation areas within their jurisdiction and are usually the principal advisers to the local planning committee in relation to proposals likely to have an impact on the historic environment. Most local authorities now have at least one conservation officer and some have small specialist teams.

In addition to its advisory role, EH administers the most important sites – as do the other UK heritage agencies (Appendix B). HS and EHS also fulfil many of the roles of the county archaeologists.

In many areas local history or amenity societies have detailed local knowledge and take active interest in anything that affects their area. Local planning authorities must consult the national amenity societies when the demolition of a listed building is proposed. In practice, the societies are also consulted when more ordinary changes are proposed, as their expertise is substantial and unique. The advisory societies are the Ancient Monuments Society, the Society for the Protection of Ancient Buildings (SPAB), the Georgian Group, the Victorian Society, the Council for British Archaeology (CBA), and the Twentieth Century Society. Local amenity societies are more ephemeral; planning authorities maintain lists of societies in their localities which they consult over changes to listed buildings. The archaeologists' professional body is the Institute of Field Archaeologists; they publish lists of their members and their specialisations. The equivalent body for conservation officers and their counterparts in English Heritage, Cadw and Historic Scotland is the Institute of Historic Building Conservation (IHBC).

7.4 Scoping and baseline studies

7.4.1 Archaeology

The aim of a baseline study is to identify and describe the nature, location and extent, period(s) and importance of the archaeological resources likely to be affected by the development. The resulting report should include:

- a summary of the archaeological context;
- an inventory of archaeological assets found both at the site and in the wider area likely to be affected by the development;
- an evaluation of these assets;
- an informed expectation of potential assets to be found in further investigation or likely to be at risk from development. Past construction activities which might have already destroyed archaeological resources should be noted;

- a map of the project area showing the location of these assets;
- a note of any inherent difficulties which may limit the study's usefulness (e.g. problems of access).

A number of sequential stages of data gathering can be identified. However, not all stages would be necessary for every EIA.

Rapid appraisal

Rapid appraisal of the archaeological resource involves the collation and review of existing and easily accessible data. This will certainly include a review of the *Sites and Monuments Record* and consultation with the county archaeologists. It may also include a site visit. This appraisal will enable a preliminary view of the likely nature and scale of the archaeological constraint. It may in itself be sufficient to meet the aims of the EIA, or may identify the need for subsequent stages of data gathering.

The main source of archaeological information in England is the *Sites and Monuments Record* (SMR). The SMR is a local archaeological database containing information about the known archaeological sites and finds in each county. The SMR has a statutory locus in that it is referred to within the General Permitted Development Order 1995; certain types of permitted development, such as mineral extraction, require permission where they affect an archaeological site registered on the SMR. The SMR information is gathered from a number of sources and in a variety of ways, from detailed surveys to chance finds. As a result there is considerable variance in the reliability of the data and the interpretation that can be placed upon it. It is often not very intelligible to non-archaeologists (and is not in fact a public document) and may need professional interpretation to assess the significance or potential of archaeological sites. The county archaeologists will usually be familiar with the nature and shortfalls of the data being considered, and will be able to advise on the appropriate interpretation of the archaeological data. It is important to note that the interpretation of archaeological data is rarely straightforward.

Whilst the SMR is a comprehensive statement of the archaeological resource as currently known, it is not a definitive statement: new archaeological information becomes available all the time. Therefore the sites on the record represent only a part of the actual archaeological resource and many archaeological sites remain as yet unlocated. This has two major implications for compiling an EIA. First, as the SMR only reflects current knowledge, there may be other important archaeological remains as yet unlocated that may be affected by a proposal. Second, if considerable time elapses between when the SMR is consulted and when that information is used, additional evidence may become available in the meantime. These unknown sites are nonetheless a material consideration and therefore should be addressed when considering a development proposal. This is recognised in PPG16:

> Where early discussions with local planning authorities or the developer's own research indicate that important archaeological remains may exist, it is reasonable for the planning authority to request the prospective developer to arrange for an archaeological field evaluation to be carried out before any decision on the planing application is taken. This sort of evaluation is quite distinct from full archaeological excavations. It is normally a rapid and inexpensive operation, involving ground survey and small-scale trial trenching, but it should be

carried out by a professionally qualified archaeological organisation or archaeologist. . . . Evaluations of this kind help to define the character and extent of
the archaeological remains that exist in the area of a proposed development,
and thus indicate the weight which ought to be attached to their preservation.
They also provide information useful for identifying potential options for minimising or avoiding damage. On this basis, an informed and reasonable planning decision can be taken. Local planning authorities can expect developers to
provide the results of such assessments and evaluations as part of their application for sites where there is good reason to believe there are remains of archaeological importance. If developers are not prepared to do so voluntarily, the
planning authority may wish to consider whether it would be appropriate to
direct the application to supply further information under the provisions of
Regulation 4 of the Town and Country Planning (Applications) Regulations
1988 and if necessary authorities will need to consider refusing permission for
proposals which are inadequately documented. In some circumstances a formal
Environmental Assessment may be necessary.

(DoE 1990)

As mentioned in §7.3.3, the county archaeologists both maintain the SMR – and
are therefore a source of initial data – and advise the local planning authority, and
are therefore an initial source of advice. They will also be able to advise on the
scope and content of the archaeological elements of the EIA. They are extremely
knowledgeable about the archaeological potential of sites in their areas, and are also
usually very realistic about development pressures. The county archaeologists will be
anxious to ensure that the archaeological content of an EIA has been properly
addressed, and will generally be happy to supply both data and advice. A charge may
be made to cover the costs incurred in supplying data (ACAO 1992). As the county
archaeologists usually advise the local planning authority regarding the acceptability
of these elements it is important to be aware of their opinions at an early stage.
Where failure to consult results in additional archaeological concerns being raised,
there is the potential for uncertainty, delay and additional costs that will negate the
benefits of having carried out the EIA.

Where a development is likely to affect a Scheduled Ancient Monument or its
setting, EH should be consulted and Scheduled Monument Consent may be required.
The need to obtain this consent is independent of the planning process and unless
identified early could introduce substantial delay or even compromise the development
altogether. Furthermore, where a development is likely to affect a monument of
national importance which, although not scheduled, may be considered for scheduling in due course, it is advisable to seek the advice and opinion of EH or equivalent.

Desk-based assessment

A desk-based assessment should identify and collate as much existing information as
possible and frequently requires some original research. Information may be retrieved
from a number of sources but the SMR is usually the most useful starting point.

Aerial photographs are an important source of data. Earthworks are often more
easily recognised and interpreted from the air than from the ground. Buried archaeological remains can also be traced from the air in certain circumstances. The buried
remains can affect the growing crop. For instance, a buried wall or road surface may

retard crop growth, or in a dry year create a parch mark. A buried pit or ditch may promote crop growth. The patterns that result can be interpreted as archaeological features or sites. Different soil colours may also reveal archaeological sites. Aerial photographs may be found in national, local authority, and possibly private collections (see Table 15.1, p. 371). The record office may contain historic maps or plans and other documents relating to the land, and it may be possible to find other data not yet assimilated into the SMR. The Victoria County Histories and Local Archaeological Societies may have additional information.

Desk assessment is usually undertaken at an early stage in project planning, so there may be an issue of commercial sensitivity. If so, it may be reasonable to use the county archaeologist and the Victoria County Histories, but not to approach the voluntary societies until later.

Field survey

A wide range of field survey techniques are available, including geophysical techniques, fieldwalking, augering, test pitting, machine trench digging and earthwork surveys. These are described below. Not all of these techniques will be applicable in all circumstances. Some can act as useful preliminaries to other techniques. A phased approach to field survey is often the most sensible and cost effective, so it is common to use a suite of techniques as the proposal develops: perhaps starting with a rapid appraisal and then a desk assessment in the earliest stages, then fieldwalking before the actual site is proposed, and machine trenching afterwards. When considering the appropriateness of the various techniques, consultation with the county archaeologists may be valuable.

The county archaeologists usually produce a brief or specification for the work when a field survey is being undertaken through the planning process. A brief is an initial statement regarding the aims and scope of the archaeological work required, identifying certain working standards. It would form the basis of any specification produced, which should be referred back to the county archaeologists to ensure that all matters in the brief have been properly addressed. Alternatively, the county archaeologists may issue a full specification which sets out in detail the works required in the field survey and would be sufficient to enable the project to be implemented and progress to be monitored.

The county archaeologists may also wish to make arrangements for monitoring the field survey to ensure that works are carried out to professional standards and to any specification that has been issued. This has benefits both for the archaeological resource and for the developer, who may have no independent means to monitor the value of the work being undertaken. It also enables the county archaeologists to keep up to date with any archaeological sites that are discovered during the fieldwork. Some county archaeologists charge for monitoring.

Geophysical techniques can be used to investigate some characteristics and properties of the ground that may be altered by previous land uses. The principal techniques used are resistivity and magnetometer surveys, although others are also available. Resistivity surveys measure the ground's resistance to the progress of an electrical current. Measuring increases and decreases in the resistance can indicate the nature and location of buried features. Magnetometer surveys measure the magnetic properties of the soil and can be used to identify locations of past human activity, particularly those that involved burning or heating.

Geophysical techniques can only be applied in suitable site conditions and an experienced geophysical operator should visit the site to assess their feasibility. Where they are appropriate, geophysical techniques have an advantage over many other field techniques in that they do not damage the archaeological resource. Because of this they are particularly appropriate for Scheduled Ancient Monuments, although Scheduled Monument Consent or a licence may still have to be obtained before surveys can be undertaken.

Although the results of geophysical techniques can sometimes be ambiguous, these techniques often successfully identify the location and extent of archaeological sites and can give some idea of their nature. The results can therefore help to focus subsequent stages of field survey to maximise data recovery. However, geophysical techniques are unlikely to provide sufficient information on their own, are not universally applicable, and are often expensive.

Fieldwalking – also known as surface artefact collection – is confined to ploughed fields. A plough breaks and turns over the surface soil. In ploughed fields there is a tendency for buried material to be brought to the surface, and where the plough intrudes into a buried archaeological site this will include archaeological artefacts. Rigorous collection and plotting of this material will enable the location, date, and extent of certain types of archaeological site to be described. The archaeological material collected can be anything that reflects human activity, like pottery sherds, worked stone, coins, building material and even stone that is not local to the area and may have been imported.

The county archaeologists will be able to suggest a fieldwalking strategy that ensures that the data gathered will be comparable to other fieldwalking data already on the SMR. The area being studied is divided up by a grid, usually based on the national grid. Artefacts are then collected from along the lines of one axis of the grid, usually the north–south axis, and stored and recorded according to where on the grid they were recovered. The size of the grid thus determines the size of the collection units, and the precision of the results. A survey on a large grid will be rapid but will represent a small sample of the available artefacts. A survey on a small grid will be more time-consuming but the results will be based on a larger sample. The size of the grid is usually determined with reference to the sorts of archaeological sites that are anticipated. For instance, a smaller grid would be required to locate small Mesolithic camps than a Roman villa. In general grid spacing is about 20 m or 25 m.

Where a site has already been located, intensive fieldwalking, called total collection, can be used to determine spatial distributions across the site. Total collection involves laying out a small grid across the site, perhaps 5 m × 5 m, and collecting all the artefacts within each grid square.

Fieldwalking is a relatively rapid and inexpensive technique that can be applied over large areas. However, the results can be ambiguous or misleading. Where a site is located by fieldwalking it is by definition being damaged. It is hard to judge from fieldwalking results alone how intact the site is, or whether it solely survives as artefacts trapped in the plough soil. A site surviving intact below the plough soil will not be represented on the surface. Certain periods do not produce artefacts which are likely to survive the ploughing action. The results of fieldwalking therefore need to be qualified by some understanding of the relationship between the depth of ploughing and the depth of the archaeology.

Augering is most frequently used in river valleys where alluvial, colluvial or peat deposits have masked the original land surface and where slightly higher ground in a wet environment may have acted as a focus for human activity. By recording the soil sequence from auger holes located over a wide area, the underlying and hidden subsurface topography can be mapped and the archaeological potential of the area can be inferred. Augering alone is unlikely to confirm the presence or absence of archaeological deposits, but can clarify the archaeological potential and so focus subsequent stages of survey. It can also be used to clarify the nature of features located by geophysical techniques, and in certain areas to assess the potential for the preservation of palaeoenvironmental data.

Test pitting involves the hand excavation of an array of small pits of a predetermined size. It provides a clear picture of the nature of the soil structure and the upper layers of the underlying geology. As with fieldwalking, the spacing and array of test pits usually reflect assumptions about the expected archaeological resource. Test pits can be varied in size and array in order to meet the requirements of the survey. They are usually 1 m × 1 m, or 1 m × 0.5 m for ease of excavation. The soil from test pits is often sieved through a wire mesh of a set size to ensure consistent artefact recovery, enabling a rigorous statement to be made regarding the number, type and depth of artefacts. Analysis of the different artefact recovery rates over an area gives an indication of the date, location and extent of archaeological sites. Test pitting is often used instead of fieldwalking where the land is pasture rather than arable, and in woodland where machine trenching may not be possible.

Machine trenching employs trenches, usually cut with a toothless ditching bucket, laid out in a pattern across the site. The trench pattern will attempt to maximise information retrieval, possibly on the basis of existing data such as aerial photographs, fieldwalking or geophysical results. The extent of trenching required is usually an agreed sample of the land. The size of the sample is currently the subject of considerable debate, but is commonly around 2%, depending on local circumstances. When archaeological deposits are encountered excavation continues by hand. The excavation is controlled by a supervising archaeologist at all times. Machine trenching quickly locates features cut into the subsoil but, where large amounts of earth are rapidly removed, there is limited opportunity to collect artefacts and the rate of artefact retrieval is low. Higher rates of retrieval can be achieved by hand-digging parts of the trench, equivalent to a test pit, and the use of metal detectors.

Trenching is very disruptive and intervenes directly into the archaeological levels. This has the advantage of producing unambiguous information but is potentially damaging to archaeological remains one might otherwise wish to protect. It is also not always possible to get a machine onto a site.

Earthwork surveys can be used for archaeological sites that are visible as earthworks such as banks, ditches, burial mounds, and sites of deserted or shrunken settlements. Sites that survive as earthworks are generally more intact than other sites. Ploughing can degrade earthworks, and the success of earthwork surveys is limited in fields that have been arable for a long time; generally, such land is more productively scanned from aerial photographs. Pasture can have visible earthworks surviving. When they are obviously visible they will often have been recorded by the Ordnance Survey (OS) or the SMR. They can also be identified through aerial photographs. Woodland, particularly ancient woodland, holds the greatest potential for producing previously unrecorded earthworks. The sites will often be obscured from the air by trees and on the ground by undergrowth, so it is best to undertake the survey during the winter or early spring.

The nature of the earthwork survey will depend on the aims of the evaluation. The survey can vary from sketch plotting the earthworks onto an OS map, through two-dimensional surveys such as plane table surveys, to a three-dimensional survey producing an accurate contour or hachure plan.

Finds are recovered artefacts. Some of the these may be subject to the laws of treasure trove; specifically all discoveries of gold or silver should be reported to the coroner, who will consider whether the items were hidden with a view to being retrieved at a later date. If this is concluded to be the case, the state may retain any of these items, paying the landowner the market value. In all other situations the artefacts are the property of the landowner. It is usually recommended that they are donated to a local authority museum, so that they can be stored in appropriate conditions and made available for future study. All finds of human bone, from any period, have to be reported to the coroner.

The developer's responsibilities arising from the destruction of the archaeological resource often continue beyond excavation. If finds are donated to the appropriate local authority museum, it is likely that the planning authority will consider the developer to have met these responsibilities. If the developer wishes to make alternative arrangements, they may need to demonstrate that this alternative is appropriate. Some museums make a charge for accepting the long-term responsibility of storing archaeological material.

Some problems with field surveys

Access to the site will not be a problem where the developer already owns the land, although there may be problems where the project has off-site implications, e.g. as a result of **dewatering**. For projects such as road schemes a field survey may not be possible until the route is finally selected and the land acquired. This is undesirably late because it does not allow a route to be chosen which would preserve important remains *in situ*.

The project timetable may constrain the fieldwork options. Fieldwalking is not possible in a standing crop, and can only be done after the fields are ploughed. Similarly, crop patterns show best in a well-grown crop and should be photographed just before the harvest.

The cost of archaeological surveys depends upon the extent and nature of the survey and the techniques employed. Surveys are frequently labour intensive and some elements can be expensive. Where the developer is liable to pay compensation to the landowner for damage arising from the evaluation, the scale of compensation will depend upon the techniques used. However, the costs should be seen against the background of the cost resulting from unexpected delay to the progress of the planning application or indeed the progress of the development if significant archaeological deposits are located at a late stage in the process.

7.4.2 Historic buildings and sites

Although listed buildings account for only some 2% of the UK's building stock, they are a fragile and valuable resource. Only a full assessment of a listed building's inherited character at the outset will allow well-informed judgements to be made about the significance of a proposed development's impacts. Both owners/developers and LPAs have their respective roles to play in such assessments.

An initial review of the listed building register will identify any **listed buildings** likely to be affected by a proposed development. Listed buildings will also normally be identified on the *Sites and Monuments Record*. If such buildings are identified, a baseline survey will be necessary, involving an audit of the buildings' special architectural and historic interest. Such a survey consists of a detailed archival search of local history libraries and other social and property record depositories. The written product should contain an evaluation of a building's particular architectural and historic significance supported by plans, sections and elevations, together with a photographic survey and diagrammatic analysis of the building's evolution over time. This information is evaluated in terms of the relative importance of the building's component parts. This survey involves specialised work and should be undertaken only by those with a qualification in historic or architectural conservation.

Baseline studies for **conservation areas** have a wider remit than those for listed buildings. An initial survey will identify characteristics of significance, including archaeological features of interest (whether buried remains or standing structures), all listed buildings with an indication of their property curtilages, building age, and geological, topographical or landscape features. Those townscape features that constitute the area's special architectural and historic interest then need to be appraised, including vernacular characteristics, indigenous building materials, spatial characteristics, sections of group coherence or special townscape value, and long-distance views within, outside or across the conservation area that are of importance in the perception of its inherited character.

The problems and policies that affect the present or future well-being of the area also need to be appraised. This consists of a statement of problems that adversely affect the physical amenity of the area (e.g. traffic intrusion, noise, visual intrusion, architectural disfigurements, decay of historic fabric, etc.), the position with respect to present and future district-wide policies for preservation and enhancement, evaluations of specific problem sites, and opportunities for area-wide enhancements and improvements, including vehicular and pedestrian movement. An increasing number of district councils have undertaken comprehensive character appraisals of conservation areas but coverage nationwide is very uneven.

7.5 Impact prediction

7.5.1 Archaeology

Prediction of archaeological impacts involves three unknowns: what the archaeological remains are (discussed in §7.4.1), what the proposed development's impacts would be, and how significant the impacts would be. Identification of impacts must include both direct and indirect impacts. The **direct impacts** are often clear, and usually involve the removal of archaeological materials. Some of the direct impacts may not be immediately obvious, however, when they result from secondary operations such as drainage and landscaping works associated with the development. A development's **indirect impacts** are often more difficult to define. For example, dewatering associated with a development may lead to the destruction of some types of archaeological deposits on adjacent undisturbed sites that had previously survived due to waterlogging. A residential development may increase recreational pressure on a nearby earthwork or affect the visual setting of an adjacent archaeological site. Positive impacts are often indirect, e.g. when a road scheme relieves congestion in a historic town centre.

The **significance** of a development's impacts depends on a number of factors linked to the interpretation archaeologists are able to put on finds given contemporary understanding. When assessing whether an ancient monument is of national importance, and thus whether it should be scheduled, the Secretary of State for National Heritage makes reference to eight 'scheduling criteria':

- period – the degree to which a monument characterises a particular period;
- rarity – the scarcity or otherwise of surviving examples of the monument;
- documentation – the significance of the monument may be enhanced by records, either of previous investigations or contemporary to the remains;
- group value – the significance of the monument may be enhanced by its association with related contemporary or non-contemporary monuments;
- condition – the condition or survival of the monument's archaeological potential;
- fragility – the resilience or otherwise of a monument to unsympathetic treatment;
- diversity – the combinations and quality of features related to the monument;
- potential – where the nature of the monument cannot be specified but where its existence and importance are likely.

These criteria are further described in Annex 4 of PPG16 (DoE 1990). They can be used to help establish the importance not only of ancient monuments but also of other archaeological remains.

Lambrick (1993) suggests that cultural impacts can be evaluated in terms of who is affected. He lists the resources: archaeological remains, palaeoenvironmental deposits, historic buildings and structures, historic landscape and townscape elements, sites of historical events or with historical associations, and the overall historical integrity of the landscape. He then gives a list of human receptors who may be affected by impacts on these resources: owners and occupiers of historic properties and monuments, visitors to sites and buildings specifically open to the public, local communities, the general public as regards general enjoyment of historic places through informal public access, and individuals or groups with special interest in the historic environment, including academic archaeologists. He then suggests:

> Perhaps the best means of considering [significance] is to say that an effect is significant if it makes an appreciable difference to the present or future opportunity for people [receptors, as defined above] to understand and appreciate the historic environment [resources] of the area and its wider context.
>
> (Lambrick 1993)

Impact significance may also be considered in geographic terms. The DoT (1993) suggested four categories of importance for archaeological remains, namely (a) sites of national importance, usually Scheduled Ancient Monuments or monuments in the process of being scheduled as such; (b) sites of regional or county importance; (c) sites of district or local importance; and (d) sites which are too badly damaged to justify their inclusion in another category.

7.5.2 Historic buildings and sites

A proposed development action can directly affect a **listed building** in a variety of ways, ranging from the minor to the extensive:

- repairs of minor elements using replacement materials;
- changes to the interiors of buildings, where decorations or other architectural features may enrich the understanding of the building's interest;
- modifications to individual elements of the building which form a significant part of its character;
- new extensions;
- partial demolitions;
- complete demolitions;
- severance of part of a property from other parts (for instance, a house from its gardens or outbuildings).

Indirect impacts to listed buildings include noise and disturbance from nearby developments leading to a loss of amenity, and *air pollution* which can lead to deterioration of buildings and damage to garden and park vegetation. Nearby developments can cause visual intrusion and change the building's original landscape setting.

Direct impacts on conservation areas from the private sector are most commonly related to proposals for development, whether new-build or refurbishment. Extensive damage can also be created by permitted development for which special directions under Article IV of the *General Permitted Development Order* are needed (DoE 1995). Public sector developments such as those by highway authorities or utility companies can affect conservation areas without reference to conservation area policies; these may be brought under the control of the Town and Country Planning Acts by specific directions under Article IV of the *General Permitted Development Order*. A conservation area can be directly affected through the loss of buildings, through **cumulative impacts** resulting in a general deterioration in the setting of the buildings, or through severance. Development can also result in the neglect of a building or site, resulting in its deterioration or destruction. More generally, development can alter or destroy open spaces and change the character of historic districts.

Any proposed development constitutes a potential intrusion into an acknowledged heritage object. Building owners, as much as government agencies and professional advisors, play a curatorial role in the building's conservation and should be involved in predicting the impacts of the proposed development. As such, impact prediction is best undertaken as a dialogue between the owner or developer and the local authority, which respectively represent the private and the public aspects of curatorial influence. The developer determines the extent of change that is expected, and thus the utilisation of the property and its financial value. The local authority makes a judgement about the extent of architectural and historic change that can be allowed, taking into account national and local policies and standards. The outcome may take the form of agreement, compromise or disagreement. This evaluation constitutes a special negotiation over and above that needed for normal building refurbishment. The LPA classes such a dialogue as an exploratory meeting. Agreement between the two parties at this stage can constitute an agreement for the later stages of design.

The **significance** of these impacts will depend on the significance of the building or site affected as well as on the magnitude of the impact. Assessing a development's impacts on a listed building involves judgements on architectural and aesthetic factors, as well as purely physical alterations to fabric. It is possible to amplify these quantitatively according to the type of impact involved. Section 7.3.2 summarised the grading systems used for listed buildings and parks and gardens, which provide an initial indication of relative importance. However, no such gradings exist for conservation areas; criteria for area-wide character and standards of amenity are needed for effective protection, and to allow judgements to be made about project impacts.

Applications for **listed building consent** should be made for any change that would affect the character of a listed building, and for planning permission to undertake development of the land. In England PPG15 gives clear guidance to LPAs and owners on the approach that should be adopted in respect of proposals affecting listed buildings. In particular paragraph 3.4 states that applicants must justify their proposals. Moreover, "they should provide the LPA with full information to enable them to assess the likely impact of their proposals on the special architectural or historic interest of the building and on its setting". UK legislation empowers an authority to seek any particulars it considers to be necessary to ensure that it has a full understanding of the impact of the proposal on the character of the building in question. In reality practice varies between authorities, some demanding impact assessments or justification statements, whilst others require less rigorous information. However, in general, increasingly detailed assessments are being called for.

An application is often a way of confirming the earlier evaluation, and for determining the full historical significance of a building and its physical condition, and the implication of any changes to the building fabric. These surveys should be undertaken only by those who are qualified in historic or architectural conservation. Most old buildings do not meet regulatory requirements governing modern building construction, but this does not necessarily make them unsafe. It takes training and experience to make judgements about their condition which obviate the destruction of the building's character. Detailed application for full planning permission and listed building consent can only be made with confidence once the initial surveys and evaluations have been successfully concluded.

7.6 Mitigation and enhancement

7.6.1 *Archaeology*

Having identified the nature of the archaeological resource and considered the development's impact upon it, a number of mitigation strategies may be recommended. For the majority of development proposals, no further archaeological activity is required because no archaeological resource has been identified, or there is no significant impact on any archaeological resource, or the scale or nature of the impact or the nature of the archaeological resource does not warrant further action.

An **archaeological watching brief** may be carried out during the relevant stages of development. These stages are likely to be earth moving, topsoil stripping, and the digging of foundations and services. The watching brief should enable any archaeological evidence encountered to be recorded, and removed if appropriate. It may be accepted that this will not cause unreasonable delay to the progress of the development; if some delay is considered likely, the circumstances which would warrant a delay should be described and agreed upon in advance.

In some circumstances the need for development may override the case for preserving an archaeological site. In this case the site should not be thoughtlessly destroyed, and the LPA may satisfy itself that appropriate provision has been made (DoE 1990). This will involve the archaeological **site-excavation** prior to the development. The developer's responsibilities also include post-excavation (e.g. the long-term storage of the excavated material and the appropriate dissemination of the results). Depending on the nature and extent of the remains, excavation, post-excavation and publication can be expensive and time-consuming.

Preservation *in situ* means leaving the archaeological site undisturbed. This is the only mitigation measure which wholly meets the EIA Directive's principle of preventing environmental harm at source. It is supported by PPG16 which states that preservation *in situ* is the preferred action. The LPA may require preservation *in situ* if the archaeological remains are important, or the developer may choose to preserve *in situ* if mitigation requirements are too expensive. Preservation *in situ* can be achieved in several ways. The development can be avoided altogether, or, if the archaeological constraint has been identified sufficiently early, by site or option selection. A common solution is to preserve the site within the design of the development, for example, as an area of open or recreational space. The LPA may attach a fencing condition to the planning permission to prevent inadvertent damage during construction work. This secures the erection of a fence around a stipulated area and prohibits work within that area. Provision may also be made for positive management of the archaeology to secure its long-term future from any indirect impact on the development. Preservation *in situ* can be achieved within the construction of a development. For instance, the less structurally demanding elements of a development, such as car parking, can be built on raised levels or rafted foundations above the archaeological deposits. Whilst these options are feasible, they can cause technical or engineering problems such as shrinkage of buried material as it dries out.

It may be possible to preserve the majority of an archaeological site by agreeing an acceptable level of destruction. For instance, a low-density pile foundation may be acceptable where the pile has been designed to avoid the most significant deposits. Ultimately, preservation *in situ* may need to be achieved by abandoning elements of the development or indeed abandoning the development entirely. Where the importance of an archaeological site merits it, the LPA can refuse an application on archaeological grounds.

7.6.2 Historic buildings and sites

Mitigation measures in EIA should include policies to highlight and strengthen the historic building's or site's inherited and intrinsic qualities and special interest, as well as to preserve them. Preservation starts with the declaration of a listed building or conservation area: all subsequent actions should strengthen and reinforce architectural characteristics and retain historic interest. Without such intent, the intrinsic qualities of a listed building or a conservation area can be diluted and destroyed.

For **conservation areas**, unlike listed buildings, the legislation specifically allows their preservation to be accompanied by enhancement measures. Proposals for area-wide preservation and enhancement may consist of programmes of building

maintenance and repair, and their implementation; programmes of building re-
storations involving the rectification of disfigurements and their implementation;
programmes of face-lift enhancements; strategies for the enhancement of floorscape
treatments and their integration into the design of public and private domains;
strategies for building materials; and new infill building developments within clearly
established building envelopes.

Proposals for the **enhancement** of conservation areas should be drawn up by LPAs
and discussed at public meetings in the localities concerned. Such proposals may be
compiled by local citizen groups with the advice and support of professionals qualified in architectural conservation or urban design, provided that the meetings at
which proposals are presented are genuinely open to all local interests and involve
elected representatives of the local authority. Where citizen groups take such initi-
atives, it still remains the province of the local authority to make formal adoptions
of the proposals presented.

7.7 Monitoring

The prediction of archaeological impacts is not an exact science, and unexpected
problems can arise. The chances of this happening are considerably reduced by a
thorough evaluation, but some contingency should still be made for the unexpected.
The planning authority has the power to revoke planning permission where an
unexpected and overriding archaeological constraint warrants it. In this circumstance compensation would have to be paid. This can prove to be an expensive
option and is one reason why local authorities are empowered to ensure, by field
survey if necessary, that the full archaeological implications of the development
have been properly identified prior to the determination of the application.

If unexpected archaeological remains are located, additional discussion between
the developer and the county archaeologist will be needed. Where agreement cannot be reached EH may be able to arbitrate between the two parties. Where these
unexpected remains warrant it the Secretary of State for Culture, Media and Sport
may schedule them and the developer would then need consent to continue work.
Developers can insure themselves against the risk of loss from encountering unexpected archaeological remains.

7.8 Conclusions

The historic environment is a specialist discipline, covering many different periods
and types of remains. Further reading on archaeological impacts includes EH (1991a,
1991b, 1992b), Lambrick (1992), Morgan Evans (1985), Ralston & Thomas (1993),
RICS (1982), and DoT (1992). Few publications exist on listed buildings and conservation areas. The most generally readable treatment is by Ross (1996), who gives
wide coverage to the rationale and evaluation of historic conservation in the UK.
Mynors (1999) provides the most comprehensive coverage of the legal provisions
affecting listed buildings, conservation areas and ancient monuments. PPG15 (DoE/
DCMS 1994) provides the most detailed official guidance. Fielden's (1982) *Conservation of Historic Buildings* is a substantial reference volume, and several other publications on techniques of repair are provided by EH and the national amenity societies.

In EIA, it is important to contact the county archaeologists in respect of archaeological sites, and the local conservation officer in respect of historic buildings and conservation areas as early as possible, since they are valuable sources of data and advice. Where consultation is left to a later stage, unexpected problems and delays are more likely to occur. The EIA should be carried out by **specialists** trained not only in survey and analysis techniques, but also in interpreting the data for the relevant period and type of remains. Specialist knowledge will be needed to interpret the relative importance of these results and suggest appropriate mitigation strategies. Using specialists in archaeology and/or historic or architectural conservation from the earliest stages of the EIA, when the data-gathering programme is first being considered, will ensure that the correct type and amount of data is obtained. The result of using inappropriately qualified staff may be that, after the EIA is completed, additional historical constraints may be identified, or additional information required, potentially introducing delay and so negating the benefits of carrying out the EIA.

Problems may arise where the developer gathers inadequate or inappropriate data for use in EIA. This frequently occurs as a result of cost-cutting on the data-gathering strategy. This can be a short-sighted saving when compared to the cost of delay to the progress of the application, or delay to the progress of the development.

References

ACAO (Association of County Archaeological Officers) 1992. *Sites and monuments record: policies for access and charging*. Chelmsford: Essex County Council (Planning Department).

DoE (Department of the Environment) 1987. Circular 8/87: *Historic buildings and conservation areas – policy and procedures*. London: HMSO.

DoE 1989. *Environmental assessment: a guide to the procedures*. London: HMSO.

DoE 1990. *Planning Policy Guidance Note 16: Archaeology and planning*. London: HMSO.

DoE 1995. *The Town and Country (General Permitted Development) Order*, Statutory Instrument No. 418. London: HMSO.

DoE/DCMS (Department for Culture, Media and Sport) 1994. *Planning Policy Guidance Note 15: Planning and the historic environment*. London: HMSO.

DoT (Department of Transport) 1992. *The good roads guide: environmental design for inter-urban roads*. London: HMSO.

DoT 1993. *Design manual for roads and bridges*. Vol. 11: *Environmental assessment*. London: HMSO.

EH (English Heritage) 1991a. *Rescue archaeology funding: a policy statement*. London: English Heritage.

EH 1991b. *Exploring our past – strategies for the archaeology of England*. London: English Heritage.

EH 1992a. *Development plan policies for archaeology*. London: English Heritage.

EH 1992b. *The management of archaeological projects*. London: English Heritage.

Fielden BM 1982. *Conservation of historic buildings*. London: Butterworth.

Lambrick GH 1992. The importance of the cultural heritage in a green world: towards the development of landscape integrity assessment. In *All natural things: archaeology and the green debate*, L Macinnes & CR Wickham-Jones (eds), 105–126. Oxford: Oxbow.

Lambrick GH 1993. Environmental assessment and the cultural heritage: principles and practice. In *Environmental assessment and archaeology*, I Ralston & R Thomas (eds), 9–19. Birmingham: Institute of Field Archaeologists.

Morgan Evans D 1985. The management of historic landscapes. In *Archaeology and nature conservation*, GH Lambrick (ed.), 89–94. Department of External Studies, Oxford University.

Mynors C 1999. *Listed buildings, conservation areas and monuments,* 3rd edn. London: Sweet & Maxwell.

Ralston I & R Thomas (eds) 1993. *Environmental assessment and archaeology.* Occasional Paper 5, Institute of Field Archaeologists, Birmingham.

RICS (Royal Institute of Chartered Surveyors) 1982. *Practitioners' companion to Ancient Monuments and Archaeological Areas Act.* London: RICS.

Ross M 1996. *Planning and heritage: policy and procedures,* 2nd edn. London: E & FN Spon.

WO (Welsh Office) 1991. Planning Policy Guidance Note 16: *Archaeology and planning.* London: HMSO.

8 Air quality and climate

Derek M Elsom

8.1 Introduction: definitions and concepts

8.1.1 Air and climate changes

A proposed development that will add pollutants to the atmosphere or alter the **weather** and **climate** may result in adverse effects on people, plants, animals, materials and buildings (Canter 1996, Colls 1997, Elsom 1992, Ortolano 1997, Turco 1997). These effects can occur at the local, regional or even global scale. Major developments, such as power stations, oil refineries, waste incinerators, chemical processing plants and roads, pose obvious potential pollution problems. In addition, even developments that emit little or no pollutants when completed and operating can create a local dust nuisance during the earth-moving and materials-handling operations of the construction stage, especially during dry weather conditions. Once completed, developments may generate additional vehicle emissions as people travel to them (e.g. edge-of-town shopping and leisure complexes).

Developments may give rise to both routine and non-routine pollutant emissions. For example, they may use one type of fuel for most of the time but on a few occasions have to switch to an alternative fuel. In the UK this can occur when an industrial plant intends to use an 'interruptible' natural gas supply. This type of supply permits the supplier the right to cease supplying gas during peak periods of national demand, during which the plant has to switch to a standby fuel such as heavy fuel oil for up to 30 days a year. Whereas natural gas produces no emissions of sulphur dioxide (SO_2), fuel oil emits significant amounts depending upon its sulphur content. Another example of non-routine emissions to consider is the possibility of an accident at a proposed development that intends to store or process toxic chemicals or nuclear fuels giving rise to the risk of the release of hazardous substances.

8.1.2 Effects of air pollutants

Air **pollutants** can affect the health of a person during inhalation and exhalation as the pollutants inflame, sensitise and even scar the airways and lungs. On reaching deep inside the lungs, they may enter the bloodstream, thus affecting organs other than the lung, and they can take up permanent residence in the body. In addition, some pollutants affect health through contact with the skin and through ingestion of contaminated foods and drinks. Pollutants affect health in varying degrees of severity, ranging from minor irritation through serious illness to premature death in extreme cases. They may produce immediate (acute) symptoms as well as longer term (chronic) effects. Health effects depend upon the type and amount of pollutants present, the duration of exposure, and the state of health, age and level of activity of the person exposed (Elsom 1996).

Pollution damage to plants and animals is caused by a combination of physical and chemical stresses that may affect the receptor's physiology. Pollutants can affect crops by causing leaf discoloration, reducing plant growth and yields, or by contaminating a crop, so making it unsafe to eat. Effects on terrestrial and aquatic ecosystems can occur locally or even regionally in the case of pollutants that contribute to **acid deposition** ('acid rain'), especially in areas where the soils and lakes lack substances to neutralise or buffer the acidic inputs (see Chapters 9, 10 & 12). Pollution problems for buildings can be short-term and reversible such as soiling by smoke (which can be removed by cleaning), whereas the effects of acid deposition can be cumulative and irreversible by causing erosion and crumbling of the stone.

8.1.3 Effects of climate changes

Weather and climate changes can occur locally when a development changes the characteristics of the area in terms of its radiation balance, surface friction and roughness, and moisture balance. Adverse microclimate changes include:

- alterations to the airflow around large structures such as office blocks, multi-storey car parks and shopping arcades, causing wind turbulence which affects the comfort and sometimes the safety of pedestrians;
- the addition of moisture from industrial cooling towers and large reservoirs, causing an increased frequency of fog or even icing on nearby roads;
- the reduction in sunlight for greenhouse crops lying beneath a persistent industrial pollution plume;
- the ponding of cold air behind physical barriers such as road and railway embankments, so increasing the incidence of frost which can damage agricultural and horticultural crops in those areas.

Macroclimatic changes can result from emissions of greenhouse gases (gases which are strong absorbers of outgoing terrestrial infra-red radiation) such as carbon dioxide (CO_2), methane (CH_4) and nitrous oxide (N_2O). These gases contribute to global warming, which is now a generally accepted trend. Because of the wide range of natural climatic variation through time, neither the significance of the human impact, nor the long-term effects of warming on global and regional climate changes can be predicted with any certainty. However, there is mounting evidence that warming is causing changes in the position and intensity of weather systems and consequent changes in regional wind, temperature and precipitation patterns.

Some regional climate changes may bring benefits, but others are likely to bring adverse impacts. Current predictions suggest that the UK may experience:

- slightly increased average rainfalls, especially in winter – resulting in increased river flows;
- increased incidence of hot, dry spells in summer – resulting in increased drought risk;

- increased variability of rainfall, and a higher proportion of intense events (higher frequency of rainstorms) – resulting in greater risks of wind damage, erosion and flooding (MAFF 2000).

Global warming is also causing global sea level to rise because of thermal expansion of the seawater and because of some melting of mountain glaciers and polar ice sheets (Elsom 1992, Hulme & Jenkins 1998, Houghton 1997). This is of particular concern in coastal areas (see §13.1).

8.2 Legislative background and interest groups

8.2.1 Air quality guidelines and standards

Epidemiological studies of community groups and laboratory-based toxicological experiments using human volunteers provide assessments of the health effects of pollutants. Consideration of these findings has enabled various national and international organisations to identify levels of air pollution concentrations (**air quality standards**) which should not be exceeded if the health of people is not to be at risk. Research studies have enabled levels to be specified to protect ecosystems too. Sometimes these levels are advisory such as the World Health Organisation (WHO) guideline values, while others, such as the UK air quality objectives and the EU limit values, are mandatory, being backed by legislation. Concentrations are expressed either as mass of the substance per unit volume of air (e.g. micrograms per cubic metre, abbreviated to $\mu g/m^3$) or as volume of the substance to the volume of air (e.g. parts per million or parts per billion, abbreviated to ppm and ppb, respectively). The units can be converted from one to another using conversion factors (published factors may vary slightly because they may be standardised to a different atmospheric pressure and temperature).

The WHO guideline values, initially issued in 1987, were revised in 1997 (Table 8.1). They are based on the lowest level a pollutant has been shown to produce adverse health effects or the level at which no observed health effect has been demonstrated plus a margin of protection to safeguard sensitive groups within the population. Sensitive groups include people with asthma, those with pre-existing heart and lung diseases, the elderly, infants and pregnant women and their unborn babies. Such groups form one-fifth of the population in the UK (Elsom 1996). Some pollutants, notably carcinogenic pollutants (e.g. arsenic, benzene, chromium, PAHs and vinyl chloride) and fine particulate matter less than 10 μm (PM_{10}) have not been given a guideline value. In the case of PM_{10}, available epidemiological data did not enable WHO to establish a level below which no health effects would be expected. Instead, exposure-effect information is provided, giving guidance to risk managers about the major health impact for short- and long-term exposure to various levels of this pollutant.

The WHO guideline values were considered by the UK and the EU when setting mandatory standards, but unlike the WHO guideline values, which are based on health considerations alone, the EU limit values and UK objectives take into account the economic costs and technological feasibility of attainment. Given the costs and problems involved in attainment, this explains why air quality standards

Table 8.1 World Health Organisation air quality guideline values

Pollutant	Value	Averaging time
Carbon monoxide	100 mg/m^3 60 mg/m^3 30 mg/m^3 10 mg/m^3	15 min 30 min 1 h 8 h
Ozone	120 µg/m^3	8 h
Nitrogen dioxide	200 µg/m^3 40 µg/m^3	1 h annual
Sulphur dioxide	500 µg/m^3 125 µg/m^3 50 µg/m^3	10 min 24 h annual
Benzene	6×10^{-6} (µg/m^3)$^{-1}$	UR/lifetime*
Dichloromethane	3 mg/m^3	24 h
Formaldehyde	0.1 mg/m^3	30 min
PAHs**	8.7×10^{-5} (ng/m^3)$^{-1}$	UR/lifetime*
Styrene	0.26 mg/m^3	1 wk
Tetrachloroethylene	0.25 mg/m^3	24 h
Toluene	0.26 mg/m^3	1 wk
Trichloethylene	4.3×10^{-7} (µg/m^3)$^{-1}$	UR/lifetime*
Arsenic	1.5×10^{-3} (µg/m^3)$^{-1}$	UR/lifetime*
Cadmium	5 ng/m^3	annual
Chromium	0.04 (µg/m^3)$^{-1}$	UR/lifetime*
Lead	0.5 µg/m^3	annual
Manganese	0.15 µg/m^3	annual
Mercury	1.0 µg/m^3	annual
Nickel	3.8×10^{-4} (µg/m^3)$^{-1}$	UR/lifetime*

* UR = excess risk of dying from cancer following lifetime exposure. Thus for benzene, 6 people in a population of 1 million will die as a result of a lifetime exposure of 1 µg/m^3; for PAHs, 87 people in a population of 1 million will die from cancer following lifetime exposure to 1 ng/m^3.
** Specifically benzo[a]pyrene.

Note: Pollutants reviewed by WHO but for which no guidelines were set, because of the lack of reliable evidence or evidence of a 'safe' level, included particulate matter, 1,3 butadiene, PCBs, PCDDs, PCDFs, fluoride and platinum.

vary nationally around the world and why they are often not as strict as the WHO guidelines (Murley 1995).

8.2.2 EU air quality limit values

From 1980 onwards, the EU began setting air quality standards in the form of mandatory health-based limit values and more stringent non-mandatory guide values to protect the environment. Guide values are intended to be long-term objectives which, when met, will protect vegetation as well as aesthetic aspects of the environment such as long-range visibility and soiling of buildings. EU limit values were initially set for sulphur dioxide and suspended particulates in 1980 (amended in 1989), lead (Pb) in 1982, nitrogen dioxide (NO_2) in 1985 and ozone (O_3) in 1992. As part of the European Community's Framework Directive on Ambient Air Quality Assessment and Management (96/62/EC), commonly referred to as the Air Quality Framework Directive, agreed in September 1996, the EU decided to review these limit values and to add values for additional pollutants. The values are specified in a series of Daughter Directives, with the first one being agreed in 1998 and covering SO_2, particulate matter (PM_{10} or $PM_{2.5}$), NO_2 and Pb. It entered into force in 2000 (Table 8.2, Elsom 1999). Subsequent Daughter Directives refer to O_3, benzene and carbon monoxide (CO), polycyclic aromatic hydrocarbons (PAHs), cadmium, arsenic, nickel and mercury.

Table 8.2 EU air quality limit values[1]

Pollutant	Target date	Measuring period	Limit value
Lead	2005	annual	0.5 µg/m³
Nitrogen dioxide	2010	hourly	105 ppb (200 µg/m³), no more than 18 exceedances per year
		annual	21 ppb (40 µg/m³)
PM_{10}	Stage 1 2005	daily	50 µg/m³, no more than 18 exceedances per year
		annual	40 µg/m³
	Stage 2 2010	daily	50 µg/m³, no more than 7 exceedances per year
		annual	20 µg/m³
$PM_{2.5}$	Action level 2005	daily	40 µg/m³, no more than 14 exceedances per year
Sulphur dioxide	2005	hourly	132 ppb (350 µg/m³), no more than 24 exceedances per year
		daily	47 ppb (125 µg/m³), no more than 3 exceedances per year

1 Further Daughter Directives will be issued to specify limit values for other pollutants.

Table 8.3 The UK National Air Quality Strategy objectives (January 2000 revision)

Pollutant	Concentration	Measuring period	Target date
C_6H_6	5 ppb (16 µg/m³)	Annual mean	31/12/2003
C_4H_6	1 ppb (2.25 µg/m³)	Annual mean	31/12/2003
CO	10 ppm (12 mg/m³)	Running 8-h mean	31/12/2003
Pb	0.5 µg/m³	Annual mean	31/12/2004
	0.25 µg/m³	Annual mean	31/12/2008
NO_2*	105 ppb (200 µg/m³) (max. 18 exceedances)	1-h mean	31/12/2005
	21 ppb (40 µg/m³)	Annual mean	31/12/2005
O_3*	50 ppb (100 µg/m³) (10 exceedances a year)	Daily max. of running 8-h mean	31/12/2005
PM_{10}	50 µg/m³ (max. 35 exceedances)	24-h mean	31/12/2004
	40 µg/m³	Annual mean	31/12/2004
SO_2	100 ppb (266 µg/m³) (max. 35 exceedances)	15-min mean	31/12/2005
	132 ppb (350 µg/m³) (max. 24 exceedances)	1-h mean	31/12/2004
	47 ppb (125 µg/m³) (max. 3 exceedances)	24-h mean	31/12/2004

* NO_2 and O_3 objectives are provisional. The O_3 objective is a national objective and not a local authority statutory responsibility.

8.2.3 UK air quality standards and objectives

Part IV of the Environment Act 1995 established the UK *National Air Quality Strategy* (NAQS) for local air quality management and specified air quality standards and objectives for key pollutants (Table 8.3, DoE 1997, DETR 1998a, 2000a,b, Elsom *et al.* 2000). The standards are derived from reviews undertaken by the independent Expert Panel on Air Quality Standards into the effects of pollutants on health (e.g. EPAQS 1996). Air quality standards are not given statutory backing and there is no timescale of attainment attached to them. Instead, the government considers the standards as reference points to be used for setting air quality objectives. These objectives represent the government's judgement of achievable air quality by specified target years between 2003 and 2008 on the evidence of costs and benefits and technical feasibility (DoE 1995, 1997, DETR 1998a, 2000a). For some pollutants the objective is identical to the standard but for others a specified number of occasions exceeding the standard is permitted (Table 8.3). As a result of the EU Framework Directive being agreed soon after the UK NAQS was implemented, the

UK air quality standards and objectives were revised to reflect EU requirements and it is these that are listed in Table 8.3. The UK has also specified public information air quality bands, classifying pollution levels into four bands: low, moderate, high and very high. If high or very high bands are experienced or are forecast to occur the next day, health advice is issued, being directed especially at sensitive groups in the community.

8.2.4 Emission standards

Air quality standards refer to the levels of air pollution to which people are exposed. Another type of legislated standard is the *emission standard*, which specifies the maximum amount or concentration of a pollutant which is allowed to be emitted from a given source. Emission standards are usually derived from consideration of the cost and effectiveness of the control technology available. The UK *Environmental Protection Act 1990* introduced the system of Integrated Pollution Control (IPC) which requires that pollution sources adopt the *Best Available Techniques Not Entailing Excessive Costs* (BATNEEC) in order to minimise pollution (DoE/Welsh Office 1990, Hughes *et al.* 1998). The act also established two groups of industrial plants and processes for regulation purposes. Major developments such as power stations belong to the schedule A group and require authorisation by the Environment Agency (EA) in England and Wales. Less-polluting industrial plants and processes constitute the schedule B group and are regulated under the system of *Local Air Pollution Control* (LAPC) by local authorities (district councils and unitary authorities).

For specific types of pollution sources the existence of an emission standard implies the type of operating process or pollution control equipment that should be employed (Process Guidance Notes have been issued by the EA). Details of emissions and emission factors from Part A and B processes are available from the National Atmospheric Emissions Inventory at http://www.aeat.co.uk/neten/airqual/naei/home.html, which is part of the National Air Quality Information Archive site. IPC and LAPC were subsequently amended to take into account the EU *Integrated Pollution Prevention and Control* (IPPC) Directive formally adopted by the European Commission in 1996 and transposed into legislation of Member States in 1999. This is similar to IPC but IPPC extends the range of processes covered by regulation (e.g. intensive animal husbandry such as large pig and poultry farms). Emission restrictions apply to air, land and water such that the *Best Practicable Environmental Option* (BPEO) must be adopted. For example, it is not appropriate to adopt a mitigation measure which removes gaseous pollutants from an industrial stack, by converting them to a sludge, if disposal of the sludge would create an even worse environmental problem in the form of landfill and/or water pollution.

Emission limits for pollutants can apply nationally. For example, the EU *Large Combustion Plants Directive* (88/609/EEC) agreed in 1988 committed the UK to reducing emissions of SO_2 from existing installations with a capacity greater than 50 MW (e.g. coal-powered power stations) by 40% by 1998 and 60% by 2003, taking 1980 emissions as the baseline. The UNECE *Second Sulphur Protocol* that was ratified by the UK in 1996 commits the Government to reducing SO_2 emissions by 80% over the period 1980 to 2010. The UNECE *Protocol to Abate Acidification, Eutrophication and Ground-level Ozone* agreed in December 1999 sets national ceilings for four acidifying, eutrophying and ozone-forming air pollutants: SO_2, NO_x,

VOCs and NH_3. Stricter national ceilings for these four pollutants for 2010 are proposed in an EU Directive on *National Emission Ceilings*. Environmental ministers of the Member States agreed in principle on the common position of this Directive in June 2000.

8.2.5 Regulations for hazardous chemicals

In the case of a proposed development that involves materials that could be harmful to people in the event of an accident, the EIA should include an indication of the preventative measures to be adopted, so that such an occurrence is not likely to have a significant effect (DoE 1989). Requirements were first included in the *Control of Industrial Major Accident Hazards* (CIMAH) regulations of 1984, which were enacted in response to the EU *'Seveso' Directive* of 1982 (82/501/EEC). This Directive was implemented as a consequence of chemical accidents at Flixborough (UK) in 1974 and Seveso (Italy) in 1976. CIMAH regulations set limits to the quantities and combinations of chemicals that can be stored at a site and require onsite and offsite plans in the case of an emergency to be drawn up to an approved standard. The EU revised the *Seveso Directive* in 1996 to place hazardous substances into risk categories and require annual inspection of top-tier sites. County Council Emergency Planning Officers can provide the latest update of the CIMAH regulations.

8.2.6 Climate standards and regulations

There are few legislated standards with regard to climate. The United States introduced regulations to ensure that visibility is protected in pristine areas such as national parks and wilderness areas. Persistent and coherent pollution plumes from industrial plants during daylight hours are considered intrusive and objectionable and mitigation measures to minimise or eliminate the plume are required. Similarly, in the UK visible plumes are regulated because they may constitute a visual nuisance (HMIP 1996).

At the global scale there are regulations concerning pollutants that contribute to global warming and those that cause stratospheric ozone depletion. In 1997 the UK and other nations agreed the *Kyoto Protocol* to the UN *Framework Convention on Climate Change*. Industrialised nations agreed to an overall emission reduction of 5.2% of 1990 levels by 2008–2012 for the three common greenhouse gases of CO_2, N_2O and methane (CH_4) and the three halocarbon substitutes, hydrofluorocarbons (HFCs), perfluorocarbons (PFCs) and sulphur hexachloride (SF_6) (a base year of 1995 can be employed for the last three pollutants). The overall 5.2% reduction is achieved by some nations taking larger cuts than others: the EU accepted a reduction of 8%, the USA 7% and Japan 6%.

The EU reduction of 8% is to be spread amongst its 15 Member States, and the UK will be required to reduce greenhouse gas emissions by 12.5%. Further, the UK Government has pledged itself to a voluntary reduction in CO_2 emissions of 20% of 1990 levels by 2010. Consequently, and because the targets are likely to be made more stringent in the future, a proposed development which will be a significant source of greenhouse gases will receive close scrutiny. Pollutants which damage the ozone layer such as chlorofluorocarbons (CFCs; they contribute to global warming too), methyl chloroform, carbon tetrachloride, hydrochlorofluorocarbons (HCFCs) and methyl bromide are subject to the Montreal Protocol on Substances that Deplete the Ozone Layer and its subsequent amendments. The Protocol requires the production and consumption of these pollutants to be reduced and eventually phased out completely.

8.2.7 Air quality and climate indicators used in EIA

Aspects of air and climate which need to be addressed in preparing an EIA are summarised by the UK guidelines (DoE 1989) as (a) level and concentration of chemical emissions and their environmental effects, (b) particulate matter, (c) offensive odours, and (d) any other climatic effects. Depending upon the development project there is a wide range of atmospheric pollutants with which an EIA may need to be concerned (Table 8.4). The existence of the NAQS objectives and EU current and proposed limit values clearly indicate the need to consider SO_2, fine particulates, CO, NO_2, Pb, benzene, 1,3 butadiene, O_3, polycyclic aromatic hydrocarbons (PAHs), cadmium, arsenic, nickel and mercury. In addition, many other health-threatening pollutants, some of which have been given WHO guideline values and others which have not simply because of insufficient evidence to be able to define an appropriate safe level, should be considered. These latter pollutants include polychlorinated biphenyls (PCBs), dioxins (PCDDs), furans (PCDFs), toxic chemicals (e.g. ammonia, fluoride, chlorine) and toxic metals (e.g. chromium, manganese, platinum). Ionising radiation (radionuclides) released from certain medical facilities and nuclear power plants should be considered too. Offensive odours could be a problem around proposed sewage treatment works, chemical plants, paint works, food processing factories and brick works. Odours often generate great annoyance when residents are subjected to them in their gardens and homes, and they may adversely affect health (e.g. ranging from discomfort, nausea and headaches through to severe respiratory illness).

Climate indicators include temperature, relative humidity, solar radiation, precipitation, wind speed and wind direction. All developments are likely to modify the microclimate to some extent, but in most cases the changes to local temperature, amount of sunlight and shade, and airflow are minor and not considered in EIA unless there are special reasons for doing so. Significant effects on sensitive environmental receptors could arise due to local changes in the frequency of weather extremes such as fog, frost, ice, precipitation and wind gusts.

8.3 Scoping and baseline studies

8.3.1 Introduction

Before the impact of a proposed development can be predicted, it is necessary to establish the current baseline conditions concerning air pollution and climate and to establish whether they are likely to change in the future, irrespective of the planned development. Knowledge of baseline pollution conditions is essential because, even when a development is likely to add only small amounts of pollution to the area, it could lead to air quality standards being exceeded if air quality in the area is already high or may become high in the future. This requires obtaining measurements of the ambient levels of the pollutants of concern at one or more locations in the study area, so as to assess the amount of pollution present.

Table 8.4 Key air pollutants and their anthropogenic sources

Pollutant	Anthropogenic sources
Sulphur dioxide (SO_2)	Coal- and oil-fired power stations, industrial boilers, waste incinerators, domestic heating, diesel vehicles, metal smelters, paper manufacturing.
Particulates (dust, smoke, PM_{10}, $PM_{2.5}$)	Coal- and oil-fired power stations, industrial boilers, waste incinerators, domestic heating, many industrial plants, diesel vehicles, construction, mining, quarrying, cement manufacturing.
Nitrogen oxides (NOx: NO, NO_2)	Coal-, oil- and gas-fired power stations, industrial boilers, waste incinerators, motor vehicles.
Carbon monoxide (CO)	Motor vehicles, fuel combustion.
Volatile organic compounds (VOCs), e.g. benzene	Petrol-engine vehicle exhausts, leakage at petrol stations, paint manufacturing.
Toxic organic micropollutants (TOMPs), e.g. PAHs, PCBs, dioxins	Waste incinerators, coke production, coal combustion.
Toxic metals, e.g. lead, cadmium	Vehicle exhausts (leaded petrol), metal processing, waste incinerators, oil and coal combustion, battery manufacturing, cement and fertiliser production.
Toxic chemicals, e.g. chlorine, ammonia, fluoride	Chemical plants, metal processing, fertiliser manufacturing.
Greenhouse gases, e.g. carbon dioxide (CO_2), methane (CH_4)	CO_2: fuel combustion, especially power stations; CH_4: coal mining, gas leakage, landfill sites.
Ozone (O_3)	Secondary pollutant formed from VOCs and nitrogen oxides.
Ionising radiation (radionuclides)	Nuclear reactors and waste storage, some medical facilities.
Odours	Sewage treatment works, landfill sites, chemical plants, oil refineries, food processing, paintworks, brickworks, plastics manufacturing.

8.3.2 *Pollution data availability*

Using information from current pollution monitors is the simplest and least expensive approach to obtaining current baseline pollution levels. There are various national monitoring networks collecting pollution data and many local authorities, universities and other organisations undertake short-term or long-term monitoring of pollutants. Pollution data from various national networks funded and/or co-ordinated by the DETR, including the Automatic Monitoring Networks and Non-automatic Monitoring Networks (e.g. the NO_2 Diffusion Tube Survey, Toxic Organic Micro Pollutants Network), are available via the National Air Quality Information Archive. This archive was established in 1997 as a comprehensive Internet site accessed directly through the site compilers, the National Environmental Technology Centre, at http://www.aeat.co.uk/netcen/airqual/home.html or via the sponsors, DETR, at http://www.environment.detr.gov.uk/airq/aqinfo.htm (DETR 2000c). Pollution monitoring sites are classified and coded by type of location, so, in the absence of a monitoring site in the vicinity of the proposed development, the data may be considered as indicative of what may be experienced at similar sites in other areas (DETR 2000c).

Expert opinion obtained from environmental consultancies and universities can advise on the validity of using pollution data from a monitoring site to represent pollution levels at a different location. Alternatively the data can be modified to reflect the location of interest by using established empirical relationships. In some cases empirical relationships enable the levels of one pollutant to indicate the likely levels of another pollutant. The National Air Quality Information Archive Internet site provides 1 km × 1 km grid square maps of background concentrations of selected pollutants for recent years.

Not all sites monitoring pollution are part of a national network such that local authority Environmental Health Officers may be able to provide information concerning their own pollution monitors. Many produce annual reports for their local authorities summarising the pollution data collected and assessing its significance in relation to air quality standards. Moreover the NAQS requires local authorities to complete a review and assessment of their air quality and these reports of air quality can be consulted. Many of these reports are available on-line from the websites of local authorities or via the University of the West of England Air Quality Management Resource Centre website (http://www.uwe.ac.uk/aqm/cente/index.html) sponsored by DETR. National reports containing air quality data are also available, such as the directory of air quality data for 249 pollutants compiled by the Meteorological Office (Bertorelli & Derwent 1995).

8.3.3 *On-site pollution monitoring*

If pollution data are not available or are insufficient, then on-site monitoring will be required and should be planned and initiated during the scoping exercise of an EIA (Harrop 1993). A baseline monitoring programme needs to consider (a) what pollutants to monitor, (b) what type of monitor to employ, (c) the number and location of sampling sites, (d) the duration of the survey, and (e) the time resolution of sampling.

Selecting the equipment to measure air pollution concentrations depends upon (a) the intended use of the data, (b) the budget allocated to purchase or hire the equipment, and (c) the expertise of personnel available to set up and maintain the equipment and, in some cases, to undertake laboratory analyses of collected samples. Setting up an automatic pollutant analyser can be costly, so hiring the equipment may be more appropriate (a list of addresses and details of companies offering equipment and consultant expertise is given in the Members Handbook available from the National Society for Clean Air and Environmental Protection (NSCA)). It is important that the equipment selected for monitoring is accredited nationally so that the data collected can be compared with UK and EU air quality standards (DETR 2000c).

Local authorities faced with the need to monitor pollution in order to assess whether air quality objectives are being attained are turning to relatively simple and inexpensive equipment such as passive diffusion tubes (DETR 2000c). Passive diffusion tubes absorb the pollutant onto a metal gauze placed at the bottom of a short cylinder open at the other end to the atmosphere. After exposure the tubes are sent for laboratory analyses. They can provide useful information for a range of pollutants including ammonia, benzene, CO, hydrogen sulphide, NO_2, O_3 and SO_2. In areas of high pollution concentrations they can produce results for daily or even three-hourly exposures although in areas with low concentrations they are usually exposed for two weeks at a time. Monthly exposure readings from these tubes can provide estimates of the annual mean concentrations. Pollution bio-indicators, types of plant that are sensitive to pollution levels (e.g. lichen for SO_2, tobacco plants for O_3) may provide supplementary information on pollution levels (Mulgrew & Williams 2000). Soil and vegetation analyses can also provide long-term levels of pollutants such as metals.

When siting monitoring equipment it is necessary to consider (a) the need to protect against vandalism, (b) access to the site, (c) the avoidance of pollution from indoor and localised sources which may make the data unrepresentative of the wider area, and (d) the availability of a power supply (if needed).

8.3.4 Projecting the baseline forwards: air pollution

Having established current baseline pollution levels, it is then necessary to consider how these levels are expected to change in the future, irrespective of the possible effects of the proposed development. If emission sources and strengths, as well as climate conditions, in the area are not expected to change in the future, then current pollution levels may be considered to approximate pollution levels in the next few years. However, changes in population and activity patterns, new industrial developments or closures, changes in fuels (e.g. decline of coal in favour of gas, the prohibition of unleaded petrol in the EU since the start of 2000) and stricter emission standards (e.g. increasing number of vehicles fitted with catalytic converters) can affect emission rates. Weather conditions that favour a build-up of pollutants (e.g. periods of calm or light winds, higher temperatures promoting increased evaporative emissions) may alter too but, in practice, these are not usually considered.

The implications of significant changes to emission rates and patterns for future pollution concentrations need to be assessed. Local, district and county authorities can usually supply information on new developments under construction as well as details of likely population and land use changes. A judgement will then have to be made as to how these and other changes (e.g. relevant UK and EU legislation) will alter emissions in the area and consequently alter baseline pollution levels. Helpfully, in support of the introduction of the NAQS, the DETR provide guidance to local authorities on projecting current pollution levels forward to future years (DETR 2000f). For example, it is suggested that background annual average CO concentration at the end of 2003 will be 0.56 times the 1996 value while the mean background concentration of NO_2 in 2005 will be 0.79 times the 1996 concentration (DETR 2000f).

If there is insufficient pollution data available in the study area it may be necessary to compile an **emissions inventory** (DETR 2000d). Taking into account the factors that may affect emissions in future years may enable emission sources and rates to be approximated for future years. These emission data then become the input into a suitable numerical dispersion model in order to predict future pollution concentrations in the area (refer to section 8.4; DETR 2000d). Emission inventories for some pollutants, compiled for the purposes of the NAQS, are available from local authorities and can save much time and effort. National 1 km × 1 km grid maps of current emissions from background sources are available from the *National Atmospheric Emissions Inventory* at http://www.aeat.co.uk/neten/airqual/naei/home.html. The maps are based on emissions from Part A industrial processes and major trunk roads and do not, at present, consider emissions from Part B processes and minor roads. Much more detailed inventories are available for major conurbations exceeding 250,000 population (DETR 2000d).

8.3.5 Projecting the baseline forwards: climate

Baseline climate conditions can be established using meteorological data readily available from hundreds of sites throughout the UK maintained by the Meteorological Office (MO), local authorities, universities, schools and individual weather enthusiasts. Some national pollution monitoring sites, especially those with multiple automatic analysers, also monitor meteorological conditions. The MO can supply hourly, daily, monthly, annual and long-term averages of temperature, relative humidity, air pressure, precipitation (including fog), wind speed and wind direction for any of its stations at a small cost. Although the meteorological site for which data are available may be some distance away from the study site, the MO and other meteorological consultants can provide expert advice concerning how local factors such as altitude, topography and nearness to the coast may lead to differences between the two locations. Future climate baseline levels are not usually predicted for the purposes of an EIA, given the major limitations of current models in predicting regional changes, let alone local changes, attributed to say, global warming due to the increase in atmospheric concentrations of greenhouse gases. Improved models may alter this situation in the future (DoE 1996, Houghton 1997, Hulme & Jenkins 1998, MAFF 2000).

8.4 Impact prediction

8.4.1 Physical models and expert opinion

There are several types of models available to predict air pollution concentrations. Physical (scale) models using wind tunnels or computer graphics are employed occasionally in situations involving complex hilly terrain or where numerical models suggest uncertainty concerning the possible effects of nearby buildings on dispersion of pollution emissions.

Predictive methods include the use of expert opinion, providing it is backed up with reasons and justification which support that opinion, such as comparison with similar existing developments or planned projects for which prediction has already been undertaken. The use of expert opinion can be justified readily on cost when a number of similar projects are being proposed in different locations.

8.4.2 Numerical dispersion models

The type of model used most frequently in predicting air pollution is the numerical dispersion model. A numerical dispersion model takes the form of a computer program run on a personal computer with a large memory. It calculates how specified emission rates are transformed by the atmospheric processes of dilution and dispersion (and sometimes chemical and photochemical processes) into ground level pollution concentrations at various distances from the source(s). Models are available for predicting pollution concentrations for emissions from a single point source (e.g. industrial stack or vent) as well as for emissions from a large number of point sources simultaneously. The basic model can be improved in accuracy by taking into account complications appropriate to the specific location under study, such as type of terrain (e.g. flat or hilly), surface roughness (e.g. urban or rural conditions), coastal influences (e.g. effects of a sea breeze) and the presence of nearby buildings which may cause building wake effects. Models are also available for area sources (e.g. construction sites, car parks, motorway service stations, industrial processes with numerous vents, urban areas, county regions), line sources (e.g. open roads, street canyons, railways) and volume sources (area sources with a vertical depth e.g. leaking gases from a group of industrial processes, take-off and landing activities at an airport).

Simple and complex (advanced) versions of numerical dispersion models are available (Table 8.5, DETR 2000e). Simple (screening) models are designed to be applied relatively easily and inexpensively as a scoping tool to identify whether or not a problem warrants further investigation. Screening models employ grossly simplified assumptions about the behaviour of pollutants in the atmosphere and are designed to calculate the worst-case pollution concentrations. As such they have pre-set meteorological conditions and the user does not usually have to input any meteorological information. If a screening model predicts that emissions from a proposed development will produce air pollution concentrations far below an air quality standard, this would indicate that it may not be necessary to obtain a more accurate estimate of the predicted concentrations using a complex model. However, if the screening model predicts that pollution concentrations are likely to approach or exceed air quality standards, then a more rigorous investigation using a complex model is needed. For major developments, regardless of how small an increase in pollution levels are caused by their emissions, the use of a complex model may be appropriate for an EIA.

Computer software and manuals for numerical dispersion models can be obtained from the United States Environmental Protection Agency (EPA) which has developed various models for regulatory purposes. These well-established models and user guides can be obtained directly for the cost of down-loading from the Internet (e.g. via Lakes Environmental Software at http://www.lakes-environmental.com/lakeepa.html). Additional software will be needed if the results are to be displayed graphically or in map form. Variations of the EPA models with user-friendly input and output routines

Table 8.5 Details of commonly used air pollution numerical dispersion models

Model name	Source type[1]	Met. data[2]	Software costs[3]	Hardware[4]	Time needed[5]	Expertise[6]
DMRB	L	N	£	N	M	A
CAR	L	wind speed only	£££	N,PC	M	L
EA Guidance	P	N	£	N	M	A
ADMS-Screen	P	N	££	PC	M	L
SCREEN	P	N	££	PC	M	L
AEOLIUS	L	S	Free	PC	M	L
DISTAR	P	L	££	N,PC	M	L
CALINE	L,A	U	££	PC	M	L
PAL	P,A,L	U	£££	PC	M	L
ISC	P,A,V	U,S,L	££££	PC+	H	L,E
ADMS-2	L,P,A,V	U,S,L	££££	PC+	H	L,E
ADMS-Urban	L,P,A,A	U,S,L	£££££	PC+	H	L,E
AERMOD	P,A,V	U,S,L	£££	PC+	M/H	L,E
INDIC Airviro	P,A,L	S,U	£££££	W	M	L,E

1 L = Line, P = point, A = Area, V = volume
2 N = none required (assumes worst scenario), U = user defined, S = sequential hourly, L = long-term statistical, S = local wind fields need to be configured by specialist software supplied
3 If purchased commercially with user-friendly input and output modules: £ = <£50, ££ = £50–£500, £££ = £500–£1500, ££££ = £1500–£10,000, £££££ = >£10,000
4 N = no hardware, PC = 486 DX with 4 MB RAM, PC+ = Pentium with 32 MB RAM, W = workstation
5 Time for setting up and running a simple scenario such as a single stack or line source: M = minutes, H = hours
6 Expertise required: A = basic maths calculator, L = understanding of air quality issues, E = expert use only

Source: modified from table A1 in DETR (2000e)

can be purchased from specialist software companies. UK models include ADMS, for which users must pay an annual licence fee, can be supplied by Cambridge Environmental Research Consultants. Environmental consultancies and some organisations have developed their own models (e.g. box models) or modified the standard ones (Barrowcliffe 1993, HMIP 1993, Street 1997). Some simple screening models are available as publications (e.g. DETR 1999b) or via the Internet (e.g. the Meteorological Office's road traffic model, AEOLIUS: Assessing the Environment Of Locations In Urban Streets at http://www.meto.gov.uk).

The most appropriate **model outputs** that should be incorporated in an EIA are predictions of short-term pollution impact (e.g. highest or 'worst-case' hourly mean concentration) and long-term impact (e.g. annual mean concentration). Outputs need to be compared with the appropriate air quality standards, objectives and guideline values and any locations which approach or exceed these concentrations must be identified. In some cases a model may not calculate pollution concentration over the averaging period used to define an air quality standard. For example, the UK SO_2 objective refers to a 15-min averaging period. In this situation it is necessary to use empirical relationships to decide whether the air quality standard is exceeded or not. It is suggested that within urban areas the 15-min sulphur dioxide is approximately 1.34 times the 1-h mean although near to tall stacks this ratio may

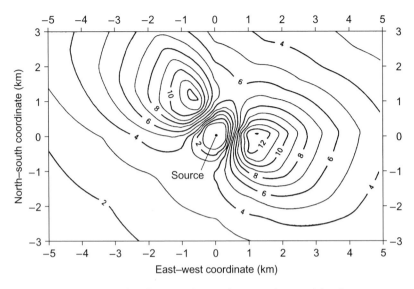

Figure 8.1 Predicted distribution of annual averaged ground-level concentrations of sulphur dioxide (μg/m³) due to emissions from a 50-m high stack using the US-EPA Industrial Source Complex model. As is often the case with UK climate data (in this example, data from Aughton, near Liverpool), the result is a distribution with two distinct peaks (to the northwest and east of the source).

increase to a factor of 2.0 (DETR 2000e). Within urban areas, the objective for PM₁₀ (a maximum of 35 exceedances of 50 μg/m³ of the daily means in any year) is unlikely to be exceeded where the annual mean concentration is less than about 28 μg/m³ (DETR 2000e). Within urban areas, the CO maximum 8-h mean is unlikely to exceed the objective if the maximum 1-h concentration is less than 14 mg/m³ (DETR 2000e). However, model developers are beginning to modify the outputs to match UK air quality objectives. For example, ADMS-3, introduced in 1999, can predict the number of exceedances for the UK 15-min SO₂ objective.

Model outputs from advanced point source models may be in the form of an **isoline map** of annual average concentrations compiled from concentrations predicted by the model for a grid spacing of say, 1 km (Fig. 8.1). This model output can be interfaced with a Geographical Information System such as ArcView (see Chapter 16) so that the pollution isolines can be overlain on an Ordnance Survey map of the area. Hourly maxima concentrations may be shown as a plot of concentration versus downwind distance for a range of specified meteorological conditions including those conditions which give rise to the highest concentration (Fig. 8.2). An EIA should seek to specify predicted concentrations at sensitive receptors such as the nearest residential housing, hospital, school, etc.

8.4.3 Models assumptions and models for point sources

For many years numerical prediction models have been based on Gaussian assumptions. The **Gaussian model** assumes that the pollutant emissions spread outwards from a source in an expanding plume aligned to the wind direction, in such a way that the distribution of pollution concentration decreases away from the plume axis in horizontal and vertical planes, according to a specific Gaussian mathematical equation, a symmetric bell-shaped distribution. Although a plume may appear irregular at any one moment, its natural tendency to meander results in a smooth cone-shaped Gaussian distribution after ten minutes of averaging time. The horizontal axis of the plume does not normally coincide with the height of the stack or point of emission, as the density and momentum of the emissions quickly carries the plume to a higher elevation, known as the "effective release height" (sometimes many times higher than the stack or point of emission). The maximum ground-level concentration experienced from a pollution plume is where the plume touches the ground.

Gaussian models assume the rate of dispersion of the plume, and consequently the pollution concentrations experienced at any location at the surface, are a function of wind speed, wind direction and atmospheric stability (Barrowcliffe 1993, DETR 2000e, Middleton 1998). Estimates of atmospheric stability for the simpler versions of the model can be obtained using a table or nomogram involving solar radiation, cloud cover and mean wind speed and expressed in the form of six or seven **Pasquill stability categories**. Stability categories range from class A (very unstable) occurring during hot, sunny conditions with light winds through category D (neutral) to class F or G (both very stable) occurring during cold, still nights with clear skies. For the purposes of the model, it is assumed that each stability class is characterised by a specified depth of boundary layer into which the pollutants are mixed. Typical mixing heights are around 1500 m for very unstable conditions through 800 m for neutral conditions to only 100 m for very stable conditions. When using the model to predict annual average pollution concentrations, the necessary summary of Pasquill stability classes for the nearest meteorological station can be obtained from the Meteorological Office Air Pollution Consultancy Group. These tables indicate the annual percentage frequencies of each stability class by 30 degree wind-direction sectors in six wind-speed bands, averaged over several years of data.

Figure 8.2 highlights that the highest ground-level concentrations from an elevated source tend to occur close to the source during light winds when the atmosphere is very unstable with substantial vertical mixing such as happens on hot summer days. It can also be seen that during light winds the peak concentration is found further from the source during conditions of increasing atmospheric stability. Where tall buildings lie adjacent to a tall stack, an occasion of strong winds is another situation that can give rise to high ground-level concentrations. This happens because buildings cause eddies to form that make the plume touch the ground much closer than would be expected otherwise. It is generally considered that building downwash problems may occur if the stack height is less than 2.5 times the height of the building upon which it protrudes. Similarly problems may occur if adjacent buildings are within about five stack heights of the release point. Other situations giving rise to high pollution concentrations may be when plumes impact directly on hillsides under certain meteorological conditions, or when valleys trap emissions during low-level inversions (DETR 2000e).

The Gaussian model most commonly used internationally is the Industrial Source Complex (ISC) model developed by the US Environmental Protection Agency (EPA). There is a long-term (ISCLT) and a short-term (ISCST) version, producing results illustrated in Figures 8.1 and 8.2, respectively. A worked example using ISCST to calculate 1-h concentrations from a single stack is given in appendix C of

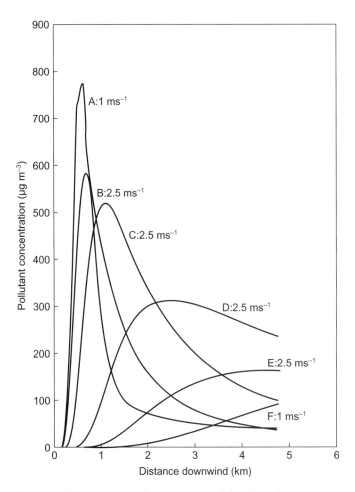

Figure 8.2 Predicted one-hour average sulphur dioxide concentrations (µg/m³) due to emissions from a 50-m high stack using the US-EPA Industrial Source Complex model for the 'worst-case' wind speed in each Pasquill atmospheric stability class.

DETR (2000e). The ISC model can be used for area and volume sources as well as for point sources. The EPA have also produced a screening model, SCREEN, which aims to predict worst-case scenarios. UK complex models include the R91 model developed originally by the nuclear industry in 1979 but available commercially as DISTAR. Another commercially available model is the INDIC Airviro model developed in Sweden. Unlike other Gaussian-based models this requires complex physiographic and meteorological configuration by the software supplier rather than relying on meteorological information from a single site.

More recently, what are termed second or new-generation models have been developed which employ atmospheric dispersion assumptions based on recent improvements in the understanding of the behaviour of pollutants released into the atmosphere (DETR 2000e, Middleton 1998). In particular they recognise that there are different turbulence and diffusion characteristics within the atmosphere at different heights and so treat the atmosphere in a more realistic way.

The ADMS (Atmospheric Dispersion Modelling System), developed in the UK by Cambridge Environmental Research Consultants, was introduced in 1993. In addition to predicting long-term concentrations it has the ability to predict short-term concentrations over averaging times of a few seconds, as is needed in the case of odours. The ADMS employs boundary layer data such as surface heat flux and boundary layer depth instead of Pasquill stability categories as its meteorological data inputs (available from the Meteorological Office). The model can treat both dry and wet deposition, building effects, terrain variations and coastal influences. Currently, the model predicts higher ground level concentrations than the Gaussian models ISC and DISTAR from tall stacks under very convective atmospheric conditions. A suite of ADMS models are available including ADMS-Screen, ADMS-1 (single source), ADMS-2 (multi-source), ADMS-3 (multi-source and capable of predicting the length of visible plumes from stacks) and ADMS-Urban. A worked example using ADMS-3 to calculate 1-hour concentrations from a single stack is presented in appendix C of DETR (2000e). In 1998 the EA issued a nomogram-based screening guide which provides precalculated ADMS results for stack emissions for 10 stack heights, four categories of surface roughness, three averaging times and three climate types (EA 1998).

In 1998 the US EPA released their new-generation model, AERMOD. It contains improved algorithms for convective and stable boundary layers, and for computing vertical profiles of wind, turbulence and temperature. In 2000 the UK Meteorological Office teamed up with Lakes Environmental to create a more user-friendly interface for this model, so its use in the UK as a competitor to ADMS may increase in future. A fluid dynamics model, PANACHE, is available which can predict concentrations for industrial (and traffic) sources and offers good treatment of very low wind speeds and wind flow patterns around uneven terrain and high-rise buildings.

8.4.4 Road traffic models

Several models have been developed specifically to predict pollution concentrations arising from emissions from road vehicles. The simplest is the Department of Transport's nomogram-based *Design Manual for Roads and Bridges* (DMRB) screening model which can be used to indicate those areas, if any, where air pollution is likely to cause concern (DoT 1994, DETR 1999b, 2000g). An Excel spreadsheet version is available from http://www.stanger.co.uk/airqual/modelhlp/dmrb/dmrb.asp. The 1994 version of DMRB was designed for motorways and trunk roads in relatively open country so had limited use in urban situations, especially street canyons where the most serious air pollution problems often occur (DoE 1994). An updated version for wider use is now available (DETR 1999b, 2000g). It is very simple to apply, and is used (a) widely by planners as a screening model, and (b) in the *New Approach to Appraisal* (NATA) (DETR 1998b) which is described in Chapter 5. The CAR-International model (Calculation of Pollution from Road Traffic, international version) developed in The Netherlands, is a Gaussian-based screening model requiring emission factors, the configuration of the road, details of the presence of trees along the road, and wind speed to be specified by the user (Eerens *et al.* 1993).

A more advanced road model is CALINE4 (the California Line Source Dispersion model, version 4). This Gaussian model can model junctions, street canyons, parking lots, bridges and underpasses and predicts 1-h concentrations of pollutants such as CO and NO_2. The model can handle up to 20 road links and 20 receptors (locations at which the pollution impact of the emissions will be predicted). The US EPA PAL (the Point, Area and Line source) model extends the CALINE algorithms to treat edge effects more accurately, which makes it useful for predicting concentrations from car parks and small areas of a city for up to 99 point, area and line sources and 99 receptors.

The US EPA CAL3QHC model (developed by extending the CALINE3 model to take into account Queuing and Highway Capacity considerations) is appropriate for traffic-congested roads and complex intersections, being able to incorporate emissions from both moving and engine-idling vehicles. It is able to predict 1-h mean concentrations for up to 120 road links and 120 receptors. Road traffic model outputs can be produced for specified locations or additional software can be used to convert the results into map form to show isolines of various pollution concentrations. The CALINE4 and CAL3QHC models and user guides can be downloaded from the Internet (e.g. Lakes Environmental Software site http://www.lakes-environmental.com/lakeepa.html).

The AEOLIUS model, developed by Doug Middleton of the UK Meteorological Office, enables the user to predict 1-h mean concentrations of pollutants from traffic flowing along a canyon-like street such as is found in city centres. Screening and full versions (AEOLIUSF, AEOLIUSQ) of this model can be downloaded free via the MO Internet site (http://www.meto.gov.uk). ADMS-Urban (Atmospheric Dispersion Modelling System, Urban module) can cope with up to 1000 road sources and includes a street canyon option (DETR, 2000e). Details of comparisons of the usefulness and limitations of many of these road traffic models (as well as point source models) were undertaken by local authorities in the trial implementation phase of the NAQS (NSCA 1998, DETR 1999a).

8.4.5 Emissions data input to models

All numerical dispersion models require emissions data either in the form of a specified emission rate for the source (e.g. the amount of pollutant released per unit of time) or a measure of the level of activity of the source (e.g. amount of fuel consumed) together with the corresponding emission factor (e.g. the quantity of pollutant emitted per unit of activity). Emission rates need not necessarily be exact, as the likely impact of a planned development could be assessed by using the highest likely emissions, such as the maximum emission limits defined for prescribed processes in the relevant IPC Process Guidance Note (DETR 2000e). If the emission rate for a proposed development is not already specified in the plant design, then an estimate may be based upon expected fuel consumption and characteristics of the fuel. Information on emissions and emission factors are available from the UK Emission Factors Database, accessible via http://www.london-research.gov.uk/emission/main.htm. Emissions factors (F) are described in terms of, for example, grams of NO_x per km driven for vehicles, grammes of NO_x per kilowatt fired for boilers, and grams of NO_x per tonne of nitric acid product for a nitric acid works. Emissions would then be calculated as $M \times F$ where M is a measure of the level of activity.

Typical emission rates can be used when calculating long-term pollution concentrations but for short-term models a number of worst-case scenarios may be needed (e.g. periods of intensive activity, during start-up, and the operation of emergency release vents). Complex models applied to a point source will require input information about the release conditions of the emissions. This may include the stack height and internal exit diameter as well as the flue-gas exit temperature and exit velocity (or volumetric flow rate). Examples of calculating emission rates for industrial sources are given in HMIP (1996) and DETR (2000d). Examples of the type of inputs and procedures for predicting 1-h concentrations from point sources using the ISC and ADMS models are detailed in DETR (2000e).

In the case of road traffic models, vehicle emission rates for a specific section of road are calculated by the model itself from input data such as vehicle flow (e.g. vehicles per day, peak hourly value), average vehicle speed, vehicle mix (e.g. fraction of heavy goods vehicles, fraction of petrol- and diesel-engine cars) and vehicle emission factors (DETR 2000e). If the model is being used to predict pollution concentrations for a future year, say 2005, then input forecast data not only for future traffic flow, speed and mix are needed but also the likely change in emission factors. Emission factors for future years, which take into account the expected effects of phasing in of cleaner technologies and fuels, are available from the UK Emission Factors Database Internet site (http://www.london-research.gov.uk/emission/main.htm) or are already embedded in some models such as the DMRB (DETR 1999b, 2000e, 2000g).

8.4.6 Model limitations

All predictions have an element of uncertainty and it is important to acknowledge this and not treat the model as a 'black box' by concentrating only on the results produced. Models are simplifications of reality and their limitations, accuracy and confidence levels should be recognised and explained (Benarie 1987, Royal Meteorological Society Policy Statement 1995). Some limitations have yet to be resolved, such as the availability of detailed and accurate meteorological and emissions input data: the quality of the input data will clearly affect the accuracy of a model. Even if accurate input data were available, the algorithms employed in the model to represent the behaviour of pollutants released into the atmosphere contain many uncertainties. Confidence in the accuracy of a model is gained by assessing its ability to predict the current baseline conditions in the study area, since the results can be verified using monitored pollution data. DETR (2000e) consider that if the predicted concentration from a numerical dispersion model lies within ±50% of the measurement, then a user would not consider that the model had behaved badly.

8.4.7 Assessing significance

The level of significance of the likely pollution impacts of a proposed development is assessed by comparing the predicted changes in the area to air quality standards, objectives or guideline values, and determining whether these are likely to be exceeded at any locations, after taking into account the existing and predicted baseline pollution levels. If the planned development is predicted to increase pollution levels in excess or close to the air quality standard, then mitigation measures need to be proposed. If the changes are well below the standard, it is useful to express the increase in ground-level pollution concentrations in a meaningful way. For example, an EIA may conclude that a proposed development is expected to increase the annual average NO_2 concentration at the location worst affected (5 km downwind) by only 3% and that this increase is well within the year-to-year variability of annual average concentration produced by meteorological fluctuations. Even when a development is likely to add only small amounts of pollution to the area, it is important that an EIA makes specific assessment of what effect (perhaps negligible) this will have on any nearby sensitive receptors such as residential areas, schools, nature reserves, SSSIs and historical buildings.

Determining the level of significance of climate changes can be difficult in some cases. A local increase in temperature, wind turbulence, fog or frost may affect people, plants and wildlife directly or indirectly (e.g. fog causing road accidents), but the level of significance of the changes may require the use of expert opinion.

8.5 Mitigation

8.5.1 *The need for mitigation measures*

Mitigation measures should aim to avoid, reduce, or remedy any significant adverse effects that a proposed development is predicted to produce. At one extreme, the prediction and evaluation of likely impacts may indicate such extreme adverse effects that abandonment or complete redesign of the proposed development is the only effective mitigating measure. More likely, modifications to the development can be suggested in order to avoid or reduce potential impacts (Wood 1989, 1990, DETR 2000h, 2000i). Some mitigation measures may be required by law for new – though not for existing – developments (e.g. fitting of specific types of pollution-control devices) but the use of others depends upon the significance of the predicted impacts.

Various mitigation measures may be suggested to solve a potential problem and it is important to assess the likely effectiveness of each measure in terms of the extent to which the problem will be reduced, as well as to indicate the costs of implementation. Whatever mitigation measures are proposed, it is important to ensure that they do not create problems of their own. Mitigation feeds back into design, so mitigation measures proposed to minimise adverse impacts of the project can be incorporated as alternatives in the project description. Subsequent proposed developments can make use of the information contained in a previous EIA in order to incorporate appropriate mitigation measures at the outset, rather than wait for its own EIA to identify potential problems.

8.5.2 *Mitigating adverse pollution impacts*

If the pollution impact from an industrial stack is predicted to approach or exceed air quality standards, this impact can be reduced by encouraging greater atmospheric dispersion and dilution of emissions by (a) raising the stack height, (b) reheating the flue gases to higher temperatures, and (c) emitting them at greater velocity. If a planned development is likely to exceed, say, maximum hourly pollution standards only during periods of poor atmospheric dispersion, then one possible mitigation measure would be to keep a cleaner standby fuel for use during those forecasted occasions or to reduce emissions by reducing production output in the case of an industrial process. Improved fuel combustion designs can reduce pollutant emissions, such as by using low nitrogen oxides burners in furnaces. In many cases the type and amount of pollutants emitted are a function of the fuel being burned, so alternative fuels can be proposed, such as fuel oil with a very low sulphur content or natural gas. Traffic-generated pollutants decrease rapidly away from roads, and this process can be enhanced by roadway trenching, embankments, walls and trees, to reduce the pollution concentrations in nearby residential areas.

The **construction stage** of most projects has the potential to cause localised wind-blown dust problems, either when excavation is taking place or when materials are being transported and stored in stockpiles. Careful design of construction operations, including the selection of haulage routes into the site and the location of stockpiles, can help to minimise dust problems in nearby residential areas. Mitigation measures can include (a) frequent spraying of stockpiles and haulage roads with water, (b) regular sweeping of access roads, (c) covering of lorries carrying materials, (d) enclosing conveyor-belt delivery systems, and (e) early planting of peripheral tree screens where they are part of the planned development.

The need for mitigation measures may not always be clear. For example, should action be taken to ensure odours from a food processing plant are not experienced by residents of a few isolated houses on several days each year when the wind blows in their direction? In such a situation, consultation with the local planning authority will be needed to agree whether the impacts are sufficiently adverse to justify the cost of mitigation measures. Alternatively the local authority may suggest the developer offers compensation to the affected residents, or offers to purchase the affected properties in order to create a buffer zone around the plant. If potential odour problems are to be tackled at source, solutions include taller stacks to encourage greater dispersion of the emissions, or removal of the pollutant completely by absorption, adsorption, oxidation or chemical conversion.

8.5.3 Mitigating adverse microclimate impacts

Adverse microclimatic changes, such as increased wind turbulence around a proposed shopping precinct, can be minimised by the widening of narrow gaps between buildings, roofing of open spaces and changing the height and layout of buildings (Oke 1987). Unwelcome high air temperatures in open shopping precincts during summer can be reduced by the choice of building materials, consideration of building layout in relation to areas of sun and shade, and the planting of trees. Frost pockets affecting agricultural and horticultural crops can be prevented by landscaping and creating openings through road or railway embankments, which allow for the passage of cold air. The frequency of icing of roads can be reduced by landscaping and choice of road surface materials. The frequency of fog forming on cold clear nights along proposed motorways can be lessened by (a) eliminating any nearby areas of standing water, (b) reducing air pollution (suspended particulates) in the vicinity, (c) raising the road onto pillars above the fog-shrouded valley floor, and (d) planting tree belts which help reduce cold air drainage and scavenge fog droplets. Water-vapour plumes from power station cooling towers, which have the potential to increase fog and icing of nearby roads, can be designed so that the banks of towers are oriented along the direction of the prevailing wind, such that the merging of individual plumes enhances buoyancy and reduces the number of occasions when plumes are brought to the ground.

8.6 Monitoring

Numerical prediction models contain uncertainties, so monitoring should be continued after completion of the development to compare predictions with those that actually occur. Confirmation of the accuracy of the predictions will provide credibility to the process of EIA. This is particularly appropriate if similar projects are likely to be proposed in the future for other locations. Continued monitoring is also necessary to assess the effectiveness of any mitigation measures proposed in an EIA and to ensure that any potential air and climate problems identified have been minimised or eliminated.

References

Barrowcliffe R 1993. The practical use of dispersion models to predict air quality impacts. *Paper presented at the IBC Technical Services Conference on Environmental Emissions: Monitoring Impacts and Remediation, London (paper available from Environmental Resources Management).*

Benarie MM 1987. The limits of air pollution modelling. *Atmospheric Environment* **21**, 1–5.

Bertorelli V & R Derwent 1995. *Air quality A to Z: a directory of air quality data for the UK in the 1990s.* Bracknell: Meteorological Office.

Canter LW 1996. *Environmental impact assessment* (Ch. 22). New York: McGraw-Hill.

Colls J 1997. *Air pollution: an introduction.* London: E & FN Spon.

DETR (Department of the Environment, Transport and the Regions) 1998a. *Review of the United Kingdom National Air Quality Strategy.* London: DETR.

DETR 1998b. *Guidance on the New Approach to Appraisal.* London: HMSO. (http://www.detr.gov.uk/itwp/appraisal/guidance/index.htm)

DETR 1999a. *The first phase review: a summary.* London: DETR (available from the Local Air Quality Management [LAQM] section of http://www.aeat.co.uk/netcen/arqual/home.html).

DETR 1999b. *Design manual for roads and bridges.* Vol. 11: *Environmental Assessment*, section 3, Part 1 Air Quality. London: TSO.

DETR 2000a. *The Air Quality Strategy for England, Scotland, Wales and Northern Ireland: working together for clean air.* London: DETR. (http://www.aeat.co.uk/netcen/arqual/home.html)

DETR 2000b. *Framework for review and assessment of air quality*, LAQM.G1(00). London: DETR. (http://www.environment.detr.gov.uk/airq/laqm.html)

DETR 2000c. *Monitoring for air quality reviews and assessments*, LAQM.TG1(00). London: DETR. (http://www.environment.detr.gov.uk/airq/laqm.html)

DETR 2000d. *Preparation and use of atmospheric emissions inventories*, LAQM.TG2(00). London: DETR. (http://www.environment.detr.gov.uk/airq/laqm.html)

DETR 2000e. *Selection and use of dispersion models*, LAQM.TG3(00). London: DETR. (http://www.environment.detr.gov.uk/airq/laqm.html)

DETR 2000f. *Review and assessment: pollutant-specific guidance*, LAQM.TG4(00). London: DETR. (http://www.environment.detr.gov.uk/airq/laqm.html)

DETR 2000g. *Design manual for roads and bridges.* Vol. 11: *Environmental Assessment*, section 3, Part 1 Air Quality Supplement 1. London: TSO.

DETR 2000h. *Air quality and transport*, LAQM.G3(00). London: DETR. (http://www.environment.detr.gov.uk/airq/laqm.html)

DETR 2000i. *Air quality and land use planning*, LAQM.G4(00). London: DETR. (http://www.environment.detr.gov.uk/airq/laqm.html)

DoE (Department of the Environment) 1989. *Environmental assessment: a guide to the procedures.* Department of the Environment, Welsh Office. London: HMSO.

DoE/Welsh Office 1990. *Integrated Pollution Control: a practical guide.* London: DoE.

DoE 1995. *Air quality: meeting the challenge*. London: DoE.

DoE 1996. *The potential effects of climate change in the United Kingdom*. 2nd Report. London: HMSO.

DoE 1997. *The United Kingdom National Air Quality Strategy*. London: The Stationery Office.

DoT 1994. *Design manual for roads and bridges*. Vol. 11: *Environmental Assessment*, section 3. London: HMSO.

EA (Environment Agency) 1998. *Guidance for estimating the air quality impact of stationary sources*. National Centre for Risk Analysis & Options Appraisal Report GN 24. London: Environment Agency.

Eerens HC, CJ Sliggers & KD van Hout 1993. The CAR model: the Dutch method to determine city street air quality. *Atmospheric Environment* **27B**, 389–399.

Elsom DM 1992. *Atmospheric pollution: a global problem*. 2nd edn. Oxford: Blackwell.

Elsom DM 1996. *Smog alert: managing urban air quality*. London: Earthscan.

Elsom DM 1999. Development and implementation of strategic frameworks for air quality management in the UK and the European Community. *Journal of Environmental Planning and Management* **42**, 103–121.

Elsom DM, JWS Longhurst & C Beattie 2000. Air quality management in the United Kingdom: development and implementation of the National Air Quality Strategy. In *Air quality management, advances in air pollution*, Vol. 7, JWS Longhurst, DM Elsom & H Power (eds). Southampton & Boston: WIT Press.

EPAQS 1996. *Nitrogen dioxide*. 7th Report. London: HMSO (details of other EPAQS reports are available from http://www.environment.detr.gov.uk/airq/aqs/index.htm).

Harrop DO 1993. Environmental impact assessment and incineration. In *Air quality impact assessment*, RM Harrison (ed.). London: Royal Society of Chemistry.

HMIP 1993. *An assessment of the effects of industrial releases of nitrogen oxides in the East Thames corridor*. London: HMSO.

HMIP 1996. *Released substances and their dispersion in the environment*. London: HMSO.

Houghton JT 1997. *Global warming: the complete briefing*. 2nd edn. London: Lion.

Hughes D, N Parpworth & J Upson 1998. *Air pollution: law and regulation*. London: Jordans.

Hulme M & Jenkins G 1998. *Climate change scenarios for the United Kingdom: Scientific Report*. UK Climate Impacts Programme Technical Report No. 1. Norwich: Climatic Research Unit, University of East Anglia. (http://www.cru.uea.ac.uk:80/link/ukcip/ukcip_report.html)

MAFF (Ministry of Agriculture, Fisheries and Food) 2000. *Climate change and agriculture in the United Kingdom*. London: MAFF. (http://www.maff.gov.uk/)

Middleton DR 1998. *Manual for modelling: a guide to local authorities*. Turbulence and Diffusion Note 241. Bracknell: Meteorological Office.

Mulgrew A & P Williams 2000. *Biomonitoring of air quality using plants*. Air Hygiene Report 10. Berlin: WHO Collaborating Centre for Air Quality Management & Air Pollution Control.

Murley L 1995. *Clean air around the world*. 3rd edn. Brighton: IUAPPA/NSCA.

NSCA (National Society for Clean Air & Environmental Protection) 1998. First Phase Authorities. *Clean Air* **28**, 40–66.

Oke TR 1987. *Boundary layer climates*. 2nd edn. London: Methuen.

Ortolano L 1997. *Environmental regulation and impact assessment* (Ch. 6). New York: Wiley.

Royal Meteorological Society Policy Statement 1995. *Atmospheric dispersion modelling: guidelines on the justification and use of models, and the communication and reporting of results*. Published in collaboration with DoE. Reading: Royal Meteorological Society.

Street E 1997. EIA and pollution control. In *Planning and environmental impact assessment in practice*, J Weston (ed.), 164–179. London: Longman.

Turco RT 1997. *Earth under siege: from air pollution to global change*. Oxford: Oxford University Press.

Wood CM 1989. *Planning pollution prevention: anticipating controls over air pollution sources*. Oxford: Heinemann Newnes.

Wood CM 1990. Air pollution control by land use planning techniques: a British–American review. *International Journal of Environmental Studies* **35**, 233–243.

9   Soils, geology and geomorphology

Martin J Hodson, Chris Stapleton and
Roy Emberton

9.1 Introduction

Much has been written about the links between soils, geology and civilisation, but considerably less is known about the impacts of human activity on soils and geology. The EU/UK EIA legislation (see §1.4) specifically identifies soil as one of the main environmental **receptors** of development impacts for which assessments must be carried out. The DoE (1989) guidance includes soil, agricultural quality, geology and geomorphology as topics in the checklist that should be included in an EIA.

Soil is defined as the top layer of the land surface within the biosphere. It is a component/subsystem of terrestrial ecosystems, providing a growing medium for flora, and a habitat for fauna (see Fig. 11.4, p. 251). From the human perspective, soil is also the basis of agricultural and forestry production for food, wood and textiles. Avoiding significant development impacts on the soil ultimately protects the whole of the ecosystem from degradation. An understanding of the local environment would be incomplete without reference to the underlying geology, but less emphasis is generally given to impacts on this, because relatively few types of development have significant impacts on geology. This chapter therefore concentrates on the assessment of significant soil impacts, although some important geological and geomorphological aspects are described briefly.

9.2 Definitions and concepts – geology and geomorphology

9.2.1 Geology

Geology is a vast and complex subject, and only a few aspects of relevance to EIA will be mentioned here. Surface geology concerns superficial deposits (e.g. drift, glacial deposits, river gravel) while solid geology only concerns pre-superficial formations. The three main groups of rock are igneous, sedimentary and metamorphic. Many igneous rocks have formed as a result of volcanic activity; they are characteristically hard and crystalline, and have crystallised from magma, a silicate melt. Sedimentary rocks are formed from pre-existing rocks by processes of denudation and sedimentation. They are relatively soft and easily eroded and include limestones, coal, evaporites and sedimentary iron ores. Sedimentary rock strata are often important as **aquifers**, and many are rich in fossils. Metamorphic rocks are formed as the result of heat, pressure and chemical activity on pre-existing solid rock.

A number of aspects of geology are of direct importance in EIA. *Earth Heritage Sites* (some of which are Sites of Special Scientific Interest (SSSIs)) are important for the conservation, protection and management of their fossils, stratigraphy, minerals or other geological interest. They have scientific and amenity value, and include exposures of value to wildlife (e.g. rocky shores, shingle structures, cliffs, screes, and limestone pavements). The underlying geology also has engineering and construction implications, and affects both geochemistry and geophysics (Ellison & Smith 1997, Bell 1999).

Some geological aspects are of more indirect importance in EIA. For example, both the storage and movement of ground and surface waters, and water geochemistry will be affected by the hard geology of an area (see Chapter 10). In addition, the physical and chemical properties of soils will be affected, as most soils are derived from bedrock or transported rock. The geology and hydrogeology of a site influences the potential for on-site and off-site *pollution*, and the extent of any pollution that may have occurred in the past. Finally, competition between mineral extraction and other land uses is also an important topic in some circumstances (Ellison & Smith 1997).

9.2.2 Geomorphology

Geomorphology can be defined as "the study of landforms, and in particular their nature, origin, processes of development and material composition" (Cooke & Doornkamp 1990). 'Material composition' includes both the geology and, where present, the soil. Geomorphology therefore includes the study of topography (the terrain) and the factors that have moulded the land to the present form (including the nature of the rock and soils in relation to the *erosion* and deposition caused by glaciers and rivers). Human impacts can include landscape/visual aspects (Chapter 6), but also consequences such as erosion (Bell 1999, Cooke & Doornkamp 1990), slope failure and subsidence, and sedimentation in aquatic systems (Chapters 10 & 12). Some aspects of geomorphology, such as soil erosion, overlap with soil studies.

9.3 Definitions and concepts – soils

The productive value of soils is determined by a number of important physical and chemical properties. An appreciation of a development's impacts on soils requires an understanding of basic soil features. The coverage of soil science here is, of necessity, brief and the reader is referred to Avery (1990), White (1997) and Brady & Weil (1999) for further information.

9.3.1 Soil composition

There are two major types of soil: mineral and organic. Typically mineral soils have four major components: mineral particles, usually derived from *weathering* of parent rock (about 45% of the volume); organic matter (about 5%); water (about 25%); and air (about 25%). Organic matter is an important component of the soil which is derived mainly from decomposing vegetation. It combines with inorganic particles and cements like iron oxides and calcium carbonate to create stable structural aggregates. The nature of the organic matter in topsoils varies according to the vegetation cover and environmental conditions. In cool wet areas, the organic matter decomposes at a relatively slower rate and tends to be more acidic. In more temperate areas, the organic matter decomposes more completely to form stable complex compounds which are collectively known as humus. Most arable agricultural topsoils contain 2–6% organic matter, and structural stability is impaired at lower organic levels.

The inorganic component of soils consists of particles that are classified into standard size ranges (gravel, clay, silt and sand). There are a number of classifications of these particles, and the following is a simplified version from the British Standards Institution (BSI):

> Gravel – particle size over 2.0 mm
> Sand – between 0.06 mm and 2.0 mm
> Silt – between 0.002 mm and 0.06 mm
> Clay – less than 0.002 mm

These categories are known as separates, and their proportions in a soil define its **texture**. Sandy soils contain at least 70% sand, and less than 15% clay; clays usually have no less than 40% clay; and loams have more equal proportions of clay, silt and sand. The texture of a soil is of great practical importance. It influences the degree of aggregation of the separates, and both the range and total volume of pore spaces, which in turn affect (a) the capacity of the soil to retain moisture, and (b) its *hydraulic conductivity* and hence the ease with which water can percolate through it.

Texture also affects the behaviour of the soil at different moisture contents (its consistency). Thus clay soils tend to be less well drained than sandy and loamy soils. They may be waterlogged in winter, show poor infiltration (see §10.2.4), and have a plastic consistency for much of the year. They are described as 'heavy' as they are difficult to cultivate. Medium to heavy loams tend to have a more friable consistency, and a greater capacity to make moisture available to plants during the summer. Sandy soils are described as 'light'. They are very friable and easy to work, but prone to drought. Loams are generally thought to have the most favourable textures for agriculture.

Soil textures often vary with depth, as a result of the mixing and redistribution of parent materials during the Ice Ages, and subsequent soil-forming processes.

9.3.2 The soil profile and soil classification

Clearly, it is important to know what type of soil is present in a study area. A pit dug into an undisturbed soil will reveal the topsoil and subsoil layers. Such a vertical section is called a **soil profile**, and each individual layer is called a **horizon**. Two different soil profiles are shown in Figures 9.1 and 9.2. Not all of the horizons are always present, and the horizons are frequently subdivided. Pedological classifications of soils are concerned with natural horizons that have formed since the last Ice Age as a result of soil-forming processes. Most natural soils have an organic rich topsoil which contains humus. A and E horizons are **eluvial** upper horizons in which the inorganic particles have become depleted of nutrients as a result of the *leaching* effect of precipitation as it percolates through the profile to groundwater and water-courses (Chapter 10). In contrast, **illuvial** B horizons are often enriched with nutri-

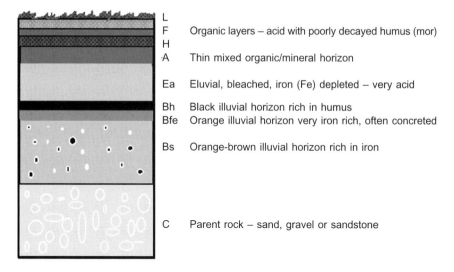

L
F   Organic layers – acid with poorly decayed humus (mor)
H
A   Thin mixed organic/mineral horizon

Ea   Eluvial, bleached, iron (Fe) depleted – very acid

Bh   Black illuvial horizon rich in humus
Bfe   Orange illuvial horizon very iron rich, often concreted

Bs   Orange-brown illuvial horizon rich in iron

C   Parent rock – sand, gravel or sandstone

Figure 9.1 Profile of a typical humus-iron podzol.
There are three superficial organic layers, L, F and H, which represent litter (leaves etc.), fermentation (where the breakdown of organic material contained in the litter largely occurs) and humus (mor). Beneath these are the eluvial A and E horizons (which are leached and often grey in colour), illuvial B horizons (rich in iron), and the parent material of the C horizon. These soils and their gleyed variants occur extensively over relatively cold and wet higher ground and some freely drained sandy parent materials in lowland areas. In these areas the main planning issues tend to be the protection of semi-natural habitats and wildlife conservation (redrawn from Bridges 1978).

ents, iron, clays or organic matter which have been leached from above and deposited in the lower subsoils. The C horizon is the weathering parent material or rock.

The soil profile is the main criterion used in **soil classifications**. This chapter concentrates on the soils likely to be found in Britain, using the classification system adopted by Avery (1990). Avery's terminology (or similar) is used in many British texts, and certainly seems to be the preferred terminology for British EIAs. There are, however, many other systems, and the classifications of the US Soil Taxonomy and FAO-UNESCO are gaining ground, even in Britain. The American textbooks (e.g. Brady & Weil 1999), use the US classification, and in Table 9.1 this terminology is compared with the equivalent British terminology for major soils of the British Isles (Avery 1990). Many EU member states including Belgium, Eire, France, Germany, Italy, The Netherlands and Portugal have their own distinctive soil classification systems, which in some cases contain elements of the US Soil Taxonomy and FAO-UNESCO classifications. An introductory account is given by Hodgson (1978).

Almost all of the soils of the British Isles have been influenced by human activity to some extent. Avery (1990) restricts the term **man-made soils** to mineral soils where present or former management of the soil has resulted in distinctive features. Outside of the hills and uplands and smaller patches of lowland **heath** (where the predominant soils are podzols which may be peaty and/or gleyed), most agricultural

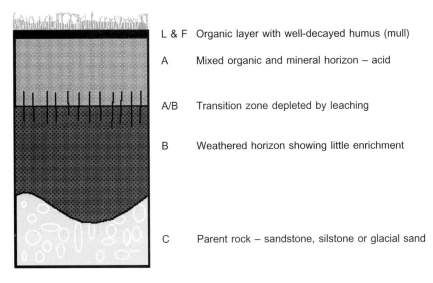

L & F	Organic layer with well-decayed humus (mull)
A	Mixed organic and mineral horizon – acid
A/B	Transition zone depleted by leaching
B	Weathered horizon showing little enrichment
C	Parent rock – sandstone, silstone or glacial sand

Figure 9.2 Profile of a typical acid brown soil.
Here the organic material is richer with well-decayed, less acid humus (mull). The soil is leached, but not nearly to the same extent as the podzol. The A and B horizons are far less distinct. These soils and their gleyed variants occur extensively over lowland areas, and the main planning issue is the protection of their productive potential, and the visual amenity of the vegetation cover which they support (redrawn from Bridges 1978).

soils consist of gleyed brown earths, brown earths and gleys. They have topsoils that extend to relatively uniform depths over subsoils, with a gradual transition into weathering parent material. Better quality soils tend to have loamy upper subsoils over lower subsoils that are generally heavier or lighter in texture, depending on the underlying parent material.

Podzols (Fig. 9.1) are typical of northern areas of Europe where they are associated with the boreal coniferous forest and heaths, and the climate is characteristically cold and wet. These soils are highly leached and acidic (*pH* often 3.0–4.5). They are little used for agriculture, but are very important for forestry. Podzols develop best on permeable sands and gravels.

Brown soils are generally associated with areas originally covered by deciduous forest and are the dominant soils of lowland Britain. There are many types of brown soil and Figure 9.2 shows one example, an acid brown soil. Brown earths are the best known and widespread category of this group, and are fairly fertile, with pH 4.5–6.5. They are generally located in warmer and drier climates than podzols, and the precipitation/*evapotranspiration* ratio (see §10.2.3) of the environments in which these soils develop is generally lower than that of the podzols. The amount of water percolating through the soil is sufficient to cause a moderate amount of leaching, but is not enough for podzol formation. Most of the original forest that grew on brown soils has been cleared for agriculture.

In some places, the profiles have distinctive features which have been imposed by the underlying rock, or a geomorphological process. For example, Carboniferous

Table 9.1 A comparison between the British soil classification of Avery (1990) and the US Soil Taxonomy

Avery (1990)	US Soil Taxonomy	Notes
Podzols	Spodosols	Humid to per-humid temperate climates. Acidic soils characterised by grey coloured A and E horizons, and the deposition of humus and/or iron in the B horizon.
Brown soils	Mostly Alfisols	Humid temperate climates. Leached and elluviated soils, but reasonably fertile. Argillic B horizon. Includes Brown Earths.
Lithomorphic soils	Mostly Entisols	Thin (30 cm) soils with no diagnostic subsurface horizon. Includes Rankers and Rendzinas.
Gley soils	Aquic soils of a great variety of types	Soils characterised by saturation with water for at least part of the time. Reducing conditions are prevalent.
Peat soils	Histosols	Organic soils, bog and fen peats, forming in humid climates often in depressions.
Man-made soils	Plaggepts and Arents	Ploughed and disturbed soils.

limestone soils tend to have very shallow soil profiles over hard rock, and gravels form impenetrable layers or pans at a range of depths, often in *alluvial* areas or on plateau surfaces. Lithomorphic soils are thin soil types where the parent rock is the dominant feature in soil development, representing an early stage in soil development. The best known lithomorphic soils are the **rendzinas**, which develop over chalk or limestone. In a typical rendzina, the A horizon, which is generally fairly thin, rests directly on the parent C horizon. The soil is very dark brown or black in colour and is alkaline (pH 7.5–8.4). In contrast, **rankers** are young, acidic soils which develop over non-calcareous rocks such as sandstones. In southern Britain the climax vegetation on rendzinas is deciduous forest (e.g. beech, oak), but the trees have often been cleared and these areas are now mostly used for agriculture.

Gley soils are hydromorphic soils, in which water stands in the profile for at least part of the year. Gleying occurs when water saturates a soil, filling most of the pore spaces and driving out air. Any remaining air is soon used up by micro-organisms, causing the development of anaerobic conditions.

Peat soils are a major soil type in some parts of the world, but cover a relatively minor fraction of the land surface of the UK (only 3% of England and Wales, but rather more in Scotland and Ireland). Pure peat is partly decayed organic (mainly plant) material that accumulates where lack of oxygen, associated with waterlogging, inhibits the activity of microbial decomposer organisms. **Mires** (peatland ecosystems) occur where there is near-permanent waterlogging and consequent peat accumulation. They provide valuable wildlife habitats, many of which are protected by statutory designations. They are also important from a global warming perspective because they contain (and hence 'lock up') a significant amount of carbon. Mires can be divided into **bogs** and **fens**, which differ largely in relation to their hydrology (see §12.2.5). According to MAFF (1988), peats contain at least 20 to 25% organic matter, depending on the clay content. The substratum of bogs is normally almost pure peat, but that in fens can contain high proportions of inorganic material such as marl (calcareous-clay mixtures). Similarly, whilst the peat in 'active' bogs is normally saturated, many peatlands have fairly free drainage, at least near the surface. However, lowering of water tables, e.g. by agricultural drainage schemes and/or water abstraction, can seriously damage peatland ecosystems (§12.5.3) and lead to soil loss by oxidation and erosion.

9.3.3 *Soil structure*

In most soils, the soil particles or separates are organised into aggregates. Soil structures, called **peds**, vary in size and shape, and generally recognised standards are described by the Soil Survey of England & Wales (SSEW 1976). Each textural soil horizon in a soil type usually contains one shape and size of structure, but structure frequently varies with depth. For example, angular and subangular blocky structures in loams become coarser (larger) with depth. In clays, there is frequently a transition from coarse angular and subangular blocky to prismatic structures with increasing depth. Sandy soils may have weakly developed angular and subangular structures in the upper subsoils, but sand particles lack cohesion, and such soils are usually devoid of structures (i.e. they are apedal) in the lower subsoil. In addition to drainage channels, soil structure provides air spaces, or pores within the aggregates or peds. These provide the air and water necessary to sustain plant roots.

9.3.4 *Soil colour*

Field observations of colour can be a clue to soil composition. A black or grey-brown soil is likely to have a high humus content. Predominant yellow or red-brown subsoil colours are due to the presence of iron oxides. A white soil may contain abundant silica, aluminium hydroxide, gypsum or calcium carbonate. The colour of subsoil horizons is an important indicator of the drainage status of the soil, and charts (Munsell Color Co. 1992) provide standard examples of the normal range of soil colours. Well-drained soils tend to have uniform brown, yellow-brown or red-brown soil colours. Colour is often inherited from the parent material (e.g. red-brown colours are associated with Triassic lithologies). In poorly drained soils the drainage channels and pore spaces are saturated and air is largely absent. Under these anaerobic conditions, iron compounds are reduced from the ferric (Fe^{3+}) to the ferrous (Fe^{2+}) state. The ferrous compounds are characterised by ochreous and blue-grey colours. Occasional waterlogging gives soils a mottled appearance, whilst more permanent waterlogging at greater depths leads to predominantly grey soil colours. These colours are known as gley morphology, and are indicative of impeded soil drainage. This feature is present in many British soils and it occurs at a range of depths. In general terms, the greater the depth at which gleying occurs, the better the drainage status and quality of the soil.

9.3.5 *Soil fertility*

This is a vast topic and the reader is referred to Brady & Weil (1999) and Cresser *et al.* (1993) for more details. Two major soil chemistry problems that are of importance in an EIA are low soil fertility and toxicity, both of which will lead to poor plant growth. **Low soil fertility** is due either to low levels of nutrients (e.g. nitrogen, phosphorus, potassium and magnesium) in the soil, or their being made unavailable for plant uptake in some way. **Soil toxicity** is caused by high levels of toxic elements or compounds being present in the soil. It can be a significant limiting factor if levels permitted by the Interdepartmental Committee for the Reclamation of Contaminated Land (DoE 1987) are exceeded. Some elements, which are essential *micronutrients* for plant growth, can be toxic at high concentrations (e.g. copper).

High levels of plant *macronutrients*, especially nitrogen and phosphorus, stimulate plant growth. However, the plant communities of *semi-natural habitats*, such as heathlands and 'unimproved' grasslands, are adapted to low nutrient levels – and can be degraded by soil *eutrophication* which favours species such as vigorous grasses at the expense of *ericoids* and *forbs*.

Soil pH *per se* rarely affects plant growth, but it strongly influences the availability of plant nutrients. Aluminium and nearly all of the *heavy metals* are much more available for plant uptake and entry to the *food chain* in acid soils than in neutral or alkaline soils.

9.3.6 *Land evaluation*

The pedological classification of soils considered above is based mainly on the nature of soil parent materials, modified by natural soil-forming processes. Land evaluation methodologies for the assessment of land quality (e.g. for agriculture or forestry) concentrate on the physical properties which cannot be altered by land management. For land use planning purposes, it is necessary to determine the relative productive value of different areas of land.

Land quality (or capability) classification systems are based on the severity of climatic, topographic and soil limitations to the agricultural or silvicultural use of the land. Climatic limitations have an overriding downgrading effect (irrespective of soil conditions) in areas which are cold and wet for most of the year (i.e. hills and uplands). In the more favourable locations (i.e. most of lowland Britain), soil drainage and liability to drought are the most common limiting factors. These are determined by both soil and climatic influences. The severity of a soil wetness limitation is determined by interactions between soil texture and structure, and the length of the period when soils are at *field capacity* in the winter. The severity of a soil drought limitation is determined by interactions between soil texture and structure, and summer *soil moisture deficits* (SMDs) in relation to selected crops. Land quality is also determined by soil depth and stone content. Shallow and stony soils are downgraded, as are sandy soils on sloping ground which are prone to water erosion, and a relatively narrow range of fine sandy and silty soils which are susceptible to wind erosion. Topographic limitations include steep slopes that preclude mechanised farm operations, and flood risk on river *floodplains* (see §10.2.7).

Table 9.2 Agricultural Land Classification (MAFF 1988)

Grade or Land Use	Quality of Land, Severity of Limitation and Cropping Capability	% Agricultural Land*
1	Excellent quality. No limitations. Very wide range of horticultural and agricultural crops.	2.3
2	Very good quality. Minor limitations. Wide range of horticultural and agricultural crops.	16.9
3a	Good quality. Moderate limitations. Wide range of agricultural crops.	19.3
3b	Moderate quality. Moderately severe limitations. Mainly cereals and grass.	35.4
4	Poor quality. Severe limitations. Mainly grass.	15.0
5	Very poor quality. Very severe limitations. Mainly semi-natural grazing and grass pasture.	11.1
Non-agricultural	Land with largely undisturbed natural soils. Includes woodland, parkland, golf courses, etc.	
Urban	Land largely devoid of soil and covered with houses and industrial development.	

* Estimates derived from MAFF News Release (277/96) dated 1 August 1996.

The quality of agricultural land in England and Wales is assessed according to a system devised by MAFF (1988), and known as the *Agricultural Land Classification* (ALC). This is the system utilised for land use planning and development control decisions, and the ALC has 5 grades (Table 9.2). Grade 1 is the best quality land which permits flexible land management and crop production and supports the full range of horticultural and arable crops. Grade 5 is so limited by severe climate, flood risk or steep slopes as to be capable of supporting only grass pasture, semi-natural vegetation and extensive grazing. Grade 3 is subject to moderate limitations, and is generally associated with cereal and grass crops. It can be subdivided into an upper category (Subgrade 3a) and a lower category (Subgrade 3b).

A reconnaissance survey of England and Wales has been carried out, based on limited field observations, and a series of provisional 1:63,360 scale maps provide a generalised indication of the distribution of land quality for use in strategic planning. The maps are not suitable for use in evaluating individual sites where development is proposed, and they have been withdrawn, to be replaced by regional maps at a more suitable scale of 1:250 k (available from MAFF). If significant soil impacts are anticipated, a detailed field survey and ALC map at a larger scale are necessary to obtain a definitive grade. In Scotland, land quality is evaluated using a similar methodology devised by the Macaulay Land Use Research Institute (MLURI 1991) and known as the *Land Use Capability Classification* (LUCC).

9.4 Legislative background and interest groups

9.4.1 Geology

The DoE (1989) suggest that an EIA of impacts on geology should consider the local geomorphology, and the "loss of, and damage to, geological, palaeontological and physiographic features". More recently published advice for planners and developers is available (DETR 1999a,b,c). In the UK, sites of geological significance (i.e. sites important for their fossils, minerals or other geological interest) are identified in the *Geological Conservation Review* (GCR) as **Earth Heritage Sites**. Most of these are protected by their designation as SSSIs, a number of which are selected largely on the basis of geological features. The selection criteria are fully described in the introduction to the GCR (Ellis *et al.* 1996).

There is also a national network of Regionally Important Geological/Geomorphological Sites (RIGs), but these do not currently enjoy statutory protection. Limestone pavements can be given special protection by *Limestone Pavement Orders* issued by LAs under the *Wildlife & Countryside Act 1981 & (Amendment) Act 1985*. The statutory consultee for a project likely to affect an Earth Heritage Site is the relevant NCCA (Appendix B). Other potential consultees or interest groups include the LA, British Geological Survey (BGS) and the local geological society.

9.4.2 Soil protection and restoration

Thompson (1990) discussed progress towards the legislative protection of soils, dating from the *European Soil Charter*, and adopted by the Council of Europe. In practice soils are protected only when they form part of a habitat or land use which is valued by the planning system for other reasons. For example, the conservation of soils in England, Scotland and Wales is achieved through policies for the protection of agricultural land from urban development, and for the restoration of mineral sites to agriculture, forestry and other soil-based land uses. In England and Wales, the policies are contained in *Planning Policy Guidance Note 7* (PPG7), issued by the DoE (1997), and *Mineral Planning Guidance Note 7* (MPG7), issued by the DoE (1996b).

As a result of the reconnaissance survey and subsequent detailed surveys, MAFF takes the view that the best and most versatile land (defined as Grades 1, 2 and 3a of the ALC) constitutes almost 40% of the agricultural land in England and Wales (Table 9.2). According to the principles of *sustainable development*, PPG7 advises that considerable weight should be given to protecting this land from development, because it is a national resource for future generations. On the other hand, outside of the hills and uplands (where lower quality land may still be important), less weight is normally given to the loss of moderate or poor quality land. Because of the national interest in protecting the best and most versatile agricultural land, LPAs, county and mineral planning authorities (MPAs) are required to consult the Farming and Rural Conservation Agency (FRCA) about any development that does not accord with **Development plans** and which involves, or is likely to lead to, the irreversible loss of more than 20 ha of agricultural land of Grades 1, 2 and 3a. The 20 ha threshold is currently under review (DETR 2000a).

For mineral sites there is the additional need to restore the land to equivalent or near equivalent quality. However, planning permissions for non-mineral developments almost always lead to the loss of the soil resource. Additional guidance on the protection of land and soil resources is given in MPG7, which states that land restoration schemes should be based upon the careful investigation of the site before it is worked for minerals, to identify the soil resources available for use in land restoration. Such pre-application site investigations are required to provide adequate information on the volumes and physical characteristics of the topsoil, subsoil, and soil-forming materials, together with a description of the original landform and drainage. It is also necessary to draw up a programme for the working and restoration of the site to include soil stripping and storage, mineral extraction, back-filling operations, soil replacement and aftercare. This represents the basis for consultations between the mineral operator and the statutory authorities over development control and land restoration conditions. Further guidance on best practice criteria is available from a number of sources, including DoE (1996c,d), DETR (1999d) and MAFF (1993).

In Scotland, policies for the protection of agricultural resources from development are contained in National Planning Policy Guidelines (NPPGs). NPPG4 *Land for Mineral Working* (SO 1994a) and NPPG15 *Rural Development* (SO 1999) are supported by Circular 25/1994 (SO 1994b). These state the need to protect prime quality land against irreversible development. Prime quality land is generally defined as Classes 1, 2 and 3.1 (the upper part of Class 3). In Northern Ireland, there is no system for the classification of agricultural land, and no comprehensive inventory of the land resource.

9.4.3 Contaminated land

Prior to 1990 there were no specific regulations related to the management of contaminated land in the UK. Authorities were restricted to using statutes and policies in related areas. These included: the *Public Health Act 1936*; the *Town and Country Planning Act 1971* and subsequent updates; the *Control of Pollution Act 1974*; and the *Derelict Land Act 1982*. The *Environmental Protection Act 1990* introduced a number of provisions that affected the management of contaminated land. Sections 78A to 78YC presented a statutory framework for dealing with waste materials and contaminated land. Section 143 of the Act introduced the requirement for local authorities to develop public registers of sites known to be contaminated. However, this section of the Act was not enacted after representation from landowners and the insurance industries, who were concerned about its effects on land values. Harris & Denner (1997) provide a good overview of the development of UK policy in this area.

The Environment Act 1995 provided a revision of the statutory framework for the assessment, management and remediation of contaminated land through Section 57, which comprised a total of 26 sections. These replaced and updated some of the general definitions of statutory nuisance given in the 1990 Environmental Protection Act, and S161 of the *Water Resources Act 1991*, and provided a procedural framework for those managing contaminated land. A detailed account of the legislation concerning contaminated land is given by Syms (1997).

The 1990 and 1995 Acts introduced the Source–Pathway–Target concept to the management of contaminated land. The use of risk assessment (Chapter 14) to assess where contamination has significant potential for causing harm has considerably assisted in the management and remediation of contaminated land (Cairney

1995). The Acts also identified those responsible for the remediation of contaminated land. As with other pollution, it is the responsibility of the polluter to decontaminate polluted sites where these are causing significant harm. However, with contaminated lands it is not always easy to ascertain the polluter, and often where the polluter can be identified the incident occurred historically and the company is no longer trading. In these circumstances, the present site owner or funding agency becomes liable for clean-up costs.

The definition for contaminated land contained in the 1990 Act was introduced under the 1995 Act. *Contaminated land is land which, because of the substances contained within it: is causing significant harm, or has the potential to do so; or affects* **controlled waters** *or has the potential to do so.* Recent case law has identified that when removed from site, contaminated soils fall under the definition of controlled waste. The Act introduced a Duty of Care for those producing, transporting and disposing of wastes. This requires the use of registered carriers, and licensed disposers, and for the producer to check and document the disposal process. Landfill tax has to be paid where contaminated soils are removed to landfill, and where the site activities fall outside the definition of 'site investigations'. The Environment Agency (EA) have powers under Section 161 of the *Water Resources Act 1991* to clean up pollution and subsequently charge the responsible party. Section 85 of the Act makes it an offence to knowingly permit any pollution of controlled waters.

There is also a range of governmental advice on contaminated land. This identifies the standards to be used, and the potential for contamination to be present. The most important are the ICRCL documents which list the trigger concentrations for specified contaminants and for specific end-uses. Unfortunately these are not comprehensive, and in many cases Action Levels or Intervention Levels developed by the Dutch or American Governments are used in the UK. In addition, a range of industry guides are published by HMIP (now the EA), which identify contaminants potentially present on sites which were subject to specified uses, such as in the chemical industry, and the iron and steel industry. The EA also publish a range of Pollution Prevention Guides, which include advice on site control and remediation. This subject area is covered by PPG23 *Planning and Pollution Control* (DoE 1994a). DETR (2000b) provides a comprehensive account of the UK Government's policy on contaminated land.

9.5 Scoping and baseline studies

9.5.1 Introduction

Both scoping and subsequent investigations will involve a desk study and consultations. At the scoping stage it is necessary to decide if these will suffice, or if a reconnaissance field survey, a detailed field survey, and laboratory analysis of soils are required. Scoping-stage site visits will normally be brief (e.g. to confirm features identified on maps), but some may involve walkover surveys. Such visits are best undertaken with other members of the EIA team, so that interactions between subject areas can be identified. For example, information on geology, geomorphology and soils may also be of relevance to other EIA components such as landscape/ visual, water and ecology (Chapters 6, 10, 11–13). Co-ordination at an early stage is therefore important to achieve an integrated approach, and to avoid duplication of effort while ensuring that key aspects are not omitted. If the use of GIS (Chapter 16) is considered appropriate for the EIA, it may be possible to include geology, geomorphology and soil layers and hence facilitate integration with other layers.

The most important scoping considerations are whether the geological or soil resources within a project's impact area are likely to be significantly affected, and if there are any practical measures which can be undertaken to mitigate anticipated impacts. Where a significant impact on soils is anticipated, it is necessary to carry out an ALC/LUCC survey to determine the grades or classes of land and the areas of best and most versatile or prime quality land. Where it is necessary to conserve the soils for land restoration (i.e. at mineral sites), or where the developer wishes to make beneficial use of this resource on the development itself (e.g. for landscaping), the field survey should also include an assessment of the volumes of topsoil and sub-soil available at the site.

9.5.2 Desk study

The desk study should make good use of existing information on geology, geo-morphology, soils and land quality, associated aspects such as site history and local climate, and contaminated land.

Information on geology and geomorphology

Information on Earth Heritage conservation, including a list of the 42 volume *Geological Conservation Review Series* is available from http://www.jncc.gov.uk/ (see also Ellis *et al.* 1996). Geological maps, published by British Geological Survey (BGS) are available for most of the British Isles. 'Solid' maps show only Pre-Quaternary rocks, and 'drift' maps also show superficial Quaternary deposits which have been laid down principally since the last Ice Age. Lithology has a big influence on soil types through the mineralogical composition and texture of the weathered rock. However, because of the erosion, mixing and redistribution of surface rocks and weathered materials during the Ice Ages, drift maps tend to give the most informative indication of the soil parent materials in a survey area.

BGS paper maps include 1:250 k regional, 1:63.36 k or 1:50 k of most areas, and 1:25 k or 1:10 k of some areas of special interest to geologists or planners. The latter include some Applied Geological Mapping (AGM) studies (e.g. within coalfields), commissioned by DETR (Ellison & Smith 1997). Digital geological maps are becoming available at the 1:10 k, 1:50 k and 1:250 k scales. BGS operates an online *Geoscience Data Index* (GDI), which is a spatial index of BGS data holdings held in an ArcView GIS (see Chapter 16). It provides the facility to zoom in on areas or place names, and gives the costs of supplying more specific information. BGS is also developing an Address Linked Geological Inventory (ALGI) which is intended to provide site-specific information on basic geology and aspects such as underground mining.

If examined in conjunction with geological and soil maps, OS topographical maps (e.g. at 1:50 k, 1:25 k and 1:10 k) will give a general idea of geomorphology. If a GIS is being used, digitised OS maps (see Table 16.1) should be useful, and it may be possible to produce Digital Terrain Models (DTMs), which may also be able to make use of remotely sensed imagery (Chapter 15). DTMs in GIS are explained in §16.2.3, and their use in geomorphology is discussed in Cooke & Doornkamp (1990). The EA holds a Geomorphology Core Survey database. Further information on sources of geological (and related) information is provided in Ellison & Smith (1997).

Information on soils and land quality

Published soil and ALC/LUCC maps provide an initial understanding of the soil types and land quality likely to be found at the sites. Soil maps of England and Wales are available from the Soil Survey and Land Research Centre (SSLRC). They include the National Soil Map in 6 regional sheets at 1:250 k, and maps at 1:50 k or 1:25 k (with reports of some areas). Soil maps of Scotland (from MLURI) include 1:250 k soil and land capability for agriculture (7 sheets); and soil maps at 1:50 k or 1:63.36 k of most areas, and at 1:25 k of some areas. In 1997 a series of soil maps covering the whole of Northern Ireland were published, and these are available from Queens University, Belfast. There are many soil memoirs (describing the soils in specific geographical areas) and monographs (describing relevant soil properties) published by MLURI and SSLRC (and listed on their websites).

The Centre for Ecology and Hydrology (CEH) holds two analytical chemistry databases. These are not computerised, but data are available from CEH Merlwood (mainly on inorganic nutrients in soils, waters and vegetation) and CEH Monkswood (mainly on pesticide and toxic chemical residues in soils, water, vegetation, and animal tissues).

Information on climate and site history

When a detailed survey of land quality is required in England and Wales, the relevant climatic information is derived from the dataset specifically produced by the Meteorological Office (MO 1989) for this purpose. The figures for each of the relevant climatic variables are available for each 5 km national OS grid intersection, and these are interpolated for the exact location and altitude of the study area. It is almost certain that, for projects which are likely to have a significant impact on soils and land quality, local climatic conditions will also have to be assessed in the field for **microclimate** and exposure.

The history of the site and estimated impact area should be investigated to identify activities or land uses that might have contaminated the land. Sources of historical information are given in Appendix C.

Information on contaminated land

If the site is contaminated, then certain additional procedures will be required. The methodology most commonly used has been developed by a number of environmental consultancies from the American ASTM method. There is no standard UK methodology, and the methods used by each consultancy differ slightly. The type and extent of any contamination that may be present on a site will depend upon the previous uses that the site has been subject to, and the management practices used to control and maintain those activities. In addition, activities on adjacent sites may also have resulted in pollution of the subsurface that may then migrate onto the subject site. It is important, therefore, to ascertain the activities which occurred on site and on adjacent sites, the management practices that occurred, and the type of chemicals used in the initial phase.

Interviews with site staff, where available, can help to determine past and present site activities that may have caused contamination of the soil or groundwater at the facility. Such staff could include the site manager, site agent, maintenance manager and the caretaker. Regulatory authorities maintain records that are very important in assessments of contaminated land. Information such as the presence of underground and above-ground storage tanks, and electrical equipment can be gathered. In addition, data on known past pollution incidents is often available. Data such as **aquifer** location, type and vulnerability should be collected (see Chapter 10). Authorities to be contacted should include the EA, the LA and the local Petroleum Officer. There is much data on the type of materials and chemicals that were used in a wide range of commercial activities. This can be gathered from published data (HMSO/TSO and governmental guidance documents, e.g. the DoE Industry Profiles), and the publications of professional bodies such as the Society of Chemical Industry. The Construction Industry Research and Information Association (CIRIA) have published 12 volumes covering all aspects of *Remedial treatment for contaminated land*. The first of these (CIRIA 1998) is a useful guide and introduction to all aspects of this topic.

9.5.3 Fieldwork

If the desk study indicates that more detailed data are required, then fieldwork will be initiated.

Geological and geomorphological surveys

The locations of Earth Heritage sites and RIGs will have been identified in the desk study, but there may be other rock exposures that are worth investigation and evaluation (e.g. for their fossil content). Where a significant geological impact is anticipated (e.g. for opencast mineral extraction), a more detailed assessment is likely to be required than can be made from existing information alone. This will usually involve field survey, e.g. sampling in **wells** to identify the extent of the mineral resource, and to understand the local hydrogeology (see Chapter 10). A topographic survey can also be carried out, for example to measure the gradients of slopes, and delineate flood risk areas.

Soil surveys

The complex geology of the British Isles and the redistribution of soil parent materials during the Ice Ages has made our soils very variable. This is a major problem for soil surveyors. Field observations are made by using a soil auger to take samples from successive horizons within a soil profile to a depth of 1.2 m. As the soils are observed only where the samples are taken, the sample network and density have to be designed to be representative of the variation in soil types within the survey area. Webster (1977) gives a very detailed mathematical account of this topic.

Generalised soil surveys of large areas are carried out by the physiographic (or free survey) technique which ensures that samples are representative of the range of geological parent materials and topography within the survey area. The results are shown on maps at intermediate scales (e.g. 1:50 k). For detailed surveys of specific development sites on undisturbed and uncontaminated agricultural land in England

and Wales, FRCA and most practitioners favour a grid sampling pattern (see Appendix F) and a minimum density of one sample per hectare, with supplementary samples as necessary to accurately delineate soil boundaries. Soil pits are dug to observe the subsoil structures and extent of crop rooting in each of the main soil types. Topsoil and subsoil resource maps are derived from the information collected during the ALC survey. These resource maps indicate the areas, thicknesses and volumes of the topsoils and subsoils. The resulting ALC and soil resource maps are usually shown at scales of 1:25 k to 1:10 k, and are capable of reasonably precise interpretation. It is important to note, however, that land classification is a field survey technique and not an exact science.

During the field survey, some soil properties like soil depth are easily measured, and other properties are either estimated by eye, or assessed using a standard technique, depending on the degree of precision required. For example, stone content can be estimated by eye or measured using a sieve and weighing scales. Some idea of texture can be gained in the field by observing it with a lens and by feeling it between the fingers. This requires much experience if an exact identification of the soil texture is needed, and occasional calibration with standard samples. However, even an inexperienced person should be able to classify the soil into the broad categories of clay, silt, sand or loam. Portable field apparatus can be used to obtain estimates of soil strength, pH and mineral status. Small hand-held penetrometers consist of a metal probe which is pushed into the soil until it reaches a certain mark. The probe is spring loaded, and the pressure required to push it into the soil is read off on a scale. Soil test kits produced for horticultural or agricultural purposes may also be used, although they require some practice before reliable results can be obtained.

Surveys of contaminated sites

Contaminated sites will require some additional measures (see Syms 1997). When deciding the layout and sampling, the assessor must be aware of what the data obtained will be used for. Assessments undertaken as part of an EIA baseline study may be supplemented by further investigations to determine appropriate remedial actions for the mitigation of impacts. *Planning Policy Guidance Note 23, Planning and Pollution Control* (DoE 1994a), clearly differentiates between the information needed at planning and subsequently for licensing. However, sufficient work should be undertaken in a single survey to answer subsequent requests of the EA, LAs and other regulatory bodies. Reconnaissance will be needed to identify areas of potential concern caused by current and obvious past activities. Such areas could include: the presence of storage tanks (underground storage tanks are often identified only by the presence of vent pipes or manholes covering filler pipes); obvious visible major classes of potentially friable building materials (lagging, spray-on fireproofing or building cladding); potentially PCB-containing electrical equipment; obvious made-up ground; blighted vegetation; and surface drainage systems and soakaways.

A commonly used method of sampling contaminated land is the grid, as recommended by DD175, and used most commonly at 50 m or 100 m centres. The number of sampling positions is increased in the vicinity of areas of potential concern. The layout may be modified due to the location of services, buildings and areas of particular environmental sensitivity. It is not uncommon for a variety of sampling methodologies to be utilised on a single investigation. Trial pits are cheapest and excavate the largest volume of material for sampling, but are restricted to less than

5 m depth, and are often used for gathering shallow soil samples for metal analysis. Window samplers are useful in areas where access is difficult, and installations for subsequent gas and groundwater sampling can be put in. Boreholes are the most expensive but are also the most permanent and can be installed considerably deeper. The types of pollutants present may affect where, when and how samples are taken. Pollutants may be in solid, liquid or gaseous form. The combustibility of the soil may also be an issue, and can be measured by loss on ignition tests or calorific value tests. Cross-contamination will affect both the design of the sampling installations and the sampling methodology used.

9.5.4 Laboratory work

The ALC/LUCC assessments set clear and quantified cut-offs between the grades and classes of land for the selected climatic, topographic and soil variables. In cases where the field observations indicate a marginal classification, it is necessary to analyse samples in the laboratory for greater precision and the definitive grading of land quality. In practice, this applies most frequently to the analysis of soil texture and stone content. It may also be necessary to determine the relationships between moisture contents and the plastic limits of topsoil and subsoil samples in the preparation of soil handling strategies for land restoration schemes.

Laboratory analysis can be expensive, and as a result it is usually undertaken on samples of soils for specific purposes only. Ball (1986) covers much of this material in detail. Soil samples may be analysed at a number of stages in the EIA process, during baseline studies for land evaluation, and as a guide to possible mitigation measures, including the treatment of contamination. Soils may also be analysed during project construction and operation for monitoring and mitigation purposes. In practice, a very wide range of analyses are selectively undertaken, but only the more common analyses are described in this account.

During baseline studies and the evaluation of undisturbed agricultural land, **topsoil texture** is often analysed in the laboratory for a definitive ALC grading. Basically the methods differentiate between the mineral fractions of soils on the basis of particle size. The usual method involves sedimentation of mineral particles in a water column. The disadvantages are that it takes a long time (several days), and at current (2001) prices each determination will cost about £25. On disturbed land that is largely devoid of natural soils, it is a matter of identifying suitable soil-forming materials for land restoration. *Soil-forming materials – their use in land reclamation* (DETR 1999d) is a useful reference. The British Standards Institute (BSI 1994) issued a specification for topsoil (BS 3882), which also refers to a number of qualifying threshold levels in soil texture and other variables. These include pH, organic matter content, **electrical conductivity**, available phosphorus, potassium, magnesium and total nitrogen. This test costs about £45. Allen *et al.* (1986) and Rowell (1994) provide detailed methods of soil analytical chemistry techniques.

Mineral extraction may be preceded by soil stripping and storage, and followed by the reinstatement of the soils. Conditions attached to planning consents by the Mineral Planning Authorities specify the **moisture content** at which soils may be moved. This is related to the **plastic limit** of the soils, and is designed to avoid damage to soil structures during soil handling. It may be determined in the field by hand, but the moisture content may have to be determined with more precision in the laboratory. Most workers use gravimetric analysis, which involves taking a sample of soil from the field and weighing it before and after heating in an oven. For soil surveys on a single site at a given time, the gravimetric method yields good results, giving information on where the dampest parts of the site are, and where soils are too wet to be moved. For other types of work where monitoring over a time period is required, more sophisticated machinery (e.g. neutron probes or time domain reflectometry) can be used (Brady & Weil 1999).

When mineral sites have been restored, a period of aftercare is instituted to recreate favourable soil conditions for a range of beneficial uses, including agriculture, forestry and wildlife, and amenity planting. As a part of this rehabilitation process, samples of soil may be taken to determine bulk density and plant nutrient status. **Bulk density** is the ratio of dry weight to total volume. It can be used indirectly to assess differences in soil structure and porosity caused by soil handling, for example. It is usually measured directly with the use of a volumetric corer. Essentially a pipe is pushed into the ground to extract a core of soil on which measurements can be made. In EIA this is a very useful measure if soil compaction is likely to be a problem. The results of these tests would be used to guide subsequent remedial cultivations and fertiliser applications.

As a result of the efforts of the DETR and bodies like the ICRCL there has been a move towards the standardisation of methods and agreement on **threshold levels of contamination**. Investigations of contaminated sites are often hindered by an incomplete understanding of the polluting activities that have taken place. There are so many potential contaminants that it would be excessively expensive to test for every possibility. Fortunately, certain suites of contaminants are associated with the main industrial processes, mining operations, and waste disposal. Tests for the more complex organic compounds are very expensive, given the need for a representative number of samples and the variable nature of most substrates.

Although each study will be unique, there are a number of screening suites that laboratories offer which can be used. These are often referred to as 'ICRCL suites' since they contain most of the components listed in the ICRCL lists. A typical 'ICRCL Screen test' may include Cu, Zn, Cd, Pb, Ni, Cr, Hg, As, total cyanide, total phenol, pH, sulphate (total), toluene extractable material (TEM) and total petroleum hydrocarbons (TPH). Other commonly undertaken tests include sulphide, ammoniacal nitrogen, boron, selenium, chemical oxygen demand (see Table 10.6, p. 225), PAH (poly-aromatic hydrocarbons – see Wilson & Jones 1993) and **biochemical oxygen demand**. Specific tests which could be used will depend upon the history of the site, and may include contaminants such as cyanide, thiocyanate, **pesticides**, PCB, chloride, mineral oils, elemental sulphur, organic acids, and the components of landfill gas.

9.6 Impact prediction

9.6.1 *Geological and geomorphological impacts*

Potential impacts on Earth Heritage sites, and other sites of conservation interest, are likely to be direct and hence relatively easy to predict. Quarrying and other forms of mining often have considerable geological impacts because they remove the geological resource and may also affect the local hydrogeological balance. Apart from possible benefits of new rock exposures (e.g. with fossil beds), this may be considered an entirely negative impact, particularly in view of (a) the finite nature of mineral resources, and (b) competition with other land uses such as agriculture or nature conservation. However, an EIA takes place within a statutory context and the government considers mineral extraction to be a valid component of sustainable development. *Mineral Planning Guidance Note 1* (MPG1) (DoE 1996a) summarises the government position. It refers to the importance of combining economic growth with care for the environment in order to attain sustainable development, and the essential contribution made by minerals to the nation's prosperity and quality of life. MPG1 also states that all of the costs and benefits of mineral extraction need to be considered, and adverse environmental impacts mitigated or controlled during the process of extraction. It is also necessary to restore worked-out mineral sites to a beneficial use, to avoid dereliction. However, this often first involves using the site for waste disposal, and, if not carefully managed, this landfill phase can result in groundwater pollution by **leachates**.

In addition to geological and geomorphological impacts, mineral extraction usually introduces a number of secondary but significant impacts on the local environment. These include noise, dust and traffic impacts, together with landscape amenity and ecological impacts in some cases. These are addressed in other chapters, but useful advice on their assessment is available in *The Environmental Effects of Surface Mineral Workings* (DoE 1991). Landfill operations may follow in the wake of worked-out opencast mineral extraction, introducing additional potential waste disposal impacts and the need to monitor operational sites for pollution.

Seismic risk is not usually a great problem in the UK, although there are occasional small earthquakes. In some parts of Europe (e.g. Italy) this may be a more serious problem. **Volcanic risk** in the UK is negligible, but in some parts of the world a section of the EIA should be devoted to this topic. Bell (1999) includes chapters on both seismic and volcanic risks. **Subsidence and slope stability** are factors which should be considered, however. Subsidence is caused by underground mining and is usually associated with traditional coalfield areas, where the subsidence extends for considerable distances around collieries (Bell 1998). It can also occur as a result of the underground extraction of salt, and in limestone areas where natural chemical solution has occurred. The risks associated with the development of land which has been disturbed by previous mining activity are addressed in PPG14, *Development on Unstable Land* (DoE 1990). Mineral Planning Guidance Note 12, *Treatment of Mine Openings and Availability of Information on Mined Ground* (DoE 1994b), gives advice with respect to previously mined sites. There are likely to be relatively few proposals for new underground workings which would create a subsidence impact, but avoiding areas of actual or potential subsidence which are a risk to the development project itself is an important part of the EIA process. Natural slope stability is a more widespread problem, and the objective of EIA is to avoid the construction of new developments in unstable areas, particularly when the development might make the area even less stable. Information on subsidence and slope stability is available from the BGS, and the Coal Authority at Burton upon Trent. The DETR (1999a,b,c) has published the findings of research projects on environmental geology in land use planning. The objective is to avoid negative interactions between development and geology. For geology, there is no overall assessment methodology, and the significance of impacts is determined in consultation with the relevant statutory authorities including the EA, English Nature and the Coal Authority.

DoT (1993) considers the impacts of **road developments** on geology and geomorphology. Such schemes can have a direct impact on geology. For example, in a mining area, they can increase the rate of collapse of underground tunnels. Indirect effects may be felt through alterations to hydrogeology (e.g. diverting streams or affecting the recharge of aquifers). The major impacts of such developments are, however, likely to be damage to geological exposures, fossil beds, stratigraphy and geomorphological systems (Anderson 1994). Not all of the impacts will be negative, and it should be remembered that about one-third of geological SSSIs have been created as a result of human activity (DoT 1993). Although road developments can create new exposures that may be of great interest to geologists, care is needed in the design of exposure angle and shape so that rock sequences can be best observed (Anderson 1994). It is more difficult to preserve geomorphological features, and the best of these (e.g. stream systems, glacial forms) should be avoided.

9.6.2 Impacts on soils

When a baseline soil survey has been carried out, the various sources of information (e.g. published and new field survey) are compiled, analysed and interpreted. The EIA has to predict the **magnitude** and **significance** of the main impacts on soil, both **temporary** and **permanent**. The DoE (1989) suggest that the following effects of a development should be taken into account: physical (e.g. changes in topography, stability and soil erosion); chemical (emissions and deposits on the soil); and land use/resource changes. The significance of these impacts is determined by the ALC land evaluation methodology in England and Wales, and the relevant legislative guidelines and standards set out in PPG7 (DoE 1997) and MPG7 (DoE 1996b). In Scotland the LUCC methodology is used, and the guidelines are to be found in NPPGs 4 and 15 (SO 1994a,b, 1999).

The UK Strategy for Sustainable Development (DoE 1994c) refers to the need to maintain the main soil functions, which include food and timber production, the support of biodiversity, a component of the hydrological and carbon cycles and a buffer and filter for pollutants. Further to this strategy, the DoE (1996e) include the following as preliminary indicators of sustainable development: the amount of land covered by urban development; critical loads for sulphur deposition and acidity in soils; agricultural productivity; soil quality (organic matter content, acidity, available nutrients and heavy metal content); mineral workings covered by restoration and aftercare conditions; and restored mineral workings. Urban development and land restoration are the more immediate concerns. The other indicators are subject to longer term and less obvious change, but these impacts also threaten the essential functions of the soil, and they must be given due consideration in the EIA process. It is unfortunate, but there is no recognised evaluation methodology beyond the ALC and LUCC systems. Although there is much information available concerning the engineering properties of soils, and on the description of soils for engineering purposes (e.g. West 1991), EIAs are more concerned with the impacts of a development on the productive potential of the soil for agriculture and forestry. These impacts are likely to be the loss of land and soil, erosion, damage to soil structure, and pollution.

Almost all developments are likely to lead to some **soil erosion** unless suitable mitigation procedures are adopted. There are two major types of erosion, by water and by wind (Hudson 1981, Bell 1999). The factors that most influence erosion by water are mean annual rainfall, storm frequency and intensity, slope, the soils infiltration capacity and vegetation cover (see §10.2.4). Rain and overland flow cause some natural erosion in most environments, but this is insignificant compared with **accelerated erosion** resulting from human activities such as the disturbance or removal of vegetation, e.g. for agriculture, mineral extraction or development (Cooke & Doornkamp 1990). Only dry soil is subject to wind erosion and so rainfall must be fairly low for it to occur (< 250–300 mm). Steady prevailing winds are generally found on large fairly level land masses, and it is these that are most susceptible to wind erosion (e.g. East Anglia).

Urbanisation can cause soil erosion by increasing **runoff** and concentrating surface water. Poor quality land restoration following mineral extraction can create unstable and unvegetated surfaces which are subject to soil erosion. Soil erosion by wind and water is considered by many to be a serious threat to the soil resource, but most soil erosion occurs as a result of agricultural land management practices which are not subject to planning controls, and which are beyond the scope of the EIA process. In the UK, erosion by water is most likely. In a new development the three factors most likely to cause erosion are the removal of vegetation, increased runoff from impermeable surfaces and creation of unstable slopes. When soil erosion by water occurs, damage will often not be restricted to the terrestrial environment. Soil removed will affect nearby water courses, causing an increase in **turbidity** and siltation. Moreover, soil erosion will also lead to an increase in soil nutrient levels in water courses. It is not uncommon for the levels of certain nutrients (particularly nitrate) to exceed legal limits in steams and rivers as a result (see Chapters 10 and 12).

Damage to soil structure can occur during soil stripping, storage and reinstatement operations at land restoration sites. This is due to the use of inappropriate methods and machinery and carrying out soil movements when the soil is too wet. Vehicles driving over soil will compact it, destroying soil structure and increasing bulk density. Topsoils also tend to become mixed with the less fertile subsoils, when they should be stripped and stored separately to facilitate the restoration of a natural soil profile upon reinstatement. Soils that have been damaged in this way lack the natural drainage channels and porosity which normally absorb precipitation and transfer it to groundwater reserves. As a result, infiltration is reduced, runoff is increased, and erosion is more likely to occur. Furthermore, soil compaction inhibits root penetration. Damaged soils also have a reduced capacity to retain moisture and to make it available to plant roots. As a result, they are prone to severe limitations by drought. A detailed account of the effects of wheel traffic on soils and the plants growing in them is provided by Voorhees (1992). In response to this threat to sustainability, the DoE has published *Guidance on Good Practice for the Reclamation of Mineral Workings to Agriculture* (1996d). Proposals for the working and restoration of mineral sites now have to conform with this guidance in order to satisfy the requirements of the MPA and FRCA (§9.4.2) concerning the protection of soil structure.

There are two types of situation where **soil pollution** is an important factor in an EIA. In the first, the site is already contaminated, and a clean-up operation is required prior to development. This will be considered in §9.7. In the second, we are concerned with predicting pollution that will happen during or after the development. Most developments pose the threat of some pollution of the local soils during the **construction phase** (e.g. oil from vehicles, dust from the building materials used). Major developments like ore smelting plants, refineries, chemical works and power stations introduce pollution to local soils during the construction phases. Soils beside new roads will receive heavy metals from exhaust fumes, motor oils and salt (from winter de-icing).

During the **operational phase**, air-borne emissions (Chapter 8) will begin to impact on the local soils. *Acidic deposition/precipitation* ('acid rain') has major effects on soil, increasing available soil aluminium and heavy metal levels, and increasing leaching of soil nutrients. Aluminium toxicity is almost certainly a major factor in the dieback that has been observed in the forests of Northern Europe and North America in the last 25 years (Huettl 1993). The problem in predicting the significance of these impacts (like sulphur dioxide from power stations, and nitrogen oxides from vehicle exhausts) is that their effects on soils are not directly visible. There are also many sources which create a cumulative and dispersed impact which is felt some distance away, and there are no recognised baseline standards. The Ecology Information Centre (EIC) at CEH Monkswood hold a **critical loads database and maps** containing estimates of the vulnerability of land to atmospheric pollution (especially acid deposition) in relation to receptor soils, geology, freshwaters and vegetation. Data correspond to 1 km national grid squares and are digital/GIS compatible. The Royal Commission on Environmental Pollution (1996) has published a comprehensive account of the sustainable use of soil, and this is a recommended source of additional information.

All of the above impacts will have serious effects on soils, but the soil types outlined in §9.3.2 will be affected to different extents by each. It is notable that podzolic soils, which occur most frequently in areas of nature conservation interest, are already acidic and have a low buffering capacity (the greater the buffering capacity, the more acid rain will be needed to change the pH of a soil). As such, these soils are the most vulnerable to acid precipitation. Podzols also suffer the greatest disruption by disturbance and soil mixing because they have distinctive layers. Gleys are also vulnerable to effects on soil hydrology. Brown earths are less susceptible to any of the above impacts. Peats are extremely sensitive soils, especially to erosion and compaction. Susceptibility to erosion is also a feature of many sandy soils.

9.7 Mitigation

Under the present legislation and planning guidance, it is not possible to mitigate the loss of land and soils arising from most types of non-mineral development. A small proportion of displaced soils may be retained for landscaping purposes, but most are lost to any productive use. Accordingly, it is necessary to ensure that the smallest area of high-quality land is lost, consistent with the sustainable functioning of the proposed development. This is a particularly important objective when land is allocated in development plans, and where there are a number of competing interests promoting alternative site locations. Site boundaries and linear developments like roads can also be adjusted to avoid better quality land and well-structured farming areas. Furthermore, development schemes can be designed to locate hard development (e.g. structures, infrastructure, and constructed surfaces) on poorer quality land, with soft development (i.e. where the soil profile remains largely undisturbed) like public open spaces on the better quality land. This is particularly effective where the soft uses are placed adjacent to agricultural land, which is not affected by urban problems like trespass and vandalism, and which remains commercially viable. Under such circumstances the soft uses have the potential to be converted back into productive use, if necessary. Similarly, they can act as **buffer zones** if placed adjacent to nature conservation sites.

The identification and conservation of soil resources and reinstatement of soil profiles described in §9.3.2 has the potential to effectively mitigate the impacts of mineral extraction. This, however, applies only to well-managed operations that conform with the best practice guidelines published by DoE (1996a–d) and MAFF (200a,b). Essentially, this is a matter of the separate handling of topsoil, subsoil and soil-forming materials using specialist machinery under appropriate weather and soil moisture conditions. If the soils are to be stored for any length of time, they may need to be grassed over to prevent erosion. In most circumstances, land evaluation methodologies assume that under normal standards of land management nutrient deficiencies can be re-medied by fertiliser applications.

Most mineral extraction occurs in rural areas, and in recent years the wider objectives of sustainable development have prompted a change from restoration for agriculture to a restructuring and diversification of land uses. As a result, FRCA generally seeks the reinstatement of best and most versatile land to a viable agricultural use, but lower quality land can be converted to other beneficial uses such as amenity and/or nature conservation wetland sites. These include broadleaved woodlands with enhanced public access and new wildlife habitats. The Forestry Commission and DETR have published some useful guidance on the reinstatement of soils for tree planting over capped wastes at landfill sites (Bending & Moffat 1997). The landfill Tax Credit Scheme, administered by ENTRUST under the *Landfill Tax Regulations 1966*, provides a compensation mechanism by which landfill operators can fund environmental projects (proposed by environmental bodies) within a ten mile radius from landfill sites.

It is not possible to cover here all of the mitigation measures necessary to prevent **erosion and compaction** problems during and after developments, but the following general guidelines are of use:

- Remove as little vegetation as possible during the development, and revegetate bare areas as soon as possible after the completion of the development.
- Where possible create gentle gradients and avoid steep slopes.
- Install suitable drainage systems to direct water away from slopes.
- Avoid creating large open expanses of bare soil. These are most susceptible to wind erosion. If such large areas are created, then windbreaks may be a useful mitigation procedure.
- If the development is near to a water body, **siltation traps** may need to be installed to trap sediment, and prevent damage to the freshwater ecosystem.
- Avoid driving over the soil or use wide tyres to spread the weight of vehicles, thereby avoiding compaction.
- Use a single or few tracks to bring vehicles to the working area.
- Cultivate the area after compaction has taken place.

The *Reclamation of Mineral Workings to Agriculture* (DoE 1996c), *Guidance on Good Practice for the Reclamation of Mineral Workings to Agriculture* (DoE 1996d) and *Good Practice Guide for Handling Soils* (MAFF 2000b) are primary references on the above. Erosion control methods are also discussed by Bell (1999).

It is important to avoid runoff of pollutants carried in a liquid form, and if this is perceived to be a major problem then procedures for the containment of the pollutants on site must be considered. The mitigation of cumulative and dispersed impacts on soil chemistry as a result of air- and water-borne pollution from developments is a matter of emission controls (see Chapters 8 and 10).

If the baseline survey shows that the site is contaminated, then remediation will have to be undertaken (some of the techniques used could also be applicable to clearing up pollution caused during and after the development phase). There are a number of remediation techniques available which include:

- **Removal of the contamination for off-site disposal** (so called shift and tip). This is the most commonly used technique, but will result in the transport of hazardous material along the public highway, and the displacement of pollution to a landfill site.
- **Excavation and on-site disposal**. This removes the need for transport, but would require a custom-designed facility, and either a waste-management licence or exemption from licensing.
- **On-site stabilisation**. These techniques remove the ability of a pollutant to move off-site.
- **In *situ* bio-remediation**. This is effective for organic pollutants and uses natural micro-organisms to break down organic pollutants. Even difficult materials, such as creosote, can now be bio-remediated (CIRIA 1995).
- **Soil washing**. Acid or solvent washing of soils is commonly undertaken in countries such as Holland, and is very effective at removing contaminants and minimising material for disposal. The end product is, however, sterile, and has to be modified if it is to be used as a growth medium.
- **Air sparging, vacuum extraction and pump and treat methods** are effective at removing a range of contaminants from groundwater.

For a review of the above techniques and a discussion of how to select the best practicable environmental option see Wood (1997). The choice of technique will depend upon the type of pollutant(s) present, the geology and hydrogeology, the development type, and the sensitivity of the surrounding environment. Even where contamination is found, it does not automatically mean that some form of remediation is required. The most common approach used in the UK at present, and the approach being promoted by the EA, is the use of risk-assessment techniques to ascertain the potential for impact on the environment (Cairney 1995, Syms 1997). The need for remediation must be proven by carrying out a risk assessment (Chapter 14).

The presence of contamination is only part of the story. To be a significant risk, the material must be in a mobile form, there must be a mechanism by which the material can move, and there must be a potential target (the Source–Pathway–Target model). The target can be human, animal, or some other component of the environment. In simple situations, the risk assessment can be undertaken using empirical methods of assessment. However, it is more usual that some form of modelling exercise will be required to quantify the potential for migration/release to occur. The modelling of transport of contaminants to groundwater is considered by Adriano *et al.* (1994). Techniques for modelling presently being used are hydrogeological models (such as MODFLOW or AQUA 3D), and air dispersion models (such as ISC3). Specialist models are currently being designed which use

Monte-Carlo simulation techniques or Lotus Hypercube techniques. Models such as LANDSIM for landfill development are specific to given end-uses.

9.8 Monitoring

The loss of soils implicit in the conversion of agricultural land to urban uses is recorded by the DETR, and other soil impacts are monitored as indicators of sustainable development. These are subject to regular review as part of the British Government's commitment to global sustainability. The DoE (1996c) has published the results of wide-ranging research into standards of land restoration following mineral extraction, and this forms the technical basis to the best practice guidelines. More recent results (DETR 1997) have been made available on a landfill site restored to agriculture and monitored since 1974, and the results are being used to further elaborate on good restoration practice.

References

Adriano DC, AK Iksandar & IP Murarka (eds) 1994. *Contamination of groundwaters*. Norwood, Essex: Science Reviews.

Allen SE, HM Grimshaw & AP Rowland 1986. Chemical analysis. In *Methods in plant ecology*, 2nd edn, PD Moore & SB Chapman (eds), 285–344. Oxford: Blackwell Scientific.

Anderson A 1994. *Roads and nature conservation: guidance on impacts, mitigation and enhancement*. Peterborough: English Nature.

Avery BW 1990. *Soils of the British Isles*. Wallingford: C.A.B. International.

Ball DF 1986. Site and Soils. In *Methods in plant ecology*, 2nd edn, PD Moore & SB Chapman (eds), 215–284. Oxford: Blackwell Scientific.

Bell FG 1998. *Environmental geology; principles and practice*. Oxford: Blackwell Science.

Bell FG 1999. *Geological hazards; their assessment, avoidance and mitigation*. London: E & FN Spon.

Bending NAD & AJ Moffat 1997. *Tree establishment on landfill sites*. London: Forestry Commission & DETR.

Brady NC & RR Weil 1999. *The nature and properties of soils*, 12th edn. New York: Macmillan.

Bridges EM 1978. *World soils*. 2nd edn. Cambridge: Cambridge University Press.

BSI (British Standards Institution) 1994. BS3882. *Specification for topsoil*. London: BSI.

Cairney T 1995. *The re-use of contaminated land: a handbook of risk assessment*. Chichester: Wiley.

CIRIA (Construction Industry Research and Information Association) 1995. *Remedial treatment for contaminated land*, Vol. IX: *In situ methods of remediation*. CIRIA Special Publication 109. London: CIRIA/DETR.

CIRIA 1998. *Remedial treatment for contaminated land*, Vol. I: *Introduction and guide*. CIRIA Special Publication 101. London: CIRIA/DETR.

Cooke RU & JC Doornkamp 1990. *Geomorphology in environmental management: a new introduction*, 2nd edn. Oxford: Clarendon Press.

Cresser M, K Killham & T Edwards 1993. *Soil chemistry and its applications*. Cambridge: Cambridge University Press.

DETR 1997. *Agricultural quality of restored land at Bush Farm*. London: DETR.

DETR 1999a. *Environmental geology in land use planning; a guide to good practice*. London: DETR.

DETR 1999b. *Environmental geology in land use planning; advice for planners and developers*. London: DETR.

DETR 1999c. *Environmental geology in land use planning; emerging issues.* London: DETR.

DETR 1999d. *Soil-forming materials – their use in land reclamation.* London: DETR.

DETR 2000a. *Our countryside: the future. A fair deal for rural England.* http://www.wildlife-countryside.detr.gov.uk/ruralwp/cm4909/index.htm

DETR 2000b. *Contaminated land (England) Regulations 2000 & Statutory Guidance: Regulatory Impact Assessment (Final).* (http://www.environment.detr.gov.uk/contaminated/land/ria/2.htm)

DoE (Department of the Environment) 1987. *ICRCL Guidance Note 59/83. Guidance on the assessment and redevelopment of contaminated land,* 2nd edn. London: HMSO.

DoE 1989. *Environmental assessment: a guide to the procedures.* London: HMSO.

DoE 1990. *Planning Policy Guidance Note 14 (PPG14) Development on unstable land.* London: HMSO.

DoE 1991. *Environmental effects of surface mineral workings.* London: HMSO.

DoE 1994a. *Planning Policy Guidance Note 23 (PPG23) Planning and pollution control.* London: HMSO.

DoE 1994b. *Mineral Planning Guidance Note 12 (MPG12) Treatment of mine openings and availability of information on mined ground.* London: HMSO.

DoE 1994c. *Sustainable development: the UK strategy.* London: HMSO.

DoE 1996a. *Mineral Planning Policy Guidance Note 1 (MPG1) General considerations and the development plan system.* London: HMSO.

DoE 1996b. *Mineral Planning Guidance Note 7 (MPG7) The reclamation of mineral workings.* London: HMSO.

DoE 1996c. *The reclamation of mineral workings to agriculture.* London: HMSO.

DoE 1996d. *Guidance on good practice for the reclamation of mineral workings to agriculture.* London: HMSO.

DoE 1996e. *Indicators of sustainable development for the United Kingdom.* London: HMSO.

DoE 1997. *Planning Policy Guidance Note 7 (PPG7) The countryside – environmental quality and economic and social development.* London: HMSO.

DoT (Department of Transport) 1993. *Design manual for roads and bridges,* Vol. 11: *Environmental assessment.* Section 3, Part 11. London: HMSO.

Ellis N, DQ Bowen, S Campbell, J Knill, A McKirdy, C Prosser, M Vincent & R Wilson 1996. *An introduction to the geological conservation review* (GCR Series, No. 1). Peterborough: JNCC.

Ellison RA & A Smith 1997. *A guide to sources of earth science information for planning and development.* British Geological Survey Technical Report WA/97/85. London: HMSO (or http://www.bgs.ac.uk/).

Harris MR & J Denner 1997. UK Government Policy and Controls. In *Contaminated land and its restoration,* RE Hester & RM Harrison (eds), 25–46. London: Thomas Telford.

Hodgson JM 1978. *Soil sampling and soil description.* Oxford: Clarendon Press.

Hudson N 1981. *Soil conservation.* London: Batsford.

Huettl RF 1993. Forest soil acidification. *Angewandte Botanik* **67**, 66–75.

MAFF 1988. *Agricultural land classification of England and Wales: revised guidelines and criteria for grading the quality of agricultural land.* London: MAFF.

MAFF 1993. *Code of good agricultural practice for the protection of soil.* London: MAFF.

MAFF 1996. News Release (277/96) dated 1 August 1996.

MAFF 2000a. *Evaluation of mineral sites restored to agriculture.* http://www.maff.gov.uk/environ/landuse/restore/

MAFF 2000b. *Good practice guide for handling soils.* http://www.maff.gov.uk/environ/landuse/soilguid/

MLURI (Macauley Land Use Research Institute) 1991. *Land capability classification for agriculture.* Aberdeen: MLURI.

MO (Meteorological Office) 1989. *Climatological data for agricultural land classification.* Bracknell: MO.

Munsell Color Co. 1992. *Soil color charts.* Macbeth Division, Kollmorgen Instruments Group, Newburgh, New York.

Rowell DL 1994. *Soil science: methods and applications.* London: Longman.

Royal Commission on Environmental Pollution 1996. Nineteenth Report – *Sustainable use of soil.* London: HMSO.

SO (Scottish Office) 1994a. National Planning Policy Guideline 4 (NPPG4). *Land for mineral working.* Edinburgh: SO. (http://www.scotland.gov.uk/library/nppg/nppg-cover.asp)

SO 1994b. Circular 25/1994. *Agricultural land.* Edinburgh: SO.

SO 1999. National Planning Policy Guideline 15 (NPPG15). *Rural development.* Edinburgh: SO. (http://www.scotland.gov.uk/library/nppg/nppg-cover.asp)

SSEW (Soil Survey of England & Wales) 1976. *Soil survey field handbook.* SSEW (now SSLRC at Cranfield University, Silsoe).

Syms P 1997. *Contaminated land. The practice and economics of redevelopment.* Oxford: Blackwell Science.

Thompson D 1990. In defence of soil. *ECOS* **11**(1), 36–39.

Voorhees WB 1992. Wheel-induced soil physical limitations to plant growth. *Advances in Soil Science* **19**, 73–95.

Webster R 1977. *Quantitative and numerical methods in soil classification and survey.* Monographs on soil survey. Oxford: Clarendon Press.

West G 1991. *The field description of engineering soils and rocks.* The Geological Society of London handbook series. Milton Keynes: Open University Press.

White RE 1997. *Principles and practice of soil science. The soil as a natural resource.* Oxford: Blackwell Science.

Wilson SC & KC Jones 1993. Bioremediation of soil contaminated with polynuclear aromatic hydrocarbons (PAHs): A review. *Environmental Pollution* **81**, 229–249.

Wood PA 1997. Remediation methods for contaminated sites. In *Contaminated land and its restoration*, RE Hester & RM Harrison (eds), 47–71. London: Thomas Telford.

10 Water

*Peter Morris, Jeremy Biggs and
Andrew Brookes*

10.1 Introduction

Water is an essential resource that sustains life on earth, and the presence of abundant liquid water is what makes the earth unique amongst known planets. However,
c.96% of the earth's free water is sea water, and c.3% is ice and snow – so liquid
fresh water only constitutes c.1%, and is a relatively scarce resource. In many parts
of the world, lack of sufficient clean water is likely to be one of the most critical
issues of the twenty-first century. In Europe, access to clean water is generally taken
for granted, and large quantities are used for domestic purposes, for cooling, rinsing
and cleaning in industry, and for irrigation in agriculture. These activities place a
heavy burden on water resources in terms of both quantity and quality.

The study of water in land areas is **hydrology**. This includes the study of precipitation (rain, snow and dew), *evapotranspiration*, surface waters (lakes, rivers, etc.), soil
water and groundwater. Hydrological systems are highly dynamic, and planning any
development that may affect them requires an understanding of variations in the
storage and flow of water (**water quantity**) and of the materials it carries (**water quality**).

Water has a pivotal role in environmental systems, and the water assessment in
an EIA is bound to overlap with most other components. For example:

- Surface waters and adjacent land often have high landscape and recreational
  value, and navigable waters can have economic importance.
- River *floodplains* usually contain valuable agricultural land, and have long
  been a focus for settlement with the result that they (a) often contain important archaeological features, and (b) are often extensively urbanised (e.g. in the
  UK c.5.8 million people now live on them (EA 2000)).
- Hydrological systems are subsystems of ecosystems – so the hydrology of an
  area is strongly influenced by local climate (Chapter 8), soils, geology and
  geomorphology (Chapter 9), and the biota (especially vegetation) – and in turn
  affect them (see Fig. 11.4, p. 251). The link between hydrology and ecology is
  particularly strong in freshwater ecosystems (Chapter 12).

Hydrology is a complex science, and a thorough hydrological assessment will
require the services of a competent hydrologist. This chapter can only provide a
brief overview of relevant aspects, together with references to sources of further
information. There is a huge quantity of literature. Texts covering most aspects
include Manning (1996), Shaw (1993), Viessman & Lewis (1996), Ward & Robinson
(1990). Texts focusing on environmental aspects include Newson (1994), Singh

(1995), Thompson (1998), Wanielista *et al.* (1996), Ward & Elliot (1995), Watson & Burnett (1995). Increasing use is being made of remote sensing and GIS (Chapters 15 & 16); texts focusing on hydrological applications of these techniques include Gurnell & Montgomery (2000), Schultz & Engman (2000), Singh & Fiorentino (1996).

10.2 Definitions and concepts of water quantity

10.2.1 Introduction

Studies of water quantity are largely concerned with the storage of water in various environmental systems and the flows of water within and between these systems. A major feature of the earth's water system is the **hydrological cycle** in which:

- water evaporates (principally from the oceans, which cover > 71% of the earth's surface) to form atmospheric water vapour;
- water vapour condenses and returns to the earth's surface as precipitation;
- water flows from the land to the seas and oceans.

This global circulation of water is a closed system with no significant gains or losses. By contrast, a site, region or land mass has an open system of water flows, with inputs (I) and outputs (O) that control the amount of water stored in it, and hence its **water budget**, which can be expressed as

$$I - O = \Delta S$$

where ΔS = change in storage (increase if I > O *or* decrease if O > I)

The only significant input to land masses is precipitation. The outputs are evapotranspiration, groundwater seepage and *runoff* (mainly in rivers).

10.2.2 Catchments

A water assessment will normally need to consider the hydrology of a *catchment*. As defined in EC (2000), this can be:

- a **river basin** (main catchment) – defined as "an area of land from which all surface runoff flows through a sequence of streams, rivers, and possibly lakes, into the sea at a single river mouth, estuary or delta"; or
- a **sub-basin** (sub-catchment) – defined as "an area of land from which all surface runoff flows through a series of streams, rivers, and possibly lakes, to a particular point in a watercourse (normally a lake or river confluence)".

A catchment is a fairly discrete system, and provides an excellent focus for scientific research, water management and EIA. It has a water budget in which:

- the main input is precipitation, although groundwater seepage can occur when a groundwater body underlies more than one catchment;
- the outputs are evapotranspiration, runoff and groundwater leakage.

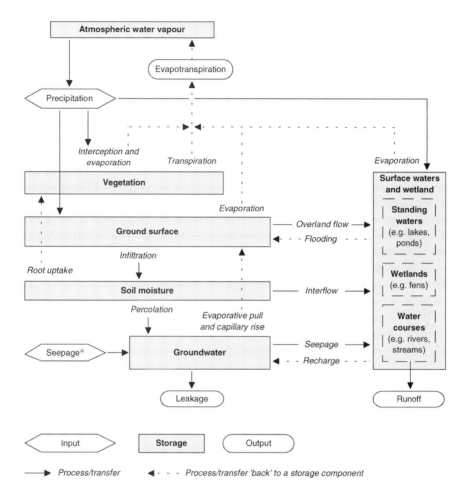

Figure 10.1 Catchment processes and storage components.

Within the catchment, various storage components and fluxes can be identified (Fig. 10.1).

A development or **receptor** site located within a catchment may receive land-phase water (surface water and/or groundwater) from higher in the catchment, and therefore has a **site catchment** and a **site water budget** with several inputs and outputs:

$$(Pn + R_s + R_g) - (ET + Q_s + Q_g) = \Delta S$$

where:
Pn = precipitation   ET = evapotranspiration
R_s = surface water recharge (run on) Q_s = surface water discharge (runoff)
R_g = groundwater recharge (seepage) Q_g = groundwater discharge (leakage)
ΔS = change in storage

The relative importance of the inputs and outputs will depend on a number of factors including the site's location in the catchment. Sites situated high in a catchment may depend largely on precipitation, and can be particularly vulnerable to water shortages in times of drought. On the other hand, low-lying sites may be susceptible to flood risk, and systems such as lowland rivers that depend on significant and sustained inputs of land-phase water (including groundwater) are vulnerable to impacts that reduce or contaminate this supply.

10.2.3 *Precipitation and evapotranspiration*

Precipitation (Pn) and evapotranspiration (ET) bring about the interchange of water between the atmospheric and land-phase water (Fig. 10.1) and a catchment water budget is markedly influenced by the balance between them. This can be expressed as the **Pn : ET ratio**, or the **meteorological water balance** (Pn – ET). When Pn > ET, there is a **water surplus** which is discharged as runoff; when Pn < ET, there is a **water deficit** which leads to a reduction in storage water and runoff.

In the long term, Pn : ET ratios are a function of the local or regional **climate**. For example, in the UK: (a) they are high in northwestern areas and lower in the south and east, (b) they show a marked seasonal pattern, all areas normally having an appreciable **winter surplus** and a **summer deficit**, which is usually slight in northwestern areas and increases to the south and east. The summer deficit normally arises from high evapotranspiration rates rather than low summer rainfall. This is because (a) evaporation increases in response to higher temperatures and lower humidities, and (b) *transpiration* increases when the vegetation is in leaf. However, evapotranspiration is often reduced because *soil moisture deficits* (SMDs) develop, especially during droughts. These inhibit transpiration and plant growth, and explain the frequent need to irrigate many crops in the drier areas.

Meteorological water balance also exhibits marked, unpredictable variation, which has been particularly apparent in recent years, with sustained deviations from normal seasonal patterns in many areas (§10.8.2). This trend is thought to be related to global warming (§8.1.3). Soil moisture levels, groundwater recharge and river flows are all very sensitive to changes in rainfall/evapotranspiration patterns, which therefore have significant knock-on effects in catchments.

Meteorological water balance is also influenced by factors other than climate. These include land cover, particularly the extent of surface waters and the extent and nature of vegetation. The latter is important because:

- **Interception** of precipitation by vegetation, and re-evaporation from the canopy, means that much precipitation water never reaches the ground (Fig. 10.1). This **interception loss** (which contributes to evapotranspiration) varies in relation to the **interception capacities** of vegetation types. For instance, it can be up to c.25% of precipitation in broadleaved woodland, higher conifer forest and tall grassland, but much lower in short swards or sparse vegetation.
- **Transpiration** can return > 50% of rainfall to the atmosphere, although it also varies with vegetation type, e.g. is higher from woodland than from grassland.

The combination of interception and transpiration can therefore account for > 75% of rainfall, leaving < 25% to become runoff – so vegetation is a major factor affecting runoff (see Baird & Wilby 1999). Both interception and transpiration are markedly reduced when vegetation is replaced by a built environment, and this is a major cause of increased runoff from urban areas. Similarly, whilst mature crops can have high interception capacities and transpiration rates, cultivated land is usually bare or sparsely vegetated for much of the year.

10.2.4 Infiltration and overland flow

Most rainfall reaching vegetated ground normally infiltrates into the soil (Fig. 10.1) where it is stored, re-evaporates, is taken up by plant roots, or percolates downwards in response to gravity. The release of infiltrated water to surface waters is normally slow. However, if precipitation exceeds the soil's **infiltration capacity** (its ability to absorb water) the excess collects in depressions or runs down inclines as **overland (sheet) flow** (Fig. 10.1).

Infiltration capacity is most likely to be exceeded under intense or sustained rainfall (especially when soils are already wet) or if heavy winter snowfall is followed by a rapid spring thaw. However, infiltration and associated surface runoff are strongly influenced by:

- **soil depth and texture**: in the US, four **hydrologic soil groups** are recognised, ranging from A, soils with good infiltration when wet, and hence low runoff potential (e.g. deep sandy soils), to D, soils with low infiltration when wet, and hence high runoff potential, e.g. heavy clays and shallow soils (see Schwab *et al.* 1993);
- **slope and vegetation cover**: overland flow tends to increase on slopes and/or where vegetation (which enhances infiltration) is sparse, and this increases the risk of soil *erosion* and flash floods.

Infiltration is also drastically reduced by factors such as soil compaction, e.g. on construction sites (§9.6.2) and is completely prevented by impervious surfaces. This increases the volume and rate of runoff from built environments, and reduces recharge to groundwater beneath them.

10.2.5 Water in the ground

The subsurface system can be divided into an **unsaturated (vadose) zone** that normally has air-filled spaces, and a **saturated zone** in which all available spaces are filled with **groundwater**. Soil is an important component of the unsaturated zone, and the properties of a soil, especially its texture and structure (§9.3.1), affect both its ability to retain water and its *hydraulic conductivity*.

If percolating water encounters an impermeable layer in the unsaturated zone, it may accumulate or move down inclines as **interflow** (Fig. 10.1), but this is usually a minor and intermittent flux compared with percolation to the saturated zone. Within the saturated zone, groundwater is usually held in strata of porous rock called *aquifers*, of which there are two main types – **confined aquifers** and **unconfined aquifers** (Fig. 10.2). Two other types occur in some areas: (a) a **'leaky' aquifer** is partially confined below a semi-pervious layer, above which an unconfined aquifer is also present, and (b) a **karst aquifer** consists of fractured rather than porous rock.

Globally, groundwater constitutes *c.*97% of all liquid fresh water. In the UK, it currently provides *c.*30% of public water demands (> 70% in southeast England),

UNCONFINED AQUIFER

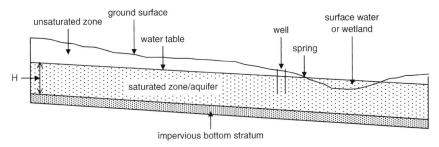

CONFINED AQUIFER

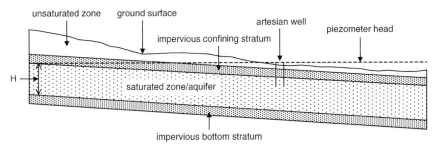

H = hydraulic head

An **unconfined aquifer** has a free **water table** (groundwater surface), and a *well* sunk into it will fill to the water table level (WTL). The **hydraulic head** is strictly the height of the groundwater body from the impervious bottom rock to the water table, but is effectively the elevation of the water table, and water will flow down the **groundwater slope** from a point where the WTL is higher to one at lower altitude. A **confined aquifer** lies beneath an impervious confining stratum, though it must fill from one or more unconfined areas. Confined groundwater is **artesian** (is under pressure) and given a break in the impervious layer, e.g. at an artesian well or spring, water will rise to the level of the **piezometer head**. This is a horizontal line drawn from the highest level of hydraulic head in the system.

Figure 10.2 Groundwater relationships in an unconfined aquifer and a confined aquifer.

and c.75% of groundwater abstracted in England is used for drinking water (EA 2000).

In Britain, groundwater can be abstracted from most 'rocks'. However, the **storage capacity** of an aquifer depends largely on its dimensions and porosity – and many strata (e.g. clays and shales) are not usually classed as aquifers because the porous material is thin (< 50 m), and the groundwater supply tends to be small and unreliable during droughts. The principal aquifers are the sandstones, limestones and chalk which underlie much of southern, eastern and midland areas of England.

Aquifer storage levels (and associated water table levels) normally follow a seasonal cycle. Storage is depleted during the summer, when output to springs and rivers continues, but (a) input is minimal because there is a meteorological water deficit, and (b) abstraction demands increase. Groundwater recharge occurs mainly during winter, when there is a meteorological water surplus. Consequently, ground-water droughts are mainly caused by a lack of winter rainfall rather than dry summers, and serious droughts occur when (a) a dry summer follows a very dry winter, as in 1975/76, or (b) winter recharge is below average for several years, as during 1988–92 in eastern England.

Groundwater flows down inclines (Fig. 10.2) but flow rates are generally slow, rarely exceeding 10 m/day and sometimes less than 1 m/yr. Moreover, the deeper in an aquifer the water is, the more slowly it moves. Water can move a short distance upwards from a water table by **capillary rise** and can be drawn further upwards by **'evaporative pull'** (exerted by evapotranspiration) (Fig. 10.1). However, groundwater cannot reach the surface from deep water tables; and in these situations (a) it is not available to vegetation, and (b) abstraction requires pumping. In some places, it reaches ground level and emerges as a spring, or seeps directly into a watercourse or water body (Fig. 10.2).

In many catchments, groundwater and surface water levels are intimately linked, and groundwater is responsible for river **baseflows** which continue when there has been little rainfall for some time. In these systems, river **low flows** during droughts can be partly due to over-abstraction of groundwater (Cook 1998).

Groundwater is also often important in supporting wetland ecosystems such as fens which are therefore threatened by groundwater depletion (see §12.5.3). Infor-mation on groundwater, with particular reference to the UK, is provided in Down-ing & Wilkinson (1992).

10.2.6 Surface waters

Apart from overland flow, which is normally transitory, surface waters can be divided into **standing waters** (lakes, reservoirs, ponds, etc.) in which there is little lateral flow, and **watercourses** (streams, rivers, etc.) in which there is appreciable flow (Fig. 10.1).

Standing water bodies occur in depressions, or in valleys with natural or artificial dams behind which the water accumulates. They range in size from small ponds to large lakes and artificial reservoirs, which are a major water resource in many areas. In spite of having little lateral flow, they are not static systems. Many have inflow and outflow streams, and typically there is movement of water between groundwater and the surface body. Consequently, levels in water bodies (a) may change within a few hours, and (b) affect local groundwater levels and streamflows. Standing waters also receive water by direct precipitation and overland flow, and lose appreciable amounts by evaporation (Fig. 10.1).

Watercourses can include slow-flowing channel systems such as canals and ditches, but most **streamflow** (the gravitational movement of water in a channel) is in streams and rivers. The rate of streamflow is influenced by channel slope, cross-sectional channel area/shape and the hydraulic roughness of the channel boundary. However, it also responds to the amount of water entering the channel, and is rarely stable for long. Water volumes and levels may rise rapidly in response to **storm rainfalls**, and because channels have a limited capacity, the water may rise above the bankfull level and spill out onto an adjacent floodplain – and **flood risk** is a principal reason for considering channel flow in EIAs. Runoff peaks (**quickflows** or **peak flows**) tend to be short-lived, but more sustained highflows can occur, e.g. in winter.

In recent years, **lowflows** have also been the subject of attention in some EIAs (Cook 1998). Prolonged dry periods can lead to markedly reduced flow in many rivers (or even drying out, e.g. of chalk rivers), especially when these have a limited natural baseflow, or where there has been significant abstraction. A report by the NRA (1993) highlighted 40 low-flow rivers in England and Wales where excessive abstraction is a problem. Lowflows can have serious consequences for river ecology, and for public water supply, particularly where this relies principally on abstraction from rivers and/or reservoirs. Without some sustained input from rivers, only the largest UK reservoirs have the capacity to meet demands through very dry summers. This applies even if they were full at the start of the summer; so whilst surface water shortages can follow the failure of winter rainfall to fill reservoirs, they are more commonly associated with low summer flows in rivers (CEH 2000a).

10.2.7 Floodplains

River floodplains constitute *c.*10% of the land area in England and Wales (EA 2000), and *c.*8% of England is at risk from river flooding (including tidal rivers and estuaries) (DETR 2000). Floodplain inundation and associated processes are natural phenomena, and (where they can occur without risk to human life) are beneficial because they provide fertile sediments to farmland, help to maintain valuable wild-life habitats, and reduce flood risk elsewhere in the catchment. A floodplain acts as a temporary store for flood waters, and facilitates their conveyance; and the release of water to the floodplain reduces (a) the flood flow in a river, and hence flood levels downstream, and (b) 'backing up' and associated increased flood levels upstream. Flooding and floodplain processes are discussed in Anderson *et al.* (1996) and Smith & Ward (1998).

Because of their locations, relatively flat topography and rich ***alluvial soils***, floodplains have long attracted a variety of human activities in spite of their natural susceptibility to flooding. For example:

- they have been utilised for farming, which has included practices to improve drainage or reduce the frequency of flooding;
- many settlements grew around river crossing points where transport routes converged;
- road and railway construction is relatively easy, and this has attracted development, the rate of which has more than doubled in some areas of the UK in the past 50 years (EA 2000).

These activities tend to restrict the capacity of rivers to accommodate large storm flows, especially in many urban areas where river channels and natural floodplains are very restricted (EA 1997). This has increased the incidence of serious flooding, and leads to further human intervention in the form of flood alleviation/defence measures such as channel modification (e.g. widening and deepening) or embankments (Brookes 1988). Such solutions can have adverse impacts on the local environment, cannot eliminate flood risk, and may increase flooding elsewhere. For example, preventing overspill to the floodplain may simultaneously increase upstream and downstream flood risks by obstructing stormflows and reducing the area available for flood water storage. Enhancing channel flow by widening and deepening can also increase downstream flood risks.

10.3 Definitions and concepts of water quality

10.3.1 Introduction

Water quality refers to the physical and chemical conditions of surface and groundwaters. Physical conditions include temperature and the presence of particulate matter; chemical conditions depend on the types and concentrations of dissolved chemicals present.

Water in the environment is never pure; natural waters always contain at least some dissolved chemicals (solutes) which originate from the atmosphere, soils or the **weathering** of bedrock. The water chemistry depends largely on the catchment climate and (especially) geology, and there is wide variation in **solute load** – the range and concentrations of solutes, including nutrients (see §12.2.1).

Natural waters also vary in the amount of particulate material present, which generally depends on the same factors as solute load. The terminology here is somewhat confusing. **Total sediment load** refers to *both* bottom *sediments* (sediments in the strict sense) *and* fine organic and inorganic particulates suspended in the water. However, while the latter are sometimes referred to as **suspended solids**, they are more commonly called **suspended 'sediments'** – and the quantity present may be called the **suspended sediment load** or just **sediment load**.

Rivers may carry large quantities of particulates. Fine suspended particles will not normally settle in a river unless the flow is slow, and even in standing waters, *silt* particles may remain in suspension for some time before sinking to the bottom to become sediment. River bed particulates are called the **bedload**, and can range in size from silts to coarse sands, gravels and boulders. Bedload materials can move, but only when the flow of water exceeds a particular energy threshold, which is related to both the channel slope and the water discharge, and increases in relation to particle size. Although the bedload is a relatively small component of total sediment load, it is a major influence on the form of the channel. The movement of bedload materials can be related to the formation and maintenance of natural features such as gravel *riffles* and point bars. These types of morphological features may also provide important habitats (see §12.2.3).

Human influences on water quality include: changes in the concentrations of naturally occurring chemicals (e.g. nitrates, phosphates, metals); the input of new synthetic substances (e.g. *pesticides*); and changes in sediment loads. In general, pollution sources can be divided into two types: *point source pollution* and *non-point source (diffuse) pollution*.

The likely effects of a development on water quality will depend not only on the development type but also on the **type and quality of the receiving waters**. For example, rivers export most of their pollutants downstream; so the effect at any one point may be transitory, but polluted water and silts may be carried considerable distances before they are sufficiently degraded or diluted to have no effect. Standing waters such as lakes and ponds are sediment sinks, and their water turnover rate is usually slow; so sediment and pollutants tend to accumulate, and impacts may intensify with time.

The aspects of water quality that are usually most relevant in EIAs are briefly discussed below. Further information can be found in standard hydrology texts and texts such as Hutchinson (1975), Kiely (1997), Moss (1998), Laws (1993), and Meybeck *et al.* (1989). Because the problems of groundwater contamination differ somewhat from those of surface water pollution, they are discussed in a separate section (§10.3.9).

10.3.2 Oxygen levels and organic pollution

The concentration of dissolved oxygen in water can have important implications for wildlife and commercial fisheries (see Table 12.2, p. 300). Oxygen levels vary naturally both within and between water bodies. Fast-flowing streams and rivers normally have constantly high levels because turbulent flow enhances oxygen absorption from the atmosphere. Levels are lower in slower-moving water, especially at night, but should never be very low in most British rivers. Still-waters, such as ponds and slow-flowing ditches, have highly variable oxygen levels which may range from supersaturated during daylight hours to zero at night. Large waterbodies, such as lakes and reservoirs, frequently stratify during the summer into an upper layer (the **epilimnion**), which is well oxygenated, and a lower layer (the **hypolimnion**), which is isolated from the atmosphere and may suffer oxygen depletion.

Oxygen depletion can occur through pollution, mainly by organic matter from sources such as sewage, soils, and agricultural or industrial *effluents*. High organic levels may be discharged from sewage treatment works, cattle yards, silage clamps, most food processing industries, and the wood and paper industry. Dissolved oxygen is consumed by the respiration of microbes that degrade the organic matter. Reduced oxygen levels can in turn lead to increased levels of potentially harmful chemicals (e.g. ammonia, methane, hydrogen sulphide, and *heavy metals*) by increasing their production or solubilities.

10.3.3 Thermal pollution

Freshwater systems have temperature regimes to which the aquatic life is adapted. Temperatures above the normal range can directly affect freshwater communities (see Table 12.2, p. 300) and can lead to oxygen starvation because increasing temperature (a) promotes oxygen consumption by increasing rates of animal and microbial respiration, and (b) reduces the amount of dissolved oxygen held by water. The main source of thermal pollution is power stations.

10.3.4 Acidification

The pH of natural waters varies considerably, and can change dramatically both seasonally and through the day. Many freshwater systems have naturally low pHs and should not be regarded as having poor water quality even if, for example, they do not support a commercial activity such as a fishery. However, acidification by *acid deposition* is now widespread, and many naturally acidic waterbodies have become more acidified during the past 100–200 years. Low pHs affect many freshwater animals directly, but a major effect is that they increase the solubility of toxic pollutants such as aluminium. Freshwater ecosystems and fisheries can be seriously affected (Table 12.2, p. 300).

10.3.5 Eutrophication

Excessive levels of nitrates and phosphates in freshwater systems can cause problems for both environmental and human health.

The principal cause of **environmental damage** is *eutrophication* of surface waters. The main source is runoff and **leaching** of fertilisers from farmland, although sewage effluent is thought to contribute *c*.5–10% of the nitrate, and detergents in waste water contribute *c*.10% to the overall phosphorous loading (DETR 1999a). Eutrophication frequently has considerable nature conservation costs (see Table 12.2). It may also bring socio-economic problems by causing fish kills, increasing drinking water treatment costs, and (by promoting **algal blooms**) decreasing the amenity value of waterbodies. It is generally perceived as a threat to standing waters, but is increasingly recognised as also having an impact on rivers.

The main **health concern** is methaemoglobinaemia (blue baby syndrome), a condition associated with nitrate. In many areas, nitrate levels in waterbodies used for drinking water (particularly rivers and aquifers) are now sufficiently high to cause concern, and have led to protective legislation such as the designation of *Nitrate Vulnerable Zones* (NVZs) and *Sensitive Areas (Nitrate)* (see Table 10.1).

10.3.6 Sediments

Sediments can be regarded as pollutants when present in unnaturally large quantities and/or when they are contaminated with chemical pollutants. Excessive sediment loads (especially of silts) can be derived from a variety of sources including agricultural land, bare urban surfaces and construction sites. Sediments from eroded soils or sewage may have a high organic content (causing deoxygenation), and where the site catchment is urbanised or intensively farmed, they may contain high levels of phosphates, metals, pathogens and pesticides.

Impacts of polluted sediments can be particularly severe in lakes and ponds, where they may become trapped and hence accumulate, with potentially damaging effects on ecosystems. These can also be adversely affected by high deposition rates, which may progressively seal waterbodies, isolating them from groundwater flows and changing the characteristics of the bottom substrate (see Table 12.2). In rivers with gravel bottoms used for spawning by fish (especially salmonids), siltation of gravels is of widespread concern as it leads to deoxygenation inside the gravels, starving the eggs and fry of oxygen.

10.3.7 Metals, micro-organics, and other harmful chemicals

Water pollution by these chemicals is largely due to accidents associated with licensed discharges to rivers, and from various difficult-to-control diffuse sources such as runoff from roads and urban or agricultural areas.

The **metals** of greatest concern in fresh waters include aluminium, chromium and heavy metals. They are normally present in the environment in low concentrations or – as in the case of aluminium – are normally not 'free'. Metals are most toxic when in solution, and metal solubility is influenced by the prevailing conditions – most are more soluble at low pH, and less soluble in hard water (with high calcium levels and normally a high pH). Consequently, different water quality standards are often set for metals in hard and soft waters. Organic compounds often remove dissolved metals from water by binding with them; but they may also release metals which would otherwise have remained insoluble, and this is the reason for concern over some water softeners such as EDTA which are added to many detergents. Metal toxicity often varies between different taxa. For example, zinc is relatively non-toxic to humans but very toxic to most fish, so levels of zinc acceptable in drinking water would be much higher than those acceptable for a fishery. Metals may also act **synergistically**.

In addition to naturally occurring toxins, between 20,000 and 70,000 compounds are estimated to be in common use worldwide (EA 1998a). These chemicals may or may not have toxic effects on organisms, or may be toxic above critical doses. An important group is the **micro-organics**, which includes most pesticides. *Environmental Quality Standards* (EQSs) have been set for many of these (see Table 10.1) but adequate toxicity data exist for only a tiny proportion of synthetic compounds, and the long-term *ecotoxicology* and environmental fate are known for only 20–30 chemicals. Recent research has shown that many chemicals have detrimental effects on organisms at levels far below those that cause immediate death, and often far below legal limits. Such sublethal effects include changes in physiology (such as hormone disruption), behaviour and reproductive rate.

Oils are commonly washed into freshwater systems from roads and industrial and development sites; and motorised pleasure boats also cause oil pollution. In addition to blanketing objects and organisms, oils can cover the water surface, reducing oxygen diffusion. They also contain many harmful chemicals (see Table 12.2).

10.3.8 Pathogens

There are four broad categories of human pathogens in temperate fresh waters – viruses, bacteria, protozoans (microscopic animals) and helminths (flatworms), although helminths are not normally a problem in Britain. Viral pathogens tend to have a limited host range, so sources are usually limited to waters containing human wastes, such as sewage. There are more potential sources of bacterial and protozoal pathogens because these tend to have less specific requirements.

10.3.9 Groundwater pollution

Porous rock has a filtering effect as water moves through it; so groundwater is generally much cleaner than surface water, and often requires little or no treatment before use. Chemicals are not completely removed, however, and there is increasing concern about groundwater pollution. Contamination can occur from a range of both urban and rural sources, and can result from point source or non-point source pollution (e.g. see Adriano *et al.* 1994, DETR 1999b, Downing & Wilkinson 1992).

Because groundwater moves very slowly (§10.2.5) pollutants take a long time to disperse naturally, and deep groundwater can remain contaminated for centuries or even millennia (EA 2000). Remedial measures, such as pollutant removal or degradation, are difficult and expensive; so it is particularly important to focus on pollution prevention. This has led to specific legislation and policies for groundwater protection.

10.4 Legislative background and interest groups

10.4.1 Legislation

The main EU Directives relevant to the water component of EIAs are listed in Table 10.1. Most of these Directives focus on one of two approaches:

Table 10.1 Key EU Directives relevant to water assessments

Surface Waters Directive 75/440/EEC[1,3] – Control of the quality of surface waters intended for abstraction of drinking water, using *Water Quality Objectives* (WQOs).

Bathing Waters Directive 76/160/EEC[1] – to protect the health of bathers, and maintain the aesthetic quality of inland and coastal bathing waters. Sets standards for 19 physical, chemical and microbiological variables, and includes requirements for monitoring and control measures to comply with the standards.

Dangerous Substances in Water Directive (DSWD) 76/464/EEC[2,3] – Control of inputs to water of dangerous substances (toxic, persistent, and likely to **bioaccumulate**). Requires member states to establish a consent system or set emission standards for two prescribed lists: those which should be prevented from entering waters (List I); and those which "should be minimised" (List II). There are related Directives for specific pollutants.

Freshwater Fish Directive 78/659/EEC and **Shellfish Waters Directive 79/923/EEC**[1,3] – to protect the health of freshwater fish and shellfish populations, by setting WQOs for **designated waters**.

Groundwater Directive 80/68/EEC[1,3] – Related to the DSWD, to protect groundwater against pollution by dangerous substances (itemised in List I and List II).

Drinking Water Directive 80/778/EEC[1] – Control of the quality of water intended for human consumption. Sets limits for total coliforms and substances such as nitrates.

Agricultural Sewage Sludge Directive 86/278/EEC[2] – Sets limits on heavy metal levels in sewage sludge applied on agricultural land.

Urban Waste Water Treatment Directive (UWWTD) 91/271/EEC[2] – Protection of surface waters by regulating the collection and treatment of urban waste water (sewage) and certain waste waters from industrial activities. Requires at least secondary **sewage** treatment for most sewage effluent, e.g. from sewage treatment works (SWTs) which have a **population equivalent** (pe) > 2 k for inland waters and estuaries or > 10 k for coastal waters. Discharges from a STW with a pe > 10 k to waters in a *Sensitive Area (Eutrophic)* or *Sensitive Area (Nitrate)* must comply with specified standards for removal of phosphorus and/or nitrogen.

Nitrates Directive 91/676/EEC[2] – Requirement to reduce nitrate pollution from agricultural sources (fertiliser and livestock manure) to safeguard drinking water, and protect fresh and marine waters from eutrophication. Sets a 50 mg/l limit and, where this is in danger of being exceeded in surface or groundwaters, requires the designation of *Nitrate Vulnerable Zones* (NVZs) within which the use of nitrate is restricted.

Integrated Pollution Prevention and Control Directive (IPPCD) 96/61/EC[2] 1996 – Pollution control for prescribed industrial installations and pollutants, using permits based on *Emission Limit Values* (ELVs), *Best Available Techniques* (BATs) and *Environmental Quality Standards* (EQSs) – concentrations of substances that should not be exceeded, based on current knowledge of their toxicities.

The Water Framework Directive (WFD) 2000/60/EC – Integrated protection and management of inland surface waters and groundwaters, estuaries and coastal waters.

---

1 Directive focusing on quality objectives for receiving waters.
2 Directive focusing on source-based controls.
3 Directive incorporated in the Water Framework Directive.

Further information can be found at http://www.europa.eu.int/comm/environment/ *or* http://www.europa.eu.int/water/

1.   **quality objectives for receiving waters** – which aim to limit cumulative pollution by setting *Environmental/Water Quality Objectives* (EQOs/WQOs);
2.   **source-based controls** – which aim to minimise pollution by setting *Emission Limit Values* (ELVs) that may be related to *Environmental Quality Standards* (EQSs) for specific pollutants.

Both these approaches have deficiencies, and the *Water Framework Directive* (WFD) (EC 2000) moves to a 'combined approach' in which WQOs and ELVs are used to reinforce each other, with the more rigorous requirements applying in any particular situation. It also aims (a) to provide a framework for integrated management of inland surface waters and groundwaters, transitional waters (e.g. estuaries), and coastal waters, (b) to maintain and enhance the status of aquatic ecosystems and dependent terrestrial eco-systems (thus integrating water management and nature conservation), and (c) to achieve long-term protection of water resources. It includes provisions for member states to:

- classify surface waters in terms of their chemical and ecological quality, set standards of 'good status', and monitor their ecological quality;
- prohibit direct discharges to groundwater, and monitor groundwater bodies;
- limit abstraction from groundwater bodies to the portion of recharge that is not needed to support connected ecosystems such as surface waters;
- produce and periodically update river basin management plans.

The main relevant UK legislation, much of which implements EU Directives, is outlined in Table 10.2. Legislation is proposed to tighten the water abstraction regulations (DETR 1999c) and implementation of the WFD can be expected in due course.

In relation to the EU/UK EIA legislation (§1.4), EIA is mandatory for six Annex I water-related project types and discretionary for 12 Annex II types, some of which only qualify if they are near **controlled waters**. However, all major projects are likely to have hydrological impacts, and the DoE (1989) guidelines prescribe screening for the water component in any EIA. Where water engineering works (including improvements to flood defences) are carried out under a General Development Order (if planning permission not required), an EIA may still be required under the EIA (Land Drainage Improvement Works) Regulations 1999 (SI 1783).

10.4.2 Policies and guidance

Because the water environment is very sensitive to impacts, it is particularly important to apply the central principles of EU/UK environmental policy outlined in §1.4, including the requirement for the polluter to pay for necessary controls (e.g. DETR 1998a). UK Government policy on water quality (DETR 1999c) includes the declaration of *designated waters*, *controlled waters*, WQOs, RQOs, NVZs and *Sensitive Areas* (*Eutrophic* and *Nitrate*) (Tables 10.1 & 10.2). In addition, the EA's policy on groundwater pollution control (EA 1998b) emphasises prevention by:

- controlling discharges;
- protecting vulnerable aquifers by the use of *groundwater vulnerability maps* (see Table 10.4);

Table 10.2 Key UK legislation relevant to water assessments

Salmon and Freshwater Fisheries Act 1975 – Regulation of inland fisheries; salmon and sea trout up to 6 miles.

Environmental Protection Act (EPA) 1990 – *Integrated Pollution Control* (IPC) system for emissions to air, land and water, which requires: EA authorisation for scheduled dangerous processes or pollutants; operators to use *Best Available Techniques Not Entailing Excessive Cost* (BATNEEC) to prevent or minimise releases and make any emissions harmless; and (when more than one medium is threatened) adoption of the *Best Practicable Environmental Option* (BPEO) to minimise damage to the environment as a whole.

Water Resources Act (WRA) 1991 – Protection of the quantity and quality of water resources and aquatic habitats. Duties and powers of the NRA (now EA) for: inland and coastal flood defences; **discharge consents** and abstraction licences; setting standards for **controlled waters**, *Water Quality Objectives* (WQOs) for inland and coastal waters, and *River Quality Objectives* (RQOs) for stretches of river; protecting groundwater; and monitoring water quality. Offences, e.g. to pollute groundwater.

Water Industry Act (WIA) 1991 – Duties of water companies; standards set for water supplies and wastewater treatment. Consents required from sewerage undertakers for discharge of trade effluents into public sewers.

Land Drainage Acts (LDA) 1991, 1994 – Powers and duties of: the NRA (now EA), mainly for flood defences and river engineering projects relating to designated **'main rivers'**; LAs, mainly for **'ordinary watercourses'** (not forming part of a main river); and Internal Drainage Boards (IDBs) for general drainage.

Environment Act 1995 – EA and SEPA established and given:
* further powers relating to flood defence and land drainage, prevention and remediation of water pollution, contaminated land, abandoned mines, and regulation of fisheries for environmental purposes.
* duties to promote the conservation of: the natural beauty and amenity of inland and coastal waters and associated land; flora and fauna which depend on an aquatic environment; geological or physiographic features of special interest, and buildings/sites/objects of archaeological, architectural, engineering or historic interest.
Regulations on mineral extraction strengthened. Duty of water companies to promote efficient water use.

The Groundwater Regulations 1998 (SI 2746) – Requirements for authorisation by the relevant EPA of direct and indirect discharges to groundwater of substances itemised in two lists (as in the Groundwater Directive).

The Pollution Prevention and Control Act (PPCA) 1999 – Implements the IPPC Directive. Replaces IPC with a *Pollution Prevention and Control* (PPC) system, which is similar but applies to a wider range of installations.

Further information can be found at the Executive Authority and EPA websites.

* protecting groundwater abstraction sites by the designation of *Groundwater Source Protection Zones* (GSPZs). For each site, three zones are defined, based on estimated groundwater travel times: Zone I (50 days); Zone II (400 days); and Zone III (the whole site catchment).

Overall policy for **land drainage** and **flood defences** is set by the relevant executive authorities (Appendix B). MAFF has produced guidance on strategies and codes of practice (MAFF/WO 1993, 1996) and a series of publications on project appraisal (see §13.3.2). Typical promoters of flood defences are **riparian** landowners or the **operating authorities** which, for inland waters, are normally the relevant EPA, LPA and Internal Drainage Board (IDB). The EA: (a) is also a developer of flood defence and certain navigation and water resources schemes, and often conducts its own EIAs; (b) takes the view that "the principles of EIA should be applied to all activities which impinge on its statutory responsibilities" (EA 1996) – and often produces or requires **informal environmental appraisals**. In Scotland, planning policy guidance is given in NPPG7 *Planning and Flooding* (SO 1995), and SEPA's flood risk assessment strategy is described in SEPA (1998).

Generally, the EPA's powers relate to river channels and flood defences, and LPAs have control over **floodplain development**. However, the EA is a statutory consultee on development plans, and seeks to persuade LPAs to follow its policies which include:

- natural floodplains (including those through settlements) should be safeguarded, and where possible restored;
- development should be resisted where it would be at risk from flooding or may cause flooding elsewhere;
- potential cumulative effects (including setting precedents) should be considered, even if the impact of a single project is small (EA 1997).

According to the DETR (2000) consultation paper, the forthcoming PPG25, *Development and flood risk*, will include the warning that flood defences can never eliminate flood risk, and will advocate:

- a much stronger presumption against new development on floodplains;
- application of the **precautionary principle** to flood-risk issues;
- recognition that "flood risk management needs to be applied on a whole-catchment basis".

The EA's policies for **catchments** are set out in *Local Environment Agency Plans* (LEAPs) (Table 10.4) which are intended as a tool for *Integrated Catchment Management* (ICM). Another potentially important management tool is *Water Level Management Plans* (WLMPs) (MAFF et al. 1994). These aim to balance and integrate the water-level needs of a range of issues including flood defence, water resources, navigation, archaeology/heritage, landscape/visual amenity, agriculture, forestry, and nature conservation – and the intention is to incorporate them in LEAPs. Under the proposed abstraction regulations (DETR 1999c), the EA will be authorised also to produce *Abstraction Management Strategies* (AMSs) which will in effect be local water resource strategies.

10.4.3 Consultees and interest groups

The principal Statutory Consultees for the water component of a UK EIA are the EPAs, which are the **competent authorities** in issuing licences and consents, such as IPC/IPPC authorisations, water abstraction licences and land drainage consents (Table 10.2).

Other interested parties will include:

- **Department of Health**, which is concerned with water quality issues affecting human health;
- the **water utilities** (see Table 10.4), which have a clear interest in potential impacts on water supply and quality;
- **riparian landowners** who own land adjoining a watercourse (and usually the river bed) and have 'riparian rights', e.g. to receive water in its 'natural' state;
- **owners of other property rights**, such as fisheries and angling associations, who also have riparian rights, and have frequently exercised these in the courts;
- **NGOs**, such as FOE, which campaign on water issues and often carry out water-quality monitoring exercises.

Where there are potential impacts on other EIA components such as heritage, landscape or wildlife: (a) the appropriate agencies (Appendix B) must be informed; and (b) relevant NGOs such as those mentioned in §11.3.3 will have an interest.

10.5 Scoping

10.5.1 Introduction

Scoping should follow the principles and procedures outlined in §1.2.2. The EA (1996) strongly advocates the use of **scoping checklists** such as Table 10.3 (which is abridged from an EA checklist, e.g. by omitting impacts on components such as traffic, landscape and heritage). Because the water environment is very susceptible to pollution, it is particularly important to make a thorough inventory of materials that will be used (and of how they will be stored and used) during both the construction and operational phases of a project (Atkinson 1999).

The water assessment is almost certain to overlap with other EIA components (§10.1), so early liaison between consultants is important. It is also essential to focus on key impacts and receptors, and a competent hydrologist should be employed at the scoping stage.

In a few cases the **impact area** may be confined to the project site and its immediate surroundings, but hydrological impacts are likely to be more widespread. This can hinder accurate determination of the impact area, and early estimates may have to be revised in the light of information obtained during the assessment process.

10.5.2 Methods and levels of study

For many hydrological variables, collection of field data is difficult, time-consuming, and requires sampling over extended periods. Consequently, the resource and time constraints in EIA often impose severe limitations on the range and depth of field survey work that can be conducted, and it is important to make maximum use of existing data by means of the desk study (§1.2.2).

Some sources of information are given in Table 10.4. The organisations referred to hold more information than that shown, and in the case of development types for which EIA is mandatory, it is obligatory for the relevant EPA to provide the developer (at a cost) with any relevant information in their possession. Other useful sources of

Table 10.3 Scoping checklist for potential hydrological impacts of construction work, with particular reference to river engineering schemes (adapted, with permission, from an Environment Agency checklist)

Issues	Sources of impact	Potential impacts
Surface water hydrology/ hydraulics	Soil excavation, removal, storage	Changed surface water runoff. Sediment contamination. Riparian drainage affected.
	Soil compaction/laying impervious surfaces (including roads)	Increased: surface runoff and velocities; magnitude, duration and frequency of flooding. Riparian drainage affected.
	Drainage	Changed flow velocities.
	In-channel works/ channel diversion	Changed flow velocities.
Channel morphology/ sediments	Riparian soil excavation/ movement/loss of trees	Changed: bank/bed stability (degradation/erosion); planform/ siltation; suspended sediment/bed loads. Sediment pollution.
	In-channel works: piling, piers, bridges, vehicle movements	Degradation/erosion of bed or banks. Disturbance to bed forms (pools, riffles). Changed: channel size; suspended sediment and bed loads.
	Channel realignment/ diversion	Changed: bank/bed stability; bed slope; planform/pattern; channel size. Disturbance to bed forms. Deposition/ siltation.
	Laying of impervious surfaces	Deposition/siltation. Degradation/ erosion of bed or banks. Changed: bank/ bed stability; suspended sediment/bed loads.
Groundwater hydraulics	Excavation	Changed flow.
	Dewatering	Changed flow. Change in water table level (**drawdown**).
	Laying of impervious surfaces	Changed: infiltration; water table level; pressure potential.
	Structures	Changed flow rates and direction.
Surface water quality	Storage and use of chemicals, fuel, oil, cement, etc., accidental spillage, vandalism, unauthorised use, site management including sanitation	Changes in quality. Chemical/organic/ microbial pollution. Rubbish/trash. Change in oxygen content. Changed *turbidity*. Changed dilution capacity. Nutrient enrichment. Change in *electrical conductivity* and pH; acidification.
	Earthworks, soil storage/ disposal	Changed turbidity. Re-suspension of contaminated sediments.
	Disturbance of contaminated land	Chemical pollution. Organic pollution. Rubbish/trash.
	Laying of impervious surfaces	Changed turbidity.

Table 10.3 (*continued*)

Issues	Sources of impact	Potential impacts
Surface water quality (continued)	Vegetation/tree removal	Change in quality and water temperature. Nutrient enrichment.
	In-channel works	Changed turbidity. Organic pollution.
	Channel realignment/ diversion	Changed dilution capacity upstream.
	Dewatering	Changed: dilution capacity; turbidity; in residence/flushing time.
	Balancing ponds	Change in quality. Changed turbidity.
Groundwater quality	Soil excavation, removal, storage	Change in quality.
	Construction below water table	Change in quality. Chemical pollution. Organic pollution.
	Storage and use of chemicals, etc.	Change in quality. Chemical pollution. Organic pollution.
	Pumping	Chemical pollution. Movement of contaminated water.
	Disturbance of contaminated land	Chemical pollution. Organic pollution.
Human related	In-channel structures	Changed flood risk. Disruption to commercial navigation.
	Dewatering	Changed water resource.
	Channel realignment	Changed flood risk. Changed abstraction rights.
Aquatic and wetland ecology	In-channel and associated works. Channel realignment/ culverting/diversion.	Altered habitat. Loss of habitat. Changes in the composition, species diversity and **biomass** of the biota, including loss of sensitive species, fish kill and effects on fish spawning.
	Sources increasing runoff, e.g. soil compaction/ impervious surfaces.	Altered habitat. Changes in the composition, species diversity and biomass of the biota, including loss of sensitive species.
	Dewatering	Altered habitat, including reduced water levels in wetlands.
	Balancing ponds	Altered habitat. Changes in the biota (as above).
	Sources affecting surface and groundwater quality	Altered habitat. Pollution through **food chains**. Changes in the biota (as above).

information include: LAs, angling clubs, local universities, previous EISs, and scientific papers. Historical information may also be relevant (Appendix C), as may information on geology and soils (§9.5.2).

Table 10.4 includes some examples of digital data. These are becoming increasingly available, and may facilitate the use of **GIS** and/or **hydrological models**. Numerous models have been developed for simulating, and predicting changes in,

Table 10.4 Some sources of information on water quantity and quality in the UK

BGS (British Geological Survey) http://www.bgs.ac.uk

Geoscience Data Index (GDI) Online spatial (GIS) index of BGS data (e.g. *well* locations, aquifer properties, streamwater chemistry/sediments, well-water chemistry). Gives costings of more specific information.

Hydrogeological Maps Various scales and information, e.g. surface water features/quality, aquifer potential.

CEH (Centre for Ecology and Hydrology) http://www.ceh.ac.uk

FEH CD-ROM – see Table 10.5.

National Water Archive (NWA) Holdings range from catchment scale data, e.g. climate and hydrology in experimental catchments, to national/international flood event data. Consists principally of:

The *National River Flow Archive (NRFA)* – includes: (a) online data for c.200 stations, e.g. catchment area and rainfall, runoff, low/high flows, abstractions/discharges affecting runoff; (b) retrieval service for other stations; (c) regional maps; (d) gauging station summary sheets; (e) hydrological trends 1961–97.

The *National Groundwater Level Archive (NGLA)* – includes: (a) online data for some observation wells; (b) a register of other sites; (c) a map showing major aquifers and gauging site locations.

Other Archives, e.g. weather station, soil moisture, flood event, and flood peak-over-threshold data.

Spatial data, e.g. digitised rivers at 1:50 k & 1:250 k; UK terrain model/map; soil types hydrology map (1 km); digital rainfall & evaporation data; flood studies report maps; floodplain/flood risk map of England & Wales.

UK Environmental Data Index (UKEDI) – Searchable database on water quantity and quality variables.

Indicators of Freshwater Quality – Results of the *Environmental Change Network* (ECN) monitoring programme for rivers and lakes.

Critical loads of acidity – Methods and results (database & and maps) for rivers and lakes.

EA (Environment Agency) http://www.environment-agency.gov.uk

Digital terrain models/maps (see §16.2.3), e.g. of flood risk areas.

Databases including: pesticides and trace organics in *controlled waters*; GQA chemistry (§10.7.1); freshwater fish (water quality); reservoirs; chemical releases inventory.

Groundwater Vulnerability maps – 1:100 k paper or digital maps of England and Wales (from TSO). A 'map picker' at the EA website gives information on each map. A 1:250 k map of N. Ireland is available from BGS.

Groundwater Source Protection Zones (GSPZs) – A national set, in digital format suitable for use with GIS, will be available soon for downloading from the EA website.

Local Environment Agency Plans (LEAPs) – (from local EA offices or online from EA-W). Assess water resources, abstraction, GQAs, groundwater quality and specific issues, and include management strategies.

Public Registers (at EA Regional Offices) – e.g. IPC; Water Quality and Pollution Control; Water Abstraction.

River Habitat Survey (RHS) database – see §12.4.2.

Table 10.4 (*continued*)

EHS (Environment and Heritage Service N. Ireland) http://www.ehsni.gov.uk

Water Quality Unit monitoring data archives – most data are available on request.

MO (Meteorological Office) http://www.met-office.gov.uk

Local climatic data including precipitation, temperature and evapotranspiration.

Met Office Rainfall and Evaporation Calculation System (MORECS) – calculates evapotranspiration and soil moisture – weekly for a 40 km nationwide grid, and at weather recording sites for hindsight data.

SEPA (Scottish Environmental Protection Agency) http://www.sepa.org.uk

Digital terrain map of mainland Scotland (1:50 k) – can show flood envelopes for the 100 year return period.

Public Registers including Integrated pollution Control (IPC), Water Quality and Pollution Control.

Reports and policies including: State of the Environment; Bathing Waters Report; Flood risk assessment.

Water UK (Association of UK water utilities) http://www.water.org.uk

Information on and links to the: water and sewerage or water-supply-only companies in England and Wales; publicly owned water operators in Scotland; and Northern Ireland Water Service.

---

hydrological systems. Reviews are provided in many hydrology texts, and the use of models in EIA is discussed by Atkinson (1999). Physical models are sometimes used, but most modelling involves the mathematical and statistical analysis of input data. Some calculations can be made using a hand calculator or computer spreadsheet (e.g. see Karvonen 1998, Thompson 1998, Wanielista *et al.* 1996), but more detailed modelling is carried out using software packages, many of which can be run on PCs (Table 10.5).

The use of models has limitations, especially in relation to the time and resource restrictions common in EIA. For example:

- some software is expensive, although much of that available from US agencies is free;
- most models need expert input by a hydrologist/hydraulic engineer, and even simple models should be used only under supervision by a competent hydrologist;
- the current capabilities of models are often limited by incomplete understanding of hydrological systems, and even complex models "necessarily neglect some factors and make simplifying assumptions about the influence of others" (Schwab *et al.* 1993);
- models can only be as good as the input data, and inadequate data can be a major source of error;
- predictions have a degree of uncertainty, and should be validated throughout the life of a project.

Table 10.5 Some hydrological modelling software available from UK and US government agencies. The US programs may relate to US databases, vegetation types, etc., but many can be adapted (with caution) for UK use.

CEH Wallingford (Centre for Ecology and Hydrology) (http://www.ceh.ac.uk/)

Micro LOWFLOWS – Estimation of: catchment characteristics (e.g. area, rainfall) and lowflow statistics from digitised river network data. Monitoring and water-use data, e.g. abstraction licences, discharge consents).

Micro-FSR – Uses rainfall-runoff methods to estimate flood magnitudes at any UK site. Hydrographs can be routed through a flood storage reservoir or balancing pond to facilitate spillway design and assessment.

PC-IHACRES – Catchment rainfall-runoff model. Requires rainfall, streamflow and temperature or evaporation data. Provides hydrographs with separation into dominant, quickflow, slowflow components.

PC-QUASAR – Comparison between present and potential water quality over time and downstream; setting of effluent consent levels; analysis of a range of water quality variables.

WINFAP – Given annual max. flood data for a site, can estimate probable events, e.g. the magnitude of an event in a give return period, or the return period of a flood of given magnitude.

WINFAP-FEH – Flood frequency analysis methods of FEH Vol. 3 (CEH 1999). Includes a dataset of > 1000 gauged sites; a range of analyses including pooled analysis of grouped sites; and input from FEH CD-ROM.

FEH CD-ROM (CEH 1999). Digital descriptors (e.g. boundaries, drainage paths) for UK catchments ≥ 0.5 km^2; rainfall depth–duration–frequency (DDF) data for catchments and 1 km grid points; a user-interface for selection of catchments/points; facility to compute design rainfalls, or estimate rainfall event rarity, from DDF data.

USDA-NRCS (Natural Resource Conservation Service) (formerly SCS) (http://www.nrcs.usda.gov/)

RUSLE2 – uses the Revised Universal Soil Loss Equation to predict annual and long-term erosion from sites (agricultural, mining, landfill, construction, etc.) where mineral soil is exposed to rainfall and overland flow.

SHPC (Soil hydraulics property calculator) – water retention (saturation and field capacities) and hydraulic conductivity from input of % sand and % clay values to a table or soil texture triangle.

TR-20 – catchment runoff hydrographs which can be combined and routed through stream reaches and reservoirs.

TR-55 Urban Hydrology for Small Watersheds (formerly SCS TR-55) – see Table 10.9.

USACE-HEC-RAS (River Analysis System) – water surface profiles in river reaches (based on channel morphometry, etc.), engineering works (e.g. bridges, **culverts** and floodways), and floodplain encroachment.

WEPP – hillslope erosion and hydrologic and erosion processes in small catchments.

USEPA (US Environmental Protection Agency) (http://www.epa.gov/)

BASINS – GIS/model for pollutants from point and nonpoint rural and urban sources.

QUAL2E – max. daily chemical streamloads in relation to dissolved oxygen.
WhAEM 2000 – wellhead groundwater capture zones.

Table 10.5 (*continued*)

USGS (US Geological Survey) (water.usgs.gov/software)

HSPF – quantity and quality (sediments and chemicals) processes on pervious/impervious surfaces and in streams and impoundments. **MODFLOW** – groundwater flow in confined, unconfined and combined-layer aquifers; **MODPATH** – flowpaths of particulates in aquifers; **MOC3D** – single solute transport in aquifers.

US NCSU Water Quality Group (http://h2osparc.wq.ncsu.edu/)

WATERSHEDSS – online package to assist in formulating mitigation/management practices for non-point source pollution. Includes information on pollutants and sources, and a linked GIS/water quality model.

US National Technical Information Service (USNTIS) (http://www.ntis.gov/fcpc/)

FLUX/PROFILE/BATHTUB – reservoir eutrophication in relation to nutrient loadings, hydrology, and morphometry. **SWMM** – management of storm water and pollutant movements in ground and surface waters.

Whilst water assessments should make maximum use of existing information, this is unlikely to be fully adequate, and it will be necessary to collect new data by field survey. Limited data are often misleading, and surveys should aim to ensure validity in terms of accuracy of measurements, number of samples, length of sampling period and frequency of sampling.

10.6 Baseline studies on water quantity

10.6.1 Introduction

This section aims to provide a brief overview of methods for obtaining new data on water quantity variables. Survey and modelling methods are described in most hydrology texts, including those referred to in §10.2.1.

10.6.2 Catchments

Most of the hydrological variables considered in an EIA will be studied in the context of the relevant catchment, and it is therefore important to obtain information on its characteristics. This should include (a) the main catchment descriptors (its boundary/area and drainage patterns) and (b) other aspects such as geomorphology (especially slopes), geology and soils, and land cover/use (including standing waters, vegetation and developments).

General information can be found in sources such as LEAPs. The main descriptors, and most other features can be determined with reasonable accuracy from Ordance Survey (OS), geological and soil survey papers or digitised maps (see §9.5.2 and Table 16.1, p. 384). Digital terrain models/maps (see §15.2.2 and §16.2.3) are becoming increasingly available (Table 10.4), and the FEH CD-ROM (Table 10.5) contains data on numerous catchments.

10.6.3 Precipitation and evapotranspiration

Precipitation data from the nearest weather station should be adequate for most EIAs, and can be obtained from the MO (Table 10.4). If rainfall–runoff modelling (§10.6.6) is envisaged, it will be necessary either to use a database, such as the FEH CD-ROM, containing rainfall depth–duration–frequency data, or to obtain long-term records from which such information can be extracted.

Occasionally, it may be desirable to obtain short-term site rainfall data, e.g. to correlate variations in streamflows to localised rainfall patterns. In such cases, rainfall can be measured using **rain gauges/recorders**. Information on these and their application can be found in most hydrology texts, and in MO (1982) and Strangeways (2000).

A complication in the estimation of evapotranspiration (ET) is that, in addition to the influence of meteorological conditions, its rate may be limited by shortages of soil water. To allow for this, distinction is drawn between **actual evapotranspiration** (AE) and **potential evapotranspiration** (PE). AE is equal to PE when the soil is saturated, but falls below PE when the soil surface dries out, and more so when SMDs develop and transpiration is inhibited (§10.2.3). Evaporation from a free water surface, and AE or PE from a vegetated surface, can be measured at point sites by using **evaporation pans**, **lysimeters** and **irrigated lysimeters**, respectively (described in Brassington 1988, Strangeways 2000, Ward & Robinson 1990). However, area ET values are usually estimated using models, such as MORECS (Table 10.4), and relevant data obtained from the MO should be adequate for most EIAs.

10.6.4 Infiltration and overland flow

Point measurements of these variables can be made (see Shaw 1993) and may be justified for small areas of particular concern, e.g. on a steep slope. However, it is not practicable to obtain direct field measurements over large areas, and resort is usually made to approximate indices (based on factors such as slope, soil properties, vegetation cover, and amount of impermeable surfaces) that can indicate runoff potential, and are incorporated in rainfall–runoff models (§10.6.6).

10.6.5 Water in the ground

The two most important aspects of water quantity in both the unsaturated (vadose) zone and the saturated zone are storage and flow (§10.2.5). For example, if a project is likely to affect **soil drainage**, it may be important to consider moisture levels and water retention and flow properties of local soils. The **soil moisture** data available from the MO (Table 10.4) should be adequate for most EIAs. If additional data are required, soil moisture contents can be measured (see §9.5.4). If the texture of a soil is known, its water retention properties (as saturation capacity and **field capacity**) and saturated **hydraulic conductivity** can be estimated (see SHPC in Table 10.5).

If the project may have a significant impact on **groundwater abstraction** rates, it will be necessary to consider the local aquifer's storage capacity and storage level patterns. It may be important also to know its **specific yield** – which is the volume of water that can be withdrawn under the influence of gravity. This is because an aquifer also has a **specific retention** – which is the proportion of water that is retained by surface tension on the solid particles, and is high in fine-grained materials.

Indicative values of specific yield for a range of geological materials are given in Brassington (1988).

General data for UK aquifers is available in the NGLA, and the locations of **wells** for which BGS holds data can be found in the GDI (Table 10.4). Methods of monitoring groundwater are described in Brassington (1988), Jones & Brassington (1992) and Wilson (1995). **Groundwater hydraulics** can be studied using (a) pumping tests in which water is pumped from wells, and groundwater flowrates are calculated from observed recharge rates, and (b) models based on the properties of the aquifers. Groundwater modelling techniques are discussed in Anderson & Woessner (1991) and some programs are listed in Table 10.5. These can be complex, but they often incorporate a simple formula known as Darcy's Law. This can provide an estimate of the flowrate in an aquifer (and the distance that water can be expected to flow in a day) on the basis that the velocity is a function of the aquifers' hydraulic conductivity and the groundwater slope. In its simplest form, Darcy's law is

$$V = K\frac{\Delta H}{L}$$

where:
V = velocity (m/day)
K = hydraulic conductivity (m/day)
ΔH = the difference in hydraulic head (Fig. 10.2) between two points in the aquifer (m)
L = the distance between the two points (m)
$\Delta H/L$ = the groundwater slope

Typical hydraulic conductivity values are given in Brassington (1988) and Atkinson (1999). The groundwater slope can be determined from aquifer maps or from field measurements of water table levels (as explained below). The simple application of Darcy's law has limitations, e.g. it assumes aquifer homogeneity (with a single hydraulic conductivity throughout) which is rarely the case.

Groundwater storage levels can be monitored by measuring water level changes in wells. Drilling new wells is expensive, but most areas contain existing monitored and/or unmonitored wells. Most of these should be shown in the GDI (Table 10.4) and there will probably be some private wells, which can be found on 1:25 k or 1:10 k OS maps. In wetland sites where the water table is normally near the surface, tubes (e.g. lengths of plastic waste pipe) can be inserted in the ground to act as mini wells.

Water level measurements can be made using continuous recorders, or more simply by weekly or monthly observations using a 'dipper'. This consists of an electric probe attached to a graduated cable, and a visual or audible signal that is activated when the probe contacts water. Because of weather-related fluctuations in water levels, monitoring should be continued for at least a year.

Measurements taken at a network of wells can also provide information on **groundwater contours**, and hence on likely flow patterns. Recorded water-level depths are subtracted from the relevant ground-level altitudes to calculate the **absolute water table elevations**. A water table contour map can then be produced to show the groundwater slope(s) and hence the likely direction(s) of flow. As illustrated by Figure 10.3, such information may be useful for assessing the vulnerability of a wetland to potential impacts such as pollution or water abstraction in its catchment. It may be beneficial also to estimate the site's water budget, and in

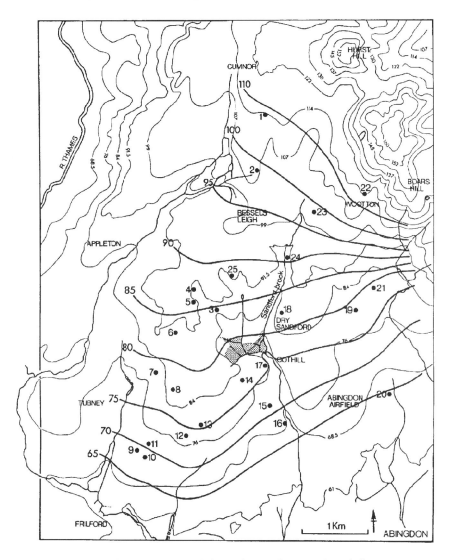

Figure 10.3 Groundwater contours (m) in the catchment of Cothill Fen SSSI in Oxfordshire.

The contours were drawn from mean absolute water table levels derived from monthly measurements over two years at 25 wells (numbered). The fen (stippled area) was thought to be threatened by a proposed extension of sand extraction workings (and subsequent land-fill) near to its western boundary. The results suggest that the groundwater flow in the area of the proposal largely by-passes the site, and that this is more vulnerable to water abstraction or pollution in the area to the north. (Morris 1988, data of Morris & Finlayson) particular the relative importance of precipitation, surface water recharge and groundwater recharge. For example, Cothill Fen was found to be largely fed by groundwater (Morris 1988). However, a site water budget (a) can only be calculated if all but one of the variables in the budget equation (§10.2.2) can be measured or neglected, and (b) requires measurements taken over at least a year.

10.6.6 *Surface waters*

The main surface-water quantity aspects likely to be important in an EIA are the current conditions of standing waters and watercourses and their vulnerability to changes in runoff, abstraction, and interference with *river corridors* and floodplains.

In order to assess the vulnerability of **standing water bodies**, it is desirable to obtain information on their size (area, depth and volume/capacity), elevation, site catchment, recharge and discharge regime, water level ranges and variability, and reservoir operating schedules. It should be possible to gather some of this information in the desk study. If necessary, recharge/discharge data for inflow/outflow streams can be measured as outlined below, but transfer between the water body and groundwater may be difficult to quantify.

An important aspect of **streams and rivers** is their flow regimes, which can have relevance to a range of issues, including water supply, pollution control, flood risk and control, and the design of bridges, etc. If an assessment is needed of a length of river, this is normally divided into **reaches** (sections of fairly uniform morphometry and flow) which are used as study units. It is particularly important to know how flows respond in times of heavy rainfall (resulting in quickflows) or drought (resulting in lowflows). Streamflows can be measured by stream gauging and/or estimated by rainfall–runoff models.

Stream gauging methods are discussed in most hydrology texts, and in particular in Boiten (2000), Gordon *et al.* (1992) and Herschy (1999). The two main methods are:

- the **velocity–area method**, which involves measuring the cross-sectional area of the channel, and flow rates (obtained with a current meter) at different points within it, with measurements repeated throughout the range of flow at the site;
- the **stream gauging structure method** in which a gauging structure (e.g. a weir or flume) is installed in the channel. This has a known stage–discharge relation (often called its rating or calibration) which permits flow rates to be calculated from water-level (**stage**) measurements. Changes in stage can be monitored by a float or sensor located in a stilling well (installed near the gauging structure), and recorded either on paper charts or by a solid-state logger.

Stream gauging results can be used to produce **hydrographs** (plots of streamflow against time). These show the frequency, magnitude and duration of events, such as highflows, which can be correlated with rainfall data, and hence can assist in flood prediction. However, stream gauging is expensive and a fairly long record is normally needed; so whilst existing data from gauged sites can be valuable, new stream gauging is unlikely to be profitable in EIAs unless monitoring is envisaged (§10.10).

In the absence of stream gauging data, streamflows can still be estimated using **rainfall–runoff models** (Table 10.5). These assume that the main factors affecting channel flow at a given location are catchment rainfall and characteristics such as area, slope and infiltration – which is affected by slope, vegetation cover, soil type and condition (including wetness), and the presence of impermeable surfaces (§10.2.4). They may include facilities for incorporating sub-catchments, runoff components such as overland flow, and flow retardance by in-channel vegetation (see Table 10.9). The input data requirements vary, depending on the sophistication of a given model and whether the software includes data for some variables. A major application of rainfall–runoff models is the estimation of flood risk at specific river locations (see §10.8.3) for which they utilise **design events**.

10.6.7 Floodplains

The limits of a river floodplain are defined in EA (1997) as the approximate extent of floods with a 1% annual probability of exceedance (1-in-100 yr flood) or the highest known level – although these "do not take account of the presence of defences or the likelihood that flood **return intervals** will be reduced by climate change" (DETR 2000). Information on **flood envelopes** (areas of recorded or design floods) is increasingly available in the form of flood studies reports, flood risk maps and digital terrain models (Table 10.4). The frequency and extent of floodplain inundation can also be estimated by computer models that utilise design floods (§10.8.3).

10.7 Baseline studies on water quality

10.7.1 Introduction

Water quality can be assessed by chemical or biological methods. Both approaches can involve a wide range of variables and techniques, or a few variables can be selected – as in the *General Quality Assessment* (GQA) method, which the EA uses for routine monitoring and assigning quality grades to stretches of rivers and canals (see Table 10.10, p. 236).

   Chemical methods involve analysing water samples for a range of variables (nitrate, oxygen, pH, etc.). They have the advantage of giving estimations of levels that can be compared with statutory standards; and apart from some microbiological techniques, they are the only available method for assessment of groundwaters. There are, however, three major disadvantages in assessing water quality from chemical data alone:

1. there are many possible pollutants in any given situation and each has to be assayed separately;
2. many pollutants (e.g. the hundreds of micro-organic compounds) are both difficult and expensive to monitor;
3. the sample will only reflect the chemical conditions at the time of sampling.

   Biological methods use living organisms as an indirect way of measuring water quality. A disadvantage of these methods is that it is not possible to determine the exact pollutant impacting a system, but they have three main advantages:

1. impacts on ecosystems are normally the primary concern of environmental protection agencies (EPAs), and surveys of biota are the most direct way of assessing ecosystem status;
2. surveys will often detect the net effects of one or more (often unknown) pollutants;

3. surveys can be used to assess long-term environmental health, e.g. pollution inputs that affect a river only occasionally may be detected, even if the pollutant is not present at the time of survey.

10.7.2 *Chemical methods of assessment*

Variables commonly measured in water quality assessments are listed in Table 10.6, which highlights those most used in relation to human health, conservation, and

Table 10.6 Common variables of water quality surveyed in water quality assessments

Variable	System	C	H	F	Notes
Nutrients					
Phosphorus	R	+	−	−	Several different forms. Much of load transported in sediment.
	L & P	+	−	+	Varies between hypolimnion and epilimnion. Detection often difficult.
Nitrate	R	+	+	+	Usually higher in late autumn/winter.
	L & P	+	+	+	Levels generally increase with amount of flow through system.
Chlorophyll a	AS	+	−	+	Used as a general index of standing crop of *algae*.
Organic matter					
Biochemical oxygen demand (BOD)	R	+	+	+	A main variable in monitoring sewage outfalls and GQAs. Can range from < 5 mg/l in clean rivers to 100,000 mg/l in industrial waste.
Chemical oxygen demand (COD)	R	+	−	+	Measures total organic matter which *could* use up oxygen. An alternative to BOD, e.g. where **non-labile organics** are suspected.
Metals					
Al, Cu, Cd, Hg, Pb, Zn	AS	+	+	+	Often serious pollutants of freshwaters. Toxicities usually increase with decreasing pH and water hardness.
Ca, Mg, Na, K	AS	+	+	+	Used to assess water type but not quality. Useful in conjunction with other variables to assess likely toxicity of other metals.
Others	AS	+	+	+	Industry-specific surveys may be needed (e.g. silver for electroplating, tin from old mines) but most not routinely covered.
Micro-organics	AS	+	+	+	Difficult to identify unless potential source suspected; so although potentially important, rarely included in standard surveys.

Table 10.6 (*continued*)

Variable	System	C	H	F	Notes
Oils					
General effects	AS	+	+	+	Most are easily detected by sight/smell. Not normally a health problem as polluted water unlikely to be imbibed. Tainting can damage fisheries.
Carcinogenic effects	AS	+	+	+	Rarely routinely done as particular carcinogen will vary with type of oil, geographic source and batch.
Others					
Ammonia	R	+	+	+	Organic decay product. Toxic to fish, and toxicity increases at high pHs.
	L & P	+	+	+	In large waterbodies, only likely to be high in intensively stocked fisheries. Small stagnant waterbodies may naturally have high levels.
Hydrogen sulphide	R	+	+	+	Generally as for ammonia.
	L & P	+	+		
Cyanide	AS	+	+	+	Very toxic but occurrence limited to particular industries.
Sediment	R	+	−	+	Part of routine monitoring, especially in relation to sewage outfalls.
	L & P	−	−	+	May be of concern in fisheries and reservoirs (may block filters).
Pathogens	AS	−	+	−	Mainly for faecal contamination, especially for water-areas.
Dissolved oxygen	R	+	−	+	A routine variable because many river animals need high levels.
	L	+	−	+	Levels vary with depth, time of day and season.
	P	−	−	+	Levels often highly variable.
pH	AS	+	+	+	Interpretation is very use-related. Used to qualify other data.
Alkalinity	AS	+	+	+	Used to qualify pH data.
Electrical conductivity	AS	+	+	+	Useful as an indication of the levels of other major variables.
Temperature	AS	+	+	+	Assessing thermal pollution, but mainly used to qualify other data.

Systems: L = lakes and reservoirs; P = ponds; R = rivers; AS = all systems (usually including groundwaters).
C, H, F = purpose: C = conservation; H = human health; F = fisheries. − = infrequently measured (but may be important in specific circumstances); + = fairly frequently measured + = frequently measured.

fisheries. The chemical component of the GQA scheme currently includes only **biochemical oxygen demand** (BOD), dissolved oxygen and ammonia, but an additional 'nutrient component' is being developed. The EPAs also monitor dangerous substances.

Levels of chemicals often vary considerably seasonally, throughout the day, and within a waterbody at a given time, sometimes over quite short distances. In addition, many elements occur in a number of different forms, only one of which may be of interest. For example, phosphorus may be measured as soluble reactive phosphorus, soluble unreactive phosphorus, particulate phosphorus, or a combination of these. Metals are often present in numerous forms, including organo-metallic forms, measurement of which is often difficult. Understanding the inherent variability of chemical variables is critical for selecting analysis and sampling programmes, and interpreting the results.

The level at which individual variables are monitored can also markedly influence the cost and extent of the survey, and care is needed to avoid selecting levels that are either too precise or too crude. For example, it would be pointless to stipulate a detection limit of 5 μg/l for monitoring nitrate in lowland rivers, where levels are never likely to fall below 1 mg/l. Conversely, there is little point in conducting a survey only to find that the assays have failed to detect the variable under study. Results of water chemistry monitoring around the world are given in Meybeck *et al.* (1989), and may help in formulating a strategy. However, water analysis will usually be carried out by an independent analyst (a public analyst if the results are to be legally accepted) who should be consulted about suitable procedures.

Assay methods are described in Golterman (1978), Hunt & Wilson (1986), Mackereth *et al.* (1978) and relevant HMSO standards (Standing Committee of Analysts). Hunt & Wilson (1986) include an extensive discussion of sampling strategies. Various samplers exist for taking samples at depth (Hellawell 1986); most other samples can be taken using a suitable bottle.

The EA has developed a predictive technique for assessing the extent to which lakes are **eutrophicated** and affected by acidification. This method 'hindcasts' the expected chemical status of lakes using equations that predict the chemical composition of runoff based on catchment geology, climate and land-use variables. Water quality is 'predicted' for the period around 1930 (which pre-dates the widespread use of chemical fertilisers, but post-dates the industrial revolution), and this is used as a baseline against which the quality of modern lakes can be judged. Decisions about the appropriateness of attempting to return a degraded lake to the 1930 conditions are then made on grounds of cost and practicability (Johnes *et al.* 1996).

10.7.3 Biological indicators of water quality

Hellawell (1986) and Rosenberg & Resh (1993) review the use of **biological indicators** in assessing water pollution, and Newman *et al.* (1992) give summary papers describing the various types of biological monitoring of river water quality throughout Europe. Most groups of freshwater organisms have been used as indicators of given pollution problems; but **macroinvertebrate** families (not species) are by far the most widely used taxa in Britain and Europe.

In Britain, the main biological assessment method in streams and rivers is the *Biological Monitoring Working Party* (BMWP) index (Hawkes 1997). This awards points to different invertebrate families according to their perceived tolerances to low oxygen levels (low points for tolerance, high for intolerance). This and the associated indices, *Number of Taxa* (TAXA) and *Average Score Per Taxon* (ASPT), are used as a broad indication of the level of water pollution.

Different rivers (or river sections) have different indices derived by this system, so it is now used in conjunction with the computer program RIVPACS (River InVertebrate Prediction And Classification System) (Wright 1995, Wright *et al.* 1998) which is available from CEH. This allows actual BMWP, TAXA and ASPT values in a river to be compared with those predicted for an unpolluted site of similar physical characteristics. BMWP scores are sometimes used incorrectly in EIAs, e.g. it is often wrongly assumed that family level macroinvertebrate data can be used to directly assess the conservation value of freshwater invertebrate communities. Hawkes (1997) gives a useful overview of the present, and probable future, use of BMWP, ASPT and RIVPACS in Britain.

Biological monitoring methods are gradually becoming available for still waters (lakes, ponds, canals, ditch systems). For example, the Predictive System for Multimetrics (PSYM), developed jointly by the EA and the Ponds Conservation Trust (PCT), assesses the ecological quality of still waters, and has been implemented for ponds and small lakes up to 5 ha (Williams *et al.* 1996, 1998), which, in Britain, represent about 98% of all discrete standing waters. PSYM operates in a similar way to RIVPACS but is based on both plant and invertebrate data and incorporates the concept of multimetric assessment for describing the overall ecological quality of waterbodies. It has been designed to fulfil the reporting requirements of the *Water Framework Directive* (§10.4.1) and also has some diagnostic potential, e.g. for identifying eutrophication impacts and poor physical habitat structure. PSYM entered a testing and development phase in 1999, prior to being made publicly available.

Several **other bioindicators** are available. Diatoms are widely used to assess river water quality and in palaeoecological studies of long-term changes in lake water quality, particularly of acidification. Many aquatic **vascular plants** are sensitive to water and sediment nutrient concentrations, and methods for assessing eutrophication in rivers using plants have recently been developed by the EA. Fish are sometimes monitored to assess incoming water quality at inlets to reservoirs (Hellawell 1986), and changes in fish populations with time can give information about long-term pollution trends such as acidification and eutrophication. Some micro-organisms, such as the bioluminescent bacterium *Photobacterium phosophoreum* have been used to assess water quality (Calow 1997). Various plant and animal species **bioaccumulate** toxins, and some are used in ecotoxicological studies using **bioassay** techniques. These are generally species-specific, however, and do not necessarily indicate the effects of pollutants on whole ecosystems. The EA and SEPA are developing *Direct Toxicity Assessment* (DTA) methods aimed at estimating the overall toxicity of effluents and receiving waters (EA 2000). Whole effluent toxicity methods used in the US can be found on the USEPA website.

Pathogens in waters can be detected by two broad methods: detection of species/strains and detection of indicator groups/species. Detection of individual species would be ideal, but there are several problems:

- there are many different pathogens in fresh water, all of which would need to be assessed;
- many species and strains of bacteria and virus require sophisticated culture and detection methods, often taking long periods of time (for some, techniques have not been developed);

- protozoan parasites are difficult or impossible to grow in culture, so large samples are often needed (e.g. a tonne of water for *Cryptosporidium*).

For these reasons, most routine monitoring involves indicator groups, and relies on two broad assumptions: (a) that the principal concern is with human faecal contamination of water, and (b) that the indicators used will be present in proportion to all pathogenic species of interest. In practice these two conditions are never fulfilled, and there has been much debate over which indicator organisms should be used and how much faith should be placed in such assessments. Nevertheless, in the absence of any other practicable method, human health limits for fresh water are set in terms of the number of indicator organisms per unit volume. The most common organisms used are coliform bacteria, some species of which are a natural (largely non-pathogenic) component of the biota of the human gut. In Britain, assessment is made for (a) total coliforms (which will include many species that are not necessarily of faecal origin), and (b) faecal coliforms (which should correspond more closely to the extent of faecal contamination of the water). Bathing waters, and surface waters used for extraction of water intended for human consumption, are also monitored for faecal streptococci and *Salmonella*.

10.8 Impact prediction

10.8.1 Introduction

Because of the complex, dynamic nature of hydrological systems, accurate prediction of impacts is often difficult, and there are bound to be uncertainties, which must be admitted in the EIS. Predictions can be assisted by the techniques referred to in Table 1.1 (p. 7); they can be qualitative, but should be quantitative where possible.

Many types of impact have already been mentioned in previous sections, and in Table 10.3. This section aims to summarise the range of impact sources and the range of impacts these generate. The impact sources can be roughly divided into those involving direct manipulation or utilisation of hydrological systems (Table 10.7), and those with less direct associations (Table 10.8). Both tables give some sources of further information.

Some projects can have potentially **positive impacts**. For example, reservoirs can provide water-based amenities and both aquatic and wetland habitats, as can mineral workings when extraction is completed – although these benefits must be weighed against construction and operational phase negative impacts.

10.8.2 Changes without the development

It is important to consider a project's potential impacts in the context of environmental changes that may occur in its absence (§1.2.3). These can be assessed in relation to past, present and predicted trends. The causes and implications of recent hydrological changes in the UK are discussed in Acreman (2000). In addition to local development trends, two potential causes of hydrological changes are ecological **succession** (see §12.5.1) and climate change.

In relation to climate change, fairly long-term records held by CEH (Table 10.4) suggest that, in areas not markedly affected by human activity, most river flows and

Table 10.7 Impacts from direct manipulation or utilisation of hydrological systems

Sources	Potential impacts
River engineering/manipulation	Brooke 1992, Brookes 1988, 1999
Resectioning/channelisation (widening, deepening, realigning/ straightening), e.g. to increase channel capacity for flood defence or drainage, or to facilitate project layout.	Loss of channel and bank habitats. Enhanced erosion and hence silt production (especially during construction, when pollution risks also increase). Increased flood risk and siltation downstream. Lowering of floodplain water table caused by deepening.
Embanking and bank protection (e.g. with concrete) usually for reasons as above.	Floodplain inundation and siltation prevented, with consequent risk of soil drought and loss of wetlands. Drainage from floodplain inhibited (unless sluices installed) with consequent waterlogging.
Clearing bank vegetation	Loss of wildlife habitats and visual/amenity value.
Fluvial dredging and deposition of dredgings, e.g. to maintain/enhance flood capacity or navigation.	Damage to channel habitats and biota at dredging sites. Increased sediment load and hence turbidity and smothering of downstream benthic and marginal ecosystems.
Diversion, e.g. to increase water supply to receptor area, or as a flood relief channel.	Decreased supply in donor area. Channelisation and evaporative loss from open channels. Risk to habitats in main river corridor.
Development on river floodplains	DETR 2000, EA 1997, Smith & Ward 1998
Use of floodplain area Construction of flood defences Laying impermeable surfaces	Increased flood risk upstream and downstream. Reduced groundwater recharge and river baseflows. Loss of ecological, heritage and visual/amenity/recreational features.
Reservoirs and dams:	Petts 1984
General	Loss of terrestrial habitats/farmland/ settlements. Local climate change and rise in water table. Visual impacts of retaining walls. Water-borne pathogens. Earthquake/landslip/ failure risks.
On-stream dams: above dam	Loss of river section; changes in flow regime; siltation.
On-stream dams: below dam	Reduced flows, oxygen levels and floodplain siltation.
On-stream dams: barrier effects	Migration of fish and invertebrates blocked.
Off-stream dams (not on a main channel)	Changes in groundwater recharge, levels and flow directions.

Table 10.7 (*continued*)

Sources	Potential impacts
Irrigation	Water abstraction (often from rivers). Increased evapotranspiration and local runoff. Risk of waterlogging and salination.
Drainage schemes	May involve channelisation. Increased soil drought risk and oxidation of organic soils. Water table lowered and wetlands lost. Increased flood/erosion risk downstream.
Water abstraction	Water resources depleted. Water table lowered. Risks of river lowflows, loss of wetlands, soil droughts and subsidence.
Sewage treatment works	Petts & Eduljee 1994 Increases in silts, nutrients (especially if treatment is poor), heavy metals, organics, and pathogens, e.g. faecal coliforms.

groundwater levels have fluctuated around a fairly stable mean. In recent years, however, rainfall, river flow and groundwater recharge have been notably variable, with sustained deviations from normal patterns in many areas. For example, exceptionally high rainfalls, especially in Scotland, and protracted dry spells in England have led, respectively, to a number of serious floods and a series of droughts. In England's drier eastern and southern areas, where water demands are greatest and much of the supply is from aquifers, groundwater levels were low for extended periods during the 1990s (CEH 2000b).

This variability is consistent with predicted climate changes associated with global warming (§8.1.3), but because of the wide range of natural climatic variation, and the influences of human activities, future hydrological changes cannot be predicted with any certainty. In addition, there is considerable variation in climate, geology, land use, and water use within the UK – so responses to climatic change will vary regionally and even locally. However, given that river flows and aquifer recharge rates are very sensitive to rainfall and evapotranspiration, increased incidence of floods (as in 2000–1) and droughts seems likely in many areas.

10.8.3 Predicting impacts on water quantity

Typical questions that should be considered in relation to water quantity are – is the project likely to significantly:

- affect river channel/corridor, standing water or wetland features because it will (a) cross or impinge on any of these, (b) involve river works, (c) need new flood defences or (d) require that a watercourse is re-routed;

Table 10.8 Impacts not directly associated with manipulation or utilisation of hydrological systems

Sources	Potential impacts
Roads	Changes in drainage systems, e.g. due to gradient changes, bridges, embankments, channel diversion or resectioning. Drawdown by dewatering when deep cutting. Increased runoff from impermeable surfaces, with risks of flash floods and erosion. Increased sediment loads from vehicles, road wear, and erosion of cuttings and embankments. Pollution of watercourses by organic content of silt, other organics (e.g. oils, bitumen, rubber), de-icing salt (and impurities), metals (mainly vehicle corrosion), plant nutrients and pesticides from verge maintenance, and accidental spillages of toxic materials. (DoT 1993)
Urban and commercial development	Changes in drainage systems due to landscaping. Abstraction. Drawdown/changes in groundwater flow, e.g. when dewatering deep foundations. Reduced groundwater recharge, and increased runoff velocities and volumes (with flood and erosion risks from rapid stormflows) due to impermeable surfaces. Pollution of watercourses and groundwaters by a wide range of pollutants which are rapidly transported to receiving waters by increased runoff. Increased sewage treatment. (Hall 1984, Shaw 1993, Walesh 1989)
Industrial development	As above but with: greater runoff effects (from a higher proportion of hard surfaces); higher pollution levels and a wider variety of pollutants including metals and micro-organics from heavy industry and refineries, pesticides from wood treatment works, and nutrient-rich or organic effluents from breweries, creameries, etc. Thermal pollution from power plants.
Mineral extraction	**Operation phase** – Removal/realignment of watercourses. Loss of floodplain storage/flow capacity. Drawdown and reduced local streamflows caused by dewatering for dry extraction, or increased runoff from process wash water or extraction methods involving water use. Increased siltation and chemical pollution downstream, e.g. from spoil heaps/vehicles/machinery/stores. **Restoration/aftercare phase** – see landfill. (Rust Consulting 1994)
Landfill	Increased runoff from raised landforms, especially if clay-capped. Reduced groundwater recharge and river baseflows if clay-sealed. Pollution of groundwater and near-surface runoff by **leachates** and by fertilisers and pesticides from restored grassland. (Petts & Eduljee 1994)
Forestry and deforestation	Reduced evapotranspiration and infiltration after felling – with consequent (a) decreased groundwater recharge, (b) increases in runoff, soil erosion, stream-sediment loads and siltation. Pollution by pesticides, especially herbicides used to prevent regrowth after clear felling.
Intensive agriculture	Enhanced runoff and erosion from bare soils. Drainage or irrigation impacts. Pollution of surface and groundwaters by: fertilisers; pesticides; organics from soil erosion, silage clamps and muck spreading; heavy metals from slurry runoff, and pathogens in animal wastes.

- increase flood risk because it will (a) constrict a river channel, (b) inhibit floodplain storage and conveyance, (c) increase channel flow directly, or (d) increase runoff;
- reduce surface and/or groundwater levels and increase the risk of river lowflows.

Physical, hydraulic and computer modelling are all used to predict the **hydraulic impacts of river works** with a reasonable degree of accuracy. For example, programs such as HEC-RAS (Table 10.5) can be used to assess the impacts of bridges and channel works on river flows and downstream flood levels. However, these require detailed information on aspects such as channel morphometry.

A major tool in the prediction of **flood frequency and magnitude** is the risk analysis technique of *design events* which can be utilised in flood–frequency models and rainfall–runoff models. The latter also require information on various catchment characteristics (§10.6.2). There are some **gauged catchments** for which data on most variables (including streamflows) are available, but most are **ungauged catchments** that lack existing data on many variables.

Most flood prediction in the UK is likely to follow the *Flood estimation handbook* (FEH) methods (CEH 1999). Importantly, this includes methods for estimating flood frequency in ungauged catchments by using techniques such as **pooled analysis** of similar sites. The FEH methods are intended for use with the accompanying software (WINFAP-FEH) and CD-ROM (Table 10.5). These are expensive, however, and since the main concern in the majority of EIAs will be to estimate the increased runoff that a development will generate, an alternative option is to use a relatively simple rainfall–runoff model such as TR-55 (Table 10.9).

Impacts of abstraction and dewatering can be estimated from the projected quantities involved and the nature of the sources (river, reservoir, aquifer). If a project is likely to contribute significantly to river lowflows, the Micro LOWFLOWS program (Table 10.5) may be applicable. However, most developments simply add to the overall demands on public water supply, and a project's requirements should be discussed with the relevant EPA. In some cases, an abstraction licence may be needed.

10.8.4 Predicting impacts on water quality

Methods for predicting changes in water quality are discussed in a number of texts including Kiely (1997) and Singh (1995). Computer models are available, but in many EIAs their application may not be appropriate or feasible.

Point source pollution is relatively easy to predict, and all point source pollutants discharged to *controlled waters* require a consent licence from the relevant EPA. In considering the application, the authority will examine the potential discharges in relation to the relevant WQOs and standards, including those for *designated waters* (Table 10.1 & 10.2). If a proposed development does not require a consent licence, but might still pose a threat, e.g. through accidental spillage, the same criteria can be applied. If adequate data can be obtained, a model such as PC-QUASER (Table 10.5) may be applicable.

Estimating the amount and effect of *non-point source pollution* is generally more difficult. There are relatively few methods, and they tend to have limited capability. Commonly used methods include the *Unit Load Method*, the *Universal Soil Loss Equation* and the *Concentration Times Flow Method*. Walesh (1989) gives a

Table 10.9 Features of the 'TF-55 Urban Hydrology for Small Watersheds' rainfall–runoff software

Procedure	Program features	User selection/data provision
Runoff curve number (CN)	Holds CNs and computes weighted CNs for sub-areas and/or the whole catchment.	Number of sub-areas (up to 10); land cover/uses and hydrologic soil groups (§10.2.4).
Time of concentration (Tc)	Up to 2 flow types (channel, sheet, shallow concentrated). Computed from L & V *or* L, S, n, A, & Wp. Limits: min. 1 h; max. 10 h.	Flow types, length (L), velocity (V), slope (S), Manning's coefficient (n); bankfull cross-section area (A) and Wetted perimeter (Wp).
Graphical peak discharge	Based on single area, Tc and CN to give peak flow rate and runoff.	Drainage area, CN, Tc, rainfall type, rainfall frequency (yr) and 24-h rain depth (in).
Tabular hydrograph	Composite flood hydrographs of sub-areas which are summed for the whole catchment.	Rainfall type, frequency (yr) and 24-h constant rain over catchment.
Detention basin storage volume	Computed from peak inflow rate (Pi), storm runoff volume and desired outflow rate (Po).	Drainage area, rainfall type, frequency and 24-h rain, runoff, CN, Pi and Po.

Runoff curve numbers (CNs) are empirical values for combinations of land cover/use and hydrologic soil groups or impermeable surfaces.

Time of concentration (Tc) is the time taken for runoff to flow from the hydrologically furthest point in the catchment (which is largely a function of distance and slope) to a point of concern (such as a river location liable to flood) and is assumed to be the shortest time for the whole catchment to contribute to flow at this point.

Manning's roughness coefficient provides values of flow retardance by various types of in-channel vegetation.

Further information is available in Schwab *et al.* (1993) and in the TR-55 manual which, together with the program, can be downloaded free from http://www.ftw.nrcs.usda.gov/tech_tools.html.

useful overview of the applications and drawbacks of these and other methods, some of which are incorporated in computer models such as RUSLE2 (Table 10.5). To guard against the uncertainty inherent in many of these methods, more than one should be employed where possible. An additional problem is that many projects will not in themselves cause significant impacts, but may contribute to the cumulative impacts, e.g. from an existing urban area.

The relevant EPA must be notified of any potential **groundwater pollution** that may require authorisation. Consideration should also be given to the project's location in relation to *groundwater vulnerability maps* and GSPZs (Table 10.4). If adequate data can be obtained, the vulnerability of other receptor sites can be assessed using Darcy's law (§10.6.5) or a computer model such as MODFLOW (Table 10.5).

10.8.5 *Significance of impacts*

Impact significance will depend on impact magnitudes and the sensitivity and value of receptors (§1.2.3). A fairly simple assessment method is provided in the water component of the *New Approach to Appraisal* (NATA) (DETR 1998b). This was designed for road schemes (see Chapter 5) but can be adapted for use with other projects. The assessment incorporates the effects of mitigation measures, and is therefore conducted in two stages:

1. Potential negative impacts are assessed using a **risk-based approach** (which only provides a neutral or negative assessment, ignoring any mitigation effects);
2. Impacts are reassessed taking into account mitigation measures and possible positive impacts.

The method focuses on two 'features' – **water quality** and **land drainage/flood defence**. For each of these, a scale of risk is produced from an assessment of the **sensitivity** of the receiving waters and the project's **potential to cause harm** (PCH).

The **sensitivity assessment** uses **nine indicators**, within each of which different categories may give a score of 'high', 'medium' or 'low' sensitivity (Table 10.10). An **overall sensitivity score** is determined as follows:

*   **overall high score** – (a) one or more 'high' score(s) for one of the **key indicators**, or (b) two or more 'high' scores for the other indicators;
*   **overall medium score** – (a) only one 'high' score for one of the non-key criteria, or (b) one or more 'medium' score(s), but no 'high' score(s);
*   **overall low score** – one or more 'low' score(s), but no 'medium' or 'high' score(s).

The DETR guidance (a) acknowledges that the PCH will be influenced by a large number of factors, but illustrates the process with two only (**traffic flows**, for potential harm to water quality; and **land take**, for potential harm to land drainage/flood defence), and (b) includes (in Annex 6E) **quantitative thresholds** for 'high', 'medium' and 'low' scores for the two factors. In practice, quantitative or qualitative thresholds should be set for all factors identified as potential causes of significant impacts on each **feature**. These should then be either combined to give an overall PCH score for the proposal or treated separately so that appropriate mitigation measures can be considered for each.

The guidance also suggests that (a) PCH scores are not totally prescriptive, but any proposed revision must be justified, and (b) scores may be upgraded if a proposal involves any of the following:

*   a route crossing or close to (250 m) a landfill site or contaminated land;
*   realignment of a watercourse;
*   major cuttings, embankments or a tunnel;
*   significant infrastructure during construction (e.g. haul roads).

For each feature, an **overall numerical assessment** (risk-based score) is derived by entering the overall environmental sensitivity and PCH scores in a matrix (Table 10.11).

Table 10.10 Assessment of environmental sensitivity in terms of NATA water quality and land drainage/flood defence indicators (after DETR 1998b, Annex 6D)

Water quality indicators	Sensitivity		
	High	Medium	Low
GQA Grade (chemical)	Grade A* (very good)	Grade B/C (good/fairly good)	Grade D/E/F (fair/poor/bad)
Freshwater Fish Directive	Designated salmonid fishery*	Designated cyprinid fishery	–
Water abstraction points	Abstraction for public water supply within critical travel time downstream*	Abstraction for other purpose within critical travel time downstream	–
Groundwater vulnerability	Major aquifer	Minor aquifer	Non-aquifer
Location of wells	Within Zone I or II of a GSPZ*	Within Zone 3 of a GSPZ	Not within a GSPZ

GQA = General Quality Assessment (§10.7.1). GSPZ = Groundwater Source Protection Zone (§10.4.2).

* Key indicators for which 'high' scores are assigned particular significance which is reflected in the Overall Sensitivity Scores.

Land drainage/flood defence indicators	Sensitivity		
	High	Medium	Low
Floodplain	Major works located in floodplain	Only minor works located in floodplain	–
Watercourses	–	Scheme crosses a watercourse	Scheme does not cross a watercourse
River Corridors – conservation value of any watercourse and corridor crossed/impacted by the scheme (see §12.6.3)	High	Medium	Low
Flood risk – increased risk of flooding upstream or downstream	Major increase in flooding risk	Minor increase in flooding risk	Current situation likely to remain

Table 10.11 Overall risk-based scores (modified from DETR 1998b)

Sensitivity of water quality *or* land drainage/ flood defence	High	−2	−3	−4
	Medium	−1	−2	−3
	Low	0	−1	−2
		Low	Medium	High
			Potential to cause harm	

Key: −4 Very large negative effects
    −3 Large negative effects
    −2 Moderate negative effects
    −1 Slight negative effects
     0 Neutral

Assuming that effective mitigation measures will be put in place, and that there may be opportunities to reduce existing impacts and/or effect enhancement, both the individual-indicator scores (Table 10.10) and overall scores (Table 10.11) are reassessed to produce final 'mitigated scores', which in some cases may be positive.

10.9 Mitigation

If a project is likely to cause a significant increase in flood risk or water pollution, there should be a strong presumption in favour of the relocate or no action altern-atives. Other mitigation and enhancement measures commonly adopted in relation to various water-impact issues are outlined in Table 10.12, together with some sources of further information.

Many mitigation measures, such as flow detention structures and storm reservoirs, will be normally designed by experts employed by the developer, but EIAs can include suggestions, e.g. of required capacities. These can be estimated using stand-ard techniques described in texts such as Schwab *et al.* (1993) and employed in some computer software (Table 10.5). For example, rainfall–runoff models such as Micro-FSR, TR-20 and TR-55 can compute desired balancing pond/storm reservoir capacities using runoff values predicted from the input data. Other software such as WATERSHEDSS can assist in the selection of suitable mitigation/management practices for diffuse source pollution. These can include the use of natural and constructed wetlands – which are discussed in §12.7, together with mitigation meas-ures relating specifically to freshwater ecosystems.

10.10 Monitoring

The sensitivity of hydrological systems, and the inevitable uncertainties associated with impact predictions, predicate that monitoring is particularly important for the water component of EIAs, and should be prescribed for both the construction and post-development phases. It can utilise baseline survey methods, and may justify the use in the baseline study of techniques and sampling programmes that would

Table 10.12 Some typical mitigation and enhancement measures relating to water-impact issues

---

Damage to riparian features and/or change in channel morphology caused by river works, etc.

Use project management and restoration techniques to minimise and repair damage (see Fig. 12.1). Create new features such as pools and riffles. Use dredgings positively, e.g. for landscaping or habitat creation. (Brooke 1992, Brookes 1988, and references cited in Table 12.1)

Increased sediment loads and turbidity caused by river channel works

Select appropriate equipment and timing, e.g. construct new channels in the dry and allow vegetation to establish before water is diverted back in (references as above).

Impacts of development on floodplains

If development is permitted: (a) steer away from wetlands and high-flood-risk areas; (b) ensure that new flood defences do not increase flood risk elsewhere; (c) take compensatory measures, e.g. floodways and flood storage areas/reservoirs to provide flood storage and flow capacity; (d) allow for failure/overtopping of defences, e.g. by creating flood routes to assist flood water discharge; (e) take opportunities for enhancement in redevelopment, especially where (as in many urban sites) existing conditions are poor, e.g. use river corridor works to restore floodplain (by removing inappropriate existing structures), enhance amenity and wildlife value, and create new floodplain wetlands. (EA 1997, Smith & Ward 1998, and references cited in Table 12.1)

Impacts of mineral workings, especially on floodplains

Operational phase – Carefully manage the use and storage of materials/spoil, and runoff from spoil heaps/earthworks. Use siltation lagoons. Route dewatering water into (a) lagoons, wells or ditches to recharge groundwater, (b) watercourses to augment streamflows. **Restoration phase** – Careful backfill and aftercare management. Enhancement, e.g. of amenity/wildlife value (see §9.7) (Rust Consulting 1994)

Impacts of new roads and bridges, or road improvement schemes

Use: careful routing; designs to minimise impacts on river corridors (not just channels); and measures to control runoff, e.g. routed to detention basins or sewage works, and not into high-quality still waters. If construction imposes river realignment, create new meandering channel with vegetated banks (see Fig. 12.2). (DoT 1993)

Impacts of dams and reservoirs

Adjust size or location (avoid sensitive areas). Minimise height and slope of embankments, and plant with trees.

Water depletion by abstraction

Promote infiltration and hence groundwater recharge in urban areas (see below). Minimise water use, e.g. by metering and the installation of water-efficient equipment/appliances.

Increased runoff from urban and industrial developments

Use **sustainable urban drainage schemes** with (a) efficient piped drainage and sewer systems and (b) **runoff source control measures**, i.e. at or near the point of rainfall, to promote infiltration and/or delay runoff before it reaches piped systems or watercourses,

Table 10.12 (*continued*)

e.g.: porous surfaces (car parks, pavements, etc.), soakaways (gravel trenches, vegetated areas); flow detention measures (grass swales, vegetated channels, stepped spillways, detention/balancing ponds/basins/storm reservoirs, and project layout/landscaping to increase runoff route). (Ferguson 1998, Hall 1984, Schwab *et al.* 1993, Shaw 1993, Walesh 1989)

Increased runoff and pollution (including sediments) from construction sites

Minimise soil compaction and erosion (see §9.7). Ensure careful storage and use of chemicals, fuel, etc. Install adequate sanitation. Guard against accidental spillage, vandalism and unauthorised use.

Chemical pollution from built environments, e.g. roads, urban/industrial areas

Control runoff (as above). Use: oil traps; siltation traps/ponds/lagoons; vegetated *buffer zones* and wetlands, e.g. constructed reed beds (see §12.7.2).

Increased sewage and/or sewage-pollutant content

Increase capacity and/or *sewage treatment level*, e.g. from primary to secondary or secondary to tertiary.

Chemical pollution from an accidental spillage

Effective contingency plans. Use booms and dispersants.

Groundwater pollution

Guide development away from GSPZs. Avoid contamination from leaking storage tanks, etc. by appropriate bunding of tanks and improved site management. Use buffer zones. (EA 1998b)

---

otherwise be excluded by time constraints. Monitoring is frequently hindered by the difficulty of isolating the effects of a project from those of other developments and activities, but aspirations for the success of a project can often be set and monitored (e.g. see Table 12.4, p. 310).

References

Acreman M (ed.) 2000. *Hydrology of the UK: a study of change*. London: Routledge.

Adriano DC, AK Iksandar & IP Murarka (eds) 1994. *Contamination of groundwaters*. Norwood, Essex: Science Reviews.

Anderson MG, DE Walling & PD Bates (eds) 1996. *Floodplain processes*. Chichester: Wiley.

Anderson MP & WW Woessner 1991. *Applied groundwater modelling*. New York: Academic Press.

Atkinson S 1999. Water impact assessment. In *Handbook of environmental impact assessment*, Vol. 1, J Petts (ed.), 273–300. Oxford: Blackwell Science.

Baird AJ & RL Wilby 1999. *Eco-hydrology: plants and water in terrestrial and aquatic environments*. London: Routledge.

Boiten W 2000. *Hydrometry*. Rotterdam: Balkema.

Brassington R 1988. *Field hydrology*. Milton Keynes: Open University Press.

Brooke JS 1992. River and coastal engineering. In *Environmental assessment: a guide to the identification, evaluation and mitigation of environmental issues in construction schemes*, CIRIA Research Project 424, Chapter 4. Birmingham: CIRIA.

Brookes A 1988. *Channelized rivers: perspectives for environmental management*. Chichester: Wiley.

Brookes A 1999. Environmental impact assessment for water projects. In *Handbook of environmental impact assessment*, Vol. 2, J Petts (ed.), 404–430. Oxford: Blackwell Science.

Calow P (ed.) 1997. *Handbook of ecotoxicology*. Oxford: Blackwell Scientific.

CEH (Centre for Ecology and Hydrology) 1999. *Flood estimation handbook: procedures for flood frequency estimation* (FEH): Vol. 1 *Overview*; Vol. 2 *Rainfall frequency estimation*; Vol. 3 *Statistical procedures for flood frequency estimation*; Vol. 4 *Restatement and application of the Flood Studies Report rainfall–runoff method*; Vol. 5 *Catchment descriptors*; FEH CD-ROM Version 1. Wallingford: CEH.

CEH 2000a. *The National Groundwater Level Archive* (NGLA): *about the data*. (http://www.ceh.ac.uk/)

CEH 2000b. *Hydrological trends – background*. (http://www.ceh.ac.uk/)

Cook HF 1998. *The protection and conservation of water resources: a British perspective*. Chichester: Wiley.

DETR 1998a. *Economic instruments for water pollution*. (http://www.detr.gov.uk)

DETR 1998b. *Guidance on the New Approach To Appraisal*. (http://www.detr.gov.uk/itwp/appraisal/ guidance/index.htm)

DETR 1999a. *Water quality: a guide to water protection in England and Wales*. (http://www.detr.gov.uk/)

DETR 1999b. *Groundwater regulations 1998: draft DETR guidance on implementation*. (http://www.detr.gov.uk/)

DETR 1999c. *Taking water responsibly*. (http://www.detr.gov.uk)

DETR 2000. Planning Policy Guidance Note 25 (PPG25) *Development and flood risk* – Consultation paper. (http://www.planning.detr.gov.uk/consult/ppg25/index.html)

DoE 1989. *Environmental assessment: a guide to the procedures*. London: HMSO.

DoT 1993. *Design manual for roads and bridges*, Vol. 11: *Environmental assessment*. London: HMSO.

Downing RA & WB Wilkinson (eds) 1992. *Applied groundwater hydrology: a British perspective*. Oxford: Oxford University Press.

EA (Environment Agency) 1996. *Environmental assessment, scoping handbook for projects*. London: HMSO.

EA 1997. *Our policy and practice for the protection of floodplains*. Bristol: Environment Agency. (http://www.environment-agency.gov.uk/)

EA 1998a. *State of the environment of England and Wales: fresh waters*. London: TSO.

EA 1998b. *Policy and practice for the protection of groundwater*. Bristol: EA.

EA 2000. *State of the environment*. (http://www.environment-agency.gov.uk)

EC 2000. Directive 2000/60/EC of the European Parliament and of the Council establishing a framework for Community action in the field of water policy. *Official Journal of the European Commission* No. OJ.L327. (http://europa.eu.int/water/water-framework/index_en.html)

EC 2000. *Water quality in the European Union. Developments of the Water Framework Directive*. (http://www.europa.eu.int/water/)

Ferguson B 1998. *Stormwater: concept, purpose, design*. New York: Wiley.

Golterman HL 1978. *Methods for chemical analysis of fresh waters*, 2nd edn. Oxford: Blackwell Scientific.

Gordon ND, TA McMahon & BL Finlayson 1992. *Streamflow hydrology: an introduction for ecologists*. Chichester: Wiley.

Gurnell AM & DR Montgomery 2000. *Hydrological applications of GIS*. New York: Wiley.

Hall MJ 1984. *Urban hydrology*. London: E & FN Spon.

Hawkes HA 1997. Origin and development of the Biological Monitoring Working Party score system. *Water Research* **32**, 964–968.

Hellawell JM 1986. *Biological indicators of freshwater pollution and environmental management*. London: Elsevier.

Herschy RW (ed.) 1999. *Hydrometry: principles and practice*, 2nd edn. Chichester: Wiley.

Hunt DTE & AL Wilson 1986. *The chemical analysis of water: general principles and techniques*, 2nd edn. London: The Royal Society of Chemistry.

Hutchinson GE 1975. *A treatise on limnology*, Vol. 1 Part 2 – *Chemistry of lakes*. Chichester: Wiley.

Johnes P, B Moss & G Phillips 1996. The determination of total nitrogen and total phosphorus concentrations in freshwaters from land use, stock headage and population data: testing of a model for use in conservation and water quality management. *Freshwater Biology* **36**, 451–473.

Jones GP & FC Brassington 1992. Data collection, storage, retrieval and interpretation. In *Applied groundwater hydrology: a British perspective*, RA Downing & WB Wilkinson (eds). Oxford: Oxford University Press.

Karvonen T 1998. *Soil and groundwater hydrology: basic theory and application of computer models*. Helsinki University of Technology (http://www.water.hut.fi/)

Kiely G 1997. *Environmental engineering*. London: McGraw-Hill.

Laws EA 1993. *Aquatic pollution: an introductory text*. Chichester: Wiley.

Mackereth FJH, J Heron & JF Talling 1978. *Water analysis: some revised methods for limnologists*. FBA Scientific Publication 36. Ambleside: Freshwater Biological Association.

MAFF/WO 1993. *Strategy for flood and coastal defence in England and Wales*, PB 1471. London: MAFF.

MAFF/WO 1996. *Code of practice on environmental procedures for flood defence operating authorities*, PB 2906. London: MAFF.

MAFF et al. 1994. *Water level management plans: a procedural guide for operating authorities*. PB 1793. London: MAFF.

Manning JC 1996. *Applied principles of hydrology*, 3rd edn. New Jersey: Prentice-Hall.

Meybeck M, D Chapman & R Helmer (eds) 1989. *Global freshwater quality – a first assessment*. Oxford: Basil Blackwell (for WHO and UNEP).

MO (Meteorological Office) 1982. *The handbook of meteorological instruments*. London: HMSO.

Morris PJ 1988. *The hydrology of Cothill Fen SSSI*. Unpublished report to Nature Conservancy Council.

Moss B 1998. *Ecology of fresh waters: man and medium, past to future*, 3rd edn. Oxford: Blackwell Scientific.

Newman PJ, MA Piavaux, & RA Sweeting 1992. *River water quality ecological assessment and control*. Luxembourg: Commission of the European Communities.

Newson M 1994. *Hydrology and the river environment*. Oxford: Oxford University Press.

NRA (National Rivers Authority) 1993. *Low flows and water resources: facts on the top 40 low flow rivers in England and Wales*. Bristol: NRA.

Petts GE 1984. *Impounded rivers: perspectives for ecological management*. Chichester: Wiley.

Petts J & G Eduljee 1994. *Environmental impact assessment for waste treatment and disposal facilities*. Chichester: Wiley.

Rosenberg DM & VH Resh 1993. *Freshwater biomonitoring and benthic macroinvertebrates*. New York: Chapman & Hall.

Rust Consulting 1994. *Hydrology and mineral workings – effects on nature conservation: Guidelines; Technical Annex*. English Nature Research Reports 106 & 107. Peterborough: EN.

Schultz GA & ET Engman (eds) 2000. *Remote sensing in hydrology and water management*. Berlin: Springer-Verlag.

Schwab GO, DD Fangmeier, WJ Elliott, RK Frevert 1993. *Soil and water conservation engineering*, 4th edn. New York: Wiley.

SEPA (Scottish Environmental Protection Agency) 1998. *Flood risk assessment strategy*. Edinburgh: SEPA. (http://www.sepa.org.uk)

Shaw EM 1993. *Hydrology in practice*, 3rd edn. London: Chapman & Hall.

Singh VP 1995. *Environmental hydrology*. Dordrecht: Kluwer Academic.

Singh VP & M Fiorentino (eds) 1996. *Geographical information systems in hydrology*. Dordrecht: Kluwer Academic.

Smith K & R Ward 1998. *Floods: physical processes and human impacts.* Chichester: Wiley.

SO (Scottish Office) 1995. National Planning Policy Guidance 7 (NPPG7). *Planning and flooding.* Edinburgh: SO. (http://www.scotland.gov.uk/library/nppg/nppg-cover.asp)

Strangeways I 2000. *Measuring the natural environment.* Cambridge: Cambridge University Press.

Thompson SA 1998. *Hydrology for water management.* Rotterdam: Balkema.

Viessman W & GL Lewis 1996. *Introduction to hydrology.* London: Addison Wesley Longman.

Walesh SG 1989. *Urban surface water management.* New York: Wiley.

Wanielista MP, R Kersten & R Eaglin 1996. *Hydrology: quantity and quality control,* 2nd edn. New York: Wiley.

Ward A & W Elliott 1995. *Environmental hydrology.* Boca Raton: Lewis (CRC Press).

Ward RC & M Robinson 1990. *Principles of hydrology,* 3rd edn. London: McGraw-Hill.

Watson I & A Burnett 1995. *Hydrology: an environmental approach.* Boca Raton; CRC Press.

Williams P, J Biggs, L Dodds, M Whitfield, A Corfield & G Fox 1996. *Biological techniques of still water quality assessment. Phase 1 Scoping Study.* R&D Technical Report E7. Bristol: Environment Agency.

Williams P, J Biggs, M Whitfield, A Corfield, G Fox & K Adare 1998. *Biological techniques of still water quality assessment. 2. Method development.* R&D Technical Report 56. Bristol: Environment Agency.

Wilson N 1995. *Soil water and ground water sampling.* Boca Raton: Lewis (CRC Press).

Wright JF 1995. Development and use of a system for predicting the macroinvertebrate fauna in flowing waters. *Australian Journal of Ecology* **20**, 181–197.

Wright JF, MT Furse & D Moss 1998. River classification using invertebrates: RIVPACS applications. *Aquatic Conservation – Marine and Freshwater Ecosystems* **8**, 617–631.

11 Ecology – overview and terrestrial systems

Peter Morris and Roy Emberton

11.1 Introduction

The EU/UK EIA legislation (§1.4) refers to the biological component of EIAs as *fauna* and *flora*. Dictionaries usually define these simply as "all the animals and plants in a given place or time, and a description of them". This is not sufficient for EIA which requires an understanding of how organisms are affected by environmental conditions and may respond to changes in these. The scientific study of the relationships between living organisms and their environments is **ecology**, so 'fauna and flora' refer to the **ecological component** of EIA.

Ecology is a broad and complex science, and the study of various aspects often requires different techniques and expertise. It is for this reason that, whilst the concepts and principles described in this chapter are generally applicable, the particular requirements for freshwater and coastal ecological assessments are discussed separately (Chapters 12 & 13).

The complex and dynamic nature of ecological systems imposes particular difficulties in obtaining adequate baseline data, making accurate impact predictions and formulating dependable mitigation measures. These problems may partly explain why reviews of the ecological assessments in EIAs (e.g. Spellerberg & Minshull 1992, Thompson *et al.* 1997, Treweek 1996, Treweek & Thompson 1997, Warnken & Buckley 1998) have shown many of them to be woefully inadequate in one or more of the following ways:

- lack of new surveys, and hence total reliance on existing (often out-dated or deficient) data;
- sources of information not cited, survey methods not adequately explained (or even mentioned), and no indication of precisely where and when field surveys were conducted;
- surveys very superficial, or inappropriate methods used, e.g.

  * information on species restricted to lists, with no quantitative data and no information of locations within sites – and faunal data often totally absent or mainly restricted to birds;
  * presence of designated sites not recorded;
  * poor (if any) description of habitats and no information on their locations and extents within sites;
  * no indication of trends;
  * field surveys carried out during inappropriate seasons, and lack of any repeat surveys to allow for seasonal variations in ecological systems;

- little or no indication of the environmental factors (including management) controlling the current ecological systems;
- prediction of impacts absent or vague, with no attempt to relate them to associated (e.g. hydrological) impacts;
- mitigation proposals absent, vague (without detailed prescriptions, and/or with little direct relevance to specific impacts identified in the EIS), or 'cosmetic', e.g. landscaping or tree planting mainly for amenity or screening purposes;
- little or no commitment to monitoring.

These findings suggest that there is considerable room for improvement in the ecological component of EIAs, and this is linked to the need for a wider understanding of the requirements for good ecological assessment and the opportunities and problems involved.

The ultimate aim of an ecological assessment is to avoid or minimise the impacts of a proposed development. They are therefore related to the aim of **nature conservation** which, in broad terms, is to maintain, and where possible increase, *biodiversity*. The importance of this was recognised by the Convention on Biological Diversity (CBD) at the 'Rio Earth Summit' (see §1.4).

Biodiversity has declined markedly during the past 50 years, and continues to decline rapidly. In the UK there have been serious losses of habitats (Fig. 11.1) and alarming declines of many, once common, species. Since 1900 there have been 154 UK species extinctions, and further species are under severe threat (WWF 1998a). Birds are currently of particular concern; of the 247 species of British breeding birds, 139 are thought to be declining, and 23 have declined by at least 50% in the past 25 years (EA 2000, RSPB 1999). Similar losses have occurred, and are continuing, in other parts of Europe, and ecosystems are being destroyed or seriously damaged at even faster rates in many parts of the world. A report by WWF (1998b) suggests that the health of the world's ecosystems declined by 30% between 1970 and 1995.

Global and regional reductions in biodiversity losses can be considered more relevant to SEA than to EIA, but it must be remembered that local losses contribute to cumulative declines. In addition, they can generate serious secondary impacts on the local environment. For example, vegetation damage or removal can lead to:

- loss of visual and noise barriers, and declines in landscape character and the context of amenity areas, heritage sites and buildings (Chapters 4, 6, 7);
- loss of slope and soil stability, and enhanced *runoff* – with consequent impacts such as increased *erosion*, flood hazard, and riverine sedimentation (Chapters 9 & 10);
- similar problems in coastal environments, e.g. reduction in land stabilisation and protection from inundation by sea water (Chapter 13);
- increased populations of pests or disease vectors through removal of natural controls.

11.2 Definitions and concepts

Ecology includes the study of species populations, biological communities, ecosystems, habitats and biotopes; and it is important to understand what these are and how they are interrelated. This section can provide only a brief explanation; further

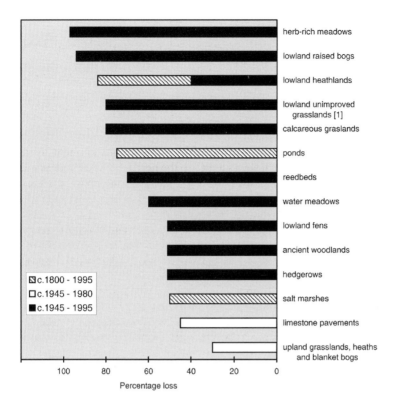

[1] The area of unimproved neutral grasslands declined by *c*.97% between 1930 and 1984 (EA 2000).

In some cases: (a) 'loss' includes degradation, e.g. of grasslands by 'improvement' or over-grazing; (b) losses have continued or accelerated in recent years, e.g. 25% of hedgerows between 1984 and 1990 (DoE 1993); so loss to date will be significantly greater than that cited.

Figure 11.1 Some estimated habitat losses in the UK.
Various sources including EA (2000), Dryden (1997), RSPB (1998).

information can be found in a wide range of ecology books, including introductory texts (e.g. Townsend *et al.* 2000) and more comprehensive texts (e.g. Begon *et al.* 1996, Krebs 2001). Canter (1996) is an EIA text that includes ecological examples, and Treweek (1999) focuses specifically on ecological EIA.

11.2.1 *Species populations*

The information presented in EISs is often restricted to lists of species that are present locally or on sites. This is rarely adequate because individuals of a species exist as members of a **species population**, and a simple presence record gives no information about (a) its *species abundance*, which may range from a few individuals to a thriving population, or (b) its population **distribution**, which could be throughout an area, or restricted to a few small patches within an area. Meaningful predictions about impacts on species frequently require abundance and distribution data, together with an understanding of why the current patterns exist and how the populations are likely to respond to the stresses imposed.

The viability of a population depends on the presence of a suitable environment with adequate resources. All organisms are constantly affected by, and interact with, a complex of **environmental factors** including:

- **abiotic** (physico-chemical) **factors** – water, temperature, light, oxygen, nutrients, toxins, *pH*, etc.;
- **biotic factors**, which involve interactions between species, i.e. competition, predation, parasitism and mutualism (mutually beneficial relationships).

Species can tolerate 'normal' short-term environmental variations, and while populations may undergo marked temporary fluctuations, they tend to remain stable in the longer term. Species may also be capable of responding to slow progressive environmental changes by evolving or changing their geographical range. However, their adaptations have evolved in response to past environmental conditions (experienced by previous generations) and they may be unable to adjust quickly enough to rapid environmental changes. '**Specialists**' (which are adapted to a narrow range of environmental conditions or food sources) are more vulnerable to such changes than '**generalists**' (which have less specific requirements). One of the greatest threats to most species is **habitat loss**, together with associated **habitat fragmentation**. A key issue in a fragmented landscape is the ability of species populations to survive in, and move between, small isolated habitat patches scattered within an urban or agricultural 'matrix' (see §11.5.4).

11.2.2 Communities

Species coexist with other species in biological communities. A **community** is an assemblage of mutually adjusted species populations in a given location at a given time. 'Mutually adjusted' means that coexistence of the species is facilitated by mechanisms such as *niche separation*, and in many cases by mutually beneficial relationships.

Strictly, a community includes all plants, animals and microbes, but although much is known about relationships between groups of species, comprehensive studies are generally limited to plant communities, which are relatively easy to study. Community studies focus on several **community attributes**. These attributes, and their relevance in EIA, are outlined in Table 11.1, and further explanation of two of them (succession and trophic structure) is provided in Figures 11.2 and 11.3.

It is important to understand that a *food chain* (Fig. 11.3) is simply a major route of energy and nutrients. In reality, a community's trophic structure consists of a **food web**, i.e. a network of feeding relationships between species. Food webs are usually complex, and few are fully understood. For example, it has taken 25 years to document the food web in a small estuary in Scotland, which has been shown to involve > 90 species and c.5500 feeding links (Gorman & Raffaelli 1993).

Community **productivity** (rate of production) varies widely, largely in relation to environmental temperature, water, and nutrient regimes. For example, tropical rainforests, swamps, estuaries and beds of marine *algae* normally have high productivities, while deserts, bogs and open oceans have low productivities. Highly productive communities have a large *biomass*, and some have high species diversities, although

Table 11.1 Major community attributes and their importance in EIA

Community attributes			Explanation and uses	Importance in EIA
Community structure	Physiognomy	Plant life-form composition	Types and proportions of plant **life forms** that make up the vegetation matrix, which is recognisable as a vegetation type, e.g. woodland, scrub, grassland.	Used in vegetation and habitat classifications.
		Vertical structure (stratification)	Characteristic of vegetation type, e.g. heaths normally have only two layers while broad-leaved woodland typically has four layers: *canopy layer* (crowns of adult large trees); *shrub layer* (shrubs/small trees/juveniles of large trees); *field layer* (mainly upright herbs); and *ground layer* (creeping herbs, bryophytes, lichens).	Useful in evaluation of woodlands (structural complexity increases with age).
		Appearance	Vegetation colour and texture; can be detected on remote sensed imagery (Chapter 15).	Useful in land cover surveys.
	Species structure*	Species composition	The species comprising a community. Can be represented by a species list, but usually includes quantitative data on **species abundance**, and often on **dominant species, keystone species** and **indicator species**.	Important in Phase 2 surveys, e.g. using the NVC (Appendix F.4).
		Species richness and species diversity	Species richness is simply the number of species in a community. Species diversity is a measure of *both* the number of species *and* the relative abundance of each – and two communities with the same number of species can have different species diversities. 'Species-rich' and 'species-poor' can refer to high and low species richness *or* species diversity.	Often used in site or community evaluation, but must be applied with caution (see Appendix D.3.2).
	Trophic structure	Food chains and food webs	The flows of nutrients and energy within communities, involving their absorption and assimilation by autotrophs and their subsequent transfer to other trophic levels (Fig. 11.3). Related attributes are community production (amount of energy utilised) and productivity (rate of production).	Knowledge of food chains and webs can assist in impact prediction and mitigation.

Table 11.1 *(continued)*

Community attributes			Explanation and uses	Importance in EIA
Community patterns	Spatial	Community gradients and mosaics	The spatial pattern of communities in a landscape or site, which tends to consist of gradients (§11.2.5) rather than discrete communities.	Interpretation of existing and new survey data.
	Temporal	Short-term (cyclical) changes	Seasonal variations; intrinsic vegetation cycles (e.g. associated with forest canopy gaps); and perturbations (e.g. caused by fire, storm, flood, drought or cold).	Implications for the timing of sampling programmes.
		Succession	Fairly rapid, progressive change that culminates in the development of a climax community (Fig. 11.2).	Prediction of change without the project.
		Long-term changes	Relatively slow post-successional changes (a) under a stable climate (e.g. by immigration/emigration, evolution and soil changes), or (b) in response to climatic change.	Relevant if project may contribute to climate change.
		Stability, sensitivity and resilience	The degree to which communities are: (a) maintained though time with little change (stability); (b) susceptible or resistant to disturbances (sensitivity); (c) capable of recovering from disturbances (resilience).	Prediction of impacts and their significance.

* Full species composition and diversity studies are usually restricted to plant communities, which are relatively easy to sample (see Appendix G).

low-productivity ecosystems are sometimes more biologically diverse than many with higher productivities (Hamblin 1998).

11.2.3 Ecosystems

An ecosystem is a self-sustaining, functional system consisting of a community and the environment in which it exists. Ecosystems are therefore total ecological systems, with innumerable components, processes and interactions – often involving delicate balances in relationships. Consequently, a change in even a single component, such as a species population or an environmental variable, tends to cause knock-on effects that are often unpredictable.

Several major subsystems of ecosystems can be recognised, all of which interact with the others and contain interacting components (Fig. 11.4). It is important to consider communities in this context. For example:

- the flows of energy and nutrients through communities are integral parts of much larger flows through ecosystems (see below);

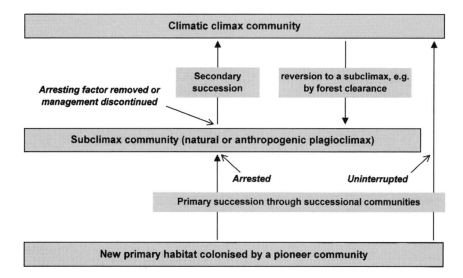

A **primary succession (prisere)** starts from a near-sterile **primary habitat**, e.g. rock (exposed by volcanic activity, glacial retreat, mineral extraction etc.) or new water body (lake, reservoir etc.) which is colonised by a **pioneer community**. This is followed by a series of **successional (or seral*) communities** (each replacing the previous one). The process (which can be quite rapid) culminates in the **climatic climax community (*biome*)** which, under an unchanging climate, will be fairly stable – although it will change slowly in response to factors such as long-term soil development.

Succession can stop at a persistent **subclimax** stage. The arresting factors can be natural, but most 'subclimaxes' (including UK **heathlands** and grasslands) are ***semi-natural*** communities maintained by human activity; and because these **anthropogenic climaxes** differ from natural subclimaxes, they are often called **plagioclimaxes**. However, they are much more natural than communities such as 'improved' grasslands. Removal of an arresting factor (including management) results in **secondary succession** which is usually rapid because features such as soil already exist.

Precise prediction of succession is difficult because (a) biomes are broad generalisations, within which there is wide variation in relation to local climates, soils etc., and (b) secondary successions are influenced by the 'stock' of potential colonisers living in the area and in the soil ***seed bank***. In most of Britain, however, abandoned land and unmanaged plagioclimax communities will revert to some form of scrub and then woodland.

* The term seral is often used as a synonym of successional, but a **sere** is strictly a particular type or example of primary succession. Recognised types include the **lithosere** (from rock) and the **hydrosere** (from open water), both of which may eventually culminate in the same climatic climax.

Figure 11.2 Simple model of ecological succession.

- community spatial and temporal patterns are associated with environmental, and hence ecosystem, patterns;
- stability, fragility and resilience are all attributes of ecosystems (not just the communities).

Ecosystems are sustained by **fluxes of energy and materials** (including nutrients). In most cases the energy source is solar radiation, only a small proportion of which (normally < 1%) is absorbed by the autotrophs and passes through the community (Fig. 11.3). However, this does not mean that the 99% is unimportant to the community or that the latter has no additional influence. For example:

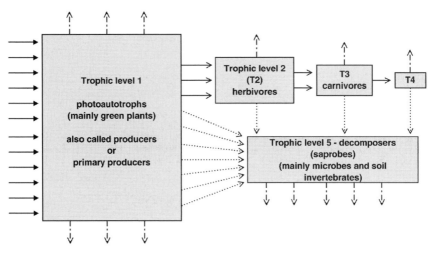

Input of light energy to trophic level 1 (autotrophs).

Transfer of energy (in organic compounds) to higher trophic levels (T2 (herbivores), T3 (carnivores) and T4 (top carnivores)) along the **consumer** (or grazing) **food chain**.

Transfer of energy to trophic level 5 (decomposers) in the form of dead plant and animal remains and animal excretory products. This route is often called the **decomposer food chain**.

Loss of energy from all trophic levels, mainly as heat generated by respiration.

The sizes of boxes and numbers of arrows are intended to indicate the relative amounts of energy entering, leaving and within the various trophic levels – but are not strictly proportional.

Communities need sustained flows of nutrients and energy. To achieve this, they rely on **autotrophs** which make organic compounds using external sources of inorganic nutrients and energy. In terrestrial (and most aquatic) communities, these are **photoautotrophs** in which the primary process is photosynthesis of carbohydrates from carbon dioxide and water using light energy absorbed by chlorophyll. All other organisms are **heterotrophs** that obtain their nutrients and energy from organic materials synthesised by autotrophs.

Energy assimilation by photosynthesis is called **primary production** (PP), and the total amount is **gross primary production** (GPP). Plants use *c*.55% of this, so *c*.45% (**net primary production** (NPP)) is available for the heterotrophs, whose utilisation of energy is called **secondary production**. In terrestrial ecosystems, only a small proportion of NPP is consumed by herbivores, and passes along the **consumer (grazing) food chain** – the bulk goes directly to the **decomposers** as dead plant remains. The decomposers also receive energy in the form of animal remains and excretory products.

All organisms carry out **respiration,** by which organic compounds are broken down to release usable energy. Much of this energy is lost to the environment as heat, so (a) energy flow through the community must be sustained by PP, and (b) less energy is available to higher trophic levels – which is why there is a **pyramid of decreasing biomass** from trophic levels 1 to 4, and why top carnivore populations are generally small.

Figure 11.3 Simple model of energy and nutrient flow through a terrestrial community. A similar model can be constructed for aquatic communities (see §13.2.5).

- the overall energy flow controls the ecosystem's thermal (temperature) regime, both directly and by providing the energy for evaporation;
- *transpiration* from vegetation can play a major role in cooling the ecosystem, both directly by its evaporative energy requirement, and indirectly by inducing the formation of a protective cloud screen (especially over tropical rainforests).

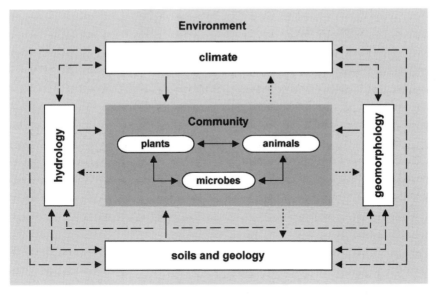

affects of environmental factors on the community

affects of the community (especially vegetation) on environmental systems

interactions between environmental systems

interactions within the community

In any given location, the *climate* affects the geomorphology, hydrology, soil and all the species of the community. Conversely (a) *microclimates* and local climates are affected by the other sub-systems, and (b) vegetation and major geomorphological features such as mountain ranges, can affect macroclimates (regional and even global).

The community is also strongly influenced by, and in turn affects, the other sub-systems – particularly the soil, which is an ecosystem in its own right, with a community consisting of plant roots, soil animals, bacteria and fungi.

In addition, there are innumerable interactions within the subsystems, e.g. between abiotic environmental factors and between species.

Figure 11.4 Simple model of interactions between subsystems of a terrestrial ecosystem.

Local and global ecosystem energy flows are minor diversions of the linear flux of solar energy from the sun to space. By contrast, there is little exchange of matter between the earth and space, and the global flows of materials are essentially cyclical. These **biogeochemical cycles** can be divided into two types:

1. **Volatile element cycles** – of elements that can exist in gaseous form or as constituents of atmospheric gases, and therefore have efficient global circulations. These are hydrogen, oxygen, carbon, nitrogen and sulphur, which are all *macronutrients*. Atmospheric pollutants can have similar global dispersions.

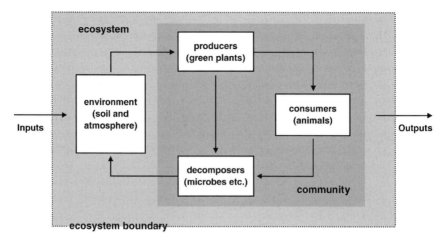

Figure 11.5 Nutrient flows within in a terrestrial ecosystem and across its boundaries. A similar model could be constructed for aquatic ecosystems, the main difference being that the environmental source of nutrients is the water body.

2. **Non-volatile element cycles** – of elements that do not have an atmospheric phase, and therefore have much less efficient cycling. They include important macronutrients (calcium, iron, magnesium, phosphorus, potassium) and *micronutrients*.

A local ecosystem has **internal cycles** in which nutrients are absorbed by the autotrophs, pass along food chains and are returned to the environment by excretion and decomposition. However, nutrients also enter and leave the ecosystem, which therefore has a **nutrient budget** (for each nutrient) that depends on the balance between inputs and outputs (Fig. 11.5).

Volatile elements can enter an ecosystem from the atmosphere, and the outputs of these are normally balanced, or even exceeded, by inputs (carbon, hydrogen and oxygen are assimilated by photosynthesis, and nitrogen is 'fixed' by lightning and nitrogen-fixing bacteria). By contrast, non-volatile elements can only enter local ecosystems by *weathering* of bedrock, or in solution or suspension in water; so their budgets are strongly influenced by the ecosystem water budget (§10.2.2). An ecosystem may receive nutrients in drainage water from higher in its *catchment*, but may lose nutrients by *leaching* and *erosion*. Consequently, if the area has a climatic water surplus, as in the UK (§10.2.3), outputs of non-volatile nutrients are likely to exceed inputs, and ecosystems tend to undergo gradual but progressive nutrient depletion.

Ecosystem nutrient regimes can be markedly affected by human activities including agriculture, forestry and development, e.g. through nutrient depletion by erosion of exposed soils, or conversely through *eutrophication*. In addition, many toxic pollutants can enter, and circulate within, ecosystems in the same ways as nutrients.

An ecosystem is a **concept** (or model) rather than a **percept**. The most perceivable elements of ecosystems are habitats, and the terms 'ecosystem' and 'habitat' are often used interchangeably.

11.2.4 Habitats

A habitat is usually defined in dictionaries as the 'home' of a species. However:

- It actually refers to the **type of environment** to which the species is adapted, and hence in which it is usually found. Strictly, this means the environment's physico-chemical conditions, but other organisms (and hence biotic factors) are effectively also components of a species' habitat.
- A given local environment also favours other coexisting species, and is thus the habitat of the whole community, although species will have specific habitat requirements, and may be restricted to *microhabitats* within a larger habitat.
- Species and communities normally inhabit particular **habitat types**, which are recognised in **habitat classifications** (Appendix F).
- Where there is appreciable vegetation, habitat types are normally identified and named largely by this (e.g. woodlands, grasslands). However, many names also reflect physico-chemical features (e.g. saltmarshes, calcareous grasslands); so the types defined in most classifications are 'broad sense habitats', characterised by *both* the abiotic environment (the habitat in the strict sense) *and* the associated community (usually represented by the vegetation).

There is a trend to replace 'broad sense habitat' by **biotope**, which is defined as a 'strict sense' habitat *and* its associated community *operating together*. Theoretically, therefore, biotopes are functional units and hence ecosystems, but in practice they are classificatory units synonymous with 'broad sense habitat types'. However, unless there are specific reasons for referring to biotopes, 'habitat' is used in this book because it is still more widely used in conservation literature and legislation.

11.2.5 Variability and inter-gradation of ecological systems

Ecological assessment tends to lean heavily on habitat and community classifications because these provide accepted categories that are meaningful to all users. However, it is important to understand that all formal classifications are intrinsically flawed because:

- Ecological systems are **infinitely variable**, and hence are not amenable to precise classification. For example, the community in a given location is the product of past and present local factors, and is therefore unique – so although it can be identified as an example of a general type, it will differ somewhat from similar examples on other sites, or even within the same site.
- Whilst a managed landscape, as in the UK, is generally characterised by sharp boundaries, these are nearly all man-made. The spatial pattern of natural communities tends to consist of **community gradients** rather than discrete entities, with attributes such as species composition adjusting progressively along environmental gradients.

Where environmental gradients are steep, there may be obvious transition zones (**ecotones**) between adjacent communities, and these are often species rich because they contain species of both communities. Mosaics of inter-grading communities may be readily apparent within sites, but changes from one community to another are often much less discernible. For example, a cursory look at a meadow may give the impression of overall homogeneity, but closer examination may reveal subtle community gradients. Indeed *semi-natural vegetation* is rarely if ever homogeneous, even within small areas. Similarly, now-isolated patches of a community type can be considered to represent 'nodes' in a continuum of variation analogous to gradients within sites.

As indicated in Table 11.1, communities and ecosystems also exhibit **temporal variation**, which is sometimes reflected by spatial patterns within sites. The three main types of temporal change are:

- intrinsic short-term cyclical variations, and longer-term progressive changes, i.e. succession and post-successional development (Fig. 11.2);
- responses to disturbances including natural perturbations (e.g. fire or severe storms) or human impacts;
- responses to progressive 'external' environmental changes, e.g. climate change.

Different ecosystems exhibit different degrees of **stability** (tendency to remain unchanged), **sensitivity/fragility** (ability to withstand disturbances) and **resilience** (ability to recover from disturbances). The potential for degradation and recovery also depends on the severity and persistence of the impacts. For example, recovery from severe soil erosion or pollution by persistent toxins is likely to be very slow.

11.3 Legislative background and interest groups

11.3.1 Legislation and international conventions

As with all EIA components, the main legal requirements governing ecological assessment are those in the EIA legislation (§1.4). An important change in the amended Directive 97/11 is the section (in Annex III) on the "location of projects" in relation to the "environmental sensitivity" of areas. However, this concept of 'sensitive environments' requires further modification if it is to rectify the widely held perception that only rare and protected habitats need be conserved.

Ecological EIA is also affected by legislation and international conventions which have one or more of the following interrelated aims:

- **protection of species** (with emphasis on high-status species[1]) – which usually also involves protection of their habitats;
- **protection of habitats** (with emphasis on high-status habitats) – usually within protected sites;
- **protection of sites** – with site designation based largely on the presence of species and/or habitats they are intended to protect;
- **countryside conservation** – which focuses on the protection of landscape and cultural features (Chapters 6 & 7) but assists nature conservation because (a) scenic features are often semi-natural habitats, and (b) the survival of many species depends on sensitive management of the wider countryside outside protected sites.

Tables 11.2 and 11.3 list the main international conventions, EU Directives and UK legislation on nature conservation. The texts of (and guidance on) many of these are available from the COE, EC-EDG, DETR, Ramsar and UNEP or UNESCO

---

1 When applied to species, habitats or sites, '**high-status**' is used to mean high conservation value in terms of the criteria described in Appendix D.

Table 11.2 International conventions/agreements and EU directives on conservation

UNESCO Man and the Biosphere (MAB) programme 1970 – established **Biosphere Reserves**.

Ramsar Convention on Wetlands of International Importance 1971 – to conserve wetlands of international importance, especially as waterfowl habitats, in **WIIs (Ramsar sites)**.

UNESCO Convention on Protection of World Cultural & Natural Heritage 1972 – to protect natural and cultural areas of outstanding international value as **World Heritage Sites**.

Council of Europe (COE) 1973 – Recommendation for establishment of European **Biogenetic Reserves**.

Convention on International Trade in Endangered Species (CITES) 1973 (strengthened in EU Directive 338/97 EEC) – Provisions to regulate the international trade of wild species of plants and animals.

(Wild) Birds Directive 79/409/EEC 1979 – Protection of all naturally occurring wild bird species and their habitats, with particular protection of rare species (listed in Annex I) in **Special Protection Areas (SPAs)**.[1]

Bonn Convention on the Conservation of Migratory Species of Wild Animals (CMS) 1979 – to protect threatened animals (listed) that migrate across national boundaries and/or the high seas.

Bern Convention on the Conservation of European Wildlife and Natural Habitats 1979 – to protect endangered species and their habitats. Amended (1989, 1996) to set up the EMERALD network of **Areas of Special Conservation Interest (ASCIs)**. Its provisions underlie the Habitats and Species Directive.

UN Conference on Environment and Development (Rio Earth Summit) 1992 – **Agenda 21** on sustainable development. Includes the **Convention on Biological Diversity (CBD)**.

Habitats & Species Directive 92/43/EEC (HSD) 1992 – (a) to protect important **natural habitats**[2] (listed in Annex I, amended in **Directive 97/62/EC**) and **species**[2] (listed in Annex II), using measures to maintain or restore their "favourable conservation status", principally by **Special Areas of Conservation (SACs)**[1], but also (through land-use and development policies) by management of landscape features of importance to wildlife outside SACs; (b) to safeguard species needing strict protection (Annex IV).

Two provisions of particular relevance to EIA are:
* Any project likely to have a "significant effect" on an SAC must go through an EIA, which should take into account the impacts of the project in relation to other projects (i.e. cumulative impacts), and will normally be accepted only if shown not to affect the "integrity of the site";
* Harmful development may be allowed if there are "imperative reasons of overriding public interest", but there must be "compensatory measures", and if the site hosts a priority habitat or a priority species, only "human health or public safety" reasons are normally acceptable.

Agreement on the Conservation of Bats in Europe (EUROBATS) 1994 – to promote the conservation of bat species across Europe.

---

1 **Potential SPAs (pSPAs)** and **Candidate SACs (cSACs)** are submitted to the EC and, when adopted, will be collectively called **Sites of Community Importance (SCIs)**, which are also known as **Natura 2000 sites** or **European sites**.
2 Some Annex I habitats and Annex II species have **priority status**.

Table 11.3 Key UK legislation on conservation

National Parks & Access to the Countryside Act 1949 – Provisions for creation of **Areas of Outstanding Natural Beauty** (AONBs), **National Parks** (NPs), **National Nature Reserves** (NNRs), **Sites of Special Scientific Interest** (SSSIs) & **Local Nature Reserves** (LNRs).

Countryside Act 1968 – Powers under the 1949 Act strengthened.

Town and Country Planning (Scotland) Act 1972 – Provisions for **National Scenic Areas** (NSAs).

Wildlife & Countryside Act 1981 (WCA) & (Amendment) Act 1985 – Increased protection of SSSIs. Enactment of the Birds Directive. Designation of **protected species** (listed in Schedules 1, 5, 8). Provisions for **Limestone Pavement Orders; Marine Nature Reserves** (MNRs) and **Areas of Special Protection** (AoSPs).

Wildlife, & Nature Conservation and Amenity Lands (N. Ireland) Orders 1985 – Declaration of **Areas of Special Scientific Interest** (ASSIs), MNRs, NNRs, and **Local Authority Nature Reserves** (LANRs).

Agriculture Act 1986 – Provision for **Environmentally Sensitive Areas** (ESAs) part-funded by the EU under Regulation 797/85/EEC.

Environmental Protection Act 1990 – New NCCA structure (Appendix B). Further protection of SSSIs.

Town & Country Planning Act 1990 – Obligation for LAs to take account of nature conservation in *Development Plans*, based on surveys to provide adequate information on species and habitats.

Planning and Compensation Act 1991 – Additions to classes of project requiring EIA. Increased powers of LAs to safeguard conservation and amenity areas. Requirement for Development Plans to include policies on conservation of natural beauty and amenity of land.

Natural Heritage (Scotland) Act 1991 – **Natural Heritage Areas** (NHAs) given special protection.

Protection of Badgers Act 1992 – Wilful harm of badgers or their setts prohibited. A licence from relevant NCCA is required to permit interference with a badger sett during development.

Conservation (Natural Habitats &c.) Regulations 1994 (The Habitats Regulations) – Implement the HSD (through provisions of the WCA 1981). Regulation 48 requires the production of an *Appropriate Assessment* to determine if a project is likely to have an adverse effect on site integrity. This can draw on information given in the EIS but must be a separate document.

Environment Act 1995 – Includes provisions for hedgerow protection, and for MAFF to run the **Countryside Stewardship Scheme** (incentives for sensitive management).

Hedgerows Regulations 1997 – Removal of most hedges of ≥ 20 m prohibited without notifying the LA, which may impose 'hedge retention notices' for 'important' hedges (≥ 30 yr old & fulfilling at least one of a set of criteria (e.g. historic; presence of protected species or prescribed numbers of woodland or woody species).

The Conservation (Natural Habitats &c.) Regulations 2000 – Extend the Habitats Regulations (1994) to include full protection to cSACs before they are adopted by the EC as SCIs (see Table 11.2).

Countryside and Rights of Way Act 2000 (CROW) – Increased protection of SSSIs, powers against wildlife crime, and access to the countryside. Improved management of AONBs. Provision of a statutory basis for the UKBAP (§11.3.2) including government duties to further the conservation of listed species and habitats.

---

websites (Appendix A). Further information on the types of protected site designated under the legislation is given in Table D.3 (p. 423), which also lists various types of non-statutory site.

The Habitats and Species Directive (HSD) is particularly important but, like most legislation, it does have deficiencies. For example,

- "Favourable conservation status", "significant effect" and "imperative reasons of overriding public interest" are not clearly defined and are hence open to interpretation.
- "Compensatory measures" may be inadequate or inappropriate.
- Emphasis on HSD species, habitats and SACs may lead to under-valuation of those that do not qualify, e.g.:

    * threatened British species and habitats that are not Annex I/II types, usually because they are well represented in the EU as a whole – which is the context of the HSD. A notable example is **ancient woodland**, which has no statutory protection *per se*, and only *c*.15% of which has any protective designation in the UK (WT 1999);
    * habitats which, because of the variability of ecological systems, do not match Annex I types precisely;
    * wider-countryside species and habitats which, in spite of the HSD provision and UK policy statements, lack statutory protection other than under the *Hedgerow Regulations 1997*;
    * sites that lack SAC (or similar international) designation.

The protection afforded to sites in the UK varies appreciably in relation to their designations. Sites with international designations have the highest level of protection, while non-statutory sites normally have little legal protection. UK statutory sites, including all SSSIs, are supposed to enjoy a high degree of protection. However, the legislation in force prior to CROW 2000 was generally recognised to be inadequate in terms of both protection from external threats and incentives for appropriate management. The new Act strengthens the protection of SSSIs, principally through the following provisions: (a) a statutory duty for public bodies to further SSSI conservation and enhancement; (b) powers for NCCAs to refuse consent for damaging activities, and to promote positive management; (c) increased penalties for deliberate or reckless damage by owners or any other party.

11.3.2 Policies and guidance

EU environmental policy includes the *Sixth Environment Action Programme* (§1.4) in which one of the key objectives is to "protect and restore the functioning of natural systems and halt the loss of biodiversity". A Pan European Biological Landscape Diversity Strategy (PEBLDS) is being developed by COE, UNEP and ECNC (which hosts a *Strategy Guide* website (http://www.strategyguide.org)).

The UK Government published its strategy for implementing the CBD (Table 11.2) in the *UK Biodiversity Action Plan* (UKBAP) (DoE 1994a), and set up the *UK Biodiversity Steering Group* (UKBSG), which made proposals: (a) to conserve key species and habitats by the production of national **species action plans (SAPs)** and **habitat action plans (HAPs)**; and (b) to promote the development of **local biodiversity action plans (LBAPs)** as a means of implementing the national plans.

It also proposed protocols for selecting key species and habitats, and produced an initial batch of SAPs and HAPs. The government endorsed the report, and set up the *UK Biodiversity Group* (UKBG) to implement the programme. This has made some revisions to the protocols, and has produced further SAPs and HAPs (see Appendix D). LBAPs are being developed by local partnerships under guidance from ULIAG (1997) – often as components of local **Agenda 21** plans.

Planning policy guidance on nature conservation is given in PPG9 (DoE 1994b). Similar guidance is given in NPPG14 *Natural Heritage* (Scotland), Technical Advice Note (Wales) 5 *Nature Conservation*, and PPS2 *Planning and Nature Conservation* (Northern Ireland). The guidance clearly states that nature conservation issues should be (a) taken into account in **Development Plans**, and (b) included in the relevant surveys to ensure that these are based on adequate ecological information, and take account of local nature conservation strategies. It also specifically refers to aspects such as the conservation of:

- designated sites, including the presumption that an EIA should be undertaken for all proposals (a) within or adjacent to designated sites of national/international importance, or (b) likely to have a significant effect on a SAC, SPA or Ramsar site;
- non-designated sites and the wider countryside, including **linear habitats** and "sites of local conservation importance".

The importance of the wider countryside is recognised in the Rural White Paper *Our Countryside: The Future* (DETR 2000). This includes commitments to: revise PPG9; protect/enhance/restore key wildlife sites/habitats; promote environmentally friendly farming; consult on (a) applying EIA to major agricultural projects that may affect wildlife, and (b) introducing 'impact fees' and 'offsetting' requirements for off-site impacts; and improve the planning and implementation of LBAPs. The wider countryside approach is also inherent in EN's *Natural Areas* programme (EN 1998), and SNH's *Natural Heritage Zone* programme (http://www.snh.org.uk/).

11.3.3 Consultees and interest groups

In the UK, the **statutory consultees** for the ecological component of EIA are the regional NCCAs (Appendix B). The relevant NCCA has several important roles:

- It must be notified by the local planning authority (LPA) about a development application and will assist in the screening and scoping procedures.
- It will hold, and has a duty to provide if requested, non-confidential information on the local ecology (and perhaps on previous EIAs).
- It will employ, and have contacts with, experienced ecologists (EN is compiling a Directory of Expertise), and will be willing to give advice on all aspects of the EIA, including appropriate mitigation measures. This will be in concept only, although it will be willing to review mitigation proposals prior to their inclusion in the EIS.
- The LPA must supply it with a copy of the EIS for comment. This may include (a) an appraisal of the EIS in terms of its scope, technical competence, validity, and proposed mitigation measures, and (b) an indication of whether it would support or oppose planning consent.

Other GOs likely to have an interest in the ecological assessment are:

- the relevant EPA (Appendix B), especially when there are concerns relating to pollution, contaminated land, freshwater ecosystems or coastal ecosystems (Chapters 9, 10, 12, 13);
- organisations such as the CEH and FC;
- the LPA(s), who will have specific policies on nature conservation and on the implementation of relevant national legislation.

NGOs that have interests which may be affected, and/or are potential sources of information and advice, include: BBCS, BBS, BCT, BLS, BSBI, BTO, CPRE, CPRW, MS, NT, NTS, Plantlife, RSPB, TWT/LWT, WT, WWF-UK, WWT and local clubs/societies (e.g. bat, birdwatching, natural history). It may be only necessary or possible to consult a few of these organisations, and a 'starting list' should be made as part of the scoping process. In addition, ecological concerns may be relevant in the context of public consultation, e.g. with parish councils, local farmers/landowners, residents and community groups.

11.4 Scoping and baseline studies

11.4.1 Introduction

Scoping should follow the principles and procedures outlined in §1.2.2. In a few cases, the **impact area** may be confined to the project site and its immediate surroundings. However, IEA (1995) recommend that a minimum 2 km radius should normally be considered for non-linear projects; and a corridor at least 1 km wide should be examined along the entire proposed route of linear projects such as roads. Moreover, impacts associated with air pollution or hydrological changes may have more widespread effects (Chapters 8 & 10). Estimating the impact area is bound to involve some educated guesswork and the initial estimate may have to be revised in the light of information that emerges during the assessment process.

The amount of ecological information that could be collected is enormous, and it is essential to **focus resources** on important aspects. This can be done by identifying *valued ecosystem components* (VECs) which can be species, habitats or other attributes such as socio-economic value (CEAA 1999, Treweek 1999). It is also important to focus on VECs that are likely *receptors*, since there is no point in using resources to study species or habitats that will not be impacted (Wathern 1999).

The most obvious VECs will be protected species, habitats and sites. However, the scoping inventory should include all species/habitats/sites that may warrant further investigation, including (a) those that may have local importance, and (b) small habitat patches and **linear habitats** (see Spellerberg & Gaywood 1993), which can be valuable in their own right, and may also act as refuges, *stepping stones*, *wildlife corridors* or *buffer zones*, often in an otherwise urban or intensively cultivated landscape. Ultimately, the baseline survey (and the evaluation of baseline conditions) should include all ecological receptors that qualify as VECs because they have 'high status' in terms of the criteria outlined in Appendix D.

11.4.2 Methods and levels of study

In selecting study methods and levels, compatibility with other ecological surveys is desirable – and there is a strong case for adopting the three-phase strategy employed by JNCC (1993) for environmental audits, and recommended by IEA (1995) for baseline surveys in EIA (although it should be noted that some standard methods, such as that prescribed in DoT (1993), use a different phasing system which is more integrated into the design process).

The JNCC phases are progressive in terms of information sought, and hence intensity of study.

- **Phase 1 survey** aims mainly to provide information on habitats. All ecological assessments should include a Phase 1 survey of the impact area.
- **Phase 2 survey** is a more detailed study of species, habitats and communities in selected areas, e.g. for the purpose of site evaluation. The majority of EIAs will require surveys at this level.
- **Phase 3 survey** involves intensive sampling to provide detailed quantitative data on species populations or communities. This level of study is rarely undertaken in EIA, and in general is necessary only when (a) adequate predictions about impacts on a high-status ecological receptor cannot be made from Phase 2 data, or (b) detailed post-development monitoring is required.

Resource and time constraints often impose severe limitations on the range and depth of field survey work that can be conducted in EIA, and it is important to make maximum use of existing information by means of the **desk study** (§1.2.2). Sources of ecological information are given in Table 11.4. The organisations listed generally hold more information than that shown, and can usually be approached directly. Some of the NGOs referred to in §11.3.3 may be particularly useful sources of local information because they often have local recording groups. However, their limited resources may restrict their ability to respond to enquiries. Sources of related information are given elsewhere in the book as follows: geological and soil maps – §9.5.2; hydrological data – Table 10.4 (p. 216); remote sensed data – Table 15.1 (p. 371); digitised (GIS compatible) data – Table 16.1 (p. 384); historical information – Appendix C (p. 416); species' conservation status, distributions and habitat requirements – Appendix E (p. 431).

Whilst the desk study is essential, much of the existing information will be sketchy or out of date. Most protected sites will have been surveyed to some extent, but even here the quantity and quality of available information is often surprisingly low; and the impact area is likely to contain other sites about which there is little or no information. Existing surveys may not be accurate because they are often conducted by teams with varied expertise. For example, the Peak District National Park Authority found 219 species-rich meadows that were not recorded in a previous Phase 1 survey (Parker 1998). Consequently, it is normally essential to undertake **new fieldwork**.

11.4.3 Resource requirements and timing

The resources needed for an ecological assessment will vary in relation to factors such as the availability of existing information and the need for Phase 2 surveys.

Table 11.4 Sources of ecological information

BSBI (Botanical Society of the British Isles) http://members.aol.com/bsbihgs

Database including species occurrence in *vice-counties*, and sources of records. Interactive map (under development) to provide information (including lists of floras) for any vice-county.

Wild flower species recommended for planting/sowing on various soils, and addresses of plant/seed suppliers.

CEH (Centre for Ecology and Hydrology) http://www.ceh.ac.uk/

BRC (Biological Records Centre) Database – Holds records of species location co-ordinates, with information on habitat and conservation status. It is used as the basis for many published distribution maps.

National *critical loads* mapping programme – Methods and results (database & maps) for assessing the vulnerability of ecosystems to atmospheric pollution (especially acid deposition and nitrogen).

Countryside Surveys (CS90 and CS2000) reports (DoE 1993; Haines-Young *et al.* 2000). CS2000 data are also available from http://www.cs2000.org.uk/, and CS90/2000 data are available from the *Countryside Information System* (CIS) (see Table 16.1, p. 384) which holds a wide range of data sets including BRC data, breeding birds atlas, designated sites/areas, UKBG broad habitats and Natural areas.

EN (English Nature) http://www.english-nature.org.uk/

Designated-area maps – small-scale maps of England showing features such as Natural Areas, AONBs, Community Forests, Heritage Coasts, NPs, National trails, protected sites, and LA boundaries.

National Map Series – 14 regional maps of England at 1:200 k, showing features such as: the 10 km National Grid; roads, railways, rivers, settlements, etc.; Natural Area boundaries; and designated areas.

Natural Areas (EN 1998) – CD-ROM. Information on Natural Areas including: habitat types; maps; designated sites; UKBAP species and habitat targets; Species Recovery Programme; and Natural Area profile.

Ancient Woodland Inventory (AWI) of sites measuring > 2 ha. The database is available online, together with a GIS digital boundary set of AWIs and SSSIs.

JNCC (Joint Nature Conservation Committee) & UKBG (Biodiversity Group) http://www.jncc.gov.uk/ukbg

Online information, e.g. habitat classifications, plant conservation status, butterfly distribution maps, bird census results, legal protection, SACs, SPAs and Ramsar sites (including locations), information services.

ISR (Invertebrate Site Register) – County-based database of important invertebrate sites. Not available online.

UKBAP database – of priority species, HAPs, SAPs and LBAPs (including contact addresses) (see §11.3.2).

NBN (National Biodiversity Network) http://www.nbn.org.uk/

NBN online database – Includes: species conservation status and current legislative protection; legislation; information on sites in relation to habitat/vegetation classifications; a catalogue of organisations/data sources.

Table 11.4 (*continued*)

NHM (Natural History Museum) http://www.nhm.ac.uk/

Postcode Plants (flora for fauna) **database** – Provides lists of plant species recorded in areas (accessed via postcodes, but based on hectads) and includes checklists of native plants, butterflies and mammals.

Other online sources of information (full names and Internet addresses are in Appendix A)

CEAA, EC, ECNC, Ecology WWW, EEA (& ETCs), ERIN, FOE, Greenchannel, IUCN, Natrurenet, RSPB, UNEP, UNESCO, USDA-NRCS, USEPA, USFWS, USNTIS, WCMC, WWF/WWF-UK.

Other sources of local information

Local habitat surveys such as Phase 1 surveys, LBAPs, and surveys conducted for Development Plans, Local Agenda 21 programmes or previous EIAs. Copies will be held by LAs, regional NCCA offices and/or LWTs.

More detailed site-specific information (including SSSI notifications) held by the NCCA, LA or NGOs.

Species distribution data including county/local-area and site-specific records. Sources include: habitat and species surveys, e.g. for LBAPs; county floras/atlases; data held by NGOs and by the local biological records centre (LBRC) – usually at the County Records Office (see Donn & Wade 1994).

Other research projects conducted independently or under contract to a GO or NGO. The local academic institution(s) and regional NCCA office may hold or know of these, but a literature search is desirable.

Mapping will be essential, and consideration should be given to the use of GIS (Chapter 16) which can facilitate impact prediction and integration of ecological information with that on other EIA components.

'Resources' must include **ecological expertise**. Employment of one or two competent ecologists is normally adequate for scoping and Phase 1 surveys, but Phase 2 surveys require experts in aspects such as species identification and the application of appropriate sampling and data analysis methods (Appendix G.1, p. 455). In addition, the work undertaken by different specialists must be co-ordinated, and the findings integrated in the EIS.

Under a seasonal climate, as in Britain, the **timing of fieldwork** can be a critical factor because it is difficult or impossible to sample many species or communities during some parts of the year (see Appendix G.1). Ideally, therefore, the survey should start a full year before the submission date of the EIS. Developers' timescales are often inconsistent with this requirement, especially (a) in the case of small projects, where the total time for design and planning application is less than a year, or (b) where a developer defers the start of an EIA until late in the development process. Under such circumstances the accuracy of the ecological assessment can be compromised, and the developer should be warned of the difficulties and potential consequences. There are many instances where planning applications have been made conditional (Section 106 agreements) on the outcome of further studies.

Phase 1 and Phase 2 surveys can be carried out sequentially, and decisions on the nature and extent of Phase 2 studies can often best be made after evaluation of the Phase 1 survey findings. However, existing Phase 2-level information can be utilised as soon as it is obtained by the desk study; and because of the time constraints, it may be necessary to start Phase 2 field surveys as soon as the need becomes apparent during scoping or the Phase 1 survey.

11.4.4 Phase 1 field surveys

Specific Phase 1 methods are available for marine habitats (§13.4.5), but the JNCC (1993) method (outlined in Table 11.5) is recommended for general use in EIA because:

- it is relatively rapid, and provides information that is easily understood by non-experts;
- it has been the standard method for environmental audit in Britain for some years, and hence:
    * it has already been applied over much of the country, and is likely to be the form of any recent survey in the impact area;
    * existing surveys are widely used by NCCAs and LPAs in formulating conservation policies and considering planning applications;
    * it is well understood, and there is no lack of qualified staff who can undertake it;
- it can be applied to relatively large areas, such as impact areas, and to smaller areas such as sites;
- it is recommended by IEA (1995) and hence is likely to be the adopted method in other EIAs.

Similarly, the JNCC habitat classification (see Appendix F.1, p. 436) is likely to remain the primary choice for UK Phase 1 surveys. However, it is beneficial to relate the results to other classifications such as UKBAP broad and priority habitats, and HSD Annex I habitats (Appendix F).

The JNCC method's emphasis on habitats, and omission of detailed faunal surveys, are appropriate strategies for the Phase 1 survey in EIA, but this should seek to extend the scope of the study somewhat. For instance, in **site surveys** it may be important to record and map the (sometimes complex) patterns of different habitats present, and to assess their relative importance, at least in terms of the proportion-of-site covered by each. This can be achieved by means of the **line intercept method**. The JNCC method recommends using this on maps or aerial photographs (Table 11.5), but it can also by applied in the field as follows:

- From one end of a selected baseline, a **line transect** (see Figure G.1, p. 458) is laid out across the study area, and the length of this that is occupied by each patch of each habitat type is recorded. At the same time, target notes and observations of species present near the transect can be made. Unless quantitative (Phase 2 level) sampling is likely to be carried out, it is also beneficial to obtain semi-quantitative records of plant species abundances, e.g. using DAFOR (see Table G.1, p. 457).

Table 11.5 Outline of the JNCC Phase 1 Habitat Survey method

Aspects	Features
Aim	The main aim is "to provide, relatively rapidly, a record of the semi-natural vegetation and wildlife habitat over large areas of countryside" (JNCC 1993).
Scope and form	Applicable to both rural and urban areas. Primarily designed to provide: • **colour maps of vegetation/habitat types** defined in the **JNCC Habitat classification** using standard colour codes and symbols, and with dominant plant species shown where possible; • **supplementary information** (largely as 'target notes'), e.g. on notable species (e.g. high-status, indicator) and environmental features (e.g. topography and substratum conditions). Because most animals are mobile, fugitive and small, large-scale faunal surveys are not considered practicable, but partial species lists can be recorded from 'casual' observations.
Use of existing information	Information is collected to facilitate and supplement the field survey, e.g.: • **OS maps**, e.g. 1:50 k (for overviews and reconnaissance), 1:25 k and 1:10 k (for field surveys). • **Geological and soil survey maps** (see §9.5.2) can be valuable aids to habitat mapping. • **Historical records and old maps** can provide valuable information. A site's history may enhance its conservation value, and knowledge of past management can be very important. • **Aerial photographs** (see Table 15.1) are considered very useful for: providing an overview prior to field survey; mapping where access is restricted; identifying some vegetation types and the locations, boundaries and areas of these or other features (e.g. hedges, roads and undeveloped urban areas) that are unclear or out of date on the OS maps. • **Satellite data** is considered useful, but not to provide "images of the quality needed for the whole range of habitats mapped" (but resolutions are increasing rapidly – see Chapter 15). • **Information on habitats and taxa** (e.g. records of species distributions and site species lists) from previous national, regional or local surveys and research studies.
Field survey and recording methods	JNCC Phase 1 habitat classification categories are recorded directly on 1:10 k or 1:25 k **OS maps** and/or **map record sheets**, using standard colour and/or alphanumeric codes, and labelled with dominant species names where possible (using standard abbreviations). **Target note record sheets** are used for additional information, e.g.: species lists; notable species (e.g. high-status, indicator); vegetation features and condition; topography and substratum (e.g. soils, geology, wetness); protection; ownership and especially management.

Table 11.5 (*continued*)

Data processing and presentation	Data are transferred from the field maps to produce **final maps** (usually 1:10 k scale for sites). Maps may be digitised, e.g. for use in a GIS. On the maps and/or aerial photographs: (a) plot areas are measured, e.g. using **Romer dot grids**, (b) the proportions (%) of different habitat types are estimated, e.g. by the **line-intercept method** which involves measuring the length of each recognised habitat type along a series of parallel line transects (e.g. map grid lines) and expressing the total length of each habitat type as percentage of total length of all transects. Map and target-note record sheets are completed, and a **written report** is usually produced.
Site evaluation	Not primarily intended for site evaluation, but: • is considered adequate for classifying sites on a three-point scale: 1 = high conservation value, 2 = lower priority for conservation, 3 = limited wildlife interest; • should provide the information required to determine the need for Phase 2 surveys.
Limitations	The maps are not 100% accurate (error estimates should be provided). Small sites may be omitted (< c.0.5 ha with 1:25 k, and < c.0.1 ha with 1:10 k scale maps). Sites are normally only visited once, so seasonal variations may be missed. Species lists may not be complete, and rarities may have been overlooked. Changes may have occurred since the survey.

- The process is repeated at suitable intervals along the baseline, using the same compass bearing for each transect so that the study area is sampled by a set of parallel transects.
- The cumulative length occupied by a habitat type, i.e. its percentage of the total length of all transects, is taken to represent the proportion of the study area that it covers (its % cover).
- The results can also be used to map the locations and extents of the habitat types, e.g. (a) on a suitable base map such as the local OS 1:10 k sheet, or (b) using a mapping program or GIS (§16.2.2) perhaps in conjunction with existing digital cartographic data (Table 16.1, p. 384).

A Phase 1 **linear habitat survey** can normally be achieved simply by walking along the feature, making notes, and recording the species seen (perhaps with DAFOR ratings). Hedgerows can be surveyed quite rapidly in this way, but their evaluation may require a more thorough Phase 2 study.

It may be necessary also to check the presence and distributions of some species within the impact area as a whole, especially when these have 'wider-countryside' distributions, and are hence unlikely to be concentrated in specific sites. Again, however, Phase 2 methods may be needed to determine distributions, especially of animals.

11.4.5 Phase 2 field surveys

IEA (1995) recommend a number of criteria for triggering Phase 2 surveys. Most of these refer to whether the project is likely to affect high-status species or habitats, and they can be encapsulated in two related questions:

1. In addition to that collected in Phase 1, what information is needed to provide a sound basis for description and evaluation of the baseline conditions, prediction of impacts and formulation of mitigation measures?
2. On what taxa, habitats and sites should Phase 2 studies be targeted, and what methods should be applied to obtain adequate information on these?

The main limitation of Phase 1 surveys is usually lack of quantitative data on species, communities and environmental factors. For example, information on plant species may be restricted to lists (with no indication of abundances or distributions) and data on animal species are likely to be even more scanty. Consequently, Phase 2 surveys normally focus on collecting quantitative (or at least more detailed) information. This requires fairly intensive sampling, which is time-consuming and usually subject to seasonal constraints; and it is essential to focus on carefully selected objectives. Principles, methods and problems of Phase 2 (and Phase 3) ecological sampling are discussed in Appendix G.

11.5 Impact prediction

11.5.1 Introduction

In addition to the general requirements listed in §1.2.3, particular requirements for ecological impact prediction are:

- predictions of the project's impacts on relevant physico-chemical environmental variables – which are often provided by other specialists (Chapters 8–10);
- adequate information about the relevant ecological receptors, and knowledge of how these may respond to the predicted environmental disturbances.

It is relatively easy to identify primary ecological impacts, but much more difficult to predict their effects, or even to itemise the numerous potential secondary (knock-on) impacts that may be generated. Three major problems are:

1. Sufficient data on impact factors and/or baseline conditions are often unavailable.
2. Ecosystems' complex interactive processes are generally poorly understood (§11.2.3) – and they frequently respond to disturbances in unexpected ways.
3. In spite of available general guidance on the assessment of cumulative impacts (see §1.2.3) it is often difficult to isolate the impacts of a specific development from the cumulative effects of others. Moreover, virtually all negative ecological impacts can be considered cumulative because they are bound to add to general pressures on ecological systems.

These problems do not excuse the vague predictions found in many EISs (§11.1), and every effort should be made to make precise, and where possible quantitative, predictions. However, they do mean that (a) predictions often have to lean heavily on expert judgement, and (b) a degree of **uncertainty** is inevitable, and should be acceptable provided that it is clearly stated in the EIS.

11.5.2 Changes without the development

In order to make valid assessments about the ecological impacts of a project, it is important to consider what ecosystem changes may occur in its absence. Such changes can result from intrinsic ecosystem processes, or from 'external' factors, whether or not these are caused by human activities.

The most significant intrinsic changes over the relevant timescales are likely to be associated with **succession** (Fig. 11.2). Precise outcomes of this are difficult to foresee, but general predictions can be made, e.g. that unmanaged grasslands and **heathlands** will give way to scrub and eventually woodland.

Human influences that may cause changes in the absence of the project include:

* other current and proposed developments in the area;
* more widespread cumulative impacts such as acid precipitation, and climate and hydrological changes associated with global warming (§8.1.3 and §10.8.2);
* management aimed at enhancing the ecological value of habitats/sites.

The predictions may have differing implications in relation to the project, e.g. that:

* it would, or would not, contribute significantly to cumulative impacts;
* its impacts on a site would not be important because the site's ecological value will decline anyway;
* whilst a site may not be considered worthy of protection in its current condition, its value will increase if impacts from the project are avoided.

11.5.3 Types of ecological impact

In relation to the various types of impact referred to in the EIA legislation (§1.4):

* There is a case for differentiating between 'indirect' and 'secondary' ecological impacts by restricting 'indirect' to those from 'secondary' external sources (e.g. increased traffic, further development) generated by a project. This is because the interactive nature of ecosystems means that almost any direct impact will have knock-on effects, and hence that the majority of ecological impacts are actually secondary.
* It may be useful to consider the duration and reversibility of ecological impacts in terms of three types of disturbance: **pulse disturbances** (short-term/temporary); **press disturbances** (sustained/chronic); and **catastrophic disturbances** (highly destructive/probably irreversible). These are explained further in §13.5.3.

Developments can have positive ecological impacts including habitat creation. For example:

	Linear projects, e.g. roads	Heavy industry & power	Light industry & urban	Mineral extraction	Landfill / waste disposal	Recreation & tourism	Agriculture and forestry
Habitat loss and fragmentation	●	●	●	●	?	•	●
Habitat damage, e.g. by trampling and erosion	○	○	○	●	•	●	•
Species disturbance (outside the built area)	●	●	•	•	•	•	•
Direct mortality, e.g. on roads	●	●	●	•	•	•	•
Pollution (including eutrophication)	●	●	•	•	•	•	●
Other physico-chemical environmental impacts	●	●	●	•	•	•	●

•	impacts usually minor and/or localised
●	impacts often major and/or extensive
○	impacts outside the final land-take area usually associated mainly with the construction phase
?	depends if the land has already been taken, e.g. for mineral extraction

The indicated magnitudes are for illustrative purposes only, and will vary in relation to specific developments and receptors.

Figure 11.6 Examples and approximate relative magnitudes of negative impacts on semi-natural terrestrial ecosystems, associated with various types of impact source.

- Sensitive redevelopment on 'brownfield' sites can improve their ecological value, e.g. by incorporating features such as 'green networks' (e.g. see Barker 1997, Harrison *et al.* 1995).
- Roadside verges can provide valuable habitats, e.g. Munguira & Thomas (1992) found that (a) sites may support up to 40% of British butterfly species – with the number increasing in relation to verge width and range of breeding habitats, and (b) butterfly mortality from vehicles was insignificant compared with other causes. However, the verges are unlikely to provide a net positive ecological impact where a new road crosses existing semi-natural habitats.

Most 'positive' impacts result from mitigation measures (§11.6) and only partially compensate for larger negative impacts. Examples of negative ecological impacts associated with major impact sources are shown in Figure 11.6, and discussed in the following sections. Publications on impacts associated with particular development types include: cross-country pipelines (DTI 1992); roads (Box & Forbes 1992, DoT 1993, ERM (1996), PAA 1994); railways (Carpenter 1994); and wastewater treatment facilities (Petts & Eduljee 1994). Impacts of particular relevance to aquatic ecosystems are discussed in Chapters 10, 12 and 13.

11.5.4 Habitat loss and fragmentation

The amount of semi-natural habitat that is directly threatened by a new development will depend largely on how much exists within the construction and final landtake areas; but some loss, if only of habitats such as hedgerows, is usually inevitable. The significance of habitat loss depends on a range of factors including:

- the area lost;
- the ecological value/conservation status of the habitat, and the degree to which high-status species depend on it;
- the degree to which displaced species can migrate to, and survive in, other suitable sites/habitat patches (which will depend on the availability of such sites, and on factors such as the existing density of the species in them, and hence the potential severity of competition);
- the degree to which overall habitat loss is accompanied by fragmentation.

Fragmentation has two primary effects: (a) it splits a habitat patch into two or more smaller patches; and (b) it creates barriers between the remaining habitat patches, resulting in (or increasing) their isolation.

Site or habitat-patch size is generally considered to be important because (a) large sites often contain greater habitat diversity than small sites, and (b) a large habitat patch is likely to support more species and larger populations than a small patch. In general, the size of undisturbed habitat fragments evidently affects the preservation or decline of species – "the smaller they are, the greater the rate of loss" (Hamblin 1998). Moreover, the more fragmented a habitat is already, the higher is the eventual number of losses (Tilman *et al.* 1994). One reason for this is that the viability of some species populations may be threatened on a small patch because it does not contain sufficient resources such as food or habitation sites – but there are additional factors including edge effects and isolation.

Edge effects are associated with the increased length of habitat edge relative to its area, i.e. of the boundary : area ratio. As a result, small isolated habitat patches are susceptible to (a) 'external' impacts such as physical damage, fly tipping, pollution and disturbance, which may be more important for some species than reductions in patch size *per se* (Kirby 1995), and (b) invasion by 'foreign' species (from neighbouring areas). In addition, habitat edges have different environmental conditions from their interiors, and a large edge : area ratio favours edge-living species at the expense of core species – which are usually more characteristic of, and dependent on, the habitat type. On the other hand, edge communities – which are ecotones (§11.2.5) – can be species-rich, and may host high-status species.

An important aspect of **isolation** is the ability (or not) of species to move between habitat patches. This has encouraged the study of **metapopulations** (Gilpin & Hanski 1991). A metapopulation consists of a group of sub-populations that exist in separate habitat patches, but are linked by dispersal between them. An important aspect is that a 'sink' sub-population in a small habitat patch (with insufficient resources to sustain it) may be augmented by immigration from a 'source' (reservoir) population living in a larger habitat patch – but only so long as the two populations are not isolated. Consequently, the viability of many species populations in fragmented landscapes may be affected by (a) their dispersal capabilities, which vary considerably between species (see below), and (b) the degree of isolation between habitat patches – and hence the distances between them and the severity of barriers. Exchange of individuals between sub-populations may also maintain their genetic diversity, and in the longer term, lack of genetic exchange may lead to the decline of isolated populations.

New barriers are created by fragmentation, the removal of interconnecting habitats, or the interposition of a 'hostile' feature (e.g. a road, fence, building or cultivated field) between existing habitat patches. A barrier may be **physical** (a species cannot cross it), **behavioural** (a species is capable but unwilling to cross it), or **hazardous** (a species may suffer high mortality in attempting to cross it – and the importance of barriers varies between species. For example:

- some species have efficient dispersal mechanisms that are not seriously affected by most local barriers;
- some species have very limited dispersal ability, and such 'low-mobility' species (a) are restricted to remaining habitat patches, and (b) are unlikely to recolonise isolated habitat patches;
- animals such as badgers, deer and otters may have range requirements that exceed the areas of remaining habitat patches (or of patches to which they have been displaced by development) and must risk crossing hazardous barriers;
- animals such as amphibians may have to cross barriers in order to reach their breeding habitats.

An additional problem that may follow from the isolation of small habitat patches is that conservation management practices, such as low-intensity grazing, may be prevented because the remaining areas are too small or are not accessible.

The increasing isolation of wildlife habitats (including protected sites) has led to fears that, in the event of their current habitats becoming unsuitable due to climate change, many species may be unable to shift their distributions, e.g. to higher latitudes or altitudes (UKBG 1999). This is adding weight to the view that more emphasis should be placed on the conservation and creation of habitats in the wider countryside in addition to their protection in designated sites (§11.3).

There is as yet no conclusive scientific evidence that fragmentation (as distinct from habitat loss) has significant effects on species' regional populations, although there is appreciable empirical evidence, and Kirby (1995) concludes that "There is sufficient evidence that it is potentially an important cause of species decline to justify opposition to further habitat fragmentation".

The heaths of east Dorset in southern England provide a good example of fragmentation effects. In 1759, the heaths consisted of 10 large blocks, separated only by rivers; by 1978 they were divided into 768 fragments (Webb & Haskins 1980) and the trend has continued since then. Webb & Rose (1994) provide evidence of associated losses of heathland *indicator species* since 1962 – especially low-mobility species (including protected species such as the sand lizard and silver-studded blue butterfly) – and especially from small, isolated patches. Much of the heathland loss has been to farming and forestry. However, in the southeast of the county, the major factor has been urban development, the effects of which are reviewed by Haskins (2000). These effects (summarised in Fig. 11.7) include (a) all the consequences of habitat loss and fragmentation outlined above, and (b) all the impact types discussed in the following sections.

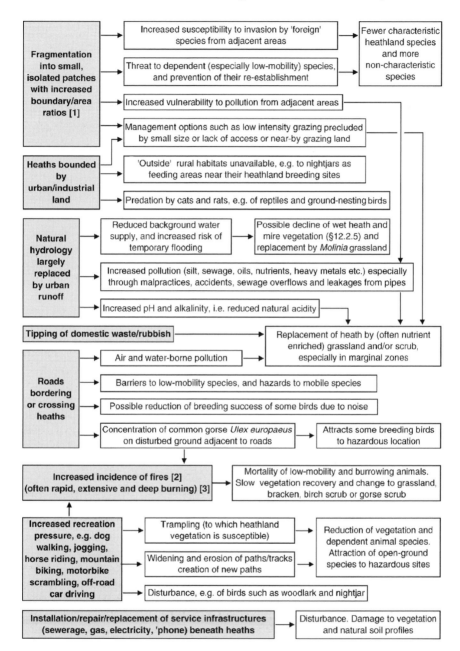

Fragmentation into small, isolated patches with increased boundary/area ratios [1]
→ Increased susceptibility to invasion by 'foreign' species from adjacent areas
→ Threat to dependent (especially low-mobility) species, and prevention of their re-establishment
→ Fewer characteristic heathland species and more non-characteristic species

→ Increased vulnerability to pollution from adjacent areas

→ Management options such as low intensity grazing precluded by small size or lack of access or near-by grazing land

Heaths bounded by urban/industrial land
→ 'Outside' rural habitats unavailable, e.g. to nightjars as feeding areas near their heathland breeding sites
→ Predation by cats and rats, e.g. of reptiles and ground-nesting birds

Natural hydrology largely replaced by urban runoff
→ Reduced background water supply, and increased risk of temporary flooding → Possible decline of wet heath and mire vegetation (§12.2.5) and replacement by *Molinia* grassland
→ Increased pollution (silt, sewage, oils, nutrients, heavy metals etc.) especially through malpractices, accidents, sewage overflows and leakages from pipes
→ Increased pH and alkalinity, i.e. reduced natural acidity

Tipping of domestic waste/rubbish → Replacement of heath by (often nutrient enriched) grassland and/or scrub, especially in marginal zones

Roads bordering or crossing heaths
→ Air and water-borne pollution
→ Barriers to low-mobility species, and hazards to mobile species
→ Possible reduction of breeding success of some birds due to noise
→ Concentration of common gorse *Ulex europaeus* on disturbed ground adjacent to roads → Attracts some breeding birds to hazardous location

Increased incidence of fires [2] (often rapid, extensive and deep burning) [3] → Mortality of low-mobility and burrowing animals. Slow vegetation recovery and change to grassland, bracken, birch scrub or gorse scrub

Increased recreation pressure, e.g. dog walking, jogging, horse riding, mountain biking, motorbike scrambling, off-road car driving
→ Trampling (to which heathland vegetation is susceptible)
→ Widening and erosion of paths/tracks creation of new paths
→ Reduction of vegetation and dependent animal species. Attraction of open-ground species to hazardous sites
→ Disturbance, e.g. of birds such as woodlark and nightjar

Installation/repair/replacement of service infrastructures (sewerage, gas, electricity, 'phone) beneath heaths → Disturbance. Damage to vegetation and natural soil profiles

[1] Very few heaths in or around the urban areas are over 100ha, and most are no more than 5ha

[2] For example, since 1990: 179 fires on Canford heath (urban); 2 on Hartland Moor (rural)

[3] Controlled fire has long been used as a management tool on heathlands, but in conditions (e.g. in winter) when it will not be rapid, extensive or deep burning.

Figure 11.7 Impacts of urban spread on the heathlands of southeast Dorset (from information in Haskins (2000)).

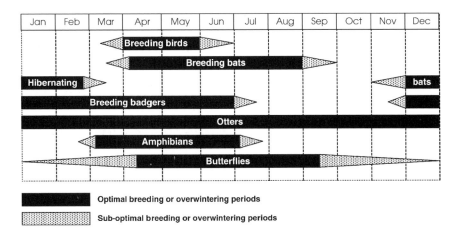

Figure 11.8 Seasonal periods in which some animal taxa are particularly sensitive to impacts.

11.5.5 Habitat damage, disturbance and direct mortality

Developments that increase the recreational use of receptor sites can generate chronic and progressive **habitat damage** by visitor pressure (Fig. 11.7). More commonly, physical damage is associated particularly with the construction phase of projects, and involves impacts such as vegetation trampling or removal, and soil compaction and erosion (§9.6.2). However, it does not follow that such impacts are temporary or reversible, and the restoration of a construction site to its original state may never be achieved, especially when long-established semi-natural habitats are affected.

Disturbance (e.g. visual, noise, trampling, night-time light pollution) can also be particularly associated with construction, but can result from many forms of increased human activity including traffic and recreation (Fig. 11.7). The susceptibility of animals to disturbance often varies seasonally (Fig. 11.8) but vulnerable periods may not be immediately obvious. For example, invertebrates with mobile adult phases (e.g. butterflies and other flying insects) may be most vulnerable when in developmental stages (eggs, larvae and pupae) because of damage to their foodplants and pupal sites. Most invertebrates are also vulnerable in winter because they are dormant and hence cannot escape. Permanent vertebrate residents are most sensitive to disturbance during the breeding season, but some species may be vulnerable also in overwintering periods, and the risk to migrant birds is during their visit period.

Effects of disturbance on birds were examined by Hockin *et al.* (1992). However, they concluded that information is limited in spite of birds being the most studied animal group. Proximity to major roads has been shown to reduce breeding densities of birds in woodlands (Reijnen *et al.* 1995) and to reduce local populations of open field species such as lapwings to a distance of up to 2 km (van der Zande *et al.* 1980). Similar effects have been found in areas of high recreational use (van der Zande *et al.* 1984). Disturbance by recreation and traffic may also affect waterfowl, especially species such as wintering geese and ducks which feed for long periods in exposed habitats.

Effects of disturbance on bats are discussed by Hutson (1993). All British bats are dependent on buildings or trees for their roost sites, and if a development is likely to affect these, the appropriate NCCA must be consulted and allowed time to advise.

Like disturbance, **direct mortality** can result from a variety of physical factors including vegetation destruction, trampling and fire (Fig. 11.7). Roads present serious long-term threats, especially to animals that need to cross them. Information on wildlife road casualties is provided by Slater (1994). Clarke *et al.* (1998) studied the effects of roads on badger populations, and concluded that badger fatalities do not necessarily increase with road size and traffic volume, probably because these discourage badgers from attempting to cross – which has the effect of restricting or modifying their movement, and the dispersal of young.

11.5.6 Pollution and other environmental impacts

Any development is bound to generate some pollution. Industrial, urban and road developments are **regular sources** of a wide range of atmospheric pollutants and water-borne pollutants (Chapters 8 & 10) that are harmful to organisms or cause community changes. It is important to remember that:

- pollutants can enter and circulate in biogeochemical cycles (§11.2.3), and that those carried in air or water can affect ecosystems far from their source;
- some pollutants accumulate in organisms, and **bioamplification** in food chains can have serious consequences, especially for top carnivores;
- some pollutants can undergo **biotransformation** in the environment or within the bodies of individuals (see Connell *et al.* 1999).

Atmospheric pollutants can affect vegetation and animals directly, or indirectly through environmental changes such as those in the chemistry of soils and waters resulting from *acid deposition* (§8.1.2, §9.6.2).

Water pollution is particularly detrimental in aquatic ecosystems and wetlands, and is therefore discussed in detail in Chapters 10, 12 and 13. However, it can also affect terrestrial ecosystems wherever polluted water periodically inundates the ground surface or soil, e.g. as **leachates** from landfill sites or surface deposits, or as runoff from urban and road surfaces. In addition, salt-rich spray regularly falls in the 'splash zones' of road verges during winter months, and influences species composition within these zones. Similarly, whilst assessment of **water quantity impacts** focuses on aquatic and wetland ecosystems, it should be remembered that terrestrial ecosystems can also be affected, e.g. by flooding, waterlogging or increased incidence of soil drought. Moreover, many sites support aquatic–wetland–terrestrial ecosystem complexes, and hydrological changes that affect any of these (e.g. as in Fig. 11.7) can degrade the conservation value of the whole site.

Ecosystem responses to **chronic environmental changes** (including pollution) depend on a complex of biotic and abiotic factors. Changes in the species composition of a community often occur because the competitive balance is altered in favour of species that are more tolerant of the new conditions. This frequently favours 'generalists' at the expense of 'specialists' (§11.2.1) which are often high-status species or species confined to high-status habitats. For example, eutrophication may induce the replacement of characteristic dry heath species by grasses (Fig. 11.7) because it favours species that thrive in more nutrient-rich conditions. Similarly, human influences evidently favour purple moor grass, *Molinia caerulea* (at the expense of other wet heath and mire species) because it is tolerant of fire, grazing, relatively dry conditions and both low and high soil pHs (Grime *et al.* 1988).

Accidental pollution can be a major threat, especially from heavy industry and transport, and developers should be asked to provide a **risk assessment** (including a worst case scenario) for this type of impact (see Chapter 14).

11.5.7 *Methods of impact prediction*

Because of the difficulties outlined in §11.5.1, predictions of ecological impacts tend to be qualitative, but should be quantitative wherever possible. Some direct impacts, such as habitat loss and fragmentation by landtake, can be readily measured and mapped; and if the distributions of species populations, habitats and communities within the sites are known, the proportionate direct losses from these can be calculated. However, most impacts are much less amenable to direct measurement, and predictions have to rely on other methods, such as those listed in Table 1.1 (p. 7).

Flowcharts and networks can be useful for identifying secondary (knock-on) effects from primary impacts, but they cannot quantify the magnitudes of impacts or their effects. The latter can be indicated by **magnitude matrices** (similar to, but more detailed than, Fig. 11.6), but these are usually simply representations of qualitative estimates.

Maps are essential, and **GIS** can be valuable (a) as a sophisticated mapping tool that can relate different variables by spatially referencing datasets ('layers'), and (b) in conjunction with an external tool, such as an expert system or simulation model (Chapter 16).

Mathematical and statistical models have been used as research tools in ecology for many years, but the limitations of using complex computerised models in hydrological EIAs (outlined in §10.5.2) are more serious in the context of ecological assessments (e.g. see North & Jeffers 1991). A major problem is the current lack of knowledge and understanding of how species and communities respond to impacts. For example:

- otters frequently exhibit *bioaccumulation* of PCBs, but the effects on otter populations are not known, and other factors are considered to be more likely to be critical;
- *acid deposition* is generally considered to be a major cause of forest decline (tree damage and death) in Europe and North America since the 1980s, but other factors (such as ozone) have been suggested and, in spite of a major research programme, no single set of causes (or single model) has been found that fully explains forest damage (Dickinson 2000).

However, with the continuing improvement of ecological knowledge/understanding and modelling techniques, the use of computer models is likely to increase, at least for major projects and in relation to developing subjects such as: **risk assessment** (Chapter 14) including estimation of minimum critical areas and minimum viable populations (e.g. see Burgman *et al.* 1993); and *ecotoxicology* (e.g. see Calow 1997, Connell *et al.* 1999) including the estimation of **critical loads** (see CEH 2000).

Expert opinion is always needed for the interpretation of data, and in the absence of definitive quantitative evidence, impact prediction has to rely on judgements based on a knowledge of impact factors and ecological systems. Such predictions may be rather general, but can still be authoritative.

11.5.8 Impact significance

As explained in §1.2.3, impact significance is a function of **impact magnitude** and the **value**, **sensitivity** and **resilience** (recoverability) of ecological receptors (EN 1994). All three receptor attributes can be assessed in relation to the criteria outlined in Appendix D, but care is needed, e.g.:

- to avoid awarding high conservation value only to sites with strong statutory designations and devaluing non-designated and small sites/habitats;
- in assessing ecosystem sensitivity and resilience, which can be one of the most important and difficult aspects of impact prediction, and will normally require expert judgement;
- when relating impact magnitude to significance. For instance, percentage-of-site affected can be misleading because the loss of even a small area may have significance if (a) it affects the integrity of the site, maintenance of which is a key requirement of the HSD (EN 1999), or (b) the habitat type is rare or declining locally or in a wider context.

The concepts of site sensitivity and integrity are incorporated in the 'biodiversity' component of the *New Approach to Appraisal* (NATA) (DETR 1998). This was designed primarily for assessment of options in road schemes (see Chapter 5) but the **option assessment scores** that are produced can be regarded as impact significance scores, and the method can be applied to projects other than roads. Assessment of a proposal/option in relation to biodiversity is based on the **nature conservation evaluation** of the **'features'** (habitats, species and geological features) and the **ecological impact** on these.

A stated intention is to adopt the "emerging environmental capital approach" (Chapter 17) in the evaluation procedure. However, the method currently evaluates **sites** into five categories using a worksheet with several **'indicators'** (Table 11.6). The initial site categories are:

A – Biogenetic and Biosphere Reserves, EDSs, Ramsar Sites, SACs, SPAs, WHSs (Table D.3), and sites hosting HSD habitats or species, or species listed in the Bonn and Bern Conventions;

B – AOSPs, LPOs, MNRs, NCR sites, NNRs, GCR sites (see §9.4.1), SSSIs, and sites hosting Red Data Book species, or species under the Wildlife Countryside Act 1981 (as amended);

C – LNRs, RIGs (see §9.4.1), SINCs, sites hosting UKBAP priority habitats or species, and other semi-natural sites of significant biodiversity importance, not referred to above;

D – sites not in the above categories, but with some biodiversity or earth heritage interest;

E – sites with little or no biodiversity or earth heritage interest.

Flexibility is recommended, e.g. to avoid undervaluing non-designated sites, or overvaluing sites that host, say, "a single individual of a widespread Bern Convention species".

Table 11.6 The site evaluation indicators used in the NATA biodiversity worksheet

Site
Location and initial category (A–E)

Features
Phase 1 habitat type or taxa, e.g. dry heath, birds, invertebrates

Scale (of importance)
International, national, regional, local

Importance
Description, including reasons for designation where appropriate

Rarity (trend in relation to target)
e.g. recent trend in relation to UKBAP targets (in HAPs and SAPs)

Substitution possibilities[1]
Assessment of whether: habitats are recreatable to sufficient quality; species can be successfully relocated; or the ecosystem services provided by the feature can be fully substituted

Evaluation
Final category (A–E) (a site can have more than one)

1 This aims to take account of the fact that the loss of an irreplaceable natural feature is often considered to be more significant than one that is replaceable, but is not regarded as a major factor in most cases.

Impacts are assessed on a seven-point textual scale (**impact categories**):

1. **Major negative** – if "the proposal (on its own or together with other proposals) may adversely affect the **integrity** of the site, in terms of the coherence of its ecological structure and function, across its whole area, that enables it to sustain the habitat, complex of habitats and/or the population levels of species for which it was classified".
2. **Intermediate negative** – "the site's integrity will not be adversely affected, but the effect is likely to be significant in terms of its ecological objectives".
3. **Minor negative** – if neither of the above apply, but some minor negative impact is evident.
4. **Neutral** – if there is no observable impact in either direction.
5. **Minor positive** – if there is a small net positive wildlife gain.
6. **Intermediate positive** – if there is "a significant gain to the biodiversity interest within the Natural Area".
7. **Major positive** – "if the net gain is of national importance".

(DETR 1998)

Table 11.7 Derivation of NATA option assessment scores from site-evaluation and impact categories

Nature conservation (site) evaluation	Impact categories		Option assessment scores (impact significance levels)
Category A	+ Major negative	=	Very large adverse[1]
	+ Intermediate negative	=	Large adverse[2]
	+ Minor negative	=	Slight adverse
Category B	+ Major negative	=	Very large adverse[1]
	+ Intermediate negative	=	Large adverse[2]
	+ Minor negative	=	Slight adverse
Category C	+ Major negative	=	Large or moderate adverse[3]
	+ Intermediate negative	=	Moderate adverse[4]
	+ Minor negative	=	Slight adverse
Category D	+ All negative categories	=	Slight adverse
Category E	+ All negative categories	=	Neutral
All categories	+ Neutral	=	Neutral
	+ Minor positive	=	Slight positive
	+ Intermediate positive	=	Moderate positive
	+ Major positive	=	Large positive

1 The option is likely to be unacceptable on nature conservation grounds alone (even with compensation).
2 There should be a strong presumption against the option, and greater than 1:1 compensation (i.e. a net gain in the Natural Area) for the very occasional case where development is allowed as a last resort.
3 The score should be "Large adverse" if the habitats or species populations are not substitutable.
4 The option should have at least 1:1 compensation (no net loss in the Natural Area) if development is allowed.

The impact categories take account of mitigation and enhancement measures that are proposed in a scheme, and are therefore assessments of **net (residual) impacts**. They do not include compensation proposals, such as habitat replacement (§11.6.3) because it is considered inappropriate to lower the impact assessment or option assessment score on this basis. However, the interpretation of some option assessment scores include possible compensation in terms of net gain or loss in Natural Areas.

Option assessment scores are derived by 'summing' the site evaluation and impact categories as shown in Table 11.7. As indicated below the table, some scores refer to 'compensation' in terms of net gain, or no net loss, in a **Natural Area**.

When more than one feature is affected, three rules are applied:

1. **Most damaging impact** – if an option affects more than one feature, the assessment score should be based on the most adverse effect.

2. **Cumulative adverse effects** – if an option affects several sites with 'slight adverse' or 'moderate adverse' scores (the cumulative effect of which is equivalent to the higher score for a single site) it should be scored in the higher category.
3. **Positive effects** – if an option has negative impacts on some sites but positive impacts (e.g. through mitigation) on others, it may be legitimate to make an overall net assessment. However, this requires careful ecological judgement rather than a simple area or number-of-sites approach.

The 'cumulative adverse effects' criterion goes some way to considering the effects of impacts on groups of receptor sites, but it does not really address issues such as the affects of habitat loss and fragmentation on metapopulations. Wathern (1999) suggests that an approximate method for assessing the **relative importance of isolated sites** within an area might be to compute (from baseline survey maps) the increase in mean inter-site distance resulting from the removal of each existing site in turn. The assumption is that the most important sites are likely to be those whose removal leads to the greatest increase. However, he also points out that various complicating factors should be considered, e.g.:

*   a site near the centre of a network will probably be more important than one at the periphery because it is likely to have sub-population interactions with a larger number of neighbouring sites;
*   removal of a population reservoir site such as a large wood, is likely to have a major impact;
*   *connectivity* between sites may be decreased by barriers and increased by linear habitats;
*   the effect of inter-site distance will vary for different species.

11.6 Mitigation

11.6.1 Introduction

As with impact prediction, the complexity of ecological systems makes the effectiveness of mitigation measures difficult to predict (which is a major reason for monitoring). However, lack of certainty should not be taken as an excuse to avoid formulating specific measures – which can aim to avoid, minimise, reverse or compensate for adverse ecological impacts.

11.6.2 Measures to avoid or minimise impacts

Most mitigation measures aimed at avoiding or minimising adverse ecological impacts are associated with a project's location, alignment, design, or construction and operating procedures.

The **location** of a project can be a key factor. It is usually determined largely by socio-economic and technical criteria rather than environmental considerations, and the choice of sites is often restricted. However, if the proposed siting will clearly cause significant impacts on high-status habitats and/or species, the relocation or no action options should be considered.

The **alignment** of linear projects such as pipelines, roads and railways can be relatively amenable to modification, and the NATA method (§11.5.8) is designed to identify the least ecologically damaging option. However, this may be precluded by technical and financial constraints, or by conflicting interests, e.g. with other EIA components.

The **project design** may be modified in various ways including:

- incorporating features to minimise pollution, soil erosion and runoff (see Chapters 8–10);
- creating site-boundary buffer zones;
- modifying site boundaries, e.g. to reduce landtake;
- providing features to reduce barrier effects, e.g. road underpasses for large mammals such as badgers (EN 1995) and small tunnels under roads for amphibians;
- modifying the within-site layout to retain semi-natural habitats and/or create 'green networks' (§11.5.3). The latter may result in enhancement, but only when the development is in an already urbanised or intensively farmed location.

Much of the ecological damage caused by developments occurs during the **construction phase**, and it is important to ensure that adequate mitigation measures are proposed and carried out. These can include:

- restricting the extent of access roads, temporary buildings and materials stores, exercising care in the routing/siting of these, and employing pollution-prevention measures;
- minimising damage to vegetation and soils, e.g. by using wide tyres on vehicles, and restricting the size of vehicles and plant;
- creating seed banks by collecting seed before vegetation is damaged;
- applying appropriate storage, handling and management of soils (see §9.7);
- where possible, avoiding major construction-phase operations during periods when taxa are particularly vulnerable to disturbance (§11.5.5);
- designating protection zones, e.g. around trees, badger setts and semi-natural habitats;
- protecting adjacent habitats by erecting boundary fences (although it should be remembered that these may act as barriers to animal movements).

Whilst mitigation during the **operational phase** will depend largely on the project design, it may also involve aspects such as maintenance procedures and the management of amenity areas.

11.6.3 Remedial and compensatory measures

When the destruction of, or serious damage to, a valuable habitat or species population is unavoidable, remedial or compensatory options include translocation, habitat restoration or recreation, and habitat creation. These can appear attractive to developers, who often assume that they are easy – which is not true.

Translocation involves 'rescuing' a species or community from a **donor site** (that will be destroyed) and moving it to a **receptor site** that already contains a suitable semi-natural habitat or (in the case of community translocation) is environmentally suitable, e.g. it has similar soil type, hydrology and climate. To have any success, therefore, translocation requires a thorough knowledge of the ecology of the species or community in question.

Translocations of species have included the marsh fritillary, great crested newt, red squirrel and bee orchid. Equipment and techniques have been developed for 'whole woodland' transplants (Buckley 1989), and translocations have occasionally involved ancient woodland (although moving the mature trees is clearly impracticable). However, community translocation has most commonly been undertaken for various types of semi-natural grassland. This is usually done by lifting turves, although transfer of rotovated topsoil and turf fragments has sometimes been used.

Some successes have been claimed (e.g. see Buckley 1989) but there have also been clear failures. A review by Gault (1997) concluded that success cannot be assumed in any of the cases she examined because (a) there has usually been insufficient time since the translocation for assessment to be made, and (b) most were poorly documented, and there had been a general lack of monitoring. A case in which monitoring was conducted – and showed that the translocated grassland deteriorated over a period of nine years – was used by EN to successfully argue that a developer's new proposal to translocate an adjacent grassland SSSI should be refused (Jefferson *et al.* 1999). Because of the uncertainties, ecologists generally argue that translocation should be considered in EIA only as a last resort.

Restoration is strictly the repair of a damaged ecosystem. It can include (a) management of the existing system, and (b) the use of 'rescued' species or community fragments (e.g. turves) to repair or enhance a community that is not impacted by a project, which Gault (1997) suggests may be one of the most useful roles of translocation. In the context of developments, however, it frequently involves the attempted reinstatement of a habitat that was destroyed, e.g. during construction – when measures such as the storage of soil or turf may have been taken to facilitate it. Consequently, 'restoration' often starts with a bare substratum, in which case it is effectively habitat **recreation**. In some cases, a 'do nothing' option can be considered because an ecologically valuable community may become re-established by natural re-colonisation from adjacent habitat patches – but this will only occur if factors such as soil conditions are still suitable.

Habitat restoration/recreation is often incorrectly taken to mean simply the reinstatement of vegetation cover (e.g. tree planting) associated with landscaping or amenity purposes. While this can have some ecological value, it is no substitute for semi-natural communities, which are much more difficult to recreate. Moreover, the more natural and complex the original community was, the less chance there is of restoring its original richness, i.e. the greater is its non-recreatability (Appendix D.3.2).

Habitat creation includes a wide variety of methods that aim to produce semi-natural habitats, starting with bare substrates or poor quality habitats, usually outside a development site. PAA (1994) suggest that planning habitat creation (including the selection of proposed habitat type) should involve several basic questions:

- Is it suited to the local geography, climate, geology and soil type (including soil fertility)?
- Is it consistent with (or will it complement) the local ecology and landscape?
- What management will it require, and will this be feasible?
- Should it be patches and/or linkages?

Appropriate methods vary according to proposed habitat type (see Buckley 1989, Gilbert & Anderson 1998). For example, grasslands may be 'sown' using rescued turf and/or native 'wild flower' seed mixes of native species that are tolerant of local climatic and soil conditions (see BSBI in Table 11.4). A similar approach can be adopted for woodland ground floras, but tree planting will normally be required. This should also be restricted to suitable native species, and to stock grown from native, and where possible local, seed (WT 1999).

Habitat patches created as mitigation measures will usually be small, but should be as large as possible because:

- small patches may not support viable populations of some species and are vulnerable to edge effects (§11.5.4);
- larger patches are likely to support larger populations;
- species richness tends to increase with area, although this may take many years.

To achieve large patches, an alternative strategy is to position new patches adjacent to existing patches. Buckley & Fraser (1998) recommend this for woodland planting, especially when the existing patches are species-rich woodland from which species may colonise the new plantation. A possible exception is when the existing habitat patch is large because:

- this is likely to be already large enough to sustain core species;
- several small patches with the same overall area as a single large patch (a) may have a better chance of hosting more species (including edge species) if they contain a wider range of habitat conditions, and (b) may increase connectivity between existing patches.

In the rare instance (in Britain) of a project being located in a highly wooded area, new woodland planting may not be an appropriate mitigation option because connectivity between woodland patches is already likely to be good (Dawson 1994), and additional woodland cover could threaten connectivity between important open habitats such as heathland or grassland (Kirby *et al.* 1999).

Although the effectiveness of linear habitats as wildlife corridors is still uncertain, it is generally agreed that they can be valuable in their own right and are likely to increase connectivity in fragmented landscapes (Dawson 1994, Kirby 1995, Spellerberg & Gaywood 1993, USDA-NRCS 1999). Dawson (1994) suggests that corridors should be as rich, wide and continuous as possible.

Habitat creation is usually proposed as a **habitat replacement** measure to compensate for the loss of a valuable habitat. Like translocation, however, this should only be undertaken as a last resort because, while a successful programme may create communities that are superficially similar to those lost, it "will never fully compensate for the destruction of high quality natural communities" (Wathern 1999). Consequently, ecologists argue that "habitat creation should never be put forward as a substitute for the conservation of semi-natural habitat" Parker (1995); and this view is endorsed in government policy, e.g. "the majority of terrestrial habitats are the result of complex events spanning many centuries which defy recreation over decades. Therefore, the priority must be to sustain the best examples of native habitats where they have survived rather than attempting to move or recreate them elsewhere" (DoE 1994a). Similarly, whilst the creation of new linear habitats may be a valuable mitigation measure, Andrews (1993) stresses that "enhancement of the existing habitat and maintaining the continuity of existing links is more important than establishing new ones".

In addition, proposals for translocation or habitat creation (including provisions for long-term management and monitoring) should be fully costed, and in many cases may be found to exceed significantly the cost of impact avoidance measures.

11.7 Monitoring

Because of the uncertainties imposed by ecosystem complexity, monitoring is particularly important in the ecological component of EIAs. Monitoring can be achieved by periodic repeat sampling of selected variables, using the methods employed for the baseline survey. Photo-sites and permanent *quadrats* can be useful to record changes (Goldsmith 1991). A difficult aspect is allowing for trends that may have occurred without the development. These can only be assessed with reasonable certainty if changes in impacted or compensatory sites are compared with those in control/reference sites (Bisset & Tomlinson 1988).

11.8 Conclusions

Several conclusions can be drawn about how ecological assessments should be carried out:

- Experienced ecologists must be employed, and (together with the statutory consultees) should be consulted early.
- It is vital to identify key impacts and receptors (if possible during scoping) and to target resources on these.
- The baseline studies should make maximum use of existing information, but this should be supplemented by new field surveys (at suitable times of year) wherever necessary.
- Whilst qualitative information is useful, quantitative data should be obtained where possible.
- The baseline conditions should be evaluated, e.g. using the criteria in Appendix D.
- Impact predictions and mitigation proposals should be as precise and quantitative as possible, although a degree of uncertainty must be accepted.
- Monitoring should be prescribed wherever necessary.
- The EIS should include clear explanations of survey methods, results, limitations, uncertainties, and relationships with other components.

References

Andrews J 1993. The reality and management of wildlife corridors. *British Wildlife* 5, 1–7.

Barker G 1997. A framework for the future: green networks with multiple uses in and around towns and cities. *English Nature Research Report No. 256*. Peterborough: English Nature.

Begon M, JL Harper & CR Townsend 1996. *Ecology: individuals, populations and communities*, 3rd edn. Oxford: Blackwell Science.

Bisset R & P Tomlinson 1988. Monitoring and auditing of impacts. In *Environmental impact assessment: theory and practice*, P Wathern (ed.), 117–128. London: Routledge.

Box JD & JE Forbes 1992. Ecological considerations in the environmental assessment of road proposals. *Journal of the Institution of Highway and Transportation* **39**, 16–22.

Buckley GP (ed.) 1989. *Biological habitat reconstruction.* London: Belhaven Press.

Buckley GP & S Fraser 1998. Locating new lowland woods. *English Nature Research Report No. 283.* Peterborough: EN.

Burgman MA, S Ferson, & HR Akçakaya 1993. *Risk assessment in conservation biology.* London: Chapman & Hall.

Calow P (ed.) 1997. *Handbook of ecotoxicology.* Oxford: Blackwell Scientific.

Canter LW 1996. *Environmental impact assessment,* 2nd edn. New York: McGraw-Hill.

Carpenter TG 1994. *The environmental impact of railways.* Chichester: Wiley.

CEAA (Canadian Environmental Assessment Agency) 1999. *Cumulative effects assessment practitioners' guide.* (http://www.ceaa.gc.ca/publications_e.htm)

CEH 2000. *The National critical loads mapping programme.* (http://www.ceh.ac.uk/)

Clarke GP, PCL White & S Harris 1998. Effects of roads on badger *Meles meles* populations in south-west England. *Biological Conservation* **86**, 117–124.

Connell DW, P Lam, B Richardson & R Wu 1999. *Introduction to ecotoxicology.* Oxford: Blackwell Science.

Dawson D 1994. Are habitat corridors conduits for animals and plants in a fragmented landscape? A review of the scientific evidence. *English Nature Research Reports No. 94.* (http://www.english-nature.org.uk/)

DETR 1998. *Guidance on the New Approach to Appraisal.* (http://www.detr.gov.uk/itwp/appraisal/ guidance/index.htm)

DETR 2000. *Our countryside: the future. A fair deal for rural England.* (http://www.wildlife-countryside.detr.gov.uk/ruralwp/cm4909/index.htm)

Dickinson NM 2000. Trees as environmental sentinels. *Biologist* **47**(4), 211–215.

DoE 1993. *Countryside 1990 Series,* Vol. 2 – *Countryside survey 1990 Main report.* London: HMSO.

DoE 1994a. *Biodiversity: the UK Action Plan.* CM2428. London: HMSO.

DoE 1994b. *Planning Policy Guidance Note 9: Nature Conservation.* (PPG9). London: HMSO.

Donn S & M Wade 1994. *The UK directory of ecological information 1994.* Chichester: Packard.

DoT (Department of Transport) 1993. *Design manual for roads and bridges.* Vol. 11: *Environmental assessment.* London: HMSO. (Sections updated periodically, e.g. ecology in 1996).

Dryden R 1997. Habitat restoration project: factsheets and bibliography. *English Nature Research Report No. 260.* Peterborough: EN.

DTI (Department of Trade and Industry) 1992. *Guidelines for the environmental assessment of cross-country pipelines.* London: HMSO.

EA (Environment Agency) 2000. *State of the Environment.* (http://www.environment-agency.gov.uk)

EN (English Nature) 1994. *Nature conservation in environmental assessment.* Peterborough: EN.

EN 1995. *Badgers: guidelines for developers.* Peterborough: EN.

EN 1998. *Natural areas: nature conservation context* (CD-ROM). Peterborough: EN.

EN 1999. *Habitats Regulations Guidance Note 3: The determination of likely significant effect under The Conservation (Natural Habitats &c.) Regulations 1994.* Peterborough: EN.

ERM (Environmental Resources Management) 1996. The significance of secondary effects from roads and road transport on nature conservation. *English Nature Research Report No. 178.* Peterborough: English Nature.

Gault C 1997. *A moving story: species and community translocation in the UK – a review of policy, principle, planning and practice.* Godalming: WWF-UK.

Gilbert OL & P Anderson 1998. *Habitat creation and repair.* Oxford: Oxford University Press.

Gilpin M & I Hanski 1991. *Metapopulation dynamics; empirical and theoretical investigations.* London: Academic Press.

Goldsmith FB (ed.) 1991. *Monitoring for conservation and ecology.* London: Chapman & Hall.

Gorman M & D Raffaelli 1993. The Ythan estuary. *Biologist* **40**(1), 10–13.

Grime JP, JG. Hodgson, & R Hunt 1988. *Comparative plant ecology.* London: Unwin Hyman.

Haines-Young RH *et al.* 2000. *Accounting for nature: assessing habitats in the UK countryside.* London: DETR. (http://www.cs2000.org.uk/report.htm) (*or* /report_pdf.htm)

Hamblin A 1998. *Environmental indicators for national State of the Environment reporting (the land).* (http://www.erin.gov.au/)

Harrison C, J Burgess, A Millward & G Dawe 1995. Accessible natural greenspace in towns and cities. A review of appropriate size and distance criteria. *English Nature Research Report No. 153.* (http://www.english-nature.org.uk/)

Haskins L 2000. Heathlands in an urban setting – effects of urban development on heathlands in south-east Dorset. *British Wildlife* **11**(4), 229–237.

Hockin D, M Ounsted, M Gorman, D Hill, V Keller & MA Barker 1992. Examination of the effects of disturbance on birds with reference to its importance in ecological assessments. *Journal of Environmental Management* **36**, 253–286.

Hutson AM 1993. *Action plan for the conservation of bats in the United Kingdom.* London: Bat Conservation Trust.

IEA (Institute of Environmental Assessment) 1995 *Guidelines for baseline ecological assessment.* London: E & F N Spon.

Jefferson RG, CWD Gibson, SL Leach, CM Pulteney, R Wolton & HJ Robertson 1999. Grassland habitat translocation: the case of Brocks Farm, Devon. *English Nature Research Report No. 304.* Peterborough: EN.

JNCC (Joint Nature Conservation Committee) 1993. *Handbook for phase 1 habitat survey – a technique for environmental audit* (separate *Field manual* also available). Peterborough: JNCC.

Kirby KJ 1995. Rebuilding the English countryside: habitat fragmentation and wildlife corridors as issues in practical conservation. *English Nature Science No. 10.* Peterborough: EN.

Kirby KJ, GP Buckley & JEG Good 1999. Maximising the value of new farm woodland biodiversity at a landscape scale. In *Farm woodlands of the future*, PJ Burgess, EDR Brierley, J Morris & J Evans (eds), 44–55. Oxford: Bios Scientific.

Krebs CJ 2001. *Ecology: the experimental analysis of distribution and abundance*, 5th edn. New York: Benjamin-Cummings (Addison-Wesley).

Munguira ML & JA Thomas 1992. Use of road verges by butterfly and burnet populations, and the effect of roads on adult disperal and mortality. *Journal of Applied Ecology* **29**, 316–329.

North PM & JNR Jeffers 1991. Modelling: a basis for management or an illusion? In *The Scientific management of temperate communities for conservation*, IF Spellerberg, FB Goldsmith & MG Morris (eds), British Ecological Society Symposium 31, 523–541. Oxford: Blackwell Scientific.

PAA (Penny Anderson Associates) 1994. *Roads and nature conservation: guidance on impacts, mitigation and enhancement.* Peterborough: EN.

Parker DM 1995. Habitat creation – a critical guide. *English Nature Science No. 21.* Peterborough: EN.

Parker K 1998. Meadows still in decline. *British Wildlife* **10**(2), 144 only.

Petts J & G Eduljee 1994. *Environmental impact assessment for waste treatment and disposal facilities.* Chichester: Wiley.

Reijnen R, R Foppen, CT Braak & J Thissen 1995. The effects of car traffic on breeding bird populations in woodland III. Reduction of density in elation to the proximity of main roads. *Journal of Applied Ecology* **32**, 187–202.

RSPB 1998. *Conservation issues: wetlands.* (http://www.rspb.org.uk/)

RSPB 1999. *The state of the UK's birds 1999.* (http://www.rspb.org.uk/)

Slater F 1994. Wildlife road casualties. *British Wildlife* **5**, 214–221.

Spellerberg IF & MJ Gaywood 1993. *Linear features: linear habitats and wildlife corridors. English Nature Research Report No. 60.* Peterborough: EN.

Spellerberg IF & A Minshull 1992. An investigation into the nature and use of ecology in environmental impact assessments. *Bulletin of the British Ecological Society*, **23**, 38–45.

Thompson S, JR Treweek & DJ Thurling 1997. The ecological component of environmental impact assessment: a critical review of British environmental statements. *Journal of Environmental Planning and Management* **40**(2), 157–171.

Tilman D, RM May, LL Lehman & MA Nowak 1994. Habitat destruction and the extinction debt. *Nature* **371**, 65–66.

Townsend CR, JL Harper & M Begon 2000. *Essentials of ecology*. Oxford: Blackwell Science.

Treweek J 1996. Ecology and environmental impact assessment. *Journal of Applied Ecology* **33**, 191–199.

Treweek J 1999. *Ecological impact assessment*. Oxford: Blackwell Science.

Treweek, J & S Thompson 1997. A review of ecological mitigation measures in UK environmental statements with respect to sustainable development. *International Journal of Sustainable Development and World Ecology* **4**, 40–50.

UKBG (UK Biodiversity Group) 1999. Climate change and conserving biodiversity. *Biodiversity News* **8**, 6 only.

ULIAG (UK Local Issues Advisory Group) 1997. *Guidance for local biodiversity action plans: Guidance Notes 1–5*. Bristol: UKBG.

USDA-NRCS 1999. *Conservation corridor planning at the landscape level: managing for wildlife habitat*. (http://www.geology.washington.edu/~nrcs-wsi/)

van der Zande AN, WJ TerKeurs & WJ van der Weijden 1980. The impact of roads on the densities of four bird species in an open field habitat – evidence of a long-distance effect. *Biological Conservation* **18**, 299–321.

van der Zande AN, JC Berkhuizen, HC van Latesteijn, WJ TerKeurs & AJ Poppelaars 1984. Impact of outdoor recreation on the density of a number of breeding bird species in woods adjacent to urban residential areas. *Biological Conservation* **30**, 1–39.

Warnken J & R Buckley 1998. Scientific quality of tourism environmental impact assessment. *Journal of Applied Ecology* **35**, 1–8.

Wathern P 1999. Ecological impact assessment. In *Handbook of environmental impact assessment*, Vol. 1, J Petts (ed.), 327–346. Oxford: Blackwell Science.

Webb NR & I Haskins 1980. An ecological survey of heathland in the Poole Basin, Dorset, England in 1978. *Biological Conservation* **17**, 281–296.

Webb NR & RJ Rose 1994. Habitat fragmentation and heathland species. *English Nature Research Report No. 95*. Peterborough: EN.

WT (Woodland Trust) 1999. Return of the natives. *Broadleaf* **52**, 10 only.

WWF (World Wide Fund for Nature) 1998a. Wildlife decline in Britain highlighted by WWF. *WWF-UK News*, 14 December 1998.

WWF 1998b. *The Living Planet Report*. (http://www.wwf-uk.org/)

12 Freshwater ecology

*Jeremy Biggs, Gill Fox, Pascale Nicolet,
Mericia Whitfield and Penny Williams*

12.1 Introduction

It is increasingly recognised that freshwater ecosystems are most likely to supply all
the services that we require from them if their ecological quality is maintained
as close to the pristine condition as possible. These services include maintenance
of aquatic **biodiversity**, provision of water for domestic, farming and industrial
use, amenity/recreational use, and commercial fisheries. Despite this, development
continues to cause much damage to fresh waters, reducing the value of water as a
resource and placing huge stresses on plant and animal communities (Meybeck *et al.*
1989, NRC 1992).

This chapter describes the main steps involved in undertaking an EiA for **freshwater ecosystems**, and highlights measures that can be used to protect them. The
chapter should be read in conjunction with Chapter 10 which provides associated
information about the physical and chemical aspects of the freshwater resource,
and with Chapter 11, which gives guidance on ecological concepts and general
approaches to ecological assessment in EIA.

12.2 Definitions and concepts

12.2.1 Introduction

Freshwater ecosystems are intrinsically dependent on their hydrology (Chapter 10)
in terms of both water quantity and water quality (including temperature and chemistry). These exhibit wide natural variation, principally in relation to the climate,
geology, geomorphology, soils and vegetation in the relevant **catchments**. Consequently, there is a huge diversity of freshwater ecosystems including, for example,
significant differences between tropical and temperate types.

This chapter focuses on temperate ecosystems and specifically on types found in
the UK. Even in this context, there is a wide natural range of water-quantity
regimes (e.g. depth and flow patterns) to which different species and communities
are adapted. The quality of natural fresh waters also varies considerably, especially in
nutrient status, **alkalinity** and **pH**. Natural water quality depends largely on the local
climate and (especially) geology. In areas of siliceous (and often hard) geological
materials, especially in the uplands of the north and west, the waters are generally
oligotrophic or even **dystrophic**. In areas of base-rich (and usually soft) materials
(mainly in the lowlands) the waters are more **eutrophic** – and in areas of chalk or

Table 12.1 Some sources of information on the ecology, conservation, management, restoration and creation of freshwater ecosystems

Ecosystem type	Sources
General	Kirby (1992), Maitland & Morgan (1997), Moss (1998), NRC (1992), USDA-NRCS (2000a).
Ponds & gravel pits	Andrews & Kinsman (1990), Williams *et al.* (1997).
Rivers & streams	Boon *et al.* (1992), Brookes (1988), Brookes & Shields (1996), Brookes *et al.* (1998), Calow & Petts (1994), De Waal *et al.* (1998), Gordon *et al.* (1992), Mainstone (1999), Petts & Calow (1996a,b), Ward *et al.* (1994).
Floodplains	Bailey *et al.* (1998), Marriott *et al.* (1999), Petts & Amoros (1997), Philippi (1996).
Wetlands	Keddy (2000), Lindsay (1995), Mitsch & Gosselink (2000), Raeymaekers (1998), Treweek *et al.* (1997).

limestone, they can be highly calcareous. These differences are used to classify British fresh waters into five main types which support very different communities (see Table F.1, p. 439). In addition to this natural variation, virtually all freshwater ecosystems are affected to some degree by human activity.

Freshwater ecology usually focuses on 'open waters', i.e. **aquatic ecosystems** – including standing waters (lakes, ponds, etc.) and watercourses (streams, rivers, etc.) – in which *macrophytes* (when present) are free floating or, if rooted, are submerged or have floating leaves. However, open waters often contain **swamp** – in which the water is shallow enough to allow the growth of emergent plants such as reeds – and which is usually classed as a type of wetland.

The term **wetland** is often taken to include aquatic ecosystems, but in this book it is restricted to ecosystems dominated by essentially terrestrial (or emergent) vegetation that is adapted to live in waterlogged conditions, at least periodically, i.e. in areas that are inundated or saturated by surface or ground water at a frequency and duration sufficient to support such vegetation. UK wetlands include swamps, mires, wet heath, marshes/marshy grasslands, and carr. Wetlands intergrade with both aquatic and terrestrial ecosystems.

Some sources of information on the ecology, conservation, management, restoration and/or creation of freshwater aquatic and wetland ecosystems are given in Table 12.1.

12.2.2 Ponds and lakes

Ponds are small natural or man-made waterbodies, less than 2 ha in area (PCG 1993). Many are important wildlife habitats and they can have economic value as fisheries and public amenities. The best ponds for wildlife generally occur in areas of *semi-natural habitat*, e.g. **heathland**, woodland, or unimproved grassland (PCT

1999). In these locations, ponds are generally protected from impacts due to water *pollution* – which is a key factor in the maintenance of high-quality ponds. These often support rich and/or distinctive communities of plants, invertebrates and amphibians, and some contain rare or endangered species (Biggs *et al.* 1993, Bratton 1990a). They typically have abundant stands of submerged and emergent plants, but in some natural habitats, such as woodland, they may be quite bare and muddy and still support valuable animal communities (Stubbs & Chandler 1978).

Seasonal ponds are a particularly neglected and internationally threatened habitat (Bratton 1990b), and where they are long established should always be treated with care. Aquifer fed, naturally fluctuating waterbodies such as turloughs can be included in this category, and are a UKBAP priority habitat (Appendix F.2).

Ponds in urban or agricultural situations (especially those which receive polluted *runoff*) or ponds stocked with large numbers of fish or ducks, are often of little wildlife interest. However, surveys are always required to confirm this.

Lakes, like ponds, may be natural or man-made. Lake communities include a wide range of generalist species, which can be found in many different freshwater habitats, together with more specialised *plankton*. The composition of these communities broadly reflects the nutrient status of the lakes. In Britain, lakes in the uplands of the north and west tend to be oligotrophic or even dystrophic (§12.2.1) and support distinctive but relatively species-poor communities. These may include rare fish like the powan (*Coregonus lavaretus*), and rare birds such as divers (*Gavia* spp.). Small oligotrophic waterbodies occur in some lowland areas, but lowland lakes are usually eutrophic, and can support a diverse fauna and flora. However, many lakes are now anthropogenically *eutrophicated* or acidified, reducing their conservation value.

In Britain, the creation of reservoirs and gravel pits during the last two centuries, has substantially increased the number of lakes in lowland areas. Many of these are now valuable wildlife habitats supporting uncommon plants and invertebrates and nationally or internationally important numbers of wildfowl.

12.2.3 Streams, rivers and floodplains

Most natural rivers and streams are characterised by three interconnected zones: the channel, the subterranean hyporheic zone, and the river floodplain. Protecting river ecosystems involves maintaining the wide range of habitats associated with all three of these zones.

The **channel** is the most obvious part of a river, and river management is often focused solely on this. River channels typically support aquatic plants and invertebrates which are restricted to flowing water habitats, but many of the wide variety of species occurring at the channel edges can be found in similar habitats in lakes and ponds (Holmes 1990). Channel habitats that are often poorly appreciated or ignored include bare mud and shingle bars, riverside cliffs and overhung or wooded sections (Bratton 1991, Kirby 1992, Shirt 1987). River channels are frequently an important fish habitat, so their management can have economic and recreational implications (Crisp 1993).

The **hyporheic zone** is, in essence, a river beneath the river: an area where water flows through the gravels or rocks below the channel base. This zone is often neglected in river assessments. However, in many rivers, particularly those with a gravel bed, it can be a major habitat for invertebrates. Research suggests that the hyporheic zone may be particularly important in temporary streams or headwaters where it can act as a refuge for invertebrates which burrow down into the river bed to escape drought (Ward 1992). In larger rivers, it is also used as a refuge during times of high flows (Marmonier & Dole 1986).

Natural and semi-natural river *floodplains* can support a wide range of habitats including temporary and permanent ponds, wetlands and wet woodland. In Europe, natural floodplains are now very uncommon, most having been targeted for agricultural improvement or urban development (§10.2.7). Remaining areas of floodplain, including river-edge strips (which, together with the channel, are often called *river corridors*), buffer watercourses from diffuse pollutants, provide habitats for riverside plants and animals, and are often considered to be valuable *wildlife corridors*. Associated backwaters, abandoned channels and floodplain ponds can add significantly to the variety of open water habitats and may provide flood refuges for fish and other animals.

12.2.4 Ditches and canals

Ditches and canals are wholly artificial in origin, but in some parts of the Britain they provide valuable open-water habitats. Canals support a predominantly still-water community whilst the flora and fauna of ditches varies from pond-like to stream-like, according to the rate of flow. The best ditch sites can support aquatic plant and invertebrate communities of exceptional nature conservation interest, sometimes including relict fen species. Important ditch systems typically occur in areas with high water tables, particularly in areas of drained wetland like the Somerset Levels. The best sites usually have water which is relatively unpolluted with nutrients, and many are non-intensively managed by grazing stock (Newbold *et al.* 1989).

From a nature conservation point of view, the most important canals are generally those with little or no boat traffic and good water quality. A few of the UK's canals have exceptionally rich freshwater plant and animal assemblages (Byfield 1990). In intensively managed arable and urban areas, ditches and canals are often degraded by runoff from surrounding land. However, even these can sometimes support important invertebrate communities (Foster *et al.* 1989) and they should always be adequately surveyed in EIAs.

12.2.5 Wetlands

Wetlands are inundated or saturated by surface or ground water with a frequency and duration sufficient to support vegetation adapted to live in waterlogged conditions, at least periodically. Like open-water ecosystems, they are dependent on their hydrology in terms of both water quantity and water quality. Conversely, in addition to their ecological value, they can have beneficial hydrological roles such as storing water, delaying runoff and filtering pollutants (§12.7.2).

British **swamps** are usually dominated by tall emergent *graminoids*, the most common type being reedbeds of common reed *Phragmites australis*.

Marshes and **marshy grasslands** are not normally inundated or saturated during the summer, so waterlogging is not sufficiently sustained to allow appreciable peat accumulation, and the substratum is predominantly mineral soil, or with peat < 0.5 m deep. The vegetation is normally dominated by grasses, sedges, rushes or *hydrophillous forbs* (see Table F.1). It is often used as *pasture* (e.g. 'grazing marsh'), but the more species-rich examples have usually been managed as **meadow** ('water meadows'/'wet meadows').

Mires (peatland ecosystems) occur where there is near-permanent waterlogging and consequent accumulation of peat (see §9.3.2) – normally > 0.5 m deep. They can be broadly divided into bogs, fens and springs/flushes.

Bogs (ombrotrophic mires) develop, and are maintained, through saturation by rainwater, with rain as the only source of nutrients (ombrotrophic) because the surface is above the influence of mineral soil water. Consequently, (a) they are restricted to high-rainfall areas, and (b) the peat is always **oligotrophic** and acid. Two types occur in the UK, **blanket bogs** and **raised bogs**, both of which normally have vegetation dominated by Sphagna (bog mosses), cotton sedges and *ericoids* (see Table F.1).

Fens (minerotrophic mires) are fed by mineral-enriched water (runoff and/or groundwater) which is also the source of their nutrients (minerotrophic). Consequently:

- they can occur in low-rainfall areas;
- their nutrient status can range from eutrophic to oligotrophic because it is strongly influenced by the local geology and soils, and by the degree of lateral water flow – ranging from **soligenous** (with appreciable lateral water flow and associated nutrient flux) to **topogenous** (with insignificant lateral flow or associated nutrient flux). They are divided accordingly into (a) **rich fens**, which are eutrophic and normally have species-rich vegetation (Wheeler 1988) and (b) **poor fens**, which are oligotrophic and have vegetation similar to that of bogs.

Fens can also be divided topographically into **valley mires, basin mires** and **floodplain mires**. Poor fen valley mires often occur in heathlands, where they intergrade with **wet heath** – which in turn intergrades with dry heath (see Table F.1).

Springs and **flushes** occur where groundwater seeps to the surface, and may have peat < 0.5 m deep. They are usually rich in **bryophytes**, sedges and rushes, and may support distinctive invertebrate communities (Kirby 1992). The best communities usually occur in areas of semi-natural habitat, especially where seepages are common.

Minerotrophic wetlands are **successional communities** that are naturally replaced (often rapidly unless managed) by terrestrial **climatic climax communities**, which in most of Britain are woodlands (see Fig. 11.2, p. 249). **Carr** (dominated by hydrophillous trees such as willows and alders) is normally a late stage in the succession, and intergrades with wet woodland. Unlike other wetlands, bogs (being dependent on rainfall) are climatic climax systems that should have long-term stability unless the climate changes.

Wetlands have long been under pressure from human activities such as draining for agriculture or forestry, peat extraction and urbanisation. For example, 99.7% of the East Anglian fens existing in 1637 have been drained (EA 2000) and there have been severe losses of lowland fens, reedbeds, lowland raised bogs and water meadows since 1945 (Fig. 11.1, p. 245). They are sensitive to impacts such as trampling and pollution (including *eutrophication*), and fens in particular are affected by lowering or increased fluctuation of the water table (see §12.5.3). Not surprisingly, the remaining wetlands have high conservation status, and the majority of types are protected under EU and UK legislation and/or are UKBAP priority habitats (see Appendices D & F).

12.3 Legislative background and interest groups

Freshwater ecosystems are indirectly protected by EU and UK legislation which is designed to protect water as a resource (see Tables 10.1 and 10.2, pp. 209 and 211, respectively), and some of this legislation (particularly the Environment Act 1995) requires the EPAs (Appendix B) to promote the conservation of flora and fauna which are dependent on an aquatic environment. In future, the EU Water Framework Directive (see §10.4.1) is likely to lead to increased protection for freshwater ecosystems. Specific protection for **high-status species and habitats**[1] is provided by the nature conservation legislation referred to in §11.3.1. However, the proportion of freshwater habitats protected in this way is small; e.g. only about 2.5% of the total river length in England has SSSI designation (EN 1997).

Policies and guidance relating to freshwater ecosystems are also covered by a combination of those for the water environment (§10.4.2) and nature conservation (§11.3.2). A potentially important development is the production of *Water Level Management Plans* (see §10.4.2). While these aim to balance the needs of a range of activities, they focus on maintaining or improving the hydrological regimes of protected wetlands such as SSSIs, SPAs, SACs and Ramsar sites (see Table D.3, p. 423). Where available, they are incorporated in the Environment Agency's LEAPs (see §10.2.4), a stated aim of which is to conserve and enhance biodiversity. MAFF (1999) has set 'High Level Targets' for flood defence operating authorities who must aim to (a) avoid damage to environmental interest, (b) ensure no net loss to habitats covered by HAPs (see §11.3.2), and (c) seek opportunities for environmental enhancement.

The **statutory consultee** for freshwater ecology assessments is the relevant NCCA (Appendix B), but in many cases, depending on the development type and possible impacts, the relevant EPA should also be consulted. Other groups who may have an interest include the Water Utilities (see Table 10.4, p. 216), and many of the NGOs, referred to in §11.3.3. Some of the best remaining freshwater habitats in Britain are now in the control of NGOs such as the RSPB and TWT, and angling clubs/associations have done much to ensure that freshwaters, especially rivers, have received protection from pollution.

12.4 Scoping and baseline studies

12.4.1 *Scoping*

Scoping for the freshwater ecology assessment should follow the principles and procedures outlined in §1.2.2 and §11.4. The use of scoping checklists should be beneficial. The example shown in Table 10.3 (p. 214) is relevant because it focuses on river engineering works and includes aquatic and wetland ecology as an issue. As in all water-environment assessments, estimation of the **impact area** can be difficult, and may need to be revised in the light of information gathered during the baseline studies.

---

1 When applied to species, habitats or sites, '**high-status**' means high conservation value in terms of the criteria referred to in Appendix D.

12.4.2 Methods and levels of study

Data gathered about freshwater communities for EIAs is mainly intended to answer the question *is the receptor site important and/or vulnerable?* This typically focuses on assessing wildlife conservation value using some of the criteria outlined in Appendix D, usually with emphasis on rarity, species richness (Table 11.1, p. 247) and community type. It must be remembered, however, that whilst species richness can be regarded as a valuable attribute of many freshwater communities, some high-status communities, such as bogs and dystrophic lakes, are intrinsically species-poor. Surveys may also involve assessments of fish stocks or gathering data about the distribution and **abundance** of high-status species.

In addition, physical, chemical and biological environmental data (like water depth, nutrient status, tree shade) may be useful to assist in understanding how the freshwater ecosystem functions, and may respond to possible impacts. Many thousands of variables could be measured for freshwater EIAs, so it is clearly important to choose those most relevant to the study area and the expected impacts. Techniques which are commonly used to study the physical and chemical aspects of the water environment are discussed in Chapter 10.

The process of gathering data for the freshwater ecology assessment can be divided into three phases, similar to those described in §11.4.2 for ecological surveys in general. Phase 1 will focus on habitats and the presence of species in these, and can utilise the JNCC method (§11.4.4). Phase 2 will involve additional (some quantitative) studies on species, communities and perhaps environmental conditions. Phase 3 (if undertaken) will require more intensive sampling to obtain detailed quantitative information. Most of the methods described in this chapter refer to Phase 1 and Phase 2 level surveys. Requirements for Phase 3 surveys are briefly discussed in Appendix G. A review of river corridor assessment methods used in the US (with similar categorisation in terms of study levels and skills required) is provided in USDA-NRCS (1999).

The **desk study** of existing information can often provide data about species and habitats of particular value, historical data about a site (see Appendix C), and information such as past and present site management regimes. Most of the organisations listed in Table 11.4 (p. 261) hold information about freshwater habitats and species; others include BTO, CEH-Windermere, EPAs, FBA and WWT. Some of these organisations hold databases relevant to freshwater ecosystems. The EA's *River Habitats Survey* (RHS) database holds data on a range of habitat features for 4500 UK reference sites, which are assigned quality grades (§12.6.3) – but the records provide little or no chemical or biological information.

Whilst the desk study is important, with the occasional exception of birds and some wetland plant communities, most data held on freshwater species or habitats will only provide background information. Additional survey work will almost certainly be needed to create adequate baseline descriptions of individual sites.

Collection of **new baseline data** needs to be carefully planned so that it can be used both for predicting the impact of a development and as a basis for monitoring post-development impacts. As in all ecological assessments, **timing** is critical. Freshwater systems are dynamic, changing seasonally and often annually, so single surveys made during any one month or year may not be representative. Surveys designed to create a reliable baseline therefore need to be undertaken over more than one season, and ideally over a number of years (Elliott 1990). Cost and time considerations mean that baseline surveys frequently fail to fulfil these requirements. However, inadequate baseline data make it difficult to (a) describe the state of the system in the absence of a development, (b) predict the effects of a development, or (c) monitor any resulting impacts.

12.4.3 Phase 1 field surveys

Phase 1 survey involves relatively brief site visits to identify freshwater habitats likely to be affected by the development. The survey can follow the JNCC Phase 1 survey method (§11.4.4) and the habitats assigned to categories of the JNCC classification (Table F.1). Care should be taken to ensure that small and apparently insignificant freshwater habitats, such as springs, seepages and temporary pools have been included. It is also essential to identify all the areas of freshwater habitat which may be impacted. For example, it is often forgotten that impacts on a river running through a development site are likely to extend some distance downstream. Similarly, some developments may bring major changes to regional groundwater levels affecting freshwater habitats many kilometres away.

12.4.4 Phase 2 field surveys

Phase 2 survey provides the main baseline information for most freshwater EIAs. It largely involves gathering data for assessment of the conservation value of species and communities in habitats identified by Phase 1 survey. There are two ways of determining whether a site supports valuable plant and animal communities: (a) by measuring environmental indicators, i.e. factors which are believed to indicate that a valuable community may occur, and (b) by recording the species present.

Environmental indicators – usually water quality but sometimes habitat diversity – are used as indirect measures of the quality of the whole system. They are based on assumptions, e.g. if water quality is high, it is assumed that the conservation value of that ecosystem will also be high, and that measures should be implemented to protect it. The advantages of this approach are: (a) it is usually relatively quick and inexpensive, because less time is spent identifying species, and (b) it treats the system as a whole, rather than perhaps focusing attention on a few rare species. The main drawbacks of adopting this approach alone are: (a) habitats with low water quality may be 'written off' as having little conservation value when valuable species may be present, and (b) it gives no indication about the species and communities present, so it is difficult to target protective or preventative mitigation measures. Typical methods of assessing the quality of fresh waters using environmental indicators such as water quality are outlined in Chapter 10.

The main disadvantage of **recording species** is that the collection of adequate data can be time consuming and expensive. However, a major advantage is that it allows direct assessments of the conservation value of ecosystems, based on the conservation status of species and community types. In general, surveys of this type are much more valuable for EIA than indicator data, because they provide more information and therefore enable more informed decision-making.

Surveys of **freshwater species** are primarily used to assess the conservation value of habitats and communities. This is normally done by looking at: (a) the range and number of species recorded and (b) the presence (and sometimes abundance) of high-status species (e.g. great crested newts, water voles) in all sites likely to be affected by the development. In addition, plant, invertebrate (and sometimes fish) communities are often assessed in terms of **community type** because this gives more information about their value and about the physico-chemical conditions which influence them. In all cases, the main aim of surveys is to provide data from which to assess whether any species or communities are of local, regional, national or international significance. If the development could physically damage a site, it may be important also to accurately locate and map important habitats or species within it.

When surveying species and communities, it is important to obtain information on species conservation status and ecology. A good introduction to the natural history of freshwater plants and animals is given by Fryer (1991). Other pulications on species conservation status, distributions and habitat requirements are listed in Appendix E. Of these, good starting points include Grime *et al.* (1988) for plants, FBA (and other) identification keys for invertebrates, Maitland & Campbell (1992) for fish, Swan & Oldham (1993) for amphibians, Cramp & Simmons (1977 *et seq.*) for birds, and Corbet & Harris (1990) for mammals.

12.4.5 *Phase 2 surveys of plants and vegetation*

Ideally, aquatic and wetland plant surveys should be undertaken in two seasons: in early summer, to catch early flowering species (such as water crowfoots) and in late summer/early autumn, when the majority of wetland plant species are readily identifiable. The definition of what constitutes an aquatic or wetland plant differs considerably, so use of a standard checklist is essential to allow comparisons of species numbers to be made between sites. A summary list, giving the national and regional status of all freshwater *vascular plants*, is available from PCTPR.

A variety of standard methods have been developed for surveying aquatic and wetland plants and communities in different habitats, and it is important that the appropriate methods are followed accurately so that plant community types can be reliably identified and compared. In the UK, The *National Vegetation Classification* (NVC) is generally recommended for Phase 2 vegetation surveys (see Appendix F.4). Volume 2 (Rodwell 1991) includes mires, and volume 4 (Rodwell 1995) includes swamps, tall-herb fens and aquatic communities. However, survey and classification methods designed specifically for standing waters, rivers or ditches are generally more appropriate for surveys of these systems, partly because they can provide a relatively rapid means of evaluating whole sites (which may contain a number of different plant communities). The main methods used in the UK are:

- The *Botanical classification of standing waters* (Palmer 1992, Palmer *et al.* 1992) which identifies ten main Site Types, with associated environmental factors such as alkalinity, pH and nutrient status. Surveys using this method should analyse open water and marginal species separately;
- The *National Pond Survey* methods for vegetation in waterbodies of up to 2 ha (Pond Action 1998);
- The *Vegetation of British Rivers* (Holmes *et al.* 1999) which includes a comprehensive classification of UK rivers. Surveys using this system must include **bryophytes**;
- The *Method for survey of ditch vegetation* (Alcock & Palmer 1985).

12.4.6 Phase 2 surveys of animals

Invertebrates make up a large proportion of the diversity of most freshwater habitats and often contribute significantly to the conservation value of a site. The main problem with invertebrate survey work is that it is only possible to record a (usually small) proportion of the species present in a habitat at any one time. To overcome this, standard survey techniques have been developed for some habitats which enable sites to be compared.

Ideally invertebrate surveys should allow (a) assessment of the value of the whole site, and (b) assessment of the value of smaller components of that site. For example, in a survey of a gravel-pit lake, samples from different habitats (such as mud or submerged plants) should be kept separate, so that the value of these habitats can be assessed. Samples should be replicated to assess whether perceived differences between habitats are likely to be real.

Aquatic invertebrate surveys should be carried out in a minimum of two seasons, and this should include an early spring visit to record mayfly and caddisfly fauna. Identification to species level should be undertaken with invertebrate groups for which keys are available (see Appendix E.2, p. 434). Given the emphasis placed on high-status species, it is essential that identifications of these are confirmed by experts.

Surveys of aquatic invertebrates have dealt mainly with species sampled using pond nets. However, there many different dredges, grabs and traps for collecting aquatic invertebrates, which may be appropriate under certain circumstances (see Elliott & Tullett 1983, New 1998, Southwood 1978). For ponds, standardised survey methods have been developed which use a three-minute hand-net sample from all significant habitats within the pond, and form the basis of the new Predictive System for Multimetrics (PSYM) system for assessing the ecological quality of ponds and small lakes (Biggs *et al.* 2000). River invertebrate communities are most frequently surveyed using methods described by Wright *et al.* (1984) (also a three-minute hand-net method), and this method forms the basis of much routine river invertebrate monitoring that takes place in Britain. Additional surveys are often conducted for adult dragonflies, either as they emerge or on the wing (Brooks 1993, Moore & Corbet 1990).

Semi-aquatic and terrestrial invertebrates associated with the margins of waterbodies are an important part of the fauna of many freshwater habitats. Commonly used survey techniques for these, and for wetland invertebrates, are outlined in Table G.2, p. 465).

Freshwater fish are often important indicators of ecosystem integrity, and are of great interest to anglers and the public. However, they are only rarely of importance in nature conservation terms, since most British native species are widespread and common. However, survey data on salmonids may be important (salmonid populations are generally declining) and fish surveys may also be relevant because of the economic and recreational importance of fish, and their significance in ecosystem function.

Fish can be surveyed by a variety of methods (e.g. see Bagenal 1978, Perrow *et al.* 1996). The main techniques are netting, electro-fishing and direct observations of breeding habitats (mainly for salmonids). Radio-tagging and counting at fish passes can be an important part of more sophisticated studies. Recently there has been considerable interest in studies of young (and hence small) fish which can be surveyed more cheaply than adults. Because of their economic and recreational importance, most fish survey work is concerned with estimating **biomass**, age structure or species diversity in order to provide data for habitat management and restoration schemes. There are two important exceptions: (a) the biology of fish is well-known enough for protection of distinctive local races of some species to be considered in conservation planning (e.g. some of the races of brown trout, *Salmo trutta*), and (b) specific surveys and measures may be undertaken to maintain populations of the few rare species (e.g. vendace, *Coregonus albula*, and powan, *Coregonus lavaretus*). Specific measures for the conservation of rare species are reviewed in Maitland & Lyle (1993).

Amphibians are usually surveyed at their breeding sites (usually ponds) during the breeding season. This varies between species and in different areas, e.g. the common frog typically spawns in late January in Cornwall but not until early April in parts of the Pennines (Swan & Oldham 1993). Juveniles and some adults remain in or near water during the summer, so summer surveys of ponds and surrounding areas can provide additional data. The main methods used are: (a) pond netting for individuals in the water, (b) 'torching' at night, (c) bottle trapping, and (d) searches for frog and toad egg masses during the breeding season. Egg searches have proved to be a quick and effective means of locating the specially protected great crested newt. Using a combination of survey methods generally proves more effective than one alone, e.g. searches for egg masses in spring, followed by summer netting for juveniles and any remaining adults.

The methods listed above really only provide information about which amphibian species are present, and cannot give more than a crude idea of population numbers (numbers collected are often 'out' by a factor of 10 or more). Collecting accurate population data for any species can be time consuming and expensive. The most frequently used method involves ring-fencing the breeding site to intercept animals moving to or from the surrounding area. More detail about amphibian survey methods can be found in Swan & Oldham (1989) and Halliday (1996).

Two of Britain's six native species are protected by law (the great crested newt, *Triturus cristatus*, and natterjack toad, *Bufo calamita*), so it is an offence to net or handle them without a licence from the relevant NCCA (Appendix B). It is also illegal to damage their habitat, including the terrestrial areas around the breeding site that they inhabit for most of the year. Great crested newts are relatively widespread in England, and so frequently feature in EIAs. Data on amphibian distribution is given by Swan and Oldham (1993). The *Herpetofauna Workers Guide* (Gibb & Foster 2000) includes information on amphibian conservation in the UK, and provides a wide range of contacts.

Reptiles such as the adder and grass snake are sometimes found in or by water, and more commonly in wetland habitats. Survey methods for reptiles are outlined in Appendix G.3.3 (p. 462).

Birds are one of the few groups where enough is known about total population sizes to make counts of individual species an important part of EIA. General survey methods are outlined in Appendix G.3.3. In freshwater habitats, the main areas of concern are likely to be whether there are (a) populations of overwintering waterfowl or waders which exceed the criteria for national or international importance (1% of population), or (b) populations of threatened breeding species. Significant overwintering populations are likely to be already monitored by the Wildfowl & Wetlands Trust (WWT). If population estimates of wetland birds using rivers and canals are required, it may be appropriate to follow the methods of the Birds of Waterways Survey, organised by BTO.

Mammals which may require specific attention in freshwater EIAs include the otter (*Lutra lutra*), bats and small mammals such as water shrew (*Neomys fodiens*), harvest mouse (*Micromys minutus*) and water vole (*Arvicola terrestris*) which may be directly associated with the margins of waterbodies. Survey methods for otters are described in NRA (1993) which can also be used a starting point for further information about otter distribution patterns and habitat requirements. Daubenton's bat (*Myotis daubentoni*) is largely reliant on waterbodies, and many other bat species use water opportunistically. General survey methods for bats and small mammals are outlined in Appendix G.3, and survey techniques for water voles are described in (Strachan 1998).

12.4.7 Analysis of baseline data

Much of the baseline analysis undertaken for a freshwater ecology assessment involves interpreting information contained in species lists collected using the methods outlined above. Interpreting these data commonly involves (a) noting the presence of high-status, uncommon and/or **indicator species**; (b) assessing the abundance of important species; (c) assessing the species richness of different sites/samples, and (d) assessing the characteristics of the habitat and community type (e.g. high-status, uncommon, rich in species, or degraded). Each of these criteria is then be described in terms of their local, regional, national or international significance. The publications cited above or in Appendix E, and the criteria outlined in Appendix D, can provide a basis for this assessment, but additional local information about the occurrence of important species/habitats may also be desirable.

Assessments of conservation value can sometimes be aided by numerically scoring sites, e.g. according to the richness or the rarity of species they host (see Appendix D.4, p. 427). Numerical methods are particularly useful where they facilitate comparisons between different sites or habitats or where they are used to combine data about different aspects of conservation value. However, simple numerical scores or indices can be misleading and lead to inappropriate conclusions; hence they should never be used in isolation.

Finally, it is clearly important to interpret wildlife data in the light of other environmental information gathered about the site (e.g. water depth, **sediment** type, habitat diversity). This is the basis for understanding freshwater communities and is an essential part of predicting impacts from the project.

12.5 Impact prediction

12.5.1 Introduction

As with all environmental components, it is important to consider the potential impacts of a project in the context of changes that may occur in its absence. In the case of freshwater ecosystems, these are likely to be changes in the water environment (§10.8.2), and changes associated with ecological **succession** (see Fig. 11.2). Many wetlands gradually become drier as they undergo succession. The likelihood of significant change occurring during the lifetime of a development can be assessed from the current conditions and management regimes. For instance, in the absence of management, a fen may be expected undergo fairly rapid succession to carr and eventually woodland.

Predicting future conditions with the development depends on (a) understanding how it will change the water environment; and (b) predicting what effect these changes will have upon the existing freshwater flora and fauna. Two inevitable difficulties with such predictions are that (a) many impacts are cumulative, and (b) the detailed effects of most impacts on most freshwater species are poorly known. EIAs therefore generally make broad predictions based on well-known tenets, backed up by more detailed work where information is available.

Freshwater communities are changed or damaged by five broad types of physical and chemical impact: (a) changes in surrounding land use (Appendix F.6, p. 452); (b) changes in water depth/level and its variability; (c) changes to the flow regime; (d) reduction of habitat size/complexity; and (e) pollution. The causes of these impacts are discussed in Chapter 10, so the following sections focus on their effects.

12.5.2 Changes in surrounding land use

Fresh waters are intimately linked to the land surrounding them. Changing the land use around a waterbody or wetland may, therefore, considerably influence its ecological quality. There are two main interactions:

1. Waterbodies surrounded by habitats such as woodlands, heathlands or meadows frequently also have distinctive aquatic and water's-edge communities. Species in these habitats (e.g. hoverflies, water beetles and amphibians) often need not only water but also specific terrestrial conditions during their life cycle. Changing the surrounding habitat may therefore eliminate these species and change the aquatic community as a whole (e.g. Fry & Lonsdale 1991).
2. Fresh waters and minerotrophic wetlands are sinks for liquids and solids which drain in from their catchments, so the quality of freshwater ecosystems usually reflects that of the surrounding land. When this has relatively non-intensive land use (e.g. moorland, deciduous woodland) they are often buffered from pollutants. If land use becomes more intensive, then the volume of pollutants such as *silt*, nutrients, organic wastes and **pesticides** draining into fresh waters can rapidly increase (e.g. Ormerod *et al.* 1993). Activities such as land drainage or urban development may also change water table levels and river flow regimes over large areas (see Chapter 10), inducing profound alterations in freshwater ecosystems, sometimes at points far away.

12.5.3 Changes in water depth or flow regime

Water level and stability are often considered to be critical for freshwater ecosystems. In fact, some water level fluctuation is natural in all open-water habitats. Typical fluctuations in still-water bodies during the year are often in the order of 0.3–0.5 m, whilst in rivers and streams, flood levels frequently rise several metres. Not all aquatic habitats are permanently wet. Temporary ponds and streams can be persistent features of natural landscapes and, particularly where they are long established, may contain specialised animals and plants of high conservation interest (Bratton 1991, Foster & Eyre 1992).

Damage to freshwater systems occurs when changes go beyond what is normal for the system, particularly if those changes are permanent or erratic. Most community damage is caused by lowering water levels but deepening can be equally devastating, especially where traditionally temporary water habitats are made permanent.

Water level and stability can be critical in wetlands such as fens (Fojt 1992), many of which been lost, degraded or are under threat as a result land drainage or abstraction, e.g. for agriculture or public water supply. For example, abstraction can result in water being pumped from protected wetland sites, and reduced water levels are estimated to adversely affect 14% of wetland SSSIs (RSPB 1999). The importance of water balance in wetland conservation is discussed in Gilman (1994).

Water flow is one of the main factors which distinguish freshwater ecosystems, and is critical to both their functioning and ecology. The effects of flow go far beyond increases in water velocity, because this is inevitably accompanied by changes in other variables such as dissolved oxygen concentration, nutrient fluxes and sediment type and volume. In general, still-water habitats like lakes and ponds accumulate sediments including organics, **heavy metals** and adsorbed nutrients like phosphates, so the ecological effect of pollutants may intensify with time. Running waters generally export materials (including pollutants) downstream. This means that the effect of pollutants on river and stream communities may be transitory at any one point, but may also affect far larger areas before becoming degraded or diluted.

Changing the flow rate of a waterbody, be it an increase or decrease, can indirectly damage communities adapted to the prevailing flow, and may irreversibly modify the physical and biological environment. For example, linking a stream and pond to 'stop the pond stagnating' will, among other impacts, change the pond's sediment characteristics and increase its infill rate. It will also introduce stream plants and animals which may considerably alter the original pond community. Similarly, creating impervious urban surfaces often causes rapid, spatey storm runoff into streams (see Chapter 10). This has been shown to physically modify stream widths, flood regimes and bottom substrates (Walesh 1989) with considerable knock-on effects for the channel and floodplain communities.

12.5.4 Reduction of habitat size and complexity

For any freshwater species there will be a critical minimum area of habitat needed to maintain a viable population. For example, the minimum area of bare gravel substrate needed by nesting little ringed plovers is around 0.2 ha (Andrews & Kinsman 1990). If a development threatens to destroy part of a water body or habitat type it is important to assess, as far as possible, whether there is sufficient area remaining to retain important species and/or communities on the site.

Habitat damage can be of particular concern for the conservation of aquatic invertebrates (Kirby 1992). Most invertebrates are very small and many rely on small-scale habitat features. Thus, small areas of bottom sediments like sand, mud, submerged wood or different complexes of plants may each support very different invertebrate communities (Harper *et al.* 1992). Habitat damage which destroys one of these small areas may eliminate an entire community. This occurs most obviously where part of a habitat is completely lost to development, but considerable damage can be caused by simplifying habitats. For example, river straightening often gives more uniform flow regimes, water depths, and bank profiles, all of which reduce habitat complexity and associated plant and invertebrate diversity (Brookes 1988).

Many species also live in (or need) more than one part of an aquatic habitat at different stages of their life cycle. For example, fish fry benefit from backwaters or bays in which they can develop (Schiemer & Waidbacher 1992) whilst nymphs of pond skaters are known to inhabit plant stands of different density as they grow. Removing any one of these habitats, or blocking the migration route between them, can therefore eliminate those species from the community.

12.5.5 Pollution

Variation in the quality of natural waters contributes to the diversity of species and habitats found in freshwaters (§12.2.1). Damage to a freshwater community is most likely to occur when human activities induce chemical changes beyond the natural range for that waterbody or wetland, e.g. by increasing phosphate to levels which are higher than would be experienced at that time of day/year in the unmodified system. Physical changes can also result in chemical impacts to freshwater systems. For example, destroying wet woodland adjacent to a river will, amongst other effects, reduce **denitrification** in the organic soils, and hence increase nitrate inputs to the river channel. The main causes of water pollution, and the effects of pollutants on water quality, are discussed in Chapter 10, and one or more of the models listed in Table 10.5 (p. 218) may be applicable in predicting the effects of pollution associated with a development. Some of the most significant ecological impacts of freshwater pollution are outlined in Table 12.2.

12.5.6 Predicting impacts using species-level information

General principles, such as those outlined above, can give a broad understanding of the impacts likely from a development. However, where species-level information is available for plants and animals in a freshwater habitat, it is usually possible to make more specific predictions about the impacts on key species or communities. For example, a development which increases the inputs of silt to a stream could cause a

Table 12.2 The effect of pollutants on freshwater flora and fauna

---

Organic matter (and associated deoxygenation of water)

Decomposition of organic matter by micro-organisms in water can lead to partial or total deoxygenation. Low oxygen levels are particularly damaging to river communities where fish and specialized river invertebrates require consistently high oxygen levels. Lakes may suffer if the bottom waters become highly deoxygenated, so (a) causing loss of bottom-dwelling biota, and (b) exacerbating the effects of eutrophication by promoting the release of phosphorus from the sediments. Small still-water bodies have highly variable oxygen levels, and support communities adapted to these conditions but may still be damaged by organic pollution if overloaded.

Thermal pollution

Temperatures above the normal range (e.g. near power stations) can: (a) exacerbate the effects of organic matter pollution by increasing decay rates and hence deoxygenation; (b) disrupt the life cycle timing of native species, (c) cause stress to cold-blooded animals by causing above-normal rates of respiration, and (d) favour, and allow to acclimatise, species not normally present in the area (including exotic and sometimes nuisance species). The first three factors may make (particularly organically polluted) waters un-inhabitable for much aquatic life.

Table 12.2 (*continued*)

Acidification

Low pH, and the toxic materials (particularly aluminium) brought into solution at low pHs, are directly injurious to many freshwater animals, and have diverse biological effects including changes in the abundance, biomass and diversity of invertebrates, plants, fish and amphibians. Effects are seen in uplands and areas of lowland heaths and bogs – wherever there is high rainfall and/or a prevalence of acidic soils.

Eutrophication

This results in enhanced plant growth (including that of macrophytes, *phytoplankton* and filamentous *algae*) with subsequent oxygen depletion of the water when this plant material decays. *Algal blooms* also increase turbidity and hence light attenuation in water, and macrophytes can clog rivers. Enrichment by nitrogen and phosphorus is often accompanied organic wastes which exacerbate the deoxygenation problem. *Anoxic* conditions may lead to the release of phosphate from bottom sediments, and hence allow the eutrophication to become self-perpetuating, irrespective of future phosphorus inputs. Most lowland lakes, and many lowland ponds, are impacted by eutrophication, as are slow-flowing and highly regulated lowland rivers. It can result in considerable loss of conservation value, including loss of species diversity, and community changes which allow dominance by a few tolerant plants (particularly algae). Fish community composition may alter, with an initial increase in fish biomass, often followed by high mortality when plant decay causes deoxygenation. Phosphorus is generally considered to be the principal eutrophicating agent in temperate freshwaters. However, once a system is rich in phosphorus, nitrates may become the main factor controlling aquatic productivity, and these conditions tend to promote the growth of nitrogen-fixing blue-green 'algae' (*cyanobacteria*) which may produce toxins.

Silt (fine organic and/or inorganic particles)

Silt may contain organic matter and hence have a high biochemical oxygen demand (so causing deoxygenation). It often carries nutrients (particularly phosphate) and adsorbed pollutants (such as micro-organics). Abrasive effects in rivers may kill fish through gill damage. Reduction of light by suspended particulates inhibits the growth of macrophytes, and may favour algal dominance. On settling, silt may (a) destroy salmon spawning beds, and the habitats of specialised plants and bottom-dwelling invertebrates, and (b) progressively seal waterbodies, isolating them from groundwater flows, and so potentially enhancing eutrophication.

Metals, micro-organics and other harmful chemicals

Polluting effects are diverse. Toxicity data are available for very few chemicals, but many have detrimental effects on aquatic life, some at levels considerably below those which cause immediate death, e.g. sub-lethal levels may enhance the risk of disease, affect reproductive capacity, or alter community structure due to changes in competitive or foraging behaviour. Some toxins may accumulate up the *food chain* or have *synergistic* effects.

Oils and grease

These deoxygenate water as they are broken down. Oil can blanket the water surface, inhibiting oxygen diffusion, and may directly coat plants and animals, causing injury and death. Oils contain many carcinogens, such as polycyclic aromatics and phenols, which mix with water and poison aquatic life.

number of generally detrimental effects including the swamping of existing gravel habitats used by aquatic invertebrates and spawning fish. A more detailed desk study of the habitat requirements of the stream's species would then aim to identify if any species were likely to be particularly vulnerable to this damage. Such predictions almost always rely on interpreting existing data from the literature (e.g. establishing habitat preferences, pollution sensitivity, breeding times). Sources of this information have been referred to in §12.4.4.

12.6 Impact significance

12.6.1 Introduction

Assessing impact significance is one of the most difficult parts of freshwater EIAs. The main problem is that, even where it is obvious that the developments will cause a change in freshwater communities, it can be very difficult to decide whether this change constitutes significant damage. The decision inevitably requires an element of judgement and should therefore draw heavily on expert opinion wherever possible.

The best practice procedure is to look at each potential impact in turn (e.g. nutrient inputs, changes in water level) and assess whether any changes to the ecosystem are likely to lie within the **natural range of perturbation** for that system. This should be considered both in the short and long term, and for all phases of the development, including construction and, if relevant, the decommissioning or restoration phase. As a rule, where impacts are likely to be within the normal range of the system, then the predicted level of change is likely to be acceptable. Where the normal range is exceeded, the significance of impacts will generally depend on (a) how much the system is likely to change from the norm, (b) how **sensitive** it is to damage, and (c) its **ecological/conservation value**.

12.6.2 Magnitudes and extents of impacts

Generally, the greater an impact's **magnitude**, the greater is its potential for damage. Thus a large area of physical damage, or the discharge of a large volume of polluting *effluent*, is likely to affect more individuals and species than a smaller impacted area or volume of effluent. Similarly, an impact is likely to have considerably more effect on a small habitat, like a stream or pond, than it would on a larger one such as a lake or river where detrimental impacts would tend to be diluted or ameliorated.

Short-term impacts, such as temporary changes in water depth or *turbidity*, generally have less significant impacts on aquatic communities than **long-term changes**. This may sometimes also be true of temporary habitat damage, but it depends on the habitat. River species, for example, are often very mobile due to downstream drift and upstream migration, and if a varied structure can be recreated or redeveloped there may be only temporary damage to aquatic communities. However, survey work is always necessary to ensure that (a) the habitats which are to be modified do not support unique species or unrecreatable features, and (b) they have good re-colonisation potential.

For many species and communities there may be **critical limits**, within which there is little change, but outside which considerable damage may ensue. Defining where these critical limits are for individual species is generally difficult. As stated above, the safest guide is to ensure that impacts lie within the range of what is already natural for the system.

Contamination of freshwater systems by non-mobile elements (like phosphorus) or non-biodegradable toxins, must be minimised since their effect may be permanent and effectively **irreversible**. Such impacts are most likely to have significant effects on systems which act as cumulative sinks for sediment, especially lakes and ponds. Although it may be possible to partially repair systems damaged in this way, practical experience suggests that the high cost of treatment makes such work extremely unlikely.

Finally, adhering to **statutory requirements** (such as maintaining water quality standards and protecting scheduled species) is clearly important, since contravening these requirements has been judged to result in a danger to health or in environmental or economic damage. Contravention may, in addition, lead to prosecution.

12.6.3 *Ecological value and sensitivity of ecosystems*

The *New Approach to Appraisal* (NATA) (see §10.8.5) suggests that river corridor sites should be assigned 'high', 'medium' or 'low' conservation/sensitivity values (Table 10.10, p. 236) using the EA's *River Habitat Survey* (RHS) scheme (Raven *et al.* 1998). This classifies sites into five grades, based on the combined score of a *Habitat Quality Index* (HQI) and a *Habitat Modification Index* (HMI). The NATA values are derived from these grades as follows: 'high' (grade 1), 'medium' (grade 2, 3 or 4) or 'low' (grade 5).

However, the RHS contains very little biological information (§12.4.2), and whether or not it is employed, river corridor and other freshwater habitats should also be evaluated using criteria selected from those listed in Appendix D. For example, sites containing **high-status species, habitats** or **communities** should be protected by the EIA process. Care should be taken not to focus on a species-centred approach and just translocate a given high-status species or manage for it alone (see §12.7.4). Species richness has already been mentioned as a valuable attribute of most freshwater communities. Two other criteria that may warrant particular attention are fragility/sensitivity and non-recreatability. For example:

- Studies have shown that where an outfall discharges urban runoff into a stream with good water quality and a sensitive community, it can significantly damage the community. The same discharges into an area of low water quality with degraded communities may cause little further deterioration.
- Once severely degraded, most freshwater systems take a long time to recover. Generally, the higher their quality the lower will be their recovery potential and some can be considered irreplaceable.

Habitats and communities which do not meet national high-status criteria, but which nevertheless are rich in species or support locally important species may be designated as of **local or regional value** (see LBAP criteria in Appendix D). Often they are characterised by clean water, and they may be buffered by areas of semi-natural habitats. Less sensitive or valuable communities are usually those which have already been extensively modified or degraded. In this case EIA should be taken as an opportunity to incorporate improvements to these ecosystems (see §12.7.3). New habitats (borrow pits, gravel pits, etc.) are also typically regarded as recreatable and of low conservation value. However, they often have areas of bare substrate and sometimes good water quality, and may support uncommon, specialist species.

It is always important to state in EIAs that the presence of lower-sensitivity communities does not mean that mitigation of development impacts is therefore unnecessary. All ecosystems are interconnected, so that severe degradation in one area can affect others. This can be seen in rivers where new pollution events may have little effect in a reach already badly damaged, but may progressively damage downstream sections, and eventually the marine environment. Without mitigation there is also the danger of subtle, persistent environmental damage even to degraded habitats. Impoverishment of local and regional biodiversity in this way is detrimental in its own right and can reduce the potential for future recovery or restoration of the area.

Finally, not all impacts are likely to be damaging to freshwater communities; some may be neutral or beneficial. However, benefits are most likely to be in the form of newly created habitats, or where change is made to these or to already highly degraded habitats. Changes to semi-natural freshwater ecosystems are much more likely to cause damage.

12.7 Mitigation

12.7.1 Introduction

As freshwater ecosystems are almost always profoundly influenced by adjacent terrestrial ecosystems, mitigation frequently involves maintaining these areas too. Mitigation is, of course, vital where it prevents damage to high-status communities. However, as stated above, it can be important even for damaged communities. The best practice for deciding upon mitigation techniques is to use the **precautionary principle**: if it is uncertain whether ecosystem damage will occur, then mitigation measures should be implemented. It also has to be recognised that full mitigation is not always possible. If residual impacts are likely to be severe and there is the potential for permanent damage to high-status (especially non-recreatable) communities, then re-planning or relocating the development should be the initial recommendation (Canter *et al.* 1991).

Examples of the broad range of mitigation measures used to minimise or prevent adverse impacts on freshwater ecosystems are given in Table 12.3 (see also Table 10.12, p. 238). In practice, deciding which method(s) will be most effective in any situation may be difficult, since there has been little monitoring of the long-term effectiveness of different mitigation techniques. As a general rule, **point source pollutants** (such as industrial effluents) should be dealt with at source, and ideally removed, recycled, or reused. **Non-point source pollution** (such as urban runoff) is best dealt with by a combination of measures, possibly including biological techniques (see below). Some mitigation measures will require periodic maintenance, such as changing the filters on oil traps or dredging siltation lagoons. Arrangements therefore need to be set in place to ensure that this is routinely undertaken. Finally it is important to consider whether mitigation measures may themselves have an adverse impact on freshwater habitats. For example, the creation of an on-stream lake to create a landscape feature or intercept sediment may have downstream implications for the flow regime, nutrient cycling and the ecology of the stream.

Table 12.3 Mitigation measures relating to impacts on freshwater ecosystems

Impact	*Mitigation*
Sediments/silt	Collect in **siltation traps**, **french drains**, or siltation basins/ponds/lagoons (maintenance is essential). Use vegetated **buffer zones** (30–100 m), including wetlands, as filters. Phase major construction periods to avoid wet seasons. Minimise disturbance during construction or operation, e.g. reduce bare areas by zoning, and install fences to protect adjacent areas. Avoid vegetation removal where possible. Revegetate bare areas rapidly, using temporary cover crops or mulches where necessary. Minimise dredging disturbance and erosion associated with bare areas, e.g. grade spoil heaps, and cover with tarpaulins.
Organic matter, nutrients and salt	Reduce silt inputs as above (P is primarily carried with silt). Reduce N inputs by minimising soil disturbance. Encourage formation of wet organic soils (i.e. create wetlands, extensive waterbody margin habitats, and wet woodland) to promote denitrification. In sewage treatment use nutrient stripping, tertiary treatments, separation of effluents, storm overflows.
Heavy metals, micro-organics, and other toxic materials	Treat or recycle industrial pollutants at source, and monitor effluents. Reduce silt inputs (as above). Reed beds may remove or manage many industrial and domestic effluents but proper design and maintenance is essential. Buffer zones (30–100 m) may give a reprieve from diffuse pollutants but can lead to long-term accumulation and/or release if these are not degradable. Minimise surface drainage from polluted areas. Reduce use where possible (e.g. of **pesticides**). Test any fill material placed in surface waters during the construction phase. Ensure isolation of waste-storage facilities and landfill sites from surface and groundwater bodies, and monitor for **leachates**. Discharge vehicle and other wash waters to foul sewers rather than surface-water drains. Guard against accidental pollution by: effective safety systems (with back-up), security systems against fire or vandalism where potential pollutants are stored or delivered; contingency plans; and education/training of personnel.
Oils	Install silt/petrol traps (gully traps) in road or parking areas and ensure a proper maintenance. Bund or dike around temporary fuel/oil storage areas during construction. Vegetated buffer zones may retain petroleum products while they degrade. Guard against accidental pollution.
Acidification	Strip power station flue gases. Control afforestation and modify forestry practices. Avoid use of liming to increase the pH of waterbodies because of adverse effects on the ecosystem.
Heat	Re-circulate and/or use to heat local buildings

Table 12.3 (*continued*)

Impact	Mitigation
Changes in flow regime and aquifer recharge	Procedures are outlined in Table 10.12. It is difficult to reproduce natural flow conditions using physical structures; so where possible, mimic natural processes by encouraging infiltration, e.g. use vegetated areas, porous artificial surfaces, or detention basins.
River engineering	Where possible, maintain natural river depths and course, bottom sediments, and floodplain/flood regimes. Use natural materials for bank protection/stabilisation, e.g. vegetation fringes and bankside trees instead of concrete or steel reinforcements. Limit damage by working from one bank and retaining vegetated areas, etc. Make new channels sinuous (not straight), and create new features such as pools, *riffles* and islands. Use dredgings for landscaping, etc.
Physical loss or other damage	Destruction or degradation of long-established semi-natural habitats should be strongly resisted, since current technology and understanding are not sufficient to allow full recreation. Whenever possible, the development should be relocated or rezoned. For other habitats, loss or damage may sometimes be minimised by retaining key areas and protecting specific species' migration routes, shelter and refuge zones. Consider habitat creation or enhancement to ameliorate loss.
Disturbance of wildlife	Create/maintain buffer zones. During construction: restrict working/access/service areas and extent of temporary roads; physically protect habitat/wildlife areas including food areas; and plan activities around critical periods (e.g. breeding, nesting). During operational phase: restrict access to valuable wildlife areas; and provide other focuses to reduce public pressure.

12.7.2 Pollution control using biological methods

Since the early 1980s, there has been considerable interest in the use of biological systems to reduce pollutant impacts through biological mitigation techniques, particularly where these are for pollutants that are difficult to mitigate by other means. The main biological systems employed are:

- *riparian buffer zones*, which usually aim to intercept diffuse pollutants such as agricultural chemicals and sediments (e.g. from soil *erosion*). If properly installed and maintained, they are said to have the capacity to remove > 50% of nutrients and pesticides, > 60% of some pathogens, and > 75% of sediment (USDA-NRCS 2000a);
- **natural** and **artificial/constructed wetlands** (especially reedbeds or ponds), which are designed mainly to intercept point source pollutants (e.g. from roads or in urban situations).

These techniques are generally a positive development with benefits that go beyond prevention of pollution by, for example, assisting in flood control from storm runoff and creating new wetland habitats. However, they are not a panacea and, although initially effective, they may create residual long-term problems. For example, a reedbed used to intercept road runoff may effectively deal with degradable pollutants such as nitrates, but have only a limited capacity to store non-degradable pollutants such as phosphates and heavy metals. Thus, it may eventually become saturated and then export most of the non-degradable pollutants subsequently received. The nature conservation benefits of biological mitigation techniques have also often been over-emphasised and used to justify other forms of damage. For example, the benefits of a creating a pond to intercept urban runoff are exaggerated; ponds filled with polluted water and sediments are unlikely to be good wildlife habitats. Further information on the use and value of natural and constructed wetlands for pollution control can be found in Cooper & Findlater (1991), Kadlec & Knight (1996), Nuttall *et al.* (1998), USDA-NRCS 2000a, USEPA (1993). Guidance on the design and use of buffer zones in available in USDA-NRCS (2000b).

12.7.3 Habitat restoration, creation, and enhancement

Where a development project is likely to lead to environmental loss or residual damage, **compensation** may be suggested in the form of the restoration or enhancement of existing habitats or the creation of new habitats. The benefits and perils of compensation and habitat creation are not always obvious, not least because monitoring projects to assess the success of existing schemes are few and far between. Some habitats such as reedbeds are relatively easy to create, but most projects which aim to accurately recreate existing waterbodies and wetlands are unlikely to be straightforward. This is particularly true of semi-natural habitats with complexes of open water and wet ground. Little is known about recreating the complex hydrological and ecological relationships in such habitats, and to date no attempts to do so have fully succeeded. Thus, justifying damage to high value communities on the basis that the same communities can be recreated elsewhere is not backed up by evidence and should be avoided.

Where damage to existing systems is unavoidable, every effort should be made to restore them (see references cited in Table 12.1) or compensate for their loss. For example, where a project necessitates re-routing a river, steps should be taken to ensure that the new section has high environmental and landscape quality. In addition, river works can sometimes be used as an opportunity to repair damage caused by earlier insensitive schemes. Two examples of mitigation and enhancement in this context are illustrated in Figures 12.1 and 12.2.

There is currently much scope to improve the quality of habitat creation proposals in EIAs. The most common form of site enhancement is simply digging a pond or lake. It would often be much better to establish a mosaic of habitats, perhaps including pools together with marsh, fen and wet grassland habitats: these not only add complexity, but some (like wet grassland), are now relatively uncommon habitats in their own right.

It is always important that attempts to restore or create aquatic and wetland habitats ensure that both the physical and chemical conditions are suitable. For instance, there is little point in trying to restore a river habitat such as a meandering stream channel unless good water quality can be assured (NRC 1992). Similarly,

Figure 12.1 Mitigation works undertaken by the NRA (now EA) as part of a comprehensive flood scheme along the River Colne, to the west of Heathrow Airport (photograph by A Brookes).
The scheme was designed to reduce flooding and provide a 1:100 yr level of protection to the adjacent buildings. Unfortunately the only option available in the constrained area was widening and deepening of the existing channel to accommodate flood flows. To mitigate the effects of widening the channel a sinuous low-flow notch (see photograph) approximating the natural width of the original channel was constructed. The higher areas within the channel were left to colonise with marginal aquatic plants, although it was recognised that a management plan would be needed to maintain the plants and therefore the hydraulic conveyance of flood flows. Islands were also retained within the reach to serve as wildlife refuges (right foreground).

provision should be made to ensure that enhancement or recreation sites can be adequately managed and maintained after their establishment. Care should also be taken to ensure that any creation sites are not already of high value, and further, that any management work to enhance the value of an existing site is actually an improvement. Temporary pools, damp hollows, wet flushes and shaded ponds or waterways are all examples of undervalued habitats which are often and easily damaged by misguided habitat enhancement work. Restoration and creation techniques for freshwater habitats are described and discussed in some general texts such as Buckley (1989) and Gilbert & Anderson (1998), and in publications focusing specifically on freshwater systems (see Table 12.1).

12.7.4 Translocation of rare species

Where species are critically endangered or have legal safeguards, translocation is sometimes undertaken. However, simply transferring rare animals or plants from one habitat to another as a means of mitigation should not be recommended except as a last resort. This is because (a) one threatened species in a habitat may well indicate the presence

Figure 12.2 Mitigation and enhancement associated with the newly constructed Blackwater Valley Road in Surrey, completed in 1995 (photograph by A Brookes).
Borrow pits used for gravel extraction (to the right of the aerial photograph) subsequently became filled with water and are important wildlife habitats and fisheries. The original river channel had been severely modified by widening, deepening and straightening over the past 100 yr or so. Whilst nearly 12 km of river channel had to be moved as a result of road construction on the floodplain, the opportunity was taken to create a more natural sinuous meandering channel on virgin floodplain, away from the road. This channel was designed to accommodate the current water and sediment discharge characteristics of the watercourse.

of others, perhaps within groups that have not been included in the baseline survey, (b) the habitat may be valuable even if no other threatened species are present, (c) the chances of success are low, and (d) there may be adverse impacts on the community of the recipient habitat. For example, attempts to transfer species such as the great crested newt into existing ponds have often proved unsuccessful, either because the ponds were unsuitable or because they were already at their maximum **carrying capacity** for that species. Translocation into new, specifically created ponds is more likely to succeed, but may not compensate for the loss of the original ecosystem.

12.8 Monitoring

Mitigation of adverse impacts from developments should not stop at the design stage, and should evolve into long-term protection and monitoring. This is essential both to check that mitigation measures are properly functioning, and to ensure that they are adequately preventing damage. Unfortunately, many aspects are difficult to monitor because it is difficult to isolate the effects of a project from those of other developments and activities. However, aspirations for the success of a project can often be set and monitored, as shown, for example, in Table 12.4.

Table 12.4 Example criteria for the success for pool-riffle reconstruction in lowland Europe (from Brookes *et al.* 1998)

Issue	Example criteria	How to measure
Surface-water hydrology	The typographic highs caused by **riffles** should not be so high that they cause overbank flooding of property.	Flood monitoring.
	The riffles should be of sufficient height to cause divergence of flow.	Mapping of flow patterns.
	The gravels forming the riffles should remain free of significant silt deposition.	Repeat topographic surveys and/or visual checks during and after construction.
Channel morphology	The gravels forming the riffles should remain *in situ*, i.e. should not erode out during moderate to high flows.	Repeat topographic surveys and/or visual checks during and after construction.
Aquatic ecology	There should be an increase in the diversity of fish, plant and invertebrate species.	Repeat ecological surveys before, during and after construction.
Visual amenity	The diversity introduced should improve the aesthetic value of the channel.	Repeat public perception surveys of the existing and improved channels.
Recreation	The addition of pools should improve the angling quality.	No standard methods exist at present.

Three requirements are essential for successful monitoring: (a) baseline data that are good enough to detect detrimental changes caused by a development; (b) funding to carry out the survey work and the monitoring of the mitigation measures; and (c) sufficient contingency funds to enable modifications to mitigation measures to be made, or faults to be rectified, if necessary. Finally, important lessons learnt, or the results produced from well-constructed monitoring projects should be made available to others wherever possible. The ability to undertake effective EIAs and to plan effective mitigation schemes is significantly hindered by lack of information. The combined efforts and experience of those involved in undertaking and monitoring environmental EIAs could do much to change this.

References

Alcock MR & MA Palmer 1985. *A standard method for the survey of ditch vegetation. CST Report No. 37.* Peterborough: Nature Conservancy Council.

Andrews J & D Kinsman 1990. *Gravel pit restoration for wildlife. A practical manual.* Sandy: RSPB.

Bagenal TB 1978. *Methods for the assessment of fish production in freshwaters.* I.B.P Handbook No. 3. Oxford: Blackwell Scientific.

Bailey RG, PV José & BR Sherwood 1998. *United Kingdom floodplains.* Otley, Yorkshire: Westbury Publishing.

Biggs J, A Corfield, D Walker, M Whitfield & P Williams 1993. The importance of ponds for wildlife. In *Proceedings of the conference 'Protecting Britain's Ponds'*, C Aistrop & J Biggs (eds), 4–7. Oxford: Pond Action and the Wildfowl & Wetlands Trust.

Biggs J, P Williams, M Whitfield, G Fox & P Nicolet 2000. Biological techniques of still water quality assessment. *R&D Technical Report E110.* Bristol: Environment Agency.

Boon PJ, P Calow & GE Petts (eds) 1992. *River conservation and management.* Chichester: Wiley.

Bratton JH 1990a. A review of the scarcer Ephemeroptera and Plecoptera of Great Britain. *Research and Survey in Nature Conservation, No. 29.* Peterborough: Nature Conservancy Council.

Bratton JH 1990b. Seasonal pools – an overlooked invertebrate habitat. *British Wildlife* 2, 22–31.

Bratton JH 1991. *British Red Data Books: 3. Invertebrates other than insects.* Peterborough: JNCC.

Brookes A 1988. *Channelized rivers: perspectives for environmental management.* Chichester: Wiley.

Brookes A & FD Shields Jr (eds) 1996. *River channel restoration: guiding principles for sustainable projects.* Chichester: Wiley.

Brookes A, P Downs & K Skinner 1998. Engineering of wildlife habitats. *Journal of the Chartered Institute of Water and Environmental Management* 12, 25–29.

Brooks SJ 1993. Review of a method to monitor adult dragonfly populations. *Journal of the British Dragonfly Society* 9, 1–4.

Buckley GP (ed.) 1989. *Biological habitat reconstruction.* London: Bellhaven Press.

Byfield A 1990. The Basingstoke Canal – Britain's richest waterway under threat. *British Wildlife* 2, 13–21.

Calow P & GE Petts (eds) 1994. *The rivers handbook:* Vol. 1 *Hydrological and ecological principles;* Vol. 2 *The science and management of river environments.* Oxford: Blackwell Scientific.

Canter LW, JM Robertson & RM Westcott 1991. Identification and evaluation of biological impact mitigation measures. *Journal of Environmental Management* 33, 35–50.

Cooper PF & BC Findlater (eds) 1991. *Constructed wetlands in water pollution.* Oxford: Pergamon.

Corbet GB & S Harris 1991. *The handbook of British mammals*, 3rd edn. Oxford: Blackwell Scientific.

Cramp S & KEL Simmons 1977 et seq. *The birds of the Western Palaearctic.* Oxford: Oxford University Press.

Crisp DT 1993. The environmental requirements of salmon and trout in fresh water. *Freshwater Forum* **3**, 176–202.

De Waal LC, ARG Large & PM Wade (eds) 1998. *Rehabilitation of rivers: principles and implementation*. Chichester: Wiley.

EA 2000. *State of the Environment*. http://www.environment-agency.gov.uk

Elliott JM 1990. The need for long-term investigations in ecology and the contribution of the Freshwater Biological Association. *Freshwater Biology* **23**, 1–5.

Elliott JM & PA Tullett 1983. A supplement to a bibliography of samplers for benthic invertebrates. *Freshwater Biological Association Occasional Publications* **4**.

EN (English Nature) 1997. *Wildlife and fresh water: an agenda for sustainable management*. Peterborough: EN.

FBA (Freshwater Biological Association) – about 60 keys (usually with ecological notes) for freshwater taxa. (listed at http://www.fba.org.uk)

Fojt W 1992. East Anglian fens and ground water abstraction. *English Nature Research Report No. 30*. Peterborough: EN.

Foster GN & MD Eyre 1992. *Classification and ranking of water beetle communities. UK Nature Conservation No. 1*. Peterborough: JNCC.

Foster GN, AP Foster, MD Eyre & DT Bilton 1989. Classification of water beetle assemblages in arable fenland and ranking of sites in relation to conservation value. *Freshwater Biology* **22**, 243–254.

Fry R & D Lonsdale et al. 1991. *Habitat conservation for insects – a neglected green issue. The Amateur Entomologist*, Vol. 21. London: Amateur Entomologists' Society.

Fryer G 1991. *A natural history of the lakes, tarns and streams of the English Lake District*. Ambleside: Freshwater Biological Association.

Gibb R & J Foster 2000. *The herpetofauna workers guide*. Halesworth: Froglife.

Gilbert OL & P Anderson 1998. *Habitat creation and repair*. Oxford: Oxford University Press.

Gilman K 1994. *Hydrology and wetland conservation*. Chichester: Wiley.

Gordon ND, TA McMahon & BL Finlayson 1992. *Streamflow hydrology: an introduction for ecologists*. Chichester: Wiley.

Grime JP, JG Hodgson & R Hunt 1988. *Comparative plant ecology*. London: Unwin Hyman.

Halliday TR 1996. Amphibians. In *Ecological census techniques: a handbook*, WJ Sutherland (ed.), 205–217. Cambridge: Cambridge University Press.

Harper D, C Smith & P Barham 1992. Habitats as the building blocks for river conservation assessment. In *River conservation and management*, PJ Boon, P Calow & GE Petts (eds), 311–320. Chichester: Wiley.

Holmes NTH 1990. British river plants – future prospects and concerns. *British Wildlife* **1**, 130–143.

Holmes NT, PJ Boon & TA Rowell 1999. *Vegetation communities of British rivers – a revised classification*. Peterborough: JNCC.

Kadlec RH & RL Knight 1996. *Treatment wetlands*. Boca Raton, Florida: Lewis (CRC Press).

Keddy PA 2000. *Wetland ecology: principles and conservation*. Cambridge: Cambridge University Press.

Kirby P 1992. *Habitat management for invertebrates: a practical manual*. Sandy: RSPB.

Lindsay RA 1995. *Bogs: the ecology, classification and conservation of ombrotrophic mires*. Edinburgh: Scottish Natural Heritage.

MAFF 1999. *High Level Targets for flood and coastal defence and elaboration of the Environment Agency's flood defence supervisory duty*. London: MAFF.

Maitland PS & RN Campbell 1992. *Freshwater fishes of the British Isles*. London: HarperCollins.

Maitland PS & AA Lyle 1993. Freshwater fish conservation in the British Isles. *British Wildlife* **5**, 8–15.

Maitland PS & NC Morgan 1997. *Conservation management of freshwater habitats: lakes, rivers and wetlands*. London: Chapman & Hall (Kluwer).

Mainstone CP 1999. *Chalk rivers: nature conservation and management*. Peterborough: EN.

Marmonier P & MJ Dole 1986. Les Amphipodes des sediments d'un bras courte-circuite du Rhone: logique de repartition et reaction aux crues. *Science del l'eau* **5**, 461–486.

Marriott S, J Alexander & R Hey 1999. *Floodplains: interdisciplinary approaches*. London: Geological Society.

Meybeck M, D Chapman & R Helmer (eds) 1989. *Global Environment Monitoring System. Global freshwater quality: a first assessment*. Oxford: Blackwell.

Mitsch WJ & JG Gosselink 2000. *Wetlands*, 3rd edn. New York: Wiley.

Moore NW & PS Corbet 1990. Guidelines for monitoring dragonfly populations. *Journal of the British Dragonfly Society* **6**, 21–23.

Moss B 1998. *Ecology of fresh waters. Man and medium, past to future*, 3rd edn. Oxford: Blackwell Science.

New TR 1998. *Invertebrate surveys for conservation*. Oxford: Oxford University Press.

Newbold C, J Honnor & K Buckley 1989. *Nature conservation and the management of drainage channels*. Peterborough: Nature Conservancy Council.

NRA (National Rivers Authority) 1993. *Otters and river habitat management*. Conservation Technical Handbook, 3. Bristol: National Rivers Authority.

NRC (National Research Council) 1992. *Restoration of aquatic ecosystems*. Washington, DC: National Academy Press.

Nuttall PM, AG Boon & MR Rowell 1998. *Review of the design and management of constructed wetlands*. London: CIRIA.

Ormerod SJ, SD Rundle, EC Lloyd & AA Douglas 1993. The influence of riparian management on the habitat structure and macroinvertebrate communities of upland streams draining plantation forests. *Journal of Applied Ecology* **30**, 13–24.

Palmer MA 1992. *A botanical classification of standing waters in Great Britain and a method for the use of macrophyte flora in assessing changes in water quality*, 2nd edn. Research & Survey in Nature Conservation No. 19. Peterborough: JNCC.

Palmer MA, SL Bell & I Butterfield 1992. A botanical classification of standing waters in Britain: applications for conservation and monitoring. *Aquatic Conservation: Marine and Freshwater Ecosystems* **2**, 125–143.

PCG (Pond Conservation Group) 1993. *A future for Britain's ponds*. Oxford: Pond Conservation Group.

PCT (Pond Conservation Trust) 1999. *The pond book*. Oxford: The Ponds Conservation Trust.

Perrow MR, IM Côté & M Evans 1996. Fish. In *Ecological census techniques: a handbook*, WJ Sutherland (ed.), 178–204. Cambridge: Cambridge University Press.

Petts GE & C Amoros (eds) 1997. *Fluvial hydrosystems: a holistic approach to river and floodplain ecosystems*. London: Chapman & Hall (Kluwer).

Petts GE & P Calow (eds) 1996a. *River biota: diversity and dynamics*. Oxford: Blackwell Science.

Petts GE & P Calow (eds) 1996b. *River restoration*. Oxford: Blackwell Science.

Philippi NS 1996. *Floodplain management: ecologic and economic prespectives*. London: Adademic Press.

Pond Action 1998. *A guide to the methods of the National Pond Survey*. Oxford: Pond Action.

Raeymaekers G 1998. *Conserving mires in the European Union*. europa.eu.int/comm/environment/

Raven PJ et al. 1998. *River habitat quality: the physical character of rivers and streams in the UK and Isle of Man*. Bristol: Environment Agency.

Rodwell JS (ed.) 1991. *British plant communities*, Vol. 2: *Mires and heaths*. Cambridge: Cambridge University Press.

Rodwell JS (ed.) 1995. *British plant communities*, Vol. 4: *Aquatic communities, swamps and tall-herb fens*. Cambridge: Cambridge University Press.

RSPB (Royal Society for the Protection of Birds) 1999. *Focusing on UK wetlands*. Sandy: RSPB.

Schiemer F & H Waidbacher 1992. Strategies for conservation of the Danubian fish fauna. In *River conservation and management*, PJ Boon, P Calow & GE Petts (eds), 363–382. Chichester: Wiley.

Shirt DB 1987. *British Red Data Books: 2*. Insects. Peterborough: Nature Conservancy Council.

Southwood TRE 1978. *Ecological methods*, 2nd edn. London: Chapman and Hall.

Strachan R (1998). *Water vole conservation handbook*. Oxford: Wildlife Conservation Research Unit.

Stubbs A & P Chandler *et al.* 1978. *A dipterist's handbook. The Amateur Entomologist*, Vol. 15. London: Amateur Entomologist's Society.

Swan MJS & RS Oldham 1989. *Amphibian communities*. Leicester Polytechnic under contract to the Nature Conservancy Council. Contract No. HF3-03-332 Year 3.

Swan MJS & RS Oldham 1993. *Herptile sites*. Vol. 1: National amphibian survey, Final Report and Appendices. *English Nature Research Reports No. 38*. Peterborough: EN.

Treweek J, M Drake, O Mountford, C Newbold, C Hawke, P José, M Self & P Benstead (eds) 1997. *The wet grassland guide: managing floodplain and coastal wet grasslands for wildlife*. Sandy: RSPB.

USDA-NRCS 1999. *Stream corridor inventory and assessment techniques: a guide to site, project and landscape approaches suitable for local conservation programs*. Watershed Science Institute Technical Report. (http://www.geology.washington.edu/~nrcs-wsi/)

USDA-NRCS 2000a. *National handbook of conservation practices*.
(http://www.ftw.nrcs.usda.gov/nhcp_2.html)

USDA-NRCS 2000b. *Buffer strips: common sense conservation*.
(http://www.nhq.nrcs.usda.gov/CCS/Buffers.html)

USEPA 1993. *Created and natural wetlands for controlling nonpoint source pollution*. Boca Raton, Florida: Lewis (CRC Press).

Walesh SG 1989. *Urban surface water management*. New York: John Wiley.

Ward D, N Holmes & P José (eds) 1994. *The new rivers and wildlife handbook*. Sandy: RSPB.

Ward JV 1992. *Aquatic insect ecology. 1. Biology and habitat*. New York: Wiley.

Wheeler BD 1988. Species richness, species rarity, and conservation evaluation of rich fen vegetation in lowland England and Wales. *Journal of Applied Ecology* **25**, 331–353.

Williams P, J Biggs, A Corfield, G Fox, D Walker & M Whitfield 1997. Designing new ponds for wildlife. *British Wildlife* **8**, 137–150.

Wright JF, D Moss, PD Armitage & MT Furse 1984. A preliminary classification of running water sites in Great Britain based on macroinvertebrate species, and the prediction of community type using environmental data. *Freshwater Biology* **14**, 221–256.

13 Coastal ecology and geomorphology

Stewart Thompson and John Lee

13.1 Introduction

The UK has more than 15,000 km of coastline and more than one-third of a million km² of territorial waters (Gubbay 1990). The coastal zone contains a variety of valuable ecosystems (§11.2.3) but is subject to considerable economic and recreational pressures. It is estimated that 70% of European coastlines are highly threatened (the highest proportion of any region), and that between 1960 and 1995 (a) a kilometre of unspoilt coastline was developed per day, and (b) most coastal habitats suffered considerable loss and decline (EUCC 1997). Pressures are exacerbated by the need to balance three conflicting requirements: meeting the demands of economic development and tourism; protecting vulnerable settlements from flooding and **erosion**; and protecting important scenic, geological and ecological systems (DoE 1993).

An additional problem is posed by rising sea levels, associated with climatic warming. Mean global sea level has risen by between 10 cm and 20 cm over the last century, and the rise is accelerating with a predicted further rise of up to 6.5 cm by 2100; and the coastal authorities in England have been warned to allow for annual increases of between 4 mm and 6 mm during the next 50 years (EA 1999a). The problem is exacerbated by the fact that many of the UK's coastal defences are in need of renovation (MAFF 1994). In addition to the risks to human life, settlements and agricultural land, rising sea levels threaten coastal habitats. For example, it has been estimated that to maintain current levels, c.8.5% of the UK's saltmarshes and c.4.3% of its tidal mudflats would have to be replaced in a 20-year period (EN 1992).

13.2 Definitions and concepts

13.2.1 The coastal zone

What precisely is meant by the term coastal zone is problematic. In terms of geomorphology (land forms and associated processes) it can be defined as the zone between the land and sea that includes the shallow waters in which waves move **sediment**, and the zone of beaches, cliffs and dunes that are affected by the movement of this sediment (Summerfield 1991). Ecologically, its inland extent can be determined approximately as the limit of influence by salt spray, which is rarely further than about 0.5 km inland (UKBG 1999). However, the EA's definition of the coastal zone includes the land within 10 km of the coast. The marine extent of the zone is usually taken to be 'inshore waters' although it can be interpreted to include 'offshore waters', which extend to the edge of the continental shelf (see §13.2.5).

All these systems are sometimes collectively referred to as maritime. More usually they are subdivided, in relation to the land–sea axis, into three zones, the **littoral** (intertidal or shore) **zone**, the **supralittoral** (or maritime) **zone**, and the **sublittoral** (or marine) **zone**. These zones are commonly used in classifications such as the UKBAP broad habitats classification (Appendix F.2, p. 441) and the BioMar biotop·; classification (§13.4.5). The communities (§11.2.2) inhabiting all three zones are profoundly influenced by the substratum type, i.e. rock or sediment type, and the substratum types define the major habitats types of the BioMar classification (Fig. 13.1). The substratum (and other habitat features) in a given location, depends on the prevalent geomorphological processes in the area.

13.2.2 Coastal geomorphology

Coastal geomorphology is important in EIA for two main reasons: (a) geomorphological formations and processes are integral components of coastal ecosystems; (b) it has direct relevance to problems such as erosion and flooding. It is a complex subject, and only a brief outline of some important aspects can be given here. Further information can be found in many texts including Carter (1988), Cooke & Doornkamp (1990), Davis (1996), Pethick (1984), and Trenhaile (1997). It involves interactions between many components, but the principal factors are waves, tidal regimes and currents.

Waves are a major erosive force, and an important ecological factor, especially in the littoral zone. Most waves are wind generated, and vary in size and force largely in relation to wind velocity, duration and *fetch* (distance of open water). Wave action is therefore greatest on coasts exposed to strong on-shore winds over extensive areas of sea. Wave effects are modified by tidal regimes.

The two main components of **tidal regimes** are rhythm and range. In most locations around Europe, the **tidal rhythm** consists of approximately two tides per day. The **tidal range** (rise and fall) varies daily, with a two-weekly cycle of large **spring tides** that advance and retreat much further than the small **neap tides** of the alternate weeks. In addition, there are larger seasonal cycles, with the largest spring tides near the spring and autumn equinoxes (in March and September). These are the extreme high-water spring-tide (EHWS) levels, and extreme low-water spring-tide (ELWS) levels. The EHWS level can be extended by waves, especially on exposed shores and during storms.

The **mean tidal range** (taken as the distance between the mean high-water spring-tide (MHWS) and mean low-water spring-tide (MLWS) levels) varies considerably between different locations. It can be *macrotidal* (as much as 12 m) in the UK, although most of the coastline has a tidal range of much less than this, and some locations have a *microtidal* regime, with a range of only 1–2 m. The tidal range affects coastal geomorphology by controlling the vertical distance over which waves and currents are effective. For instance, tidal sand ridges develop in macrotidal environments (Reading & Collinson 1998). The tidal rhythm also affects the intensity of currents.

Coastal currents are important, both as a means of sediment transport and as agents of erosion. They result from the interaction of climate, tides, wave regime and coastal morphology. Two important types are:

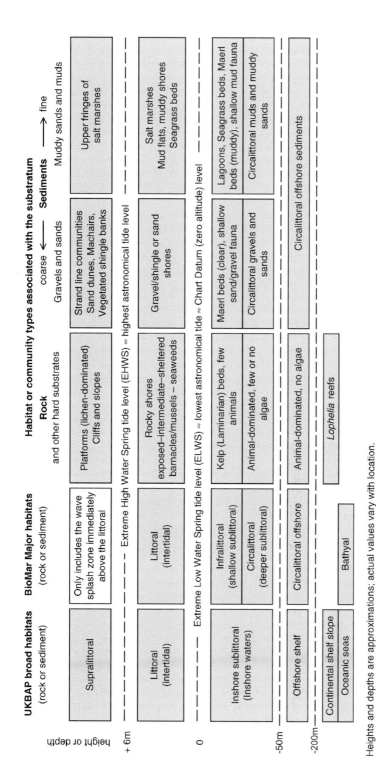

Figure 13.1 Outline of maritime and marine habitats defined in the UKBAP habitats classification and the BioMar biotopes classification.

Heights and depths are approximations; actual values vary with location.

Coarse sediments occur on fairly exposed coasts, and fine sediments in more sheltered locations. The BioMar classification also includes "mixed sediments". Shingle is coarse gravel (of particle size 2–200mm) but the term is most commonly used for supralittoral banks which become at least partially vegetated.

- **longshore currents**, which are commonly caused by oblique waves, tend to run parallel to the shore line, and result in lateral movement of sediment (**longshore drift**) along the coast (Carter 1988);
- **tidal currents**, which are important where coastal morphology funnels tides, e.g. in narrow straights and inlets, and in extreme cases become **tidal rapids**.

Under certain conditions (usually a combination of low atmospheric pressure and high winds) funnelling of shallow coastal seas by coastlines can cause **storm surges** or, when combined with high tides, **tidal surges**. These pose serious flooding risk. For example, at the southern end of the North Sea, surges can raise predicted tide levels by over 2 m. The bathymetric (sea floor) topography is also important in influencing the pattern of currents.

The complex interaction of tides, currents and waves creates either an erosive or a depositional (and hence constructive) regime in a given area.

Erosive processes are widespread, e.g. it has been estimated that 70% of the world's sandy coastline is being eroded (Bird 1985), with waves being the most important agent. The predominant erosional landform types are sea cliffs and shore platforms. Cliffs suffer minimal destruction in deeper water but from severe erosion at their base when water is sufficiently shallow for waves to break. As a cliff retreats a shore platform is left which protects the cliff base by dissipating wave energy. The stability of cliffs may also be affected by groundwater seepage and frost action above sea level, and by cliff geology. Soft cliffs, consisting of unconsolidated materials such as boulder clay, are often subject to quite rapid erosion, as along many stretches of the East coast of England.

Depositional processes use sediment from inland (mainly in rivers), the sublittoral or elsewhere along the coast. Coastal erosion sources include cliff sediment from exposed coasts, but more commonly the unconsolidated materials of dunes and beaches. Deposition leads to a variety of landforms the most important of which are:

- **beaches**, which hold the greatest amounts of deposited coastal material, usually sand or pebbles;
- **spits**, which are formed by longshore drift of sediments (usually shingle) along fairly exposed coastlines;
- **dunes**, which are formed by wind-blown sand, usually from beaches and sand banks exposed at low tide, although their development and maintenance depends on the vegetation, which facilitates accretion and stabilisation of the sand (see Carter *et al.* 1992, Nordstrom & Carter 1991);
- **tidal flats**, which occur in estuaries and sheltered inlets, and consist largely of muddy sediments. In estuaries, accretion is enhanced by the mixing of fresh water and sea water, which causes flocculation and hence settling of water-borne sediments (see Dyer 1998). Further accretion occurs if mudflats are colonised and stabilised by saltmarsh plants.

All these systems depend on a continued supply of sediment, which can be interrupted by activities such as coastal protection works and dredging (§13.5.2). Changes in sediment supply are the commonest cause of **downdrift effects** (impacts on the lee side of coastal activities) including downdrift erosion.

Along the coastline of England and Wales, 11 major **coastal sediment cells** have been identified, which are evidently largely self-contained in terms of the movement of coarse sediments (HR 1993). Consequently, these cells (or smaller sub-cells which have also been identified) are considered to be suitable units for study, and for the development of *Shoreline Management Plans* (§13.3.2).

The upper levels of estuaries and inlets can be regarded as parts of coastal **floodplains**, which act as buffers against flooding of the hinterland. However, much coastal floodplains are now protected by man-made flood defences, and have a variety of land uses ranging from grazing marsh to urban and industrial development. They are susceptible to flooding from poor drainage and overspilling of watercourses, or breaching or overtopping of coastal defences; and estuarine tidal floodplains can be flooded from raised sea levels, river floodwaters, or a combination of both (EA 1997).

13.2.3 Littoral habitats

The **littoral zone** can be very narrow (when the slope of the land is steep) or quite extensive, e.g. on mudflats. Its approximate boundaries are the EHWS and ELWS tide levels (Fig. 13.1). In most littoral habitats the resident species are essentially marine. However, they are adapted to the regime of immersion and emersion associated with tidal cycles. Organisms living near the EHWS or ELWS levels are only submerged or emersed, respectively, for short periods during the year. Between these extremes, the communities usually exhibit clear zonation along the land–sea axis, although this is controlled by elevation rather than distance.

Littoral habitats vary in substratum type depending on their exposure to wave action and currents, e.g. mud only accumulates in sheltered locations, and rocky shores occur where exposure prevents any sediment deposition (see Table 13.1).

Rocky shores provide a generally impenetrable substratum that precludes burrowing or penetration by roots, but supports seaweeds and animals that adhere to rock surfaces (**epibiota**). There is wide variation in wave exposure, and associated community types. Sheltered shores are normally dominated by seaweeds such as fucoids (species of *Fucus* and similar 'shrubby' brown algae) but host many animal species and generally have high species diversity (Table 11.1, p. 247). By contrast,

Table 13.1 Typical locations of, and relationships between, littoral and supralittoral habitats in relation wave exposure and currents

Locations	Littoral zone	Supralittoral zone
Exposed coastlines and headlands	Rocky shores	Sea cliffs and slopes
Fairly exposed coastlines, usually where lateral currents drag the material (longshore drift) to the deposition locations	Shingle beaches	Vegetated shingle banks
Exposed or fairly exposed coastlines, often in bays or at the mouths of estuaries	Sandy shores	Sand dunes and machairs
Estuaries and sheltered inlets (sometimes behind sand dunes or shingle banks)	Mudflats	Saltmarshes[1]
Depressions partially cut off from sea water, usually by barriers of sand or shingle	Saline lagoons	

1 Saltmarshes are littoral habitats, but are sometimes classed as 'maritime'.

most seaweeds and animal species are excluded from very exposed rocky shores, which are usually dominated by barnacles and/or mussels. Green algal beds sometimes occur, e.g. where fresh water crosses the shore. Rocky shore ecology is discussed in Lewis (1976) and Moore & Seed (1985).

Shingle beaches are a hostile environment in which (because of the constant grinding action) few resident species can survive. However, supralittoral shingle is a valuable habitat (§13.2.4).

Sand and mud are soft, unstable substrates that do not provide adequate anchorage for epibiota, but are suitable for animals such as burrowing shellfish and marine worms that live in the substratum (**infauna**). Sandy shores are a relatively hostile environment (see Brown & McLachlan 1990) and most are too unstable for plant growth. However, *Zostera* (seagrass) beds occur on some muddy sands in the lower littoral and sublittoral (see UKBG 1999). Mudflats are usually coated with a film of microscopic **algae** (Coles 1979) and normally contain a rich invertebrate infauna, especially in estuaries.

Saltmarshes develop on mudflats where there is sufficient shelter, and the mud is sufficiently stable to permit colonisation by **vascular plants**, which further stabilise and generate accumulation of the substrate. Saltmarshes are often classed as maritime habitats because (a) they are largely restricted to the zone between mean high-water neap tides and mean high-water spring tides, so only the lower fringes are submerged by the daily tidal cycle throughout the year, and the upper levels are only subject to inundation at EHWS tides, and (b) they are dominated by essentially terrestrial vegetation, although the plants must be halophytes (species adapted to live in saline conditions). The communities usually exhibit zonation, along the land–sea axis, in relation to the frequency of inundation (see Table F.1, p. 440). Saltmarshes are often located in **estuaries**, which are unique ecosystems in which the mixing of fresh and salt water is a major ecological factor. Information on saltmarsh and estuarine ecology is provided in texts such as Adam (1993), McLusky 1981 and Packham & Willis (1997).

Coastal lagoons are bodies of saline or brackish water, that are partially separated from the sea, but retain some sea water at low tide (see Downie 1996, UKBG 1999). They often contain unusual communities that include algae, vascular plants, and invertebrates that rarely occur elsewhere (see Barnes 1994).

13.2.4 Supralittoral (maritime) habitats

Supralittoral habitats lie above the limits of the EHWS tides, and support terrestrial vegetation. Some salt tolerance is needed, however, especially in near-littoral locations which are affected by wave splash and spray. The zone includes several important habitats (Fig. 13.1). As in littoral habitats, an important factor is substratum type; indeed, the different types of littoral habitat are usually backed by supralittoral habitats on similar substrates (Table 13.1).

Maritime cliffs and slopes vary in relation to their geology and local landforms, and can have faces ranging from vertical to gently sloping. They are considered to extend inland to at least the limit of salt spray deposition, and hence sometimes encompass whole headlands or islands (UKBG 1999). They include a variety of habitats such as rock crevices and ledges, and coastal grasslands and **heathlands** (see Table F.1).

Vegetated shingle banks sometimes support scrubby vegetation or a grass sward, but more exposed areas have open vegetation with scattered vascular plants and lichens. They are sensitive to disturbance, and are slow to recover (see Packham & Willis 1997, UKBG 1999).

Sand dune systems usually consist principally of several dunes (aligned approximately parallel to the coastline and increasing in age along the sea–land axis) interspersed with depressions (dune slacks). They are complex systems that include a range of habitats (see Table F.1), and are very sensitive to disturbance. Information on sand dune ecology is provided in Carter *et al.* (1992), Crawford (1998), Gimmingham *et al.* (1989), and Packham & Willis (1997).

Machairs are distinctive systems on wind-blown calcareous sand, and are confined to northwest Scotland and western Ireland where strong onshore winds prevail. They have a long history of traditional management, e.g. by seasonal grazing and rotational cropping. The vegetation is usually predominantly short grassland, but can include dunes, fen and swamp (see Bassett & Curtis 1985, Owen *et al.* 1996, UKBG 1999).

British maritime habitats are usually backed by agricultural land, urbanised areas, or recreational developments such as caravan sites or golf courses. In the absence of these, they would intergrade with terrestrial habitats such as woodland. Consequently, the landward extension of the coastal zone is less well defined than it might seem at first sight.

13.2.5 Sublittoral habitats

The upper limit of the **sublittoral zone** is the ELWS tide level in a given location, which is a fairly discrete boundary. The seaward limit less is clear. It can be taken to include all the shallow seas which extend to the edge of the **continental shelf** that fringes the European land mass. The extent of this varies, but it includes the English Channel, the Irish Sea, and most of the North Sea. It also varies in depth, but generally slopes gently to a depth of about 200 m before the **continental shelf slope** falls steeply to the deep ocean floor. However, classifications such as BioMar restrict the sublittoral zone to **inshore waters** (usually up to *c*.5 km from the shore and *c*.50 m deep) and define the remaining area as the **offshore shelf** or **circalittoral offshore zone** (Fig. 13.1). Legally, inshore waters are defined as within six nautical miles of the shoreline (where the UK has authority to exercise unilateral protection of fish stocks). For EIA purposes, 'sublittoral' should at least include these waters, but *pollution* of the entire North Sea demonstrates that land-based developments can have significant impacts on the whole continental shelf area.

As in all ecosystems, sublittoral and open sea communities depend on flows of energy and nutrients along *food chains* and – except for deep ocean hydrothermal vents which have chemoautotrophs that utilise chemical energy – they rely on **photoautotrophs** (see Fig. 11.3, p. 250). However, apart from kelp and seagrass beds that may occur in the infralittoral (Fig. 13.1), the primary producers (Fig. 11.3) are *phytoplankton* which form the basis of food chains in both **pelagic** (free floating and swimming) and **benthic** (seabed) communities. A difference from terrestrial communities is that a much larger proportion (up to 80%) of the 'primary pasturage' (the phytoplankton) is consumed by the *zooplankton*, and passes along grazing food chains.

Most EIAs are likely to focus on the benthic communities of inshore waters, the ecology of which is discussed in Earle & Erwin (1983) and Hiscock (1998). The environment here is less widely fluctuating than that of the littoral zone, and the seabed is dominated by soft sediments – rocky substrates being normally restricted to narrow zones adjacent to coastlines, and isolated features such as reefs (UKBG 1999). However, there is appreciable variation for the following reasons.

- There is still a range of substrates that largely control the benthic communities, and to a lesser degree those in the water above (which include organisms that depend on the seafloor for food, shelter or reproduction).
- There is considerable variation in water movement, including turbulence, currents and tidal movement.
- Near the mouths of rivers and estuaries, salinity may be reduced by fresh water, and **turbidity** increased by suspended sediments, especially in wet weather. Turbidity tends to be high also in areas with a muddy or sandy seabed, especially when the sediments are disturbed during storms.

13.2.6 *The ecological value of British coastal ecosystems*

The coast and seas around northwest Europe are amongst the most productive wildlife habitats in the world. They are home to a range of flora and fauna which are often present in numbers of international importance. A major factor in their presence is the high nutrient status of the water. This supports large numbers of primary producers which form the basis of large and often complex **food webs** (e.g. see §11.2.2). However, many coastal communities are susceptible to, and slow to recover from, disturbances.

The British coastal zone is particularly special because of its location. Temperate, warm temperate, and Arctic species are all found around the shores. The coastline is geologically and topographically varied, heavily indented, and subject to a wide range of wave activity and tidal regimes. These features provide a wide variety of habitats, often within a small stretch of coast.

The UK's coastal areas are particularly important to birds, including many rare species, and British seabird colonies are of global importance. Of the 261 internationally important bird areas in the UK, 28 qualify because they hold over 1% of the world population of a seabird species, and 61 qualify because they hold over 1% of the EU population total (RSPB 1991).

Two particularly valuable UK habitats are maritime cliffs, which often support internationally important populations of breeding seabirds, and estuaries, which are internationally important. Because of the indented coastline and large tidal ranges, Britain has the highest proportion of estuarine habitats in Europe (Davidson *et al.* 1991). In addition, they are among the most biologically productive ecosystems in the world. This is because factors such as the warm seas, mild winters, and nutrient inputs from the land, promote large, species-rich invertebrate communities in their intertidal mudflats (Rothwell & Housden 1990). Consequently, they provide rich feeding grounds for birds and, in particular, form vital links between the breeding and overwintering grounds of migratory waders and wildfowl. Britain hosts about 20% of the migratory populations each spring and autumn, and over 33% of the overwintering waders on the European Atlantic coast (EA 1999a).

The importance and threatened nature of UK coastal habitats is reflected by the facts that (a) all the major maritime types, and 12 marine types are UKBAP priority habitats (see §11.3.2, and Table F.2, p. 442), and (b) the coastal zone hosts many UKBAP priority species, including 29 marine species (UKBG 1999). In addition, many species, habitats and sites are afforded legislative protection.

13.3 Legislative background and interest groups

13.3.1 Legislation

Coastal zone legislation in the UK is complex, and reflects a fragmentation of responsibility and management, e.g. DoE (1993) noted 40 separate government bodies with varying responsibilities and limits of jurisdiction. There have been moves towards greater unification of responsibility. For instance, MAFF and NAW now set the overall policy for **coastal defence** (against erosion or flooding) in England and Wales, and the EA and Maritime District Councils (MDCs) are the **operating authorities**, responsible for policy implementation. (The EPAs' responsibilities for pollution control also extend to coastal waters.) However, other organisations have responsibilities in coastal management, and may have powers to pass by-laws for specific purposes. These include the NCCAs (Appendix B), LPAs, harbour authorities, and landowners (including NGOs such as the RSPB and NT).

Legislation and international agreements containing specific references to the coast are listed in Table 13.2. As can be seen in the table, much of this legislation refers to both inland and coastal waters. Similarly, the *Water Framework Directive* (see §10.4.1) includes inland fresh waters, estuaries and coastal waters. In addition, EU directives such as the DSWD and IPPCD (Table 10.1, p. 209) and related UK legislation such as the EPA and PPCA (Table 10.2, p. 211) aim to control pollution of all surface waters, including coastal waters. Similarly, whilst most legislation on nature conservation (Tables 11.2 and 11.3, pp. 255 and 256, respectively) is not specific to the coast, it is highly important for the zone. For instance, Annex I of the HSD lists 24 coastal habitats (including five with priority status) that occur in the UK (see Table F.3, p. 444).

As shown in Table D.3 (p. 423), the only specifically coastal sites with statutory protection are MNRs, of which there are currently only three. Similarly, the only non-designated coastal sites are Heritage Coasts, PCZs, and SMAs/MCAs. However, much of the zone has some degree of protection under the general designations listed in the table. As always, the greatest protection is afforded to 'international' sites, with which the zone is well endowed. For example, of the 155 recognised estuaries in the UK, 68 are Ramsar sites, SPAs or cSACs. Many other coastal sites have national (e.g. 'ordinary' SSSI) designations, or are non-statutory sites, and significant stretches of coastline are owned or managed by NGOs such as the RSPB, NT or NTS. The geological/geomorphological importance of the coast is also recognised by a number of *Earth Heritage Sites* and RIGS (see §9.4.1).

In spite of the stringent obligations imposed by many designations, they frequently afford little protection. For example, many UK estuaries recognised as internationally important wildlife sites are still being subjected to severe development pressures.

13.3.2 Policies and guidance

The EC *Fifth Environmental Action Programme* (see §1.4) calls for **sustainable development** of coastal zones in accordance with the **carrying capacity** of the coastal environments; and the development of Integrated Coastal Zone Management (ICZM) has been called for by several UN and international conferences, including the *Earth Summit '92* (see §1.4) and the *World Coastal Conference* (WCC) 1993 (available online from CZMC – see Table 13.3, p. 327).

Table 13.2 International agreements and EU or UK legislation specific to the coastal zone

---

Coast Protection Act 1949 – Powers of coastal protection authorities, control structures below low-tide level.

Salmon and Freshwater Fisheries Act 1975 – Regulations for inland fisheries and for salmon and sea trout within a six-mile zone.

Bathing Water Directive (BWD) (76/160/EEC) – see Table 10.1.

Shellfish Waters Directive (SWD) (79/923/EEC) – to protect coastal and brackish shellfish waters by setting water quality standards and requiring member states to reduce pollution where necessary. Standards are set for a number of parameters including salinity, dissolved oxygen and nine metals in *designated waters* (see §10.4.2).

Food and Environment Protection Act 1985 – Pollution control in coastal waters. Licences required for construction works, dumping at sea (including dredged materials); use of herbicides affecting tidal waters.

Water Act 1989 – Defines coastal waters as those which are within the area which extends landwards from baselines from which the breadth of the territorial sea is measured as far as the limit of the highest tide or tidal limit of the river.

North Sea Conference 1990 – signatories agreed to reduce inputs of dangerous substances (including *heavy metals*, *pesticides* and PCBs) to the North Sea, and of nutrients where these may cause problems. Extended by the UK Government to include all UK coastal waters.

Urban Waste Water Treatment Directive (UWWT) (91/271/EEC) – see Table 10.1.

Water Resources Act (WRA) 1991 – see Table 10.2.

Land Drainage Act (LDA) 1991 – includes provisions for coastal (as well as inland) flood defences.

Agreement on the Conservation of Small Cetaceans of the Baltic and North Seas (ASCOBANS) 1991

Environment Act 1995 – see Table 10.2.

Agreement for the Conservation of Cetaceans (whales and dolphins) **in the Black Sea, Mediterranean Sea and Contiguous Atlantic Area (ACCOBAMS) 1996**

Oslo & Paris Commission – Convention for the Protection of the Marine Environment of the North-East Atlantic (OSPAR), accepted in EU Council Decision 98/249/EC – signatories agreed to continually reduce emissions of hazardous substances, with the aim of achieving near background levels of naturally occurring substances and near zero concentrations of synthetic substances by 2020 (see EA 1999a).

The EIA Regulations 1999 (implementing Directive 97/11/EC) (see §1.4) – include (a) in Schedule 1: large ports and piers (except ferry piers), and (b) in Schedule 2: coast protection works (other than maintenance or reconstruction); large fish farms; reclamation; shipyards; marinas > 0.5 ha; and construction of harbours and ports > 1 ha (unless included in Schedule 1). Other particularly relevant projects include oil or gas extraction plants and pipelines, and extraction of minerals by fluvial dredging (but not marine dredging).

In Britain, the general policies on nature conservation, outlined in §11.3.2, apply to the coastal zone. In addition, specific coastal zone planning and management are now receiving a higher profile than in the past, as can be seen in government publications such as the following:

- DoE (1990) refers to development on unstable land, including that near erod-ing cliffs.
- DoE (1992a, 1993, 1995, 1996) and SO (1997) recommend clearer definition of the coastal zone, greater integration of responsibilities, and a more strategic approach to coastal management.
- MAFF/WO (1993, 1995, 1996) state similar aims with particular reference to coastal defences, including the need to consider impacts in the context of **coastal sediment cells** (§13.2.2) or at least sub-cells, rather than administrative boundaries.
- MAFF (1999a) set 'High Level Targets' for coastal defence operating author-ities in relation to **biodiversity**, i.e. they must aim to (a) avoid damage to environmental interest, (b) ensure no net loss to habitats covered by HAPs (§11.3.2), and (c) seek opportunities for environmental enhancement.
- MAFF (1999b; 2000a,b; in prep.a,b,c) provide guidance on appraisal of coastal defence projects, and incorporate current policies including the need for envir-onmental appraisals (when formal EIAs are not required).

Probably the most important of these publications are PPG20 (DoE 1992a) and NPPG13 (SO 1997), both entitled *Coastal Planning* – which provide a framework for a comprehensive coastal management strategy. Of particular note are the recom-mendations that:

- the coastal zone be extended landward and seaward with its limits defined by the geographical extent of the natural coastal processes and the human activities affecting them;
- only those developments which require a coastal location should be approved;
- new projects be directed to areas already developed, and away from areas that may be affected by flooding or instability;
- consideration should be given to the off-shore impacts of on-shore developments;
- "Policies should seek to minimise development in areas at risk from flooding, erosion and land instability. The degree of risk involved will have to be care-fully considered and policies will specifically be needed to control or restrict development in low lying coastal areas . . ." (DoE 1992a)

Perhaps the most important aspect of PPG20 is the recognition that coastal planning is a strategic issue and therefore relies upon close co-operation between all those interested in coastal zone management. However, this is inhibited by the large number of statutory authorities and other interested parties that have responsibil-ities in coastal areas.

MAFF is encouraging a more strategic approach through the development of *Shoreline Management Plans*, *Estuary Plans* and *Coastal Zone Management Plans*, which should include 'strategic environmental appraisals' (MAFF/WO 1995, 1996). In England, the operating authority and EN are also to develop *Coastal Habitat Man-agement Plans* (CHaMPs). In addition to providing a framework of conditions to which proposed projects should comply, these are intended to identify the flood and coastal defence works likely to be needed to conserve the nature conservation interest of SACs, SPAs and Ramsar sites, especially where the current defence line may be unsustainable (EN 2000, MAFF 2000b).

Such plans are intended to form integral components of MAFF's **managed re-alignment/setback** strategy (the deliberate setting back of a flood or sea defence line) in areas where, in the face of rising sea levels, existing sea walls (a) are unlikely to cope, or (b) prevent intertidal habitats from migrating inland, so causing substantial loss of these habitats – a process known as **coastal squeeze**. The strategy may involve (a) allowing the old wall to breach or deliberately breaching it, (b) providing wave breaks behind which saltmarsh can develop, and (c) building a new (shorter) defence behind the marsh (MAFF 2000b, MAFF/WO 1996). It is consistent with EN's **managed retreat** strategy which aims to reverse the recent trend of saltmarsh loss (EN 1995). A potential ecological problem is that, in some cases, there may be conflict between conserving intertidal habitats and others (such as freshwater wetlands) located behind the current defence line.

13.3.3 Consultees and interest groups

In Britain, the statutory consultees for coastal zone ecology and geomorphology are the relevant NCCA and EPA (Appendix B). Other potential consultees and interested parties will include (a) those referred to in §11.3.3 for ecological assessment in general, (b) organisations such as the Marine Conservation Society, port and harbour boards, sea angling clubs, and commercial fishing firms.

13.4 Scoping and baseline studies

13.4.1 Introduction and scoping

Coastal EIAs should employ the scoping procedures outlined in §11.4, including the strategy of phases (study levels) for ecological surveys. Much of the ecological interest of the coastal zone is linked to the geomorphology, and ecological studies must take this into account. Moreover, geomorphological impacts can have important implications for coastal defence.

Establishment of the **impact area** may be difficult because of the indeterminate boundaries, especially of the sublittoral zone, and the original estimate may have to be revised in the light of information that emerges during the study. The lateral extent of most geomorphological impacts should be confined to **coastal sediment cells** (§13.2.2), and MAFF (2000b) suggests that, where available, *Shoreline Management Plans* and *Estuary Plans* (§13.3.2) should be the starting points for project design and appraisal.

The coastal zone is also affected by developments in associated catchments (§13.5.2); so another important aspect may be *catchment* hydrology (Chapter 10). Since many of a project's impacts are likely to be cumulative, it is also important to seek information on predicted trends such as further development and recreational use. Time and resources permitting, the use of GIS (Chapter 16) should be beneficial, e.g. for facilitating integration between different aspects.

13.4.2 Use of existing information

Much of the information required for a coastal assessment can be compiled from a desk study. General sources of ecological information are given in §11.4.2. Aerial photographs and satellite data (Chapter 15), topographic maps, and bathymetric charts can provide information on the current and recent form of the coast, and may reveal any substantial changes such as coastal erosion. In some cases, it may be beneficial to consult old maps or other historical information (Appendix C, p. 416), although historic records of coastal erosion are usually scarce (MAFF 2000a).

An increasing amount of information is available in the form of inventories and databases (Table 13.3). Although many of the data are unlikely to refer to the immediate vicinity of a project, they can still be useful. For example, tidal regimes can be calculated from data for the nearest ports, available from the Tidal Prediction Service at CCMS-POL. The organisations hold, and may be willing to supply, information other than that listed in the table. (The EPAs have a duty to supply relevant information, on request, for EIAs.) Their websites also provide links to other sites, often worldwide.

In spite of the increasing range and extent of existing information, much of it is likely to be out of date or inadequate in terms of quality or resolution, and new surveys should be conducted wherever necessary.

13.4.3 Geomorphological surveys

Geomorphological parameters can be measured by a variety of methods, using *in situ* recording instruments and remote sensing techniques (see §15.2.2 and §15.4.3). A review of methods for measuring littoral-zone processes, beach morphology and

Table 13.3 Inventories and databases on the coastal zone

British Geological Survey (BGS) (http://www.bgs.ac.uk)

Geoscience Data Index (GDI) – online spatial index of BGS data holdings (e.g. seabed datasets (sediment particle size and geochemistry (including contaminants), saline intrusion of aquifers). It is held in a GIS, can be zoomed to small areas, and gives costings for the supply of more specific information.

British Oceanographic Data Centre (BODC) (http://www.bodc.ac.uk/)[1]

United Kingdom digital marine atlas (UKDMAP), 3rd edn (1998) – CD ROM containing maps and databases, e.g. geomorphology, protected areas, JNCC coastal & marine data, species distributions (including seabirds and mammals), plankton, benthos, fisheries, currents, tides, waves, weather, chemical distributions.

UK Directory of Marine Environmental data (UKDMED) and **European Directory for Marine Environmental Data (EDMED)** – online searchable directories of datasets relating to the marine environment.

Coastal Zone Management Centre (CZMC) (http://www.minvenw.nl/projects/netcoast/info/czmc.htm)

NetCoast – computer program to provide online access to information on ICZM; links to relevant websites.

Table 13.3 (*continued*)

Environment Agency (EA) (http://www.environment-agency.gov.uk)

Bathing Waters Directive database for 464 coastal sites; **National Marine Monitoring Programme (NMMP)** database of significant contaminants, benthic biology and biological effects in estuarine and coastal waters for 87 sites (held at the EA's National Centre for Environmental Data and Surveillance, Bath.

Environment and Heritage Service Northern Ireland (EHS) (http://www.ehsni.gov.uk/)

Water Quality Unit monitoring data archives – most data is available on request.

European Union for Coastal Conservation (EUCC) (http://www.coastalguide.org/)

Coastal Guide – online information on topics such as coastal typology, tidal ranges, threats and management.

Joint Nature Conservancy Committee (JNCC) (http://www.jncc.gov.uk)

National inventories of coastal systems – estuaries (JNCC 1993–97), saltmarshes (Burd 1989), vegetated shingle (Sneddon & Randall 1993–94) and sand dunes (JNCC 1993–95).

Marine Nature Conservation Review (MNCR) series (JNCC 1996–99) – focuses on benthic habitats.

Coastal Directory Series (JNCC 1995–98) – focuses on environmental and human-use information.

MNCR Mermaid (http://www.jncc.gov.uk/mermaid) – online database with good search and distribution-map facility for species, sites, BioMar biotopes (complete hierarchy), MNCR sectors and marine cSACs.

Marine Biological Association (MBA) (http://www.marlin.ac.uk)

Marine Life Information Network for Britain and Ireland (MarLIN) – includes: (a) species listed in Conventions and EU/UK legislation: (b) information on species' identification, biology, habitat preferences, distributions, sensitivity (to a wide range of factors) recoverability, and importance; (c) information on BioMar biotopes; (d) links to other UK datasets (see also Tyler-Walters & Jackson 1999).

Proudman Oceanographic Laboratory (POL) (http://www.pol.ac.uk/[1])

Tidal prediction Service and software (see Table 13.4); **Archived data** on physico-chemical variables, bathymetry, waves, currents, sea levels, extreme tide estimates, storm surges, etc.

Scottish Environmental Protection Agency (SEPA) (http://www.sepa.org.uk)

Public Registers including: Integrated pollution Control (IPC), Water quality Pollution Control.

Reports and policies including: State of the Environment; Bathing Waters Report; Flood risk assessment.

---

1 BODC is housed at POL, which is a component of CCMS (Centre for Coastal and Marine Sciences).

coastal erosion is provided by Dugdale (1990). However, the methods are generally time-consuming and expensive. Moreover, although coastal geomorphology is very dynamic, changes occur relatively slowly; so many methods require repeat measurements over extended periods. Consequently, (a) assessment of trends will normally have to rely on existing information, and (b) new surveys for EIA are likely to be restricted to large projects and post-development monitoring programmes (in which case it may be beneficial to initiate appropriate studies at the baseline survey stage). In making decisions about the need for new data, and the selection of appropriate methods, advice should be sought from agencies such as the EPAs, CCMS-POL and CEFAS.

13.4.4 Problems of ecological field surveys

The coastal zone presents special problems for ecological sampling, especially of the sublittoral zone. However, this is not a good reason to exclude new fieldwork. The study should include as many habitat types and taxa as possible, but as in all ecological assessments, sampling and identification of many taxa can be difficult, time consuming and expensive, so surveys must be carefully targeted, e.g. on high-status species. Identification books and keys are listed in Appendix E, but experts in both sampling methods and identification will usually be needed.

The timing of field surveys and (where possible) repeat sampling, are particularly important in a coastal zone assessment because many of the ecosystems have a high degree of seasonality. Whilst some animals can be found all year round, many fish and bird populations change in relation to breeding and overwintering strategies. In particular, many waders and other migratory seabirds are resident only during the winter, and some have shorter 'stop over' periods, e.g. in the spring and autumn. Saltmarsh vegetation grows and flowers relatively late in the summer, and sand dune animals (and some annual plants) should be sampled earlier than the most suitable period for a general vegetation survey (Fig. G.2, p. 459). Resident shore communities can be sampled at most times of year, but neap tides do not expose the lower shore, and sampling is best conducted during the large spring tide periods in March or September.

13.4.5 Phase 1 ecological field surveys

Phase 1 surveys of maritime and littoral habitats can employ the JNCC Phase 1 habitat survey method (§11.4.4 and Appendix F.1, p. 436), and for highly developed coastlines, it may be beneficial to include additional land use categories (see Appendix F.6, p. 452). However, The JNCC classification does not cover the sublittoral zone, and future surveys of both sublittoral and littoral habitats are likely to use the MNCR BioMar marine (benthic) biotopes classification (Connor *et al.* 1997a,b; Picton & Costello 1997).

The **biotope** (§11.2.4) was chosen as a fundamental unit in the classification because: (a) there is a strong relationship between benthic marine communities and abiotic habitat factors such as substratum type, water depth and exposure to waves or currents; (b) many marine habitats, especially in deeper water, lack **macrophytes**. Consequently "more significant use of the habitat is made than for many terrestrial classifications, where vegetation is often the prime determinant of the classification's structure" (Connor *et al.* 1997a).

The classification has five levels:

1. **Major habitats** – very broad divisions based on substratum type (rock or sediment) and major zones (see column two of Fig. 13.1);
2. **Habitat complexes** – broad divisions of major habitats, defined by (a) exposure of rock to wave action or currents (exposed, moderately exposed, sheltered) or (b) sediment types (Fig. 13.1);
3. **Biotope complexes** – groups of biotopes with similar general character that can be recognised by the dominant *life forms*, e.g. (a) lichen crusts on supralittoral (splash zone) rock; (b) brown algal shrubs (dense fucoids) on sheltered littoral rock, (c) faunal crust (barnacles and/or mussels) on exposed littoral rock;
4. **Biotopes** – typically characterised by *dominant species* or assemblages of conspicuous species;
5. **Sub-biotopes** – typically characterised by "less obvious differences in species composition, minor geographical and temporal variations, more subtle variations in the habitat or disturbed and polluted variations of a natural biotope" (Connor *et al.* 1997a).

The three top levels are readily applicable to Phase 1 surveys because information on levels 1 and 2 can usually be obtained by a desk study, and biotope complexes can be identified by non-experts or subtidal video. Because they involve species identification, biotopes and sub-biotopes may be considered to require surveys at the Phase 2 level. However, dominant and conspicuous species are not usually difficult to identify; and guidance on Phase 1 survey and mapping (Bunker & Foster-Smith 1996, Richards *et al.* 1995) suggests that Phase 1 surveys should include biotopes except in cases of uncertainty.

13.4.6 Phase 2 surveys of maritime and benthic species and communities

In general, Phase 2 fauna and flora surveys of maritime (supralittoral) habitats can follow the procedures described in Appendix G and Chapter 12 for terrestrial or freshwater systems.

Coastal birds are included here because they are most frequently surveyed from the land. In addition to the general census techniques referred to in Appendix G.3.3 (p. 462), a number of specific methods have been developed for seabirds, e.g. see: Tasker *et al.* (1984) for seabirds at sea; Lloyd *et al.* (1991) and Walsh *et al.* (1995) for seabird breeding colonies; BTO (undated) for breeding waders; and BTO (1992) for low-tide counts. Information on seabird distributions and numbers is available in a number of publications, including Gibbons *et al.* (1993), JNCC (1992–98), Lloyd *et al.* (1991), and Stone *et al.* (1995). Consequently, data will already exist for many sites and species, so: unless there are reasons to suspect that the information is out of date, a new survey may not be needed, or can focus on assemblages of local or regional importance; at least for some species, it should make it possible to determine if an area supports a nationally or internationally important population.

Vegetation surveys can employ the NVC (Appendix F.4, p. 450). Rodwell (2000) contains the relevant maritime communities, and those of the two vegetated littoral habitats – saltmarshes and seagrass beds. The latter are also included in the BioMar classification, levels 4 and 5 of which are suitable for Phase 2 surveys of other littoral and sublittoral biotopes. The BioMar system for measuring **species abundances** is SACFOR ratings (Superabundant, Abundant, Common, Frequent, Occasional, Rare) which are based on ranges of % cover *or* density (Table G.1, p. 457) depending on the species being sampled. This is because some species, such as seaweeds and encrusting animals, are best sampled by % cover, while most animals are best sampled as density. Details of the system, and guidance on survey and mapping methods are given in Connor *et al.* (1997a,b).

Quadrat sampling can be employed on **rocky shores** because seaweeds and most animal residents are immobile and easily visible at low tide (see Baker & Crothers 1987). Rocky shore communities normally show clear zonations along the land–sea axis, so the use of transects along this axis is usually a suitable sampling pattern (Appendix G.1, p. 455).

A similar sampling pattern may be suitable for **sandy and muddy shores and mudflats**, but sampling the **infauna** of these requires different techniques. Subsurface **macroinvertebrates** are an important group in these habitats because they are at the base of the food chain. They can be surveyed by a number of methods ranging from a simple inspection of the sediment (e.g. to estimate the densities of lugworms from their castes) to methods which employ the use of corers and grabs to estimate densities and **biomass** (see New 1998, Wolff 1987).

A problem affecting **sublittoral benthic** surveys is the need for specialist equipment and personnel (e.g. boats and/or divers), and EIA baseline studies may have to rely on existing information.

In both Phase 1 and Phase 2 surveys, attention should also be paid to the presence of BAP priority habitats and HSD Annex I habitats. Approximate correspondences of these with NVC communities and BioMar categories are given in Table F.3 (p. 444).

13.4.7 Phase 2 surveys of pelagic species and communities

Like those of sublittoral benthic habitats, pelagic (free swimming or floating) species and communities are relatively inaccessible and difficult to survey.

Plankton present problems because (a) they are very small and diverse, (b) they are widespread over large areas of sea, and (c) concentrations fluctuate in time and space (e.g. in relation to currents). Satellite and airborne sensors that respond to chlorophyll-a fluorescence may provide detailed distribution maps for phytoplankton, and are used in **eutrophication** studies, e.g. by the EA (see §15.2.2 and §15.4.3). However, the method is very expensive and therefore generally unrealistic for EIA, and most plankton sampling employs nets and samplers that can be filled with seawater at prescribed depths. These methods, and techniques for analysing the samples, are explained by Tett (1987), who suggests that for survey purposes it is convenient to adopt categories based mainly on ecological rather than taxonomic criteria.

Fish survey techniques are numerous and variable in their level of complexity. They are influenced by various characteristics of the fish populations and communities, including: distribution (vertical and horizontal); size and mobility; and population and community dynamics, e.g. single or mixed species shoals, and seasonal migration and breeding patterns. Reviews of methods are provided in Bagenal (1978), Blower *et al.* (1981), Perrow *et al.* (1996), Pitcher & Hart (1982), and Potts & Reay (1987). They can be grouped under two broad headings:

- **observation**, e.g. aerial, direct underwater, underwater photography and acoustic surveys;
- **capture**, e.g. by traps, hook and line, hand nets, set nets, seines, trawls, lift, drop and push nets – most of which can provide specimens for mark-recapture programmes (Table G.1, p. 457).

The samples obtained by these methods can be analysed to provide information on species abundance, age structure, fish health, dietary requirements and site productivity (see Potts & Reay 1987). This information can indicate the relative worth of a site to fish stocks and hence the significance of a development's potential impact.

Marine mammals can prove difficult to survey. It is relatively easy to estimate numbers in colonies of common seal and grey seal because these are easily recognised, are faithful to particular stretches of coast, and come ashore (especially at pupping time and during the seals' moult) – when aerial and boat surveys can be conducted (Hiby *et al.* 1988, Thompson & Harwood 1990, Ward *et al.* 1988). Numbers of Cetaceans (whales, dolphins and porpoises) can be estimated by aerial, ship and land-based sightings (Hammond 1987, Hammond & Thompson 1991, Hiby & Hammond 1989). However, precise estimations of marine mammal populations involves the use of time-consuming and often expensive field techniques such as mark-recapture and radio telemetry, and are therefore unlikely to be considered in EIA.

13.4.8 Evaluation of the baseline conditions

When evaluating the baseline conditions, particular attention should be paid to sensitive geomorphological systems and high-status species, habitats and sites.[1] UK Government guidelines tend to focus on SACs, SPAs and Ramsar sites, but this should not preclude the thorough evaluation of 'less important' and small sites, especially if these host high-status species or habitats.

In evaluating habitats, consideration should also be given to 'secondary' attributes. For example, in addition to their high ecological value, sand dunes and saltmarshes are natural coastal defence systems. Sand dunes also help to prevent saline intrusion by maintaining water table levels, and saltmarshes can act as 'oil traps' for stranded spills. Baker (1979) classified British saltmarsh plants into five groups which reflect their susceptibility to, and ability to recover from, oil spillage damage. This can be used to identify sites supporting saltmarsh vegetation that is the most likely to recover from accidental spillage, e.g. from a proposed oil terminal. However, the classification is very simple and must not be viewed in isolation, as many other factors need to be considered when siting an oil terminal.

13.5   Impact prediction

13.5.1 Introduction

The difficulty of accurately predicting impacts in the face of the ecosystem complexity (§11.5.1) particularly applies to the coastal zone because of its diversity. In

---

1 When applied to species, habitats or sites, **'high-status'** means high conservation value in terms of the criteria referred to in Appendix D.

addition, each type of development brings with it a suite of potential problems which are peculiar to that type, and no two development types are the same. For example, the potential impacts of a salmon farm on an inshore sea loch are very different from those of a nuclear power station or barrage scheme. However, because of the value and fragility of many coastal ecosystems, any development which has the potential to disrupt the fine balance of interacting processes on which they depend must be viewed with concern.

Coastal ecosystems are dynamic, and a combination of natural trends and human impacts will lead to changes in the absence of a project. For example: some soft cliffs are currently suffering rapid erosion; some estuaries are changing through progressive sedimentation; and systems such as sand dunes are intrinsically unstable, and can be affected by severe storms in addition to human impacts. A new project must be considered in this context, and in the knowledge that many of its impacts may be cumulative (contributing to impacts of other developments and pressures).

13.5.2 Sources and types of impact

Major causes and associated types of impact in the coastal zone are shown in Figure 13.2. This does not indicate relationships between development types or all the possible relationships of these with impacts. As in all ecosystems, primary impacts inevitably lead to secondary and cumulative impacts.

Urban, industrial and commercial development is considerable, and is the greatest source of impacts, in the coastal zone. In England and Wales: about 31% of the coastline (and 11% of land within 10 km) is developed; and the 10 km zone is heavily populated, with about 33% of the total population (EA 1999a). About 40% of UK industry is also situated at or near the coast. Much of this is *heavy* (including chemical) *industry*, and many of the developments are very large. As indicated in Figure 13.2, major impacts include habitat loss and fragmentation, and pollution.

A principal reason for industrial siting at the coast is access to ports, but another has been the easy disposal of unwanted products by simply discharging them into the sea or river. Industrial discharges are now under tighter control, but many water bodies still bear the scars of years of uncontrolled discharge, and pollution incidents still occur. Moreover, urban and industrial developments are still the main sources of coastal water pollution, e.g. in Scotland, *point source pollution* by sewage *effluent* is the most important; and industrial effluent is second, affecting 43% of polluted estuarine waters and 11% of polluted coastal waters (SEPA 1999). The North Sea Conference 1990 declared 36 dangerous substances that the signatories agreed to reduce (EA 1999a). **Bioaccumulation** is a serious problem, especially in shellfish and top carnivores (see Brouwer *et al.* 1990, Davies & McKie 1987, NERC 1983, Walker 1990).

Urban development can also cause marine **eutrophication**, especially near sewage outfalls. This can have various consequences, including contamination of shellfish by toxins from **algal blooms**. Coastal waters also suffer pollution by *garbage* from land-based sources, ships and pleasure craft. It is estimated that 1 million seabirds worldwide die each year from entanglement or swallowing garbage, especially plastic (RSPB 2000).

Tourism and recreation pressures are increasing both onshore and in the use of inshore waters. They have indirect impacts such as adding to urban development. Direct impacts include visitor pressure on sensitive maritime systems such as sand

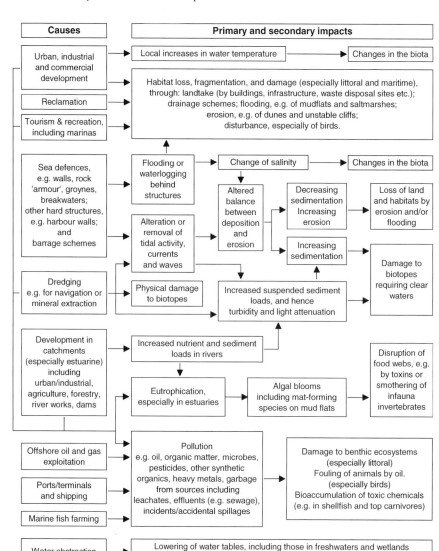

Figure 13.2 Causes and types of impact in the coastal zone.

dunes. Once the vegetation cover is damaged, dunes are very susceptible to wind erosion, which can severely damage the young frontal dunes and cause large *blowouts* even in mature dunes. A major effect of developments such as marinas is disturbance of wildlife, especially birds, which are heavily reliant upon undisturbed feeding sites. Consequently, the growth of marinas is cause for concern; by 1990, 154 existed in

UK estuaries, with a further 78 proposed (Davidson *et al.* 1991), and the problem is exacerbated by their concentration along the popular stretches of coast.

Reclamation has a long history and is an on-going threat. About one-third of all British intertidal estuarine habitat and about half the saltmarsh area have been reclaimed since Roman times (Thornton & Kite 1990), and the intertidal area of the Tees estuary has been reduced by around 90% in the past 100 years (Rothwell & Housden 1990). Moreover, the pressure has increased in recent years, e.g. in 1989 at least 50 UK estuaries were subject to one or more proposals involving land claim (Davidson *et al.* 1991) – and this has resulted in extensive losses especially on the south and east coast of England where the largest tracts were once found. Estuarine habitat loss has also been widespread in other countries. Reclamation also reconfigures the morphology of the coastline, and hence may alter sedimentation and erosion patterns (Cooke & Doornkamp 1990). Drainage schemes can cause further problems such as destruction or fragmentation of freshwater wetland habitats (RSPB 1991).

Barrage schemes fall into two basic categories: permeable and impermeable. Some *impermeable barrages*, such as the Thames Barrier, are flood defences against high tides and tidal surges, but many are intended for total exclusion of tides, primarily for amenity purposes such as water sports or providing pleasing views for waterside developments (Therivel *et al.* 1992). The immediate impacts of the latter include the replacement of marine habitats, such as mudflats, by freshwater bodies. Long-term consequences are not well known, but (a) EA (1999b) have identified a range of potential impacts of the Cardiff Bay Barrage (mainly within the impounded area and its catchment), and (b) barrages are known to have profound effects on sedimentation regimes for many kilometres along the coastline, often enhancing erosion at susceptible sites. *Permeable barrages* are intended to harness tidal power for generating hydro-electricity. These may also change sedimentation patterns and enhance eutrophication by inhibiting tidal activity 'upstream'. By 1990, 22 estuarine sites had been subject to preliminary investigation for this type of barrage scheme (Rothwell & Housden 1990). The design and environmental impacts of barriers are discussed in Burt & Watts (1996).

Sea defences are essential to protect many coastal settlements and agricultural land from flooding. However, 'hard' defences, and structures such as harbour walls, can cause serious geomorphological impacts. They fail to dissipate wave energy and, by deflecting waves and currents, affect deposition and erosion processes (Carter 1988). For example, sea walls can cause erosion of the protecting beach, and deprive a coastal system of sediment which may be vital in the replenishment of beaches further along the coast, thus causing *downdrift erosion* (Komar 1983). On the other hand, changes in *longshore drift* (§13.2.2) can lead to enhanced sedimentation in calmer waters. Such problems are particularly likely if coastal defences are managed in relation to administrative rather than geomorphological boundaries, when, for example, one district's erosion can be another's deposition (Clayton 1993). An additional problem associated with hard sea defences is *coastal squeeze* of intertidal habitats in the face of rising sea levels (§13.3.2).

Dredging is carried out for various purposes including: (a) maintenance of navigable waterways, e.g. to ports; (b) harbour and marina creation; and (c) provision of sand and gravel, e.g. for *beach replenishment* (replacement of eroded material), other coastal defence work, or use by the construction industry. In 1996, 11 million tonnes of aggregates were dredged from the sea (EA 1999a). Conversely, it is estimated that c.40 million tonnes of dredged material are dumped annually at sea (MAFF/WO 1993). Impacts associated with dredging include:

- physical damage to the site, and associated habitat loss;
- deepening of inshore waters, increasing shoreface slopes and allowing larger waves to break closer to the shore, thus increasing the risk of shoreline erosion (Carter 1988). Mineral extraction from the foreshore or dunes can also enhance erosion (MAFF/WO 1995);
- creation of turbid conditions in and around the extraction site, thus lowering lower light levels and causing problems for flora and fauna requiring clear water conditions;
- disruption of natural sedimentation patterns (by extraction or dumping), with consequences such as smothering of benthic communities;
- possible release of toxins and nutrients which normally remain locked up in the sediment, thus creating toxic pollution or eutrophication problems;
- landtake by disposal sites which, partly because of the high water content and poor settling qualities of dredged material, can require large areas (Yell & Riddell 1995).

Catchment development is important because most coastal sites lie within catchments (§10.2.2) and coastal zone parameters such as river flow, groundwater levels, and water quality (including nutrient, sediment and toxic pollutant loadings) can be affected by developments anywhere in the catchment – often many kilometres inland. Water abstraction, and developments such as dams and irrigation schemes, can reduce (a) groundwater levels, and (b) river flows and sediment loads, leading, for example, to lower sediment accretion rates in estuaries. Conversely, urban development, river works and agro-forestry (including deforestation) can increase:

- *runoff* (including flash floods during storm periods);
- soil erosion and consequent suspended sediment loads, which can lead to increased sedimentation and associated consequences such as the need for dredging to maintain navigation;
- nutrient and toxic pollutant loadings in estuarine and marine waters.

Most runoff from the land is in rivers, so estuaries are particularly affected. For example, *diffuse pollution* from agriculture affects 7% of the polluted estuarine area in Scotland (SEPA 1999). *Eutrophication* can be particularly problematic in small estuaries or those in which tidal flushing has been reduced by other activities, e.g. in the Ythan estuary, mat-forming algae now cover extensive areas of mudflat during the summer, smothering invertebrate species and, since these form the basis of the food web, threatening to affect the whole community including birds and fish (Gorman & Raffaelli 1993).

Catchments are not the only source of pollutants from outside the coastal zone; airborne and sea-borne pollutants can come from distant sources, and contamination of areas such as the North Sea is caused by discharges from all the bordering countries.

Oil and gas exploitation involves exploration, laying of pipelines, construction of offshore rigs and onshore terminals, and eventual decommissioning. All of these activities cause at least local disturbance of marine species and ecosystems. However, the greatest hazard is probably accidental pollution by oil. This also applies to

ports/terminals and associated **shipping** which can cause additional pollution problems by the accidental spillage of other toxic cargoes or the illegal disposal of bilge waters.

Marine fish farming such as salmon farms are, in the UK, usually located in the sheltered waters of sea lochs. These farms have a high potential to lower water quality in and around the rearing cages as they have a heavy reliance upon chemicals to control pest outbreaks. Further pollution results from the high loadings of organic and nitrogenous compounds in faecal material and uneaten food (Thompson *et al.* 1995). This may reduce the environmental quality of the sea lochs, and hence their ability to support viable populations of characteristic wild species. Additional concerns include the disturbance caused by fish farm operational activities, the excessive use of wild stocks of fish to feed the captive fish, and the effects on the genetic constitutions of wild salmon as they breed with captive bred stock.

Water abstraction within the coastal zone is an important issue because many coastal communities rely on groundwater for their drinking water supply, and by depleting the groundwater in *aquifers*, abstraction can lead to intrusion by sea water. The main result is *saline intrusion*, but the groundwater can also be contaminated by pollutants present in the sea water. Removal or alteration of certain habitat types such as sand dunes can have a similar effect because these maintain the water table at an elevated level. Saline intrusion can also affect the biota of maritime fresh or brackish water habitats. In some areas, the combination of abstraction and the weight of development has caused the land to sink relative to sea level (MAFF/WO 1993).

Over fishing, including that of small species for purposes such as animal feed or fertilisers, can disrupt food chains, with particularly serious consequences for top carnivores, including large fish, birds and marine mammals.

Many of the development types and activities referred to above tend to be concentrated in estuaries, and this has serious ecological consequences. Any significant loss or alteration to the estuarine habitat means that the diversity and biomass of invertebrates will substantially decrease, with inevitable effects on the whole resident and migratory biota. The cumulative implications of estuary development should therefore be a primary consideration when viewing estuarine development proposals. A survey by Rothwell & Housden (1990) concluded that of the 123 UK estuaries surveyed, 80 were under some degree of threat, with 30 in imminent danger of permanent damage. Some of the impacts are irreversible and seriously reduce the extent of inter-tidal habitat.

13.5.3 *Methods of impact prediction*

Potential changes (with or without the development) can be assessed in relation to the baseline conditions and information on past, current and predicted trends. Most information on trends will have to be sought through the desk study, although comparison of field survey data with previous data can help to elucidate recent trends.

Some geomorphological impacts can be predicted using standard risk assessment methods such as the calculation of *return periods*. For example, for the purpose of designing flood defences, the limits of a coastal floodplain are defined, in DoE (1992b), by the peak water level of an appropriate peak flood event, which is normally the greater of the 1 in 200 yr return period flood or the highest known water level (although defences are often designed to a higher standard). The use of return periods is discussed in MAFF (2000a) and Penning-Rowsell *et al.* (1988).

As with all ecological assessments, prediction of some **primary impacts**, such as habitat loss by land take, is a relatively easy task. Similarly, it may be possible to state that, in order to maintain a viable population, or a population of given size, a particular species requires a certain set of environmental variables operating within a given area. However, prediction of **secondary impacts** on environmental-factor systems (and hence of the integrity of remaining habitat patches) is much more difficult, as is prediction of gradual and cumulative impacts.

A number of **computer models** have been developed for predicting changes in coastal systems and/or for coastal management. Some examples are given in Table 13.4. However, the software can be expensive, may not be suitable for 'off the peg' use, and can only be as good as the data input. Moreover, predictive models of geomorphological and ecological processes have a high degree of uncertainty. For example, coastal sedimentation and erosion are very uncertain process, and even

Table 13.4 Some modelling software for coastal systems

CZMC (Coastal Zone Management Centre)
(http://www.minvenw.nl/projects/netcoast/info/czmc.htm)

Rapid Assessment Module for Coasts (RAMCO) and **Coastal Zone Simulation Model (COSMO)** – which may facilitate integration of information (including socio-economic) for impact predictions.

Coastal Zone Biodiversity Simulation Model (COSMO-BIO) – which is based on a *population risk factor* for selected 'key' species.

Hydraulics Research Ltd (HR Wallingford) (http://www.hrwallingford.co.uk/)

SeaWorks – includes modules for: wind–wave prediction, wave transformation (deep water to nearshore), wave energy dissipation, waves and currents and sediment transport in the surf zone, long-term shoreline changes.

SandCalc – contains methods for calculating sediment transport dynamics in rivers, reservoirs, estuaries, the coastal zone and offshore, e.g. water density and viscosity, velocity profiles, bed shear stress, threshold of sediment motion, bed forms, sediment concentration profiles, sediment transport rates, settling velocity (sand or mud), mud erosion and deposition.

TELEMAC – includes modules for: environmental impact of reclamation and dredging schemes, dredged material disposal, strategic water quality planning, outfall design and pollutant dispersion, coastal defence design, port and harbour design, wave activity including harbour resonance, failure of dams or dykes.

Proudman Oceanographic laboratory (POL) (http://www.pol.ac.uk/)

COAMES (COAstal Management Expert System) – under development.

POLPRED for Windows – will compute sea elevations and currents for a specified location or area (e.g. UK Continental Shelf model (12 km) and finer resolution UK models) for any specified point in time. It is one of a suite of **POL Hydrodynamic Numerical Models**, some of which have been developed by commercial firms for specific purposes such as predicting the transport and spreading of oil spills.

POLTIPS (Tidal Information and Prediction System) for Windows – tidal predictions for nearly 700 ports.

where good historical records exist, it is often dangerous to assume that the same conditions will continue to apply (MAFF 2000a). Advice on available models, and on the feasibility of utilising them in an EIA, can be sought from organisations such as those listed in Table 13.4. In most cases, predictions will have to rely on relatively simple methods such as those outlined in §11.5.7.

In assessing **impact significance**, particular attention should be paid to (a) the sensitivity and vulnerability of important geomorphological features and processes, and of high-status species, habitats and sites, and (b) how these are likely to respond to particular impacts, including whether the effects will be temporary, long-term, reversible or permanent. In this context it is useful to differentiate between pulse, press and catastrophic disturbance types (Glasby & Underwood 1996).

A **pulse disturbance** is a short-term disturbance, of high intensity, which may result in a temporary response in a population. Examples might be (a) the short-term impacts associated with the construction of a building near a coastal waterway which results in disposal of spoil to that waterway, or (b) the temporary bathymetric changes associated with the disposal of dredged sediment at sea.

A **press disturbance** is a sustained or chronic disturbance to the environment which may cause a long-term response. For example, any permanent development such as a coastal defence scheme will cause long term changes to the sediment balance, perhaps enhancing erosion or sediment accretion (which may have positive or negative consequences). Other examples could be (a) the long-term discharge of a thermal plume from a nuclear power station, causing changes in the distribution of littoral biota, or (b) the increased presence of fish near the intake screens of a water-cooling system, and their subsequent entrapment on the sieve system.

A **catastrophic disturbance** is a major habitat destruction from which populations are unlikely to recover. An example is the permanent flooding of inter-tidal mudflats by a static barrage scheme. Similarly, cliff collapse caused by the construction of buildings on unstable cliffs might result in the permanent loss of valuable geological or geomorphological features (Baird 1994).

Although these definitions are clear, in practice (a) a project may generate combinations of the disturbance types, and (b) responses to them may vary between organisms (Glasby & Underwood 1996). For example, a pulse disturbance to a population of very long-lived organisms may be a press disturbance to a population of organisms with a short lifespan (Lincoln-Smith 1998). Similarly, a local geomorphological pulse disturbance, such as the dumping of dredged material, may upset the sediment balance, and lead to catastrophic disturbances elsewhere in the coastal sediment cell.

These complexities illustrate how inadequate data and/or understanding of the coastal system hamper impact prediction, and explain why scientists are often loath to make concrete statements regarding changes or losses that a project will generate. Coastal ecosystems involve the interaction of numerous processes and factors that are poorly understood, with the result that impact prediction is an inexact science.

13.6 Mitigation

Ecologists and geomorphologists involved in the EIA of coastal developments should have the formulation of appropriate mitigation measures as one of their main objectives. They should provide detailed descriptions of proposed measures, indicate how they should actually be put in place, and propose how they might be modified in the light of unforeseen impacts. The last point is particularly relevant to coastal zone developments because of the relative lack of knowledge about their impacts, and the dynamic nature of the ecosystems they affect. Attention should be paid to aspects such as water abstraction and sewage production, which may be minor in the context of the project alone, but may require mitigation in the context of the severe stresses already present in the coastal zone.

Wherever possible, proposed mitigation measures should emphasise the need to minimise or avoid:

- potentially harmful geomorphological changes;
- pollution, including eutrophication;
- habitat loss or fragmentation;
- disturbance of species and communities.

Means of mitigating against potential geomorphological impacts are based largely on **coastal engineering techniques** (a good review is provided by Fleming 1992). For example, it is now generally recognised that if a project requires the construction or modification of sea defences, it is desirable that these are 'soft' rather than 'hard' (§13.5.2). Options include:

- construction or replenishment of shallow sloping beaches, which are more effective at dissipating wave energy and maintaining the erosion/deposition regime (Brampton 1992, West 1992);
- using groynes to stabilise beaches where replenishment is not an option, e.g. due to a lack of suitable material. Groynes are usually effective in the short term, but they disrupt deposition patterns. This can be reduced by minimising their encroachment onto the littoral zone or, for many beaches, by placing them at intervals along the coast (Cooke & Doornkamp 1990);
- encouraging the maintenance and development of natural barriers such as saltmarshes and sand dunes, which also have positive ecological impacts. This is implicit in *managed realignment* strategies (§13.3.2).

Some mitigation measures can involve 'sensitive' construction methods. For example, during the construction phase of projects such as barrage schemes, impacts on sediment balance can be minimised by conducting the work on the leeward side of existing structures, and/or by the use of floating platforms for the construction machines.

Impacts of **dredging** can be reduced by carefully planned extraction programmes and controlled techniques. Operations can be confined to ebb tide periods, and can seek to avoid areas where they are likely to generate impacts, e.g. (a) areas with high nutrient and toxin loadings in the sediments, and (b) areas where tidal movement and existing sediment loadings can result in turbidities and sedimentation rates high enough to have serious impacts on the biota. In addition, the need for extraction dredging (to provide materials for coastal defences, etc.) can be reduced by using navigation dredgings (MAFF 2000b). Guidelines on the useful disposal of dredged material are provided in CPMP (1993).

Point source pollution (including eutrophication) from sewage outfalls can be minimised by (a) nutrient stripping at the sewage treatment works, or (b) good planning, including predicting the levels of sewage the stretch of coastline can accept in relation to factors such as water circulation (tides and currents) which buffer the environmental stress imposed by sewage disposal (Carter 1988). Diffuse pollution, e.g. from agriculture, is more difficult to control.

In addition to maintaining and enhancing natural features, such as sand dunes, that maintain water table levels, mitigation against groundwater contamination by seawater intrusion can be achieved by methods such as **artificial recharge** of the aquifer, e.g. by importing fresh water from outside the catchment or by re-routing streams or storm runoff into infiltration pits, which reduce *evapotranspiration* (Carter 1988). However, care is needed to ensure that such measures do not generate other impacts on the freshwater systems involved.

It is important to avoid or minimise **habitat loss or fragmentation** on both the landward and seaward sides of a project. Together with disturbance of wildlife, these impacts depend largely on project location and design, including infrastructure such as new roads, so mitigation measures must focus on sensitive siting and design. If loss of valuable habitat is unavoidable, compensation may be considered as an alternative, but this should be seen as a last resort since it is rarely successful or adequate (see §11.6.3).

Apart from protection in reserves that are closed to the public, **damage to fragile habitats** such as sand dunes by visitor pressure can be limited by measures such as exclusion of vehicles, provision of boardwalks and management procedures to control or repair wind erosion. These may include the use of netting or brushwood fencing or, more effectively, replanting and protecting vegetation, especially marram grass (see Carter *et al.* 1992, Doody 1985, Houston 1997, Ranwell & Boar 1986).

13.7 Monitoring

Given the importance of the coast it is essential that a monitoring system be in place to measure residual impacts, and hence to assess the effectiveness of mitigation measures and alert the interested parties to any development-linked ecological or geomorphological problems (see §1.2.6). Monitoring has been neglected in UK EIAs generally, and the coastal zone is no exception. MAFF has stated that monitoring (during and after construction) is an essential element of any scheme, and should be in place to avoid harmful environmental impacts (MAFF/WO 1993, 1996). However, this only refers to flood and coastal defence works. *Shoreline Management Plans* (§13.3.2) are also supposed to incorporate monitoring programmes, which should assist in monitoring the effects of future developments.

Monitoring should be undertaken by experts and in consultation with the statutory/regulatory authorities and relevant NGOs (§13.3.3). Some geomorphological parameters can be monitored using fairly simple techniques. For example:

- On rocky coasts, cliff recession can be measured with pegs driven into the rock, and beach profiles can be measured using conventional field surveying techniques. Other methods of measuring processes such as coastal erosion are reviewed in Dugdale (1990).
- At a constructional coast, Chorley *et al.* (1984) suggest that rates of deposition can be monitored by indicators such as accumulation/erosion at breakwaters and groynes, dilution rates of particles in sediment of known source, or the use of sediment traps or tracers such as dyes.

- Rates of mud accretion (e.g. on salt marsh) can be measured using standard levelling techniques or sediment traps (Hargrave & Burns 1979).
- Sediment transport can be monitored (a) directly by sampling water, or (b) indirectly by beach profile and groyne height exposure measurements, benthic sampling, or remote sensing (§15.4.3).
- Photographic or video records can be made, e.g. of beach profiles and sand dune erosion or recovery.

Advice on the use of other methods, e.g. for sublittoral monitoring and the use of models, can be sought from organisations such as POL. Most biological monitoring will require repeat sampling using the same methods as in baseline surveys.

13.8 Conclusions

There is much room for improvement of current EIA practice in the coastal zone. For example:

- Most coastal zone EIAs fail to address the cumulative impacts that coastal developments generate.
- There is a tendency to focus on individual species, sites or zonal components, while impacts should be considered in relation to the total resident and migratory biota of the whole zone.
- There is a need to ensure that developers allow adequate time for baseline studies, and allocate sufficient resources for appropriate fieldwork and post-development monitoring.

Project-level EIA often fails to quantify the overall impact of developments on biodiversity (e.g. Thompson *et al.* 1995, Treweek *et al.* 1998), and there is also a need for a more integrated and strategic approach to coastal zone planning and management. The production of *Shoreline Management* Plans, etc. (§13.3.2) is a promising development. However, these are in their infancy, appear to be largely focused on coastal defence works rather than on developments as a whole, and are not statutory. It is important that LA structure plans adopt the same principles and seriously consider limiting the development of projects which affect coastal ecosystems.

In addition, a national strategic planning framework for the coast is needed. EN promotes SEA as an effective way of helping to achieve sustainable development and promoting the maintenance and enhancement of biodiversity (Therivel & Thompson 1996). Adoption of SEA along the coastal zone would lead to the removal of piecemeal development and would provide an arena in which to bring together conflicting and overlapping interests. Strategic assessment, planning and management will not remove the need for project-based EIAs. Indeed, it should facilitate their execution and effectiveness.

Central government should also adhere to the **precautionary principle**, which is advocated in European environmental policy and was recommended at the North Sea Ministers Conference of 1990 for policies and activities affecting the marine environment. This approach acknowledges the current lack of knowledge about the biology and ecology of marine systems.

The coastal zone is an outstanding area for wildlife in the UK, and EIA provides a means of providing checks on development activities which undermine its eco-logical worth. Both ecological and geomorphological science have an obvious role in the process. However, ecology is often under-resourced or ignored (Treweek 1996), and the importance of geomorphology is only now being recognised. If coastal EIA is to develop as a tool for environmental management, which helps to realise the goals of conservation and sustainability, it is important that ecologists and geomorphologists have a greater input to the process.

References

Adam P 1993. *Saltmarsh ecology*. Cambridge: Cambridge University Press.

Bagenal TB (ed.) 1978. *Methods of assessment of fish production*, 3rd edn. IPB Handbook No. 3. Oxford: Blackwell.

Baird WJ 1994. Naked rock and the fear of exposure. In *Geological and landscape conservation*, D O'Halloran, C Green, M Harley, M Stanley & J Knill (eds), 335–336. London: The Geological Society.

Baker JM 1979. Responses of salt marsh vegetation to oil spills. In *Ecological processes in coastal environments*, RL Jefferies & AJ Davy (eds), 529–542. Oxford: Blackwell.

Baker JM & JH Crothers 1987. Intertidal rock. In *Biological surveys of estuaries and coasts*, JM Baker & WJ Wolff (eds), 157–197. Cambridge: Cambridge University Press.

Barnes RSK 1994. *The brackish-water fauna of Northwest Europe. An identification guide to brackish-water habitats, ecology and macrofauna for field workers, naturalists and students*. Cambridge: Cambridge University Press.

Bassett JA & TGF Curtis 1985. The nature and occurrence of sand dune machair in Ireland. *Proceedings of the Royal Irish Academy* **85B**, 1–20.

Bird ECF 1985. *Coastal changes: a global review*. Chichester: Wiley.

Blower JG, LM Cook & JA Bishop 1981. *Estimating the size of animal populations*. London: Allen & Unwin.

Brampton AH 1992. Beaches – the natural way to coastal defence. In *Coastal zone planning and management*, MG Barrett (ed.), 221–229. London: Thomas Telford.

Brouwer A, AJ Murk & JH Koeman 1990. Biochemical and physiological approaches in ecotoxicology. *Functional Ecology* **4**, 275–281.

Brown AC & A McLachlan 1990. *Ecology of sandy shores*. Amsterdam: Elsevier.

BTO (British Trust for Ornithology) 1992. *National low tide counts: detailed instructions*. Thetford: British Trust for Ornithology.

BTO (undated). *Instructions to counters: breeding waders of wet meadow survey*. Thetford: British Trust for Ornithology.

Bunker F & RL Foster-Smith 1996. *A Field guide for seashore mapping*. Peterborough: EN/SNH/CCW/JNCC.

Burd F 1989. *The saltmarsh survey of Great Britain: an inventory of British saltmarshes*. Research and Nature Conservation No. 17. Peterborough: NCC.

Burt N & J Watts 1996. *Barrages: engineering, designs & environmental impacts*. Chichester: Wiley.

Carter RWG 1988. *Coastal environments: an introduction to the physical, ecological and cultural systems of coastlines*. London: Academic Press.

Carter RWG, TGF Curtis & MJ Sheehy-Skeffington (eds) 1992. *Coastal dunes: geomorphology, ecology and management for conservation*. Proceedings 3rd European Dune Congress. Galway June 1992. Rotterdam: Balkema

Chorley RJ, SA Schumm & DE Sugden 1984. *Geomorphology*. London: Methuen.

Clayton KM 1993. *Coastal processes & coastal management*. Cheltenham: CC.

Coles SM 1979. Benthic microalgal populations on intertidal sediments and their role as precursors to salt marsh development. In *Ecological processes in coastal environments*, RL Jefferies & AJ Davy (eds), 25–42. Oxford: Blackwell Scientific.

Connor DW, DP Brazier, TO Hill & KO Northern 1997a. MNCR *marine biotope classification for Britain and Ireland*. Vol. 1. *Littoral biotopes*, Version 97.06. JNCC Research Report No. 229. Peterborough: JNCC.

Connor DW, MJ Dalkin, TO Hill, RHF Holt & WG Sanderson 1997b. MNCR *marine biotope classification for Britain and Ireland*. Vol. 2. *Sublittoral biotopes*, Version 97.06. JNCC Research Report No. 230. Peterborough: JNCC.

Cooke RU & JC Doornkamp 1990. *Geomorphology in environmental management*, 2nd edition. Oxford: Oxford University Press.

CPMP (Oslo and Paris Conventions for the Prevention of Marine Pollution) 1993. *Oslo Commission guidelines for the management of dredged material*. Fifteenth meeting, Annex 1. Berlin: Oslo and Paris Commissions.

Crawford RMM 1998. Shifting sands: plant survival in the dunes. *Biologist* **45**(1), 27–32.

Davidson NC, D d'A Loffoley, JP Doody, LS Way, J Gordon, R Key, CM Drake, MW Pienkowski, R Mitchell & KL Duff 1991. *Nature conservation and estuaries in Great Britain*. Peterborough: NCC.

Davies IM & JC McKie 1987. Accumulation of total tin and tributyl tin in muscle tissue of farmed Atlantic salmon. *Marine Pollution Bulletin* **18**(7), 405–407.

Davis RA 1996. *Coasts*. Hemel Hempstead: Prentice-Hall.

DoE (Department of the Environment) 1990. *Planning and Policy Guidance Note 14: Development on unstable land* (PPG14). London: HMSO.

DoE 1992a. *Planning and Policy Guidance Note 20. Coastal Planning* (PPG20). London: HMSO.

DoE 1992b. *Development and Flood Risk*, Circular 30/92 (Welsh Office Circular 68/92, MAFF Circular FD 1/92). London: HMSO.

DoE 1993. *Coastal planning and management: a review*. London HMSO.

DoE 1995. *Policy guidelines for the coast*. London: HMSO.

DoE 1996. *Coastal zone management – towards best practice*. London: HMSO.

Doody P (ed.) 1985. *Sand dunes and their management*. Focus on Nature Conservation No. 13. Peterborough: JNCC.

Downie AJ 1996. *Saline lagoons and lagoon-like saline ponds in England*. English Nature Science No. 29. Peterborough: EN.

Dugdale R 1990. Coastal processes. In *Geomorphological techniques*, 2nd edn, A Goudie, M Anderson, T Burt *et al.* (eds), 351–364. London: Unwin Hyman.

Dyer KR 1998. *Estuaries: a physical introduction*, 2nd edn. Chichester: Wiley.

EA (Environment Agency) 1997. *Our policy and practice for the protection of floodplains*. Bristol: Environment Agency. (http://www.environment-agency.gov.uk/)

EA 1999a. *State of the environment in England and Wales: coasts*. London: TSO. (Extracts available from http://www.environment-agency.gov.uk/)

EA 1999b. *The Environment Agency's latest position on the Cardiff Bay Barrage*. (http://www.environment-agency.gov.uk/)

Earle R & DG Erwin (eds) 1983. *Sublittoral ecology, the ecology of the shallow sublittoral benthos*. Oxford: Oxford Scientific.

EN (English Nature) 1992. *Targets for habitat recreation*. Report by Cambridge Environmental Research Consultants. Peterborough: EN.

EN 1995. *Managed retreat: a practical guide*. Peterborough: EN.

EN 2000. *Coastal Habitat Management Plans: an interim guide to content and structure*. Peterborough: EN/EA/CCMS.

EUCC (European Union for Coastal Conservation) 1997. *Threats and opportunities in the coastal areas of the European Union*. Report for the Dutch Ministry for Housing, Spatial Planning and Environment. Leiden, The Netherlands: EUCC.

Fleming CA 1992. The development of coastal engineering. In *Coastal zone planning and management*, MG Barett (ed.), 5–20. London: Thomas Telford.

Gibbons DW, JB Reid & RA Chapman (eds) 1993. *The new atlas of breeding birds in Britain and Ireland*. Calton: T & A Poyser.

Gimmingham CH, W Ritchie, BB Wiletts & AJ Willis (eds) 1989. *Coastal sand dunes*. Edinburgh: RS.

Glasby TM & AJ Underwood 1996. Sampling to differentiate between press and pulse disturbances. *Environmental Monitoring and Assessment* **42**, 241–252.

Gorman M & D Raffaelli 1993. The Ythan estuary. *Biologist* **40**(1), 10–13.

Gubbay S 1990. *A future for the coast: proposals for a UK Coastal Zone Management Plan*. Ross-on-Wye: Marine Conservation Society.

Hammond PS 1987. Techniques for estimating the size of whale populations. *Symposium of the Zoological Society of London* **58**, 225–245.

Hammond PS & PM Thompson 1991. Minimum estimation of the number of bottlenose dolphins *Tursiops truncatus* in the Moray Firth, N.E. Scotland. *Biological Conservation* **56**, 79–87.

Hargrave BT, & NM Burns 1979. Assessment of sediment trap collection. *Limnological Oceanography* **24**, 1124–1135.

Hiby AR & PS Hammond 1989. Survey techniques for estimating abundance of cetaceans. *Report to the International Whaling Commision (Special Issue No. 11)*, 47–80. Cambridge: International Whaling Commission.

Hiby AR, D Thompson & AJ Ward 1988. Census of grey seals by aerial photography. *Photogrammetric Record* **12**, 589–94.

Hiscock K (ed.) 1998. *Benthic marine ecosystems of Great Britain and the north-east Atlantic*. Peterborough: JNCC.

Houston J 1997. Conservation and management on British dune systems. *British Wildlife* **8**, 297–307.

HR (Hydraulics Research) 1993. *Coastal management: mapping of littoral cells*. Report SR 328. Wallingford: HR.

JNCC (Joint Nature Conservation Committee) (various authors) 1992–98. *Seabird numbers and breeding success in Britain and Ireland*. Peterborough: JNCC.

JNCC 1993–95. *Sand dune vegetation survey of Great Britain: Part 1 England* (GP Radley 1994); *Part 2 Scotland* (TCD Dargie 1993); *Part 3 Wales* (TCD Dargie 1995). Peterborough: JNCC.

JNCC (Buck AL et al.) 1993–97. *An inventory of UK estuaries: Vol. 1 Introduction and methodology* (1997); Vol. 2 *South-west Britain* (1993); Vol. 3 *North-west Britain* (1993); Vol. 4 *North and east Scotland* (1993); Vol. 5 *Eastern England* (1997); Vol. 6 *Southern England* (1997), Vol. 7 *Northern Ireland* (1996). Peterborough: JNCC.

JNCC (Barne JH et al.) 1995–98. *Coasts and seas of the United Kingdom – coastal directory series* (16 regional volumes). Peterborough: JNCC (also available (1999) on CD ROM from CIRIA).

JNCC 1996–99. *Coasts and seas of the United Kingdom – Marine Nature Conservation Review series*: Rationale and methods (K Hiscock (ed.) 1996); 15 regional sectors (various authors 1996–99). (See also Hiscock 1998). Peterborough: JNCC.

Komar PD 1983. Coastal erosion in response to the construction of jetties and breakwaters. In *Handbook of coastal process and erosion*, PD Komar (ed.), 191–204. Boco Raton, Florida: CRC Press.

Lewis JR 1976. *The ecology of rocky shores*, 2nd edn. London: Hodder & Stoughton.

Lincoln-Smith MP 1998. *Guidelines for assessment of aquatic ecology in EIA*. Unpublished report to the NSW Department of Urban Affairs and Planning.

Lloyd CS, ML Tasker & KE Partridge 1991. *The status of seabirds in Britain and Ireland*. London: T & A Poyser.

MAFF (Ministry of Agriculture Fisheries and Food) 1994. *Coast protection survey of England: summary survey report*, PB 1667. London: MAFF.

MAFF 1999a. *High Level Targets for flood and coastal defence and elaboration of the Environment Agency's flood defence supervisory duty.* London: MAFF.

MAFF 1999b. *Flood and coastal defence project appraisal guidance: economic appraisal*, FCDPAG3. London: MAFF.

MAFF 2000a. *Flood and coastal defence project appraisal guidance: approaches to risk*, FCDPAG4. London: MAFF.

MAFF 2000b. *Flood and coastal defence project appraisal guidance: environmental appraisal*, FCDPAG5. London: MAFF.

MAFF (in prep. a, b, c). *Flood and coastal defence project appraisal guidance: overview*, FCDPAG1; *strategic planning and appraisal*, FCDPAG2; *post project evaluation*, FCDPAG6. London: MAFF.

MAFF/WO (MAFF & Welsh Office) 1993. *Strategy for flood and coastal defence in England and Wales*, PB 1471. London: MAFF.

MAFF/WO et al. 1995. *Shoreline management plans: a guide for coastal defence authorities*, PB 2197. London: MAFF.

MAFF/WO 1996. *Code of practice on environmental procedures for flood defence operating authorities*, PB 2906. London: MAFF.

McLusky DS 1981. *The estuarine ecosystem.* Glasgow: Blackie.

Moore PG & R Seed (eds) 1985. *The ecology of rocky coasts.* London: Hodder & Stoughton.

NERC (National Environment Research Council) 1983. *Contaminants in marine top predators.* Report to Marine Pollution Monitoring Management Group. London: DoE.

New TR 1998. *Invertebrate surveys for conservation.* Oxford: Oxford University Press.

Nordstrom, K & W Carter 1991. *Coatsal dunes: form and process.* Chichester: Wiley.

Owen N, M Kent & P Dale 1996. The machair vegetation of the Outer Hebrides: a review. In *The Outer Hebrides: the last 14,000 years*, D Gilbertson, M Kent & J Grattan (eds), 123–131. Sheffield: Sheffield University Press.

Packham JR & AJ Willis 1997. *Ecology of dunes, saltmarsh and shingle.* London: Chapman & Hall (Klewer Academic).

Penning-Rowsell E, P Thompson & D Parker 1988. Coastal erosion and flood control: changing institutions, policies and research needs. In *Geomorphology in environmental planning*, JM Hooke (ed.), 211–230. Chichester: Wiley.

Pethick J 1984. *An introduction to coastal geomorphology.* London: Edward Arnold.

Perrow MR, IM Côté & M Evans 1996. Fish. In *Ecological census techniques: a handbook*, WJ Sutherland (ed.),178–204. Cambridge: Cambridge University Press.

Picton BE & MJ Costello (eds) (1997). *BioMar biotope viewer: a guide to marine habitats, fauna and flora of Britain and Ireland* (CD-ROM). Dublin: Environmental Sciences Unit, Trinity College.

Pitcher TJ & PJB Hart 1982. *Fisheries ecology.* London: Croom Helm.

Potts GW & PJ Reay 1987. Fish. In *Biological surveys of estuaries and coastal habitats*, JM Baker & WJ Wolff (eds), 342–373. Cambridge: Cambridge University Press.

Ranwell DS & R Boar 1986. *Coast dune management guide.* London: HMSO.

Reading HG & JD Collinson 1998. Clastic coasts. In *Sedimentary environments: processes, facies and stratigraphy*, 3rd edn, HG Reading (ed.), 154–231. Oxford: Blackwell Science.

Richards A, F Bunker & R Foster-Smith 1995. *Handbook for marine Phase 1 survey and mapping.* (CCW Report No. 95/6/1). Bangor: Countryside Council for Wales.

Rodwell JS (ed.) 2000. *British plant communities*, Vol. 5: *Maritime communities and vegetation of open habitats.* Cambridge: Cambridge University Press.

Rothwell PI & SD Housden 1990. *Turning the tide: a future for estuaries.* Sandy: RSPB.

RSPB 1991. *Coastal zone planning: evidence to the House of Commons Select Committee on the Environment.* Sandy: RSPB.

RSPB 2000. *Conservation issues: marine action.* (http://www.rspb.org.uk)

SEPA (Scottish Environmental Protection Agency) 1999. *Improving Scotland's water environment.* Stirling: SEPA. (http://www.sepa.org.uk/)

Sneddon P & RE Randall 1993–94. *The coastal vegetated shingle structures of Great Britain:* Main Report (1993); Appendix 1 *Shingle sites in Wales* (1993); Appendix 2 *Shingle sites in Scotland* (1994); Appendix 3 *Shingle sites in England* (1994). Peterborough: JNCC.

SO (Scottish Office) 1997. *National Planning Policy Guidance Note 13* (NPPG13). *Coastal Planning.* Edinburgh: SO. (http://www.scotland.gov.uk/library/nppg/nppg-cover.asp)

Stone CJ, A Webb, C Barton, N Ratcliffe, TC Reed, ML Tasker & MW Pienkowski 1995. An *atlas of seabird distribution in north-west European waters.* Peterborough: JNCC.

Summerfield M 1991. *Global geomorphology.* Harlow: Longman.

Tasker ML, PH Jones, TJ Dixon & BF Blake 1984. Counting seabirds from ships: a review of methods employed and a suggestion for a standardised approach. *AUK* **101**, 567–577.

Tett PB 1987. Plankton. In *Biological surveys of estuaries and coastal habitats,* JM Baker & WJ Wolff (eds), 280–341. Cambridge: Cambridge University Press.

Therivel R & S Thompson 1996. *Strategic environmental assessment and nature conservation.* Peterborough: EN.

Therivel R, E Wilson, S Thompson, D Heaney & D Pritchard 1992. *Strategic environmental assessment.* London: Earthscan.

Thompson PM & J Harwood 1990. Methods for estimating the population size of common seals, *Phoca vitulina. Journal of Applied Ecology* **27**, 924–938.

Thompson S, JR Treweek & DJ Thurling 1995. The potential application of Strategic Environmental Assessment (SEA) to the farming of Atlantic Salmon (*Salmo salar* L.) in Scotland. *Journal of Environmental Management* **45**, 219–229.

Thornton D & DJ Kite 1990. *Changes in the extent of the Thames estuary grazing marshes.* Peterborough: NCC.

Trenhaile AS 1997. *Coastal dynamics and landforms.* Oxford: Clarendon Press.

Treweek JR 1996. Ecology and environmental impact assessment. *Journal of Applied Ecology* **33**, 191–199.

Treweek JR, P Hankard, DB Roy, H Arnold & S Thompson 1998. Scope for strategic ecological assessment of trunk road development in England with respect to potential impacts on lowland heathland, the Dartford warbler (*Sylvia undulata*) and the sand lizard (*Lacerta agilis*). *Journal of Environmental Management* **53**, 147–163.

Tyler-Walters H & A Jackson 1999. *Assessing seabed species and ecosystem sensitivities. Rationale and user guide.* MarLIN Report No. 4. (January 2000 edn). Plymouth: MBA. (http://www.marlin.ac.uk)

UKBG (UK Biodiversity Group) 1999. *Tranche 2 Action Plans:* Vol. V. *Maritime species and habitats.* Peterborough: English Nature. (The priority species list, HAPs and SAPs, are available from http://www.jncc.gov.uk/ukbg).

Walker CH 1990. Kinetic models to predict bioaccumulation of pollutants. *Functional Ecology* **4**, 295–301.

Walsh PM et al. 1995. *Seabird monitoring handbook for Britain and Ireland: a compendium of methods for survey and monitoring of breeding seabirds.* Peterborough: JNCC.

Ward AJ, D Thompson, AR Hiby 1988. Census techniques for grey seal populations. *Symposia of the Zoological Society of London* **58**, 181–191.

West GM 1992. Engineering the beaches. In *Coastal zone planning and management,* MG Barrett (ed.), 231–236. London: Thomas Telford.

Wolff WJ 1987. Identification. In *Biological surveys of estuaries and coastal habitats,* JM Baker & WJ Wolff (eds), 404–423. Cambridge: Cambridge University Press.

Yell D & J Ridell 1995. *ICE design and practice guide: dredging.* London: Thomas Telford.

Part II

Shared and integrative methods

14 Environmental risk assessment and risk management

Andrew Brookes

14.1 Introduction

Risk provides the answer to three key questions: what can go wrong; how likely is it; and what are the consequences? (Kaplan & Garrick 1981). Risk assessment and management as applied to environmental and ecological issues is a rapidly growing discipline within its own right. There is now a wealth of publications which range from the provision of guiding principles set by governments for public domain risk analyses (e.g. USEPA 1992, DoE 1995) to handbooks which prescribe more detailed approaches to particular aspects of risk assessment (e.g. Calow 1998). Decisions are increasingly being made on a risk footing and some government agencies, for example, have acquired specialist expertise in risk analysis (e.g. EA 1997a). EIA practitioners also need to familiarise themselves with risk assessment as a complementary and powerful tool for analysis (e.g. Petts & Eduljee 1994, Carpenter 1995).

There are problems with such a new and rapidly evolving discipline, not least the need for clarification of terminology. There are also many instances where the term 'impact' is used, rightly or wrongly, in an interchangeable way with 'risk'. This chapter is written with the needs of the EIA practitioner in mind, rather than a risk specialist, and seeks to demonstrate the considerable benefits of following a risk-based approach. From an EIA perspective, risk assessment has conventionally been used as a tool for prediction and evaluation, but this chapter also seeks to explore its role as a complementary approach in its own right.

14.2 Definitions and concepts

14.2.1 Overview

Risk assessment is well established in the fields of banking, insurance and engineering as a management tool for dealing with uncertainty. It is also well used as a tool for improving occupational safety and setting priorities for the allocation of resources. This experience stretches back several decades. People use risk assessment, either consciously or subconsciously, in their everyday lives such as in negotiating a busy road as a pedestrian or placing a bet on a horse. There is a wealth of information concerned with human health risk assessment methods (Carpenter 1995). However, it is only relatively recently that risk assessment techniques have been extended to wider environmental considerations.

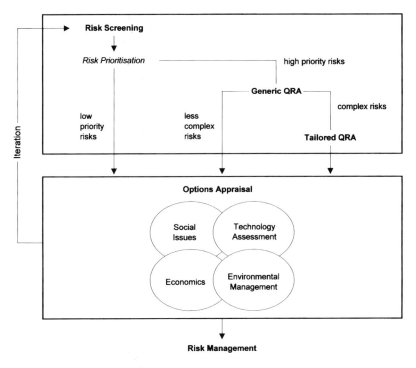

Figure 14.1 A framework for environmental risk assessment.

What does the term 'environmental risk assessment' (ERA) mean? It is a relatively new, emerging and exciting technique concerned with the structured gathering of available information about environmental risks and then the formation of a judgement about them (DoE 1995, DETR 2000). Risk management involves reaching decisions on a range of options that balance these risks against the costs and benefits (specifically including the environmental costs and benefits). Communicating the nature and scale of risk and the options is also a key part of the process. Figure 14.1 sketches out the basic elements of a framework within which environmental risk assessment may be carried out, including the options of generic and tailored Quantitative Risk Assessment (QRA).

14.2.2 Environmental risk assessment in the context of EIA

Uncertainty is an inherent and unavoidable aspect of EIA and is a characteristic of all natural systems (see Holling 1978). Uncertainties arise from a variety of sources, including the available data and in the decision-making process itself. Previous literature has largely failed to address this issue with a consequence that EIAs have often included sweeping statements about impacts and the effectiveness of untried mitigation measures (see Brookes 1999, Brookes *et al.* 1998). Where numerical values are used in EIA a single representative number is chosen which is typically either an average value or the worst case scenario. This can be very misleading, particularly where there are considerable uncertainties about an outcome and it may be totally inappropriate to use a single number (Harrop & Pollard 1998). By contrast approaches for managing uncertainty have been developed in parallel with risk assessment techniques (De Jongh 1988) and as a consequence uncertainty is explicit. ERA is a practical tool that can be used to express the likelihood of an outcome.

EIA and ERA are very similar concepts in that they broadly have the same goals and are tools that can inform decision-makers about the frequency and magnitude of adverse environmental consequences arising from activities or planned interventions. A response to such predictions might be that the manager wishes to mitigate or eliminate a particular impact or reduce the risk. Alternative sites or technological options or risk management may be desirable (Fig. 14.1). A major additional aspect provided by environmental risk assessment is that it can give probabilities to predicted impacts (Suter 1993). EIA and risk assessment often overlap and are mutually supportive of each other: they both deal with uncertainty, are essentially multifunctional in approach and seek to predict impacts to improve policy, programme, plan and project decisions.

Traditionally EIA practitioners have perhaps generally regarded risk assessment as a costly tool and have used this as a reason for limiting its use. For example, within the UK the results from the assessment of risk arising from landfill liner failure have sometimes been incorporated within EIAs of potential landfill sites. Risk assessment has also had limited, although increasing, use in a number of EIAs relating to waste-to-energy plants (Harrop & Pollard 1998). However, in general there has been limited wider application of the tool. Legal requirements also have a direct bearing on its use but the premise of this chapter is that risk assessment applied to particular problems may not need to progress as far as the detailed quantitative stage and therefore may not involve large costs. If applied as a tool for best practice there are considerable advantages to be gained by informing decision-makers of the potential risks of particular projects or proposals. Understanding the risks facing the environment, and the factors that govern whether such risks occur, is essential to a proactive approach to environmental management.

14.2.3 Problems with the terminology

One of the difficulties with the concept of risk is that it has been developed and applied across a broad range of disciplines and activities, leading to different terminologies. However the Royal Society (1992) attempted to provide more consistent definitions and these are followed in this chapter:

- **Hazard**: a property or situation with the potential to cause harm
- **Risk**: a combination of the probability, or frequency of the occurrence of a particular hazard and the magnitude of the adverse effects or harm arising to the quality of human health or the environment
- **Probability**: the occurrence of a particular event in a given period of time or as one amongst a number of possible events
- **Risk Management**: the process of implementing decisions about accepting or altering risks

In addition ERA is taken to be a comprehensive term including both human health and wider ecological aspects (see Calow 1998). Ecological risk assessment is seen as a sub-component of ERA.

A particular issue for ERA is the lack of a definable measure of harm to the environment. In dealing with ecosystems (§11.2.3) there are no equivalent end-points to the premature death of a human used in health risk assessment. Species extinction is a definable end-point but also whole communities of many species and their habitats are of interest (Carpenter 1995). Although there are some definitions laid down in law, appropriate criteria will need to be chosen in other circumstances to reflect both scientific information and social judgements.

14.3 Legislative and policy background and interest groups

14.3.1 Legislative and policy background

By comparison to EIA it is only recently that policies for consistent approaches to risk assessment have been developed for environmental protection. Many current regulations and proposed legislation require human health risk assessment. The Environment Act of 1995 specifically requests local authorities to carry out risk assessment and maintain registers of contaminated land (King 1998). MAFF's Control of Pesticides Regulations (1986) requires environmental risks to be assessed and to some extent the Health and Safety Executive, which is responsible for enforcing legislation on workplace safety, includes elements of environmental protection (e.g. Control of Substances Hazardous to Health (COSHH) Regulations, 1994). Generally, however, risk assessment concerned with ecosystems is not specifically defined in legislation and, unlike EIA, it is not a process that has been tied to the planning system.

14.3.2 Interest groups and sources of information

In recent years much progress has been made in the UK in harmonising the approaches to risk assessment advocated or used by government (e.g. DoE 1995). Considerable efforts are being made nationally to extend the use and acceptability of environmental risk assessment as a tool. Much is being done to promote it as a best practice tool and a principal reason for undertaking risk assessment and risk management is a commitment to *sustainable development*, i.e. meeting the needs of the present without compromising the ability of future generations to meet their own needs (World Commission on Environment and Development 1987). The Environment Agency, through its National Centre for Risk Analysis and Options Appraisal (EA 1997a) is an example of a specific group tasked with the development of tools and techniques. Since ERA is an emerging discipline there are relatively few 'how to do it' manuals. Whilst it is beneficial to refer to examples of practice, such as previous EIAs with risk assessments of incinerators or landfill sites, at the prescriptive level it may be wise to employ a risk specialist.

14.4 Key steps in performing an Environment Risk Assessment

ERA attempts to analyse the risks to human health and ecosystems from both human activities and natural phenomena. There are several basic steps (outlined below) which should be followed in a process that is iterative.

14.4.1 Hazard identification and analysis

The set of hazards to be identified needs to be clearly defined. For a hazard to result in harm there must be a way in which it can affect a receptor. If this is not the case then a risk is non-existent. Some risk specialists use the term Source–Pathway–Receptor to describe the process. An example for a flood defence scheme might be: how likely is it that the scheme will be over-topped with flood water? (Source of Hazard); how might people living on the neighbouring floodplain be exposed? (Pathway) and what effects might be experienced by an exposed individual? (Receptor). For a sewage treatment works the hazard might be the likelihood per year of the exceedance of Environmental Quality Standards (EQSs) to an adjacent river; a pathway would be how fauna and flora are exposed; and a receptor might be the effects on a single exposed organism.

Identification of the routes by which a hazardous event may occur is exemplified by the example of a lined landfill site with a leachate collection system and an associated treatment plant. Since the concern is the escape of leachate to groundwater, then it is not adequate to consider only the possibility of the liner being punctured. It is equally important to look at the possibility of failure of the leachate treatment plant. Techniques are available for the identification of hazards. However, **event tree analysis** is an accepted means of undertaking hazard analysis. Figure 14.2 shows a typical *event tree* for an accidental spillage. Event trees (also called *decision trees*) can be relatively simple as in the example shown and it is important not to make them too detailed.

Hazard analysis also involves estimating the probability or chance of occurrence of a particular hazard. This involves the collection and analysis of data. The more data that are available the better, particularly those which are relevant to the local circumstances under consideration. For example, real data on actual crashes of road tankers on British roads would be far more relevant to analysis of the risk of a chemical spill from a motorway in Britain than would be worldwide data on past road accidents. In putting numbers or scores on event trees it is important not to be too precise. Precision to one decimal place may have little credibility.

14.4.2 Exposure assessment

The next step is to examine the potential consequences associated with exposure to a hazardous event. A chemical spill, for example, could have a wide range of impacts on the built and natural environment. Factors to take into account would include:

- a clear definition of the nature of the hazard (e.g. quantity and rate of spill). This should be relatively straightforward;
- the characteristics of the local environment (e.g. sensitivity of the local environment; presence of rare species, etc.). Determining this can be problematic – a detailed site survey over a considerable area could be costly;
- behaviour of the hazard (e.g. infiltration rates, stream dilution, air dispersion);
- specific 'dose-response' relationships which might be known for particular species or environmental attributes being considered.

Determining the first factor is a relatively straightforward process, but the other three are much more difficult and complex, and demonstrate some of the difficulties

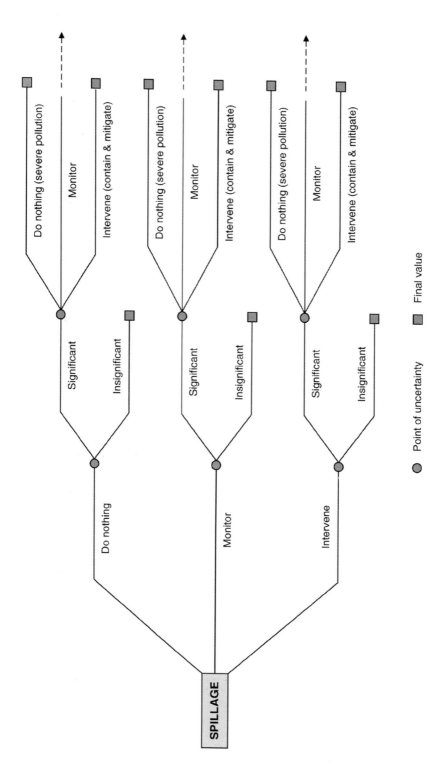

Figure 14.2 An event or decision tree for accidental spillages and pollution risk.

Table 14.1 Example table of consequences

Type of consequence	Description
Very high risk	Ecosystem irreversibly altered; no recovery. Over 100 km^2 affected.
High risk	Ecosystem altered, but not irreversibly; recovery may take as long as 50 yrs. 50–100 km^2 affected.
Moderate risk	Only one component of the ecosystem affected; 10 yr recovery period.
Low risk	Temporary alteration; effects confined to less than 0.5 km^2; recovery in less than 5 yrs.
Very low risk	Temporary alteration; very localised and minor consequences.

surrounding environmental risk assessment. Table 14.1 lists some descriptors that might be used to describe various levels of consequence.

14.4.3 Risk estimation

Risk can be determined by combining the results of hazard and consequence analyses and the simplest form of risk estimation is a matrix (Table 14.2).

Such matrices can be designed to be as simple or as complex as appropriate. Approaches to completing a matrix can be qualitative, quantitative, or a combination of both. More complex (and perhaps more controversial) approaches include the use of multi-criteria analysis (MCA) which can involve ranking, scoring and weighting methods to attain an overall risk score. Such methods have now been successfully used to examine risks due to genetically modified organisms (see DoE 1995) and road transport (EA 1997b).

Finally, it is possible to present risk results in numerical terms, e.g. that there is a 20% chance that the use of pesticides will lead to the loss of 50% of butterflies.

14.4.4 Risk evaluation/options appraisal

The importance of this step is in the judgement of the acceptability of the risk. In terms of human health this risk might be expressed in terms of the number of

Table 14.2 Simplified risk matrix

Probability or likelihood	Magnitude		
	High	Medium	Low
High	Very high risk	High risk	Moderate risk
Medium	High risk	Moderate risk	Low risk
Low	Moderate risk	Low risk	Very low risk

additional deaths per million people arising from a lifetime of exposure or the probability of the frequency of events causing fatalities. From an environmental perspective the preferred option is likely to be the one with the lowest risk. However, risk acceptability depends on a complex set of psychological factors.

The communication of the ERA results should take the form of an *Options Appraisal*, i.e. for each option what are the risks, costs and benefits. Effective communication can change a lay person's preconceived assessment of risks. This leads to more rational decision-making based less on emotions.

14.4.5 Risk management

Since all risk assessments systematically examine the causes and consequences of potential failures, then it is usually possible to pinpoint where improvements could be made. Risk management uses the results of ERA to mitigate or eliminate unacceptable risks. It is, however, important to consider whether or not a particular risk management measure leads to a secondary consequence. It is also important to ensure that the appropriate level of resources is directed to the level of risk reduction warranted in a particular circumstance. It is clearly not sensible to direct huge funds at a minor risk. There is a need to iterate between risk management and hazard analysis. Table 14.3 lists the types of options that could be evaluated in relation to road transport and the environment (EA 1997b).

Table 14.3 Risk management options that might be addressed in consideration of road transport impacts on the environment

Type of option	Examples of risk management
Policy level	Developing a multi-modal approach to transport, e.g. consideration of investment in forms of transport other than roads
Programme	Consideration of the roads programme for the whole country: rejecting schemes at an early stage with the potential for significant environment impact
Plan	Integrating land use and transport plans, e.g. to consider options for reducing traffic congestion in urban areas
Project level	Improved road design for minimising environmental impact: noise reduction using newer types of road surface; improved safety
Technology	New technology fitted to cars to reduce emissions; using techniques for the secondary treatment of road runoff to remove sediments and other pollutants
Economic	Mechanisms for charging for road use (e.g. in selected city areas; increased taxation on fuel, etc.)
Education	Improved driver training to minimise accidents but also to instruct the relevant services of what to do in an emergency situation to minimise pollution to the environment

14.5 Different levels of risk analysis

One way of describing the application of risk analysis is 'different horses for different courses'. A traditionally held perception is that risk assessment, perhaps as applied to an operational failure, is a very complex, involved and hence costly process. This may very well be the case where the circumstances warrant such a detailed level of analysis. There are various levels of sophistication for risk assessment (see Pollard *et al.* 1995). It is important to recognise the value of different stages in the environmental risk assessment process. It may be that in many circumstances there is no justification to progress beyond the initial stages that may be relatively low cost. The degree of sophistication should be determined by: the magnitude and significance of the risks being studied; the sensitivities of receptors; the quality of available data; and the means by which risks are to be communicated and the outputs utilised (Pollard *et al.* 1995). Figure 14.3 shows the different levels of sophistication that might be used with increasing risk and cost. It is important to adopt the most appropriate techniques to suit the issue under consideration. A global problem such as the depletion of the ozone layer is likely to require a different approach to remediation of an old gas works site for housing development.

The different levels of risk assessment can be described as follows (see EA 1997a):

- **Risk Screening and prioritisation** – the process used (a) to determine the range of risks, and the factors that control whether they will result in environmental damage, and (b) to describe the most important risks. It may be based on available data and substantially on professional judgement. If the decision is made to progress further with analysis, then monies can be invested in these key risks rather than looking in detail at all risks.

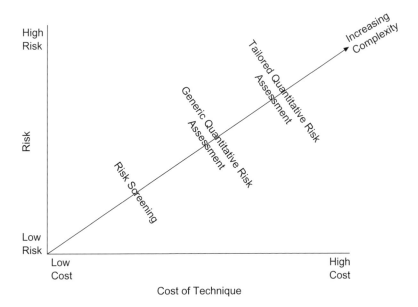

Figure 14.3 Levels of sophistication that might be used with increasing risk and cost.

- **Generic Quantitative Risk Assessment** – the use of generally available and tested models to provide simple quantification of the risks;
- **Tailored Quantitative Risk Assessment** – the development of specific models to meet a particular purpose. Usually complex and costly (e.g. the disposal of radioactive waste).

14.6 Parallels between EIA and ERA

This section attempts to draw some of the parallels between EIA and ERA and also to demonstrate the value of ERA as a tool in its own right (see Table 14.4).

Table 14.4 Comparison between EIA and ERA

Framework for EIA	*Framework for ERA*
Screening of the project or proposal and preliminary assessment of the existing environment to decide whether to carry out a full blown EIA followed by *Scoping* of the key environmental issues likely to be affected by the project or proposal.	**Screening** to determine the range of risks, and the factors that control whether they are likely to result in damage to the environment. When all risks have been identified **prioritisation** or ranking is conducted to ensure that resources for further work are targeted at the highest priority risks. Defining the problem is also known as **hazard identification**.
Baseline studies – collection of existing information.	
Impact prediction – determining the magnitude, spatial extent and probability of impacts, including direct and indirect effects.	**Hazard analysis** involves identification of the routes by which hazardous events could occur and estimation of the probability or chance of occurrence. **Consequence analysis** involves determining the potential consequences of a hazard. **Risk determination** combines the results of hazard and consequence analysis.
Assessment of the relative importance of the predicted effects, taking into account the present condition and the future condition that would result, as well as any measures of **mitigation**.	Judging the significance of the estimated risk is known as **Risk evaluation**, i.e. whether the environment is likely to withstand the effects. It may well be right for decisions to be taken partly in
Evaluation of the overall acceptability of the proposal or project and each of its alternatives, leading to selection of one or more **preferred options**.	response to pressures generated by **risk perceptions**. **Risk management** options may be concerned with tolerating or altering risks.
Monitoring and audit, e.g. leading to confirmation or rejection of predicted effects.	**Monitoring and audit**. Confirmation or rejection of predicted effects.

Both EIA and ERA are structured tools leading to recommendations concerning the environment that can assist decision-takers. Whilst there are clear parallels to be drawn, there are also fundamental differences: for example, EIA typically involves consideration of development alternatives whilst ERA does not. Both are essentially iterative processes and it is important that as a final stage after implementation of a project or proposal that monitoring and audit be considered. It is only through learning by experiences and mistakes that decisions can be improved 'next time'. Whilst public consultation and participation are talked about widely in EIA circles, risk perception and risk communication and similar concepts are recognised in ERA. Both EIA and ERA have been developed initially for application at the project level but the processes can be extended to strategic levels of decision-taking.

14.7 Opportunities and challenges for ERA

Environmental risk assessment should be regarded as a tool which allows the 'what if' question to be systematically addressed. It is far better to base decisions on the available evidence and in a structured way, rather than relying simply on the 'gut feel' of an individual. However, there should not be a preoccupation with precision and a quantitative output. Rather the process should be seen as tool for assisting decision-takers; it should be transparent, recording the assumptions made and uncertainties in the estimates; and it should be regarded as an iterative process, leading to future refinement. It is important to recognise that risk assessment and management is necessarily affected by considerable uncertainties. In established areas of risk assessment such as occupational health and safety evaluations there is a common denominator, namely human exposure. However, ERA is much wider in scope and therefore complex with far greater uncertainty (see Wright 1993). Some factors leading to uncertainty in ERA are:

- Ecosystems are open, dynamic and complex.
- Ecosystems have built-in variability and recoverability.
- Adjustment or recovery from particular impacts may be over a time span longer than a human life.
- It is inherently difficult to measure causal relationships.
- Release of certain persistent materials may cause irreversible change.
- Synergistic effects may arise, e.g. when two chemical pollutants interact and the combined effect is greater than the sum of their separate effects.
- Individual subsystems may be inter-dependent.
- Perceived risk may be just as important (if not more so) than real risk.

However, if the best available information at the time is used, and erroneous data discounted, then gross errors can be avoided.

It is important that those who use risk assessment as a tool do not profess more objectivity and confidence for their probability estimates and subsequent management decisions than is warranted. Commonly risk predictions are based more on subjectivity rather than objectivity (Kaplan & Garrick 1981) and it is essential that for the purposes of transparency gross assumptions and limitations are recorded. Risk assessment should be seen as an aid to informed decision-taking. It cannot itself make decisions and may not even be able to provide a preferred option. It is also a misconception to think that more prescriptive and detailed forms of risk assessment will make decisions clearer: difficult choices and trade-offs will still have to be made.

14.8 Risk communication

It is necessary to consider the way in which risk information is communicated. Common pitfalls include unrealistic levels of precision in estimates of risk and the portrayal of a zero-risk option. Risk information has often required interpretation by middle management before use by senior decision-takers. It is important that communication between risk experts and decision-takers is appropriate: there needs to be a common understanding of the precise meaning in a particular situation of terms such as significance and inference.

Just as there are calls for closer public involvement in EIA, there are those who advocate communication between the risk expert and the public. It is wrong to believe that public consultation exercises to inform after decisions have already been made will suffice. Unfortunately, all too often an assumption is made that the 'expert is right' and that there need not be a dialogue to ascertain the public's risk perceptions. A more open approach is to create a dialogue that ascertains what the public already knows about a risk and to take on board the public's insight and views on particular management options. In communicating risk it is important for risk experts to convey that there is no such thing as a risk-free world and that there are considerable uncertainties in scientific knowledge.

14.9 Concluding issues

"Risk assessment can avoid giving wrong answers, but it cannot give uniquely right answers" (Hrudey 1996).

Both ERA and EIA are similar forms of impact assessment. Whilst EIA currently remains the predominantly used tool for assessing the impacts of projects and proposals, not least because of its definition in legislation, it is clear that ERA has much to offer both as a supportive and as a complementary technique. It is often better at attempting to estimate the certainty, timing and magnitude of impacts than EIA. There are an increasing number of examples of ERA being used as part of an EIA (and vice versa) to provide information that is combined with information from other sources to contribute to an overall decision. Whilst there are still sceptics of ERA as a tool in its own right, this should not be through ignorance of its usefulness. This chapter has sought to demonstrate that, as long as the assumptions and limitations of ERA are made transparent, then it can be a very useful and credible tool to assist the decision-taker. As practice increases and the benefits are realised, then ERA should become more widely accepted. Traditionally risk assessment has been regarded as a highly quantitative tool, fraught with uncertainties and costly (see Thomas 1996). However, this chapter demonstrates current thinking on ERA tools ranging from relatively simple checklists and matrices to more complex models tailored to specific problems. It is therefore important not to disregard the application of ERA on grounds of cost: it is a highly adaptive and flexible tool.

Table 14.5 Some issues for EIA/ERA cross-fertilisation

Issue	EIA	ERA
Objective process	Development need: EIS reviews often give a high score to grammatical and procedural elements of a report rather than objectively assessing the technical credibility	Considerable experience: although not professing to be a very objective process, scientific information is considered systematically
Recognition of uncertainties	Development need: many EISs profess that 'all will be well'. May contain unqualified statements about the effectiveness of new technologies for mitigation	Considerable experience: consideration of uncertainty is fundamental to risk assessment
Consideration of alternatives	Considerable experience: implicit that development alternatives are considered early in the process	Development need: more consideration could be given to consideration of alternatives early in the process
Public involvement	Development need: calls for public participation in the EIA process	Considerable experience: enormous literature on the value of, and procedures for, evaluating risk perception and communicating risk
Strategic levels of appraisal	Considerable experience: theory and some practical examples of the EIA process at policy, programme and plan levels	Development need: considerable potential to translate what has been learned in Strategic EIA to Strategic ERA

It is also important to appreciate that ERA and EIA have developed largely in parallel and isolation. There is great scope for cross-fertilisation of experience and procedures between the two processes (Petts & Eduljee 1994). Since the concepts are so similar there is, for example, much that EIA practitioners can learn from ERA and risk assessment in general (Table 14.5). Not least, learning lessons might arise from a wider acceptance that uncertainty is a fact of life and that risks perceived by the public may be just as important as 'real risks'.

References

Brookes A 1999. Environmental impact assessment for water projects. In *Handbook of environmental impacts assessment*, Vol. 2. J Petts (ed.), 404–430. Oxford: Blackwell Science.

Brookes A, P Downs & K Skinner 1998. Uncertainty in the engineering of wildlife habitats, water and environmental management. *Journal of the Chartered Institute of Water and Environmental Management* 12(1), 25–29.

Calow P (ed.) 1998. *Handbook of environmental risk assessment and management*. Oxford: Blackwell Science.

Carpenter RA 1995. Risk Assessment. In *Environmental and social impact assessment*, F Vanclay & DA Bronstein (eds), 193–219. Chichester: Wiley.

De Jongh P 1988. Uncertainty in EIA. In *Environmental impact assessment: theory and practice*, P Wathern (ed.), 62–84. London: Unwin Hyman.

DETR 2000. *Guidelines for environmental risk assessment and management*. London: TSO.

DoE 1995. *A guide to risk assessment and risk management for environmental protection*. London: HMSO.

EA (Environment Agency) 1997a. *A guide to the National Centre for Risk Analysis and Options Appraisal*. Bristol: Environment Agency.

EA 1997b. *Road transport and the environment, Risk profile No.1*. Bristol: Environment Agency.

Harrop DO & SJT Pollard 1998. Quantitative risk assessment for incineration: is it appropriate for the UK? *Journal of the Chartered Institute of Water and Environmental Management* **12**(1), 48–53.

Holling CS 1978. *Adaptive environmental assessment and management*. Chichester: Wiley.

Hrudey SE 1996. *A critical review of current issues in risk assessment and management*, Eco Research Chair in Environmental Risk Management, published as a paper by Environmental Health Program, Department of Public Health Sciences, University of Alberta, Canada, 16pp.

Kaplan S & B Garrick 1981. On the quantitative definition of risk. *Risk Analysis* **1**, 1–27.

King NJ 1998. Application of risk assessment in policy and legislation in the European Union and in the United Kingdom. In *Handbook of environmental risk assessment and management*, P. Calow (ed.), 249–260. Oxford: Blackwell Science.

Petts J & G Eduljee 1994. *Environmental impacts assessment for waste treatment and disposal facilities*. Chichester: Wiley.

Pollard SJ, DO Harrop, P Crowcroft, SH Mallett, SR Jeffries & PJ Young 1995. Risk assessment for environmental management: approaches and applications. *Journal of the Chartered Institute of Water and Environmental Management* **9**, 621–628.

Royal Society 1992. *Risk analysis, perception and management*. London: Royal Society.

Suter GW II (ed.) 1993. *Ecological risk assessment*, Boca Raton, Florida: Lewis.

Thomas I 1996. *Environmental impacts assessment in Australia: theory and practice*. New South Wales: The Federation Press.

USEPA (US Environmental Protection Agency) 1992. *Framework for ecological risk assessment*, Risk Assessment Forum, Report EPA/630/R-92/001. Washington, DC: EPA.

World Commission on Environment and Development 1987. *Our common future, The Brundtland Commission*. Oxford: Oxford University Press.

Wright NH 1993. *Development of environmental risk assessment (ERA) in Norway*. Norway: Norske Shell Exploration and Production.

15 Environmental Remote Sensing (RS)

Kelly P Davis and John Lee

15.1 Introduction

Campbell (1996) suggests that:

> Remote sensing is the practice of deriving information about the Earth's land and water services using images acquired from an overhead perspective, using electromagnetic radiation in one or more regions of the electromagnetic spectrum, reflected or emitted from the earth's surface.

Over the last twenty-five years, remote sensing (RS) has allowed the collection of great quantities of environmental information, including data from previously inaccessible areas, in a systematic and reliable manner. In addition to optical images, RS extends the way humans can see substances on the earth by using parts of the electromagnetic spectrum which go beyond our normal vision.

Remote sensing data can be obtained using three types of platforms: ground based, aircraft and satellites. **Aerial photography** has a long history of use in providing overviews of landscape, and details of patterns and features that cannot be readily or speedily ascertained by traditional surveys. Most modern aircraft can be equipped with sensors or conventional cameras. **Ground-based instruments**, such as spectrometers, are providing scientists with reflected radiation information that help create spectral libraries for many plant species and mineral types. These libraries are used to assist with classifying ortho-imagery and satellite data. **Satellite imagery** is a relatively recent development that has revolutionised the ability to scan large areas repeatedly. New satellites are providing data that have spatial resolution under 1 metre, making remote sensing a valuable tool in both quantitative as well as qualitative studies.

Airborne and satellite data can both have important roles in EIA and SEA. They can be used to produce Digital Terrain Models (DTMs) (§16.2.3) and to obtain and interpret information on geology, geomorphology, soils, hydrology, vegetation, land cover and land use.

15.2 Definitions and concepts

This section only provides an overview of the basic concepts of RS. Readers interested in a more in-depth understanding of the subject are referred to texts such as Barrett & Curtis (1992), Campbell (1996), Danson & Plummer (1995), Lillesand & Kiefer (1994) and Vincent (1997).

15.2.1 Electromagnetic energy

The source of remotely sensed information is **electromagnetic energy** whose fundamental makeup is the **photon**. This subatomic particle comprises radiation given off
by matter when it is excited thermally, by nuclear processes, or by other radiation
(Short 1998). On earth, the main supply of photons is solar radiation from the sun.

Electromagnetic energy travels through space in **waves**. The size of each wave
can be measured by calculating the distance between successive crests, known as
wavelength. Waves may also be expressed as the total number of equivalent crests
that pass a reference point in a second. Known as **frequency**, this measurement has
a direct correlation with wavelength. As wavelength decreases, frequency increases.
The distribution of all radiation incident on the earth can be plotted using wavelength or frequency in a chart called the **electromagnetic spectrum** (ES).

The human eye can gather information from the visible portion of the ES. The
extent of our vision is limited to blue (0.4–0.5 µm), green (0.5–0.6 µm) and red
(0.6–0.7 µm). Remote sensing instruments can extend our human limitation by
collecting information below the blue portion of the ES into **ultraviolet** (0.3–
0.4 µm) and smaller to **X-rays**. Likewise, wavelengths larger than red light such as
near infrared (0.7–3.0 µm), known as NIR, and **far infrared** (c.3–100 µm), known
as FIR, can be recorded and provide valuable information on vegetation patterns.
Beyond FIR, remote sensing devices collect data in the microwave portion of the
ES. **Microwaves** can be broken down into two unique sections: passive and active
(see §15.2.4). Active microwave remote sensing is becoming an attractive source
of data for scientists in many different disciplines. Its popularity rests in the ability
of microwaves to 'see through' clouds and as a result making it valuable in tropical
regions.

15.2.2 Photography and optical data

The images produced by both photography and optical sensing instruments rely
mainly on light reflected from the earth's surface. However, they may also utilise
wavelengths beyond the visible band, principally the near infrared. This is partly
because shorter wavelengths, such as blue, tend to be scattered by dust and water
vapour in the atmosphere. Consequently, images recorded with blue light energy
can appear blurred or hazy and are not ideal for detailed resource observation. Green
light is less affected by the atmosphere and provides useful images, especially of
vegetation. The same is true with red light, and many studies use both green and red
energy in combination with wavelengths beyond visible light (e.g. infrared) to give
a better overall understanding of vegetation trends.

Photography (the earliest form of RS) provides images by means of a camera and
photographic film. A large selection of film types are available, each exploiting
different sections of the electromagnetic spectrum. The most common and inexpensive of these is black-and-white panchromatic film that has a spectral sensitivity
covering most of the visible portion of the ES. In general, however, colour photographs allow better discrimination of features such as soil, rock and vegetation
types. The limiting factor may be the costs associated with colour photography over
panchromatic, which can be considerable.

Black-and-white infrared film, originally developed by the military to separate
camouflage from the natural surroundings, has been widely used for vegetation studies.

Again, the more expensive option is infrared colour photography, which is "less affected by atmospheric haze than colour photography and can allow clearer discrimination of vegetation types and soil moisture variations" (Ellison & Smith 1998).

The principal platform for photography is still aircraft, although Russian satellite photography is available. This ranges in ground resolution from 3 m to 30 m and can be used at scales up to 1:5000.

Optical sensors convert electromagnetic radiation into a digital signal that can be recorded and subsequently displayed. The most common form is a Multi Spectral Scanner (MSS). An oscillating mirror focuses energy through a filter which separates the different bands of information (blue, green, red, etc.). The light passing through the filters reaches detectors where it is digitally recorded. The resulting **optical data** must be processed by an image analyser to produce the image. MSS has some inherent advantages over photography, particularly the ability to collect information from a larger portion of the ES (0.3–14 μm rather than the 0.3–0.9 μm available with photographic film).

The ability of sensors to collect and transmit information in digital format has proved very successful on space-borne platforms. It is possible for sensors to collect data over a single area for a prolonged period of time or, conversely, to collect global data for the lifetime of the satellite. This depends on orbital characters, which vary from satellite to satellite. The ability to continually collect or 'revisit' an area can be useful in hazard monitoring such as for forest fires or flooding.

The most widely used optical data are those provided by the American Landsat and French SPOT satellites. The only optical instrument carried in early Landsats was MSS, which has limited spectral bands and low (80 m) spatial resolution, but the current satellite, Landsat-7, also carries the Extended Thematic Mapper (TM). This has a higher-resolution 15 m panchromatic band and seven spectral channels covering the visible and NIR. It also has a thermal sensor, for one band in the FIR, allowing emitted heat energy from the earth's surface to be recorded at 60 m spatial resolution (§15.2.3). TM data can be used at scales up to 1:50,000.

The French SPOT ("Systeme Propatoire de l'Observation de la Terre") satellite has two main advantages over Landsat:

- It has better ground resolution, of 10 m for one band in the visible (which can be used at scales up to 1:10,000) and of 20 m for two in the visible and one in the NIR.
- It can produce off-centre images, which allows the production of stereo images.

On the other hand, SPOT has reduced spectral coverage; with four bands it covers only part of the visible spectrum and only extends as far as the NIR.

The late 1980s and 1990s have seen an influx of new non-military satellites being launched for Earth observation. The Indian IRS series have sensors similar to four of the Landsat TM bands, and improved ground resolution of up to 5 m. Some Russian satellites have similar ground resolution but more limited spectral resolution. The Japanese JERS-1 sensor has good spectral resolution with additional bands in the IR.

Optical sensors can also be mounted on aircraft. These include:

- DETR's Airborne Thematic Mapper (ATM), which can achieve ground resolutions up to 1 m, using a similar spectral range to Landsat TM;

- NASA's Airborne Visible Infrared Imaging Spectrometer (AVIRIS), which has high resolution and a wide range of spectral bands;
- the Compact Airborne Spectrographic Imager (CASI), which also has a wide spectral range of 400–1000 nm (hyperspectral) and can gather information at a spatial resolution of 40 cm depending on altitude.

The Environment Agency (EA) has found the CASI useful for the following applications:

- land cover and vegetation visualisation and mapping;
- identification and tracking of dissimilar water bodies and mapping of mixing zones, e.g. in estuaries;
- detection and monitoring of water pollution, including suspended solids concentrations and eutrophication (by chlorophyll-a estimation);
- estimation of and changes in coastal morphology.

The success of CASI has prompted the development of CASI-2 which is an extremely compact and power-efficient instrument. This allows the unit to collect visible and near infrared information from aircraft, land vehicles and terrestrial-based platforms. CASI and CASI-2 are being used by government and educational institutes, private service companies, international space agencies, and the military.

The EA also employs an airborne Light Detection and Ranging system (LIDAR) which uses a laser to measure the distance between the aircraft and the ground. LIDAR applies the same principals as RADAR remote sensing. The device transmits light to a ground-based target. Radiation scattered by the target is collected by the instrument and processed to provide information about the target and/or the path to the target.

There are three types of LIDAR utilisation in remote sensing:

1. LIDAR rangefinder methods can be used to calculate distances that can be converted to accurate digital terrain models (DTMs);
2. Differential Absorption LIDAR (DIAL) uses two different laser wavelengths to record information about chemical concentrations;
3. Doppler LIDAR can be used to determine the velocity of an object.

The commonest application of LIDAR is as a tool for creating extremely accurate terrain maps, e.g. for assessing flood risk (EA 2000). However, it is also gaining popularity as a method for determining pollutants in the atmosphere and specifically, toxic emissions from large factory stacks.

15.2.3 Thermal imagery

Thermal imagery uses measurements of temperature and hence heat, and is commonly affiliated with infrared energy (IR). However, this is only partially true because there is a difference between Near Infrared (NIR) energy and Far Infrared (FIR) energy. NIR, like light, is energy that is initially part of incoming solar radiation, and is reflected from objects on the earth's surface. For example, inbound NIR energy is reflected by chlorophyll, making it an ideal portion of the spectrum to use for vegetation studies. In contrast, FIR is energy that has been absorbed by the earth, or a water body, and re-emitted as heat, which is why FIR is often referred to as thermal IR.

The absorption and re-emittance of FIR occurs on a daily cycle, e.g. during the day a lake absorbs sunlight and therefore appears dark on a FIR image. At night, however, the lake emits the stored energy and shows up very brightly, usually providing a clear view of the shoreline due to the comparatively colder land surrounding it.

The difference between the two forms of IR energy marks a distinct change in the way sensors record information, and thermal imagery, which measures temperature, strictly refers to FIR. Some satellites, such as Landsat, carry thermal sensors, but airborne sensors, such as NASA's Thermal Infrared Multi spectral Scanner (TIMS), can provide more information because it is nearer to the ground and is designed specifically for recording thermal data in six separate channels.

During the past decade thermal imagery has moved from being military data only, to a valuable information source for environmental applications. The most popular of these is monitoring of open waters and near-surface waters. Temperature patterns associated with variations in ground moisture can be easily found, making the data useful for detecting soil moisture gradients, spring lines, leaks in pipelines or water-bore pollution. Thermal pollution in open waters from large factories and refineries is another practical application for thermal information.

15.2.4 Radar

In recent years the use of microwave energy has added a new dimension to remote sensing. For the purposes of RS, the use of microwaves is separated into two major divisions, passive and active. **Passive microwave remote sensing** concentrates on the emission of microwave energy from the earth. In this respect, it is has similarities to FIR as both record temperature information. In contrast, **active microwave remote sensing**, or radar (RAdio Detecting And Ranging) uses artificially created energy produced by the RS instrument itself. This usually produces pulses of microwaves at a predefined wavelength. The pulses are reflected by objects on the ground and recorded as digital information by the sensor. Radar can provide images during the day or night, and is unaffected by weather conditions because it can penetrate cloud cover. This is particularly valuable in areas such as the rain forests where cloud free days are rare.

One of the main conditions affecting the response of radar signals is surface roughness. In general, the rougher the surface, the higher the backscatter and the brighter that portion of the image will be. The angle at which the energy strikes the ground, called incidence angle, also plays an important role in the backscatter of microwave pulses. Generally, as the incidence angle increases, so does the expected backscatter. Consequently, radar can be used to provide information on surface roughness and topography, and to produce detailed DTMs. Backscatter is also influenced by the ability of substances to conduct electrical energy; and since this includes water, radar can be very beneficial to projects focusing on the presence of water in vegetation or soils.

Four satellites are currently providing daily coverage of the earth's surface, using synthetic aperture radar (SAR) sensors. These are the European ERS-1 and ERS-2, the Canadian Radarsat and the Japanese JERS-1. Radarsat is the most flexible system; it can provide data at various spatial resolution (9 m to 100 m) and has the ability to change the incidence angle (between 10° and 60°). Other recent radar missions include the Shuttle Radar Topography Mission (SRTM) that used two radar systems mounted on the space shuttle and a technique called radar interferometry to create a detailed DTM for the entire planet.

Airborne radar sensors are also used, the most comprehensive coverage currently being that provided by the commercial firm Intera (Ellison & Smith 1998). Airborne radar can be broken down into two groups: Real Aperture Radar (RAR) and Synthetic Aperture Radar (SAR). Real aperture radar is often associated with Side Looking Airborne Radar (SLAR), which is a sensor used for displaying back-scatter from surficial objects. SLAR works by emitting and receiving microwave energy from a high-powered antenna mounted on the aircraft. As the antenna size is increased, the spatial resolution is improved. To achieve greater spatial resolution, without having an antenna too large to mount on an aircraft, synthetic aperture radar systems have been designed. A small SAR antenna can achieve the same or greater spatial resolution as a SLAR, with less power requirements. This makes SAR ideal for space-borne platforms also.

15.2.5 Satellite orbits and new developments

There are two types of orbit, equatorial and polar. Satellites in equatorial orbits circle the earth near the plane of the equator. They are often referred to as geostationary because their orbital period is equal to the rotating period of the earth, and so the orbiting sensors appear to be motionless. Unfortunately, to achieve this the satellite must maintain an altitude of 35,000–36,000 km, which severely degrades spatial resolution. Consequently, they are mainly meteorological and communication satellites.

RS satellites are usually in **polar orbit**, circling the earth from pole to pole and hence at right angles to the earth's rotation. This has the following advantages:

- By offsetting the orbit slightly (oblique to the lines of longitude) the local sun time of each point along the orbital track will be the same (Legg 1992). Most RS satellites, including Landsat and SPOT, are in this type of sun-synchronous orbit, and those with optical sensors collect information between 9:30 am and 11:00 am, which is normally the period of least cloud cover.
- The earth's rotation at right angles to the orbit of the satellite allows complete coverage of the earth's surface during consecutive satellite orbits.
- Polar orbiting satellites usually travel at relatively low altitudes (between 700 km and 900 km) which allows better spatial resolution, ranging from approximately 10 m to 30 m.

New high-resolution systems will orbit at altitudes as low as 300 km, similar to the space shuttle, providing sub-metre resolution. Unfortunately, atmospheric drag becomes a serious problem at such low altitudes, and the life span of 'low orbiters' will be limited by the need to maintain altitude, and hence to carry fuel which will eventually burn out.

New breeds of high-resolution satellites are sure to make an impact on earth observation in the future. Unlike traditional missions, these are run by private-sector consortia. The spatial resolutions of these systems make them competitors for aerial photography. Satellites such as IKONOS can provide images that should be able to distinguish objects as small as 4 m. Other projects such as Quickbird and Orbview-3 are promising spatial resolutions of less than 2 m. This is very exciting for those interested in using remotely sensed data for environmental applications.

15.3 Sources of remote sensing information, software and data

Examples of organisations that provide RS information, software and data are given in Table 15.1. Examples of digital maps using RS data are listed in Table 16.1, p. 384. Landsat TM images of the UK, digitally registered to the National Grid, are available from the Remote Sensing Group at BGS.

Table 15.1 Organisations providing remote sensing information, software and data

Aerial photographs

UK Local Authorities and **Government agencies** (usually the 'Air Photographs Officer/Unit at'), e.g. those listed in Appendix B. Organisations such as RCAHMS, RCAHMW and RCHME. Photographs can usually be examined at the relevant offices, and some organisations are willing to make searches. **Commercial firms**, some of which offer digitally scanned photographs at 1:10 k scale, with display at up to 1:2 k (see NAPLIB 2000).

Government agencies of other countries, e.g. ERIN, USDA-NRCS, USGS-EROS

Satellite imagery

General information:
Canadian Centre for Remote Sensing (CCRS) – http://www.ccrs.nrcan.gc.ca/ccrs/
ITC Remote Sensing Tutorial Pages – http://www.itc.nl/pors/
National Remote Sensing Centre, UK (NRSC) – http://www.nrsc.co.uk
Remote Sensing Frequently Asked Questions (RSFAQ) –
    http://www.geog.nottingham.ac.uk/remote/satfaq.htm
Remote Sensing Speciality Group (RSSG) –
    http://www.earthsensing.com/rssg/web_rsrc.html
The Remote Sensing Tutorial – http://www.rst.gsfc.nasa.gov/
Technical Research Centre of Finland (VTT) remote sensing group –
    http://www.vtt.fi/aut/rs/virtual/

Online sample imagery:
Carterra Image Archive – http://www.mapserver.esri.com/si/html/main.htm
European Space Agency (ESA) – http://www.shark1.esrin.esa.it/
Jet Propulsion Laboratory (JPL) – http://www.jpl.nasa.gov/earth/land/
Johns Hopkins University AVHRR Image Gallery –
    http://www.fermi.jhuapl.edu/avhrr/gallery/index.html
MIT + MassGIS Orthophoto Project – http://www.ortho.mit.edu/nsdi/index.html
SPIN-2 High resolution imagery – http://www.spin-2.com/
Terra Images – http://www.earthobservatory.nasa.gov/Newsroom/NewImages/
    images_index.php3

Data providers:
Australia Centre for Remote Sensing (ACRES) –
    http://www.auslig.gov.au/acres/index.htm
British Geological Survey (BGS) – http://www.bgs.ac.uk

Table 15.1 (*continued*)

Canadian Centre for Remote Sensing – http://www.ccrs.nrcan.gc.ca/
China Remote Sensing Satellite Ground Station – http://www.rsgs.ac.cn/
Danish Centre for Remote Sensing (DCRS) – http://www.emi.dtu.dk/research/DCRS/
Engesat Imagens de Satélite, Brazil (ENGESAT) – http://www.engesat.com.br/
German Remote Sensing Data Center – http://www.dfd.dlr.de/
Indian Space Research Organisation (ISRO) – http://www.isro.org/sat.htm#irs
National Remote Sensing Agency, India (NRSA) – http://www.nrsa.gov.in/
National Remote Sensing Centre, UK (NRSC) – http://www.nrsc.co.uk
National Space Development Agency, Japan (NASDA) –
   http://www.nasda.go.jp/index_e.html
Netherlands Earth Observation Network – http://www.neonet.nl/
Radarsat, Canada – http://www.rsi.com
Russian Space Agency – http://www.arc.iki.rssi.ru/
South Africa Satellite Application Centre – http://www.sac.co.za/
TELSAT (Belgium) – http://telsat.belspo.be/
USGS-EROS – http://edcwww.cr.usgs.gov/eros-home.html

Satellite information
ERS (Europe) – http://www.earth1.esrin.esa.it/ERS/
Ikonos – Space Imaging (USA) – http://www.spaceimaging.com/
Landsat 7 – NASA (USA) – http://www.landsat.gsfc.nasa.gov/
Orbview – Orbimage (USA) – http://www.orbimage.com/
Radarsat – Radarsat International (Canada) – http://www.rsi.ca/
Spot – Spot Image (France) – http://www.spotimage.fr/
Terra – NASA (USA) – http://www.terra.nasa.gov/

15.4 Applications of remote sensing with particular reference to EIA

15.4.1 Introduction

Remote sensing has been applied at a range of spatial and temporal scales reflecting its diversity. Uses range from the monitoring of global climatic change (e.g. Taylor 1996), by, for example, measuring the extent of polar pack-ice (Piwowar & Ledrew 1995), to the assessment of the habitat requirements of individual species (e.g. Austin *et al.* 1996).

RS allows for the possibility of gathering data over large areas in a relatively short time frame which compares favourably with time-consuming and therefore expensive fieldwork, although this is often carried out to supplement or verify the RS data. Many applications use ground-based sampling and/or GIS in conjunction with RS, and GIS generated DTMs are often especially useful (Chapter 16).

RS is being used increasingly in EIA, particularly in the scoping, baseline and monitoring stages – although RS data integrated with a GIS can also be used to model potential scenarios as an aid to impact prediction and mitigation.

There are differences in the suitability of airborne or satellite data in relation to the aims and scales of particular studies, including those undertaken in EIA. Some of these differences are listed in Table 15.2, from which it can be seen that, in general, satellite data are most useful for SEA and linear projects, whilst airborne

Table 15.2 Comparative features of airborne data and **satellite data (shaded boxes)**

Can be taken at given times of day or season, and bad weather can be avoided.

Times and dates are predetermined, and cloud cover can impede optical or thermal images.

Sensors can be readily changed and configured.

Configuration of the sensors is predetermined.

Ground resolution can be varied by flying at different altitudes.

Ground resolution is predetermined.

Images are susceptible to distortion because at low altitudes (a) a wide viewing-angle is needed to image reasonably large areas, (b) the platform is relatively unstable in most weather conditions. However, digitally scanned aerial photographs can be corrected by computer so that they can be accurately overlaid on conventional 1:10 k scale maps.

Images are much less susceptible to distortion because satellites orbit at very high altitude and hence (a) can use a small viewing-angle to image a large area, (b) provide a stable platform because they are above the earth's weather systems. Consequently they can provide consistent regional data sets that can be readily corrected to map systems such as the UK national grid (Ellison & Smith 1998).

Images are relatively inexpensive, and can be selected for specific small areas.

Images are more expensive, and selection of specific small areas is less available.

Can provide very high spatial resolution data that allow good discrimination (e.g. between different soil, rock or vegetation types) and are particularly suitable for detailed studies of sites or small areas including impact areas of site-based projects.

Very high spatial resolutions are not yet available, but data provide a synoptic view of large areas that is particularly suitable for SEA or assessing linear projects (e.g. roads, railways and pipelines) and linear features (e.g. coastlines).

data are likely to be more useful for EIAs of non-linear projects. The application of airborne and satellite RS to habitat and vegetation mapping is reviewed by Pooley & Jones (1996).

15.4.2 Linear projects

In a survey of the ecological component of British EISs, Thompson *et al.* (1997) identified 12 single-category development types and two mixed categories. The most common of the former were **road developments** which, due to their linear extent, present particular problems to the EIA process. Roads affect habitats at a range of spatial scales, and usually impinge upon a number of habitats across a region as well as fragmenting individual habitat patches and as acting as barriers to species' movement (Mader 1984). The impact of roads is thus difficult to assess using conventional techniques. Consequently, it is generally only the chosen route that is investigated rather than any of the alternative proposals. However, since data can be gathered for large areas in a short time, RS can be used to predict the impacts of a proposed road development, and alternative routes can be compared (e.g. in relation to the locations of known habitats) to assess the one which has the least cost to the environment (Rao 1996, Treweek & Veitch 1996).

Other types of linear development pose similar problems in EIA because of their large spatial extent. In particular, **pipeline construction** commonly extends over international boundaries, which presents difficulties not only with the scale of the development but also with the co-ordination and logistics of ground-based surveys. Remote-sensed imagery can be used for impact prediction along a series of proposed routes by considering the effects on land use, topography, geology, hydrology and existing infrastructure (Feldman *et al.* 1995). Finer-scale RS can be used for monitoring the effectiveness of mitigation measures for linear projects, for example in the monitoring of the re-vegetation of a pipeline route by videography (Um & Wright 1996, 1998).

15.4.3 Coastal zone studies

The coastal zone has a similar linearity to that of roads and pipelines and poses the associated problems with measurement and monitoring. Coastal processes occur at large spatial scales and over both short and long temporal scales, and consequently, conventional monitoring techniques are difficult to apply. Furthermore, the environment often precludes direct measurement of the seabed and inferences must be made based on both terrestrial and marine surface conditions (Chapter 13). Given the increasing pressure on the coastal zone for both commercial and recreational uses, orbital RS (particularly in the visible and IR spectral regions) is being used to monitor the effects of a range of impacts such as fluvial discharges, marine processes, weathering and waste disposal on land and at sea (Barale & Folving 1996).

Remote sensing can monitor both long- and short-term changes at the coast. For example, historical aerial photographs can be used to measure changes to **coastal geomorphology** resulting from variation in the long-term sediment balance, and satellite imagery can be used in conjunction with current nautical charts and tide tables to detect gross sediment transport (Culshaw 1995). More dynamic water-based processes can be analysed by studying the evolution of temperature patterns and their relationship to the spatial distribution of plankton or to currents (Barale & Folving 1996, Baban 1997). The seabed form and the patterns of water movement are inter-related and strongly influence biological diversity in the coastal zone. RS can be used to monitor the effects of developments by mapping the **marine biodiversity** (Davies *et al.* 1997).

Estuaries can also be monitored for changes in their sedimentation regime. Both the gross sediment load (Culshaw 1995) and more subtle indices such as the distribution of suspended solids, turbidity, temperature, salinity, and amounts of chlorophyll and phosphorous (Baban 1997) can be measured to check the health of an estuary.

Some coastal developments have a direct effect on smaller sections of the coastline, but RS is still an invaluable data gathering tool due to the extensive nature of coastal processes. For example, the emission into the sea from nuclear power stations can be measured using high-resolution visible and absorption IR imagery to detect the suspended sediment levels and surface debris associated with the thermal discharge (Davies & Mofar 1993). For submarine pipelines, the usefulness of RS in gathering data from a large inaccessible area is of primary importance (Yuksel *et al.* 1995) and for proposed large-scale waste-water disposal schemes RS enables a wide area to be considered cost effectively and with benefits to the regional environment (Gonenc *et al.* 1995).

15.4.4 Freshwater studies

In addition to their large spatial extent, RS is particularly useful for measuring coastal zone waters because of their inaccessibility to measurement. Freshwater bodies such as lakes, reservoirs and waste-water stores are similarly difficult to measure other than by time consuming and costly direct sampling. For example, RS can be used to:

- measure waste-water bodies, which can be classified by combining RS with laboratory analysis of water samples (Braude *et al.* 1995);
- estimate the volumes of waste water held in reservoirs (Rigol & Chica-Olmo 1998);
- monitor reservoirs for effluent quality by studying water transparency and colour due to the quantities of organic matter, suspended solids and chlorophyll in the water (Dor & Benyosef 1996, Oron & Gitelson 1996, Gitelson *et al.* 1997);
- assess macrophyte distributions in reservoirs (Malthus & George 1997);
- target the optimum location for mitigation measures such as establishing riparian buffer zones of vegetation to control runoff (Narumalani *et al.* 1997).

The estimation and monitoring of **runoff** from RS imagery is a useful index of land-use change. Remote sensing enables large tracts of land to be sampled and land-use data can be combined with a DTM generated from contour or spot height data within a GIS. Estimates can be made for the amount of runoff from a single grid cell (the size of which will depend on the image resolution) of any land-use type and the direction and cumulative runoff can be calculated from the DTM (Drayton *et al.* 1992). Similar methods can be applied to the movement of subsurface water or in pipeline construction projects (Florinsky 1998).

Runoff is partly a function of the quantity of **soil moisture**, which can be estimated from RS using SAR (§15.2.4) imagery (Ragab 1995, Griffiths & Wooding 1996). Similar techniques can be applied to **watercourses** whereby field sampling is used in conjunction with remotely-sensed land-use data and climatic, hydro-geological and geomorphological data to monitor their chemical composition (Simpson *et al.* 1993). Aerial photography is also used in the US for rapid assessment of the condition (in terms of hydrology, soils and biology) of riparian-wetland corridors (river channels and **floodplains**) in large areas (USDA-NRCS 1999).

15.4.5 Land use and land cover studies

Landfills have been extensively classified and their status assessed using current and historical aerial photographs (Barnaba *et al.* 1991, Pope *et al.* 1996), visible-FIR imagery (Johnson *et al.* 1993) and multi-spectral images and DTMs (Vincent 1994). This classification allows for the prediction of impacts from new or expanding sites, but more specific scoping is possible, such as the analysis of the surrounding geology for water-bearing fractures (Frohlich *et al.* 1996). Monitoring of post-development conditions is also possible either by direct observation of old subsurface structures which may have held oils (Rugge & Ahlert 1992) or by identifying areas of surface vegetation chlorosis or dieback due to movements of landfill gases towards the root zone (Jones & Elgy 1994).

The scoping and monitoring of **wastelands** is important since they are commonly the site of development projects. Wasteland can be identified with thermal RS as having relatively high soil moisture content and associated low temperature (Irvine *et al.* 1997). Remote sensing is also ideally suited to mapping and classification of wasteland sites at regional scale (Rao *et al.* 1991, Foody & Embashi 1995, Zellmer & Eastman 1997), enabling the targeting of sites for remediation. Monitoring of hazardous sites can be undertaken using high-resolution, multi-temporal video RS and colour and IR photography (Marsh *et al.* 1991) and thermal-IR imagery and ground-penetrating radar (Weil *et al.* 1994).

In the **rural–urban fringe** remotely-sensed mapping is especially useful, since information becomes rapidly outdated (Kenny 1996). This allows for scoping over large areas in zones where there is conflict between the demands of urban or industrial development, and those of agriculture or nature conservation (Li & Yeh 1998). Urban expansion is generally complex, rendering precise impact prediction difficult. However, data acquisition over large areas and for different time periods allows for the development of transport plans and predictions of land requirements for expansion (Pathan *et al.* 1993). The expansion of urban areas can be monitored by analysis of the heat island effects using thermal RS (Lo *et al.* 1997).

RS is widely applicable to **land use and land cover mapping**, and standard classifications and datasets have been used in national and pan-continental applications (see Table 16.1, p. 384 and Appendix F.6, p. 452). However, there are problems in using non-customised imagery because land classification schemes:

- may be biased towards a particular area different from the one under investigation (Cruickshank & Tomlinson 1996);
- usually consist of relatively few categories that can be readily identified using RS data, and mapped at low resolutions; and these often correspond poorly with habitats defined in habitat classifications (Appendix F.6).

Furthermore, it is essential that the RS data are at a scale both temporally (frequency of repeat observations) and spatially (both extent and resolution) appropriate to the subject under investigation.

Changes in land cover over large areas, and in response to developments, can be effectively monitored by RS (Dimyati *et al.* 1996, Gillespie *et al.* 1996, Sommer *et al.* 1998). Indeed, using Landsat data to monitor land cover change is a major component of the UK CS90 and CS2000 surveys (Table 16.1). Similarly, the state of crops can be monitored by visible and IR imagery and related to developments adjacent to the agricultural areas (Akiyama *et al.* 1996, Kondratev *et al.* 1996). Crop health can be adversely affected by pollution from both industrial and agricultural development, and this may manifest itself in irrigation waters which can be monitored by multi-temporal RS (Abderrahman & Bader 1992, Adinarayana *et al.* 1994).

15.5 Conclusion

Remote sensing has been used in EIA for a range of development types. Its usage will increase as imagery becomes more widely available and at a low price. RS allows data to be collected for large tracts of the earth's surface in a short period of time. Additionally, data can be gathered from otherwise inhospitable or inaccessible environments such as seas, lakes and mountainous areas. Care must be taken, however, that appropriate scales and techniques are used which reflect the spatial and temporal extent of the development and its type. RS is a valuable tool at each stage of the EIA process and is especially valuable when used in conjunction with ground and GIS-based methods.

References

Abderrahman WA & TA Bader 1992. Remote-sensing application to the management of agricultural drainage water in severely arid region – a case-study. *Remote Sensing of Environment* **42**(3), 239–246.

Adinarayana J, JD Flach & WG Collins 1994. Mapping land-use patterns in a river catchment using Geographical Information Systems. *Journal of Environmental Management* **42**(1), 55–61.

Akiyama T, Y Inoue, M Shibayama, Y Awaya & N Tanaka 1996. Monitoring and predicting crop growth and analyzing agricultural ecosystems by remote-sensing. *Agricultural and Food Science in Finland* **5**(3), 367–376.

Austin GE, CJ Thomas, DC Houston & DBA Thompson 1996. Predicting the spatial distribution of buzzard *Buteo buteo* nesting areas using a Geographical Information System and remote sensing. *Journal of Applied Ecology* **33**(6), 1541–1550.

Baban SMJ 1997. Environmental monitoring of estuaries estimating and mapping various environmental indicators in Breydon Water Estuary, UK, using Landsat TM imagery. *Estuarine, Coastal and Shelf Science* **44**(5), 589–598.

Barale V & S Folving 1996. Remote-sensing of coastal interactions in the Mediterranean region. *Ocean & Coastal Management* **30**(2–3), 217–233.

Barnaba EM, WR Philipson, AW Ingram & J Pim 1991. The use of aerial photographs in county inventories of waste-disposal sites. *Photogrammetric Engineering and Remote Sensing* **57**(10), 1289–1296.

Barrett EC & LF Curtis 1992. *Introduction to environmental remote sensing*. London: Chapman & Hall.

Braude C, N Benyosef & I Dor 1995. Satellite remote-sensing of waste-water reservoirs. *International Journal of Remote Sensing* **16**(16), 3087–3114.

Campbell JB 1996. *Introduction to remote sensing*. London: Taylor & Francis.

Cruickshank MM & RW Tomlinson 1996. Application of CORINE land cover methodology to the UK – some issues raised from Northern Ireland. *Global Ecology and Biogeography Letters* **5**(4–5), 235–248.

Culshaw ST 1995. A visual interpretation of an ERS-1 SAR image of the Thames Estuary. *Journal of Navigation* **48**(1), 97–104.

Danson FM & SE Plummer (eds) 1995. *Advances in environmental remote sensing*. Chichester: Wiley.

Davies J, R Foster-Smith & IS Sotheran 1997. Marine biological mapping for environment management using acoustic ground discrimination systems and Geographic Information Systems. *Underwater Technology* **22**(4), 167–172.

Davies PA & LA Mofor 1993. Remote sensing observations and analyses of cooling water discharges from a coastal power-station. *International Journal of Remote Sensing* **14**(2), 253–273.

Dor I & N Benyosef 1996. Monitoring effluent quality in hypertrophic waste-water reservoirs using remote-sensing. *Water Science and Technology* **33**(8), 23–29.

Drayton RS, BM Wilde & JKH Harris 1992. Geographical Information-System approach to distributed modeling. *Hydrological Processes* **6**(3), 361–368.

Dimyati M, K Mizuno, S Kobayashi & T Kitamura 1996. An analysis of land use/cover change using the combination of Mss Landsat and land-use map – a case-study in Yogyakarta, Indonesia. *International Journal of Remote Sensing* **17**(5), 931–944.

EA 2000. *State of the Environment*. (http://www.environment-agency.gov.uk)

Ellison RA & A Smith 1998. *A guide to sources of earth science information for planning and development*. British Geological Survey Technical Report WA/97/85. London: HMSO. (http://www.bgs.ac.uk/)

Feldman SC, RE Pelletier, E Walser, JC Smoot & D Ahl 1995. A prototype for pipeline routing using remotely-sensed data and Geographic Information-System analysis. *Remote Sensing of Environment* **53**(2), 123–131.

Florinsky IV 1998. Combined analysis of digital terrain models and remotely sensed data in landscape investigations. *Progress in Physical Geography* **22**(1), 33–60.

Foody GM & MRM Embashi 1995. Mapping despoiled land-cover from Landsat Thematic Mapper imagery. *Computers Environment and Urban Systems* **19**(4), 249–260.

Frohlich RK, JJ Fisher & E Summerly 1996. Electric-hydraulic conductivity correlation in fractured crystalline bedrock – central landfill, Rhode-Island, USA. *Journal of Applied Geophysics* **35**(4), 249–259.

Gillespie MK, DC Howard, MJ Ness, RM Fuller 1996. Linking satellite and field survey data, through the use of GIS, as implemented in Great-Britain in the Countryside-Survey 1990 Project. *Environmental Monitoring and Assessment* **39**(1–3), 385–398.

Gitelson A, R Stark & I Dor 1997. Quantitative near surface remote sensing of wastewater quality in oxidation ponds and reservoirs: a case study of the Naan system. *Water Environment Research* **69**(7), 1263–1271.

Griffiths GH & MG Wooding 1996. Temporal monitoring of soil moisture using ERS-1 SAR data. *Hydrological Processes* **10**(9), 1127–1138.

Gonenc IE, O Muftuoglu, BB Baykal, E Dogan, H Yuce & M Gurel 1995. The Black-Sea factor influencing the waste-water disposal strategy for Istanbul. *Water Science and Technology* **32**(7), 63–70.

Irvine JM, TK Evers, JL Smyre, D Huff, AL King, G Stahl & J Odenweller 1997. The detection and mapping of buried waste. *International Journal of Remote Sensing* **18**(7), 1583–1595.

Johnson E, M Klein & K Mickus 1993. Assessment of the feasibility of utilizing Landsat for detection and monitoring of landfills in a Statewide GIS. *Environmental Geology* **22**(2), 129–140.

Jones HK & J Elgy 1994. Remote-sensing to assess landfill gas migration. *Waste Management & Research* **12**(4), 327–337.

Kenny FM 1996. Geographic Information-Systems and remote-sensing techniques in environmental assessment. *Geoscience Canada* **23**(1), 41–53.

Kondratev KY, PP Fedchenko & VA Grekov 1996. Determination of the state of agricultural crops by remote-sensing. *Earth Observation and Remote Sensing* **13**(5), 789–797.

Legg CA 1992. *Remote Sensing and Geographic Information Systems: geological mapping, mineral exploration and mining*. New York: Ellis Horwood.

Li X & AGO Yeh 1998. Principal component analysis of stacked multi-temporal images for the monitoring of rapid urban expansion in the Pearl River Delta. *International Journal of Remote Sensing* **19**(8), 1501–1518.

Lillesand TM & RW Kiefer 1994. *Remote sensing and image interpretation*. Toronto: Wiley.

Lo CP, DA Quattrochi & JC Luvall 1997. Application of high-resolution thermal infrared remote sensing and GIS to assess the urban heat island effect. *International Journal of Remote Sensing* **18**(2), 287–304.

Mader HJ 1984. Animal habitat isolation by roads and agricultural fields. *Biological Conservation* **4**, 82–90.

Malthus TJ & DG George 1997. Airborne remote sensing of macrophytes in Cefni Reservoir, Anglesey, UK. *Aquatic Botany* **58**(3–4), 317–332.

Marsh SE, JL Walsh, CT Lee & LA Graham 1991. Multitemporal analysis of hazardous-waste sites through the use of a new bi-spectral video remote-sensing system and standard color-ir photography. *Photogrammetric Engineering and Remote Sensing* **57**(9), 1221–1226.

NAPLIB (National Association of Aerial Photographic Libraries) 2000. *NAPLIB Directory of Aerial Photographic Collections in the United Kingdom*, 2nd edn. (http://www.naplib.org.uk/)

Narumalani S, YC Zhou & JR Jensen 1997. Application of remote sensing and geographic information systems to the delineation and analysis of riparian buffer zones. *Aquatic Botany* **58**, 393–409.

Oron G & A Gitelson 1996. Real-time quality monitoring by remote sensing of contaminated water-bodies: waste stabilization pond effluent. *Water Research* **30**(12), 3106–3114.

Pathan SK, SVC Sastry, PS Dhinwa, M Rao, KL Majumdar, DS Kumar, VN Patkar & VN Phatak 1993. Urban-growth trend analysis using GIS techniques – a case-study of the Bombay Metropolitan Region. *International Journal of Remote Sensing* **14**(17), 3169–3179.

Piwowar JM & EF Ledrew 1995. Hypertemporal analysis of remotely-sensed sea-ice data for climate-change studies. *Progress in Physical Geography* **19**(2), 216–242.

Pooley R & MM Jones 1996. Application of remote sensing to habitat mapping and monitoring. *Scottish Natural Heritage Review No. 57*. Battleby: SNH.

Pope P, E Vaneeckhout & C Rofer 1996. Waste site characterization through digital analysis of historical aerial photographs. *Photogrammetric Engineering and Remote Sensing* **62**(12), 1387–1394.

Ragab R 1995. Towards a continuous operational system to estimate the root-zone soil-moisture from intermittent remotely-sensed surface moisture. *Journal of Hydrology* **173**, 1–25.

Rao DP, NC Gautam & B Sahai 1991. Irs-1a application for wasteland mapping. *Current Science* **61**(3–4), 193–197.

Rao KML 1996. Regional network planning and the development of rural transportation using remote-sensing techniques. *International Journal of Remote Sensing* **17**(17), 3453–3466.

Rigol JP & M Chica-Olmo 1998. Merging remote-sensing images for geological-environmental mapping: application to the Cabo de Gata-Nijar Natural Park, Spain. *Environmental Geology* **34**(2–3), 194–202.

Rugge CD & RC Ahlert 1992. Ground and aerial survey of a peninsular landfill. *Water Research* **26**(4), 519–526.

Short NM 1998. (http://code935.gsfc.nasa.gov/Tutorial/Start.html)

Simpson PR, WM Edmunds, N Breward, JM Cook, D Flight, GEM Hall & TR Lister 1993. Geochemical mapping of stream water for environmental-studies and mineral exploration in the UK. *Journal of Geochemical Exploration* **49**(1–2), 63–88.

Sommer S, J Hill & J Megier 1998. The potential of remote sensing for monitoring rural land use changes and their effects on soil conditions. *Agriculture Ecosystems & Environment* **67**(2–3), 197–209.

Taylor FW 1996. Remote sensing of the Earth from space. *Contemporary Physics* **37**(5), 391–405.

Thompson S, JR Treweek & DJ Thurling 1997. The ecological component of environmental impact assessment: a critical review of British environmental statements. *Journal of Environmental Planning and Management* **40**(2), 157–171.

Treweek J & N Veitch 1996. The potential application of GIS and remotely sensed data to the ecological assessment of proposed new road schemes. *Global Ecology and Biogeography Letters* **5**(4–5), 249–257.

Um JS & R Wright 1996. Pipeline construction and reinstatement monitoring – current practice, limitations and the value of airborne videography. *Science of the Total Environment* **186**(3), 221–230.

Um JS & R Wright 1998. A comparative evaluation of video remote sensing and field survey for revegetation monitoring of a pipeline route. *Science of the Total Environment* **215**(3), 189–207.

USDA-NRCS 1999. *Assessing conditions of riparian-wetland corridors at the areawide level using proper functioning condition (PFC) methodology – an interdisciplinary assessment tool.* (Watershed Science Institute Technical Report). (http://www.geology.washington.edu/~nrcs-wsi/)

Vincent RK 1994. Remote-sensing for solid-waste landfills and hazardous-waste sites. *Photogrammetric Engineering and Remote Sensing* **60**(8), 979–982.

Vincent RK 1997. *Fundamentals of geological and environmental remote sensing.* Upper Saddle River: Prentice-Hall.

Weil GJ, RJ Graf & LM Forister 1994. Investigations of hazardous-waste sites using thermal ir and ground-penetrating radar. *Photogrammetric Engineering and Remote Sensing* **60**(8), 999–1005.

Yuksel Y, D Maktav & S Kapdasli 1995. Application of remote-sensing technology to the relation of submarine pipelines and coastal morphology. *Water Science and Technology* **32**(2), 77–83.

Zellmer JT & SM Eastman 1997. Incorporation of historical aerial photographs and land-use information into environmental site assessments. *Environmental & Engineering Geoscience* **3**(3), 431–441.

16 Geographical Information Systems (GIS) and EIA

Agustin Rodriguez-Bachiller and Graham Wood

16.1 Introduction

Geographical Information Systems (GIS) are databases with powerful mapping capabilities and, for this reason among others, they are becoming increasingly associated with environmental studies of all kinds, including EIA. The definition of GIS has been the subject of some debate (Maguire 1991), and although GIS can be simply described as databases where the information is spatially referenced, what has made GIS so popular is the fact that the spatial referencing of information is related to 'maps'. It is the manipulation and analysis of the spatial database and the display of maps with relative speed and ease that is the trademark of GIS.

The conceptual and technical origins of GIS can be traced back to the late 1960s and early 1970s, but developing that ease and speed of map combination and display beyond the research environment into commercially viable off-the-shelf systems has taken 20 years of development of computer technology. Today there are thousands of commercial firms worldwide engaged in GIS, in a market worth several billion dollars/year and growing fast, with forecasts suggesting a global worth of over $8 billion by the year 2000 (Antenucci 1992).

The main benefits of GIS seem to be associated with long-term cost-savings in map-production, as well as with extending the use of the GIS to other areas that improve the overall performance of organisations. When GIS technology started to be widely available in the late 1980s, the initial costs of GIS map-production were found to be about twice those of traditional mapping. With time, the two tend to converge so that after about 7–8 years the costs of GIS mapping start to be less than those of traditional mapping and the returns of the (sometimes considerable) initial investment in the new system begin to materialise.

Today, the comparison must be made with map production using other (non-GIS) mapping technologies now available rather than with traditional 'manual' mapping, but the question about the time lapse between any investment in GIS technology and the returns it generates still remains. More generally, the problems associated with GIS have changed over the years, and so have the costs associated with these issues. After initial technical problems with the *development* of GIS in the 1970s, the lack of available *expertise* to use the systems became the greatest issue in the 1980s, and in the 1990s and into the new millennium it is mainly the availability of *data* at affordable prices that has become the greatest bottleneck in GIS use.

In this chapter we shall first introduce and discuss some basic technical aspects of GIS, and then go on to discuss its potential and applications in EIA. The literature on GIS is vast, but there are two benchmark publications (Maguire *et al.* 1991,

Longley *et al.* 1999)[1] which summarise most of the research and development issues in this field. A very good introduction to all GIS issues is provided by Antenucci *et al.* (1991). On environment-related GIS, Melli & Zanetti (1992) report on an IBM-sponsored meeting on computer assisted environmental modelling in 1990 with many references to GIS, and Goodchild *et al.* (1993, 1996a, 1996b) contain papers from three subsequent seminal conferences on environmental modelling and GIS in the US in 1991, 1993 and 1996, respectively. Overviews of GIS in ecology are provided by Johnston (1998) and Wadsworth & Treweek (1999).

The best known journal on GIS is the *International Journal of Geographical Information Science* (formerly *International Journal of Geographical Information Systems*), and very useful magazines containing up-to-date information about the GIS industry and its applications are *Mapping Awareness* (for the UK), *GIS Europe*, and *GeoWorld* (formerly *GIS World*). There are regular conferences on GIS, e.g. in the US (the GIS/LIS Conference), Canada (the Canada GIS Conference), Australia (the Australian GIS Conference), and Europe (the EGIS conference). In addition a range of other conferences on environmental or other subjects frequently have a section on GIS. Specifically related to EIA, Guariso & Page (1994) report on a conference in 1993 on Information Technology for impact assessment with many references to the potential of GIS. More recently, the conferences of the International Association for Impact Assessment (IAIA) have included GIS-related papers (Portugal 1996, New Orleans 1997, New Zealand 1998, Glasgow 1999), albeit with decreasing prominence after the interest shown at the Portugal meeting in 1996.

16.2   GIS concepts and techniques

To understand the potential and limitations of GIS, it is important to remember that in essence these systems are just a combination of a computer-cartography system that stores map-data, and a database-management system that stores attribute-data (an attribute being a characteristic of a map-feature, like the land use of an area (see Appendix F.6, p. 452) or the slope of a stretch of road). Hence, GIS share the issues and problems these two types of systems – or, for that matter, any information system – have, namely: data capture and storage; data manipulation and analysis; and presentation of results.

16.2.1   Data capture

The technology for GIS map-data capture is quite varied, and changing rapidly, but the techniques can be divided into three categories that can be called *primary*, *secondary* and *tertiary* data capture.

Primary data capture techniques (from the real world) include: (a) ground surveys based on sampling, the traditional source of cartographic data; (b) remote-sensing based on classifying the 'pixels' in a satellite infra-red picture (see Chapter 15); and (c) Global Positioning System (GPS), a hand-held system that can register instantly

---

1 Although Longley *et al.* (1999) is presented as a "second edition" of Maguire *et al.* (1991), it is an entirely new publication, with different authors and chapters, so the two should really be taken *together* as a quite complete and excellent source on GIS.

on a computer – using a network of satellites to pinpoint the location – the co-ordinates of a point with errors of less than 1 m. GPS is used today for all kinds of cartographic and navigation applications, and is probably the most important advance of recent times in the field of cartographic data input.

Secondary data capture techniques (from paper-maps or aerial photographs) include: (a) digitising (tracing) on a magnetic table the points on a map as well as the 'caricature' of its lines (lines broken down into straight segments) – labour-intensive and expensive; (b) scanning maps, using refined versions of the type of technology used in fax machines – cheaper but still prone to errors in the form of gaps in the scanned lines; and (c) so-called overlay-digitising (or 'heads-up' screen digitising) which combines the advantages of both: a map is scanned cheaply and easily and then displayed on the screen, and vector map-features are derived from the image using a 'mouse' as a digitiser.

Tertiary data capture is based on 'importing' data from existing sources already in digital form, provided by public or private organisations. That this is currently an area of fast growth is not surprising, given the cost and difficulty of obtaining primary data and the labour-intensive nature of secondary data capture. A first attempt at a guide to digital data sources in the UK is provided by O'Carroll *et al.* (1994), although the accelerating growth and diversification of this type of data-source indicates that a guide of this nature requires constant updating. Digital data from airborne and satellite sensors is becoming increasingly available (see Table 15.1, p. 371), and many national cartographic and environmental agencies are now providing digital cartographic information which is GIS-compatible, although this proliferation of data sources is increasingly raising the question of format-compatibility between them. For example, satellite data, such as that used for the *Land Cover Map of Great Britain* (LCMGB) may have a lower resolution than ground survey data. However, methods of integrating such data have been demonstrated (e.g. Mack *et al.* 1997). Some examples of digital products and sources of data are given in Table 16.1.

16.2.2 Data storage

Raw map-data become information when interpreted by conceptual *data-models*, and the type of model used to store GIS maps is one of the clearest dividing lines between different types of systems (see Fig. 16.1).

Regular-tessellation 'raster' models store maps using more or less simplified versions of a matrix-file, where the different square cells (*rasters*) are stored with the value of their attributes. The advantage of a file of this kind is that it simultaneously defines the map (*where* features are) and the *values* of particular attributes (one for each map) for every feature. First-generation GIS belonged to this kind. They are relatively easy to program and simple in terms of file-structure, but are wasteful of space (they repeat the same information many times, once for every cell); and their greatest drawback is that their accuracy is ultimately determined by the *size* of the cells they use. Well-known raster-based GIS are:

- GRASS – free, by the US Army (http://www.geog.uni-hannover.de/grass/ *or* http://www.baylor.edu/~grass/);
- IDRISI – very cheap, by Clark University (http://www.clarklabs.org/);
- SPANS – analytically powerful, by TYDAC (http://www.tydac.com/).

Table 16.1 Some digital products and sources of digital data

Ordnance Survey (OS) (http://www.ordsvy.gov.uk/)

Land-Line series – digitised at 1:1250 for urban areas, 1:2500 for rural areas, or 1:10,000 for 'moorland' areas (e.g. moors, mountains, estuaries), and has up to 36 feature codes (roads, buildings, land plots, etc.)

Land-Line Plus – has an extra 27 feature codes including landform features and vegetation types

Land-Form series – contains topographic contour information at 1:50,000

Raster data – from printed maps at 1:10,000 (intended for background maps)

'Oscar' – the road centre-lines for the whole country

Historical maps – dating from 1843, e.g. at 1:10,000 or 1:2500

Centre for Ecology and Hydrology (CEH) (http://www.ceh.ac.uk/)

Data from the **Countryside Survey 1990 (CS90)** (Table 11.4) including: (a) the **Land Cover Map of Great Britain (LCMGB)** which is based on Landsat satellite data (§15.2.2), uses 17 land cover classes, and has raster format; (b) the **Ecological Surveys database** which holds data on vegetation, soils and rivers. The system and data have been updated for the **Countryside Survey 2000 (CS2000)** and a new land cover map **(LC2000)** with vector-based mapping using the UKBG broad habitat classification (Appendix F.2) will be available in 2001 (http://www.cs2000.org.uk/). CS90/2000 data are also available for the *Countryside Information System* (CIS) (http://www.cis-web.org.uk/). The CIS software can generate maps, charts, etc. (based on 1 km grid squares) and has some facilities for data analysis and data input by the user.

English Nature (EN) (http://www.english-nature.gov.uk/)

Boundary maps of designated conservation sites and ancient woodland inventory (AWI) sites.

British Geological Survey (BGS) (http://www.bgs.ac.uk/)

Some **geological maps** at 1:250,000 and 1:1:50,000 scales are becoming available in digital form.

MAFF-FRCA (http://www.maff.gov.uk/)

National Environmental Database (NED) – Uses a range of software including GIS (Arc/Info) and satellite image processing (ERDAS), and holds datasets on features such as soils, climate, land cover and boundaries of National Parks, Environmentally Sensitive Areas (ESAs) and **Natural Areas**.

European Environment Agency (EEA) (http://www.eea.eu.int/)

CORINE databases and maps (EU-funded), e.g.:

   CORINE land cover database (EEA-ETC-LC) with digital map (using 44 land cover classes) at 1:100,000 and minimum mapable units of 25 ha, from satellite data.

   CORINE biotopes database and map (EEA-ETC-NC & CEH Monkswood) > 7500 nature conservation sites. Records include site location and extent, habitats and species descriptions, and protection status.

Other sources (full names and Internet addresses of organisations are given in Appendix A)

Data and online links, e.g. provided by ECNC, ERIN, FAO, UNEP-GRID, USEPA, USGS-EROS, USNTIS

Base maps, etc. from **GIS software vendors** and other **commercial firms.**

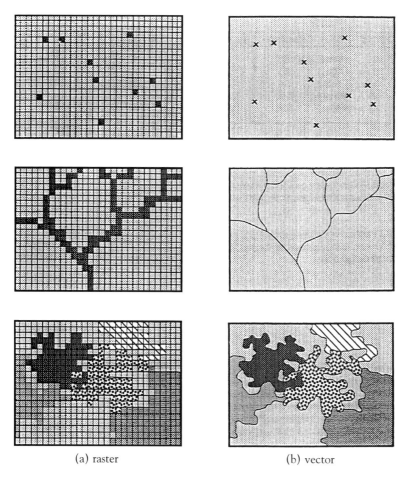

(a) raster   (b) vector

Figure 16.1 Map-features represented using different data-models (*source*: ESRI, Arc/Info reference manual for GRID).

Irregular-tessellation 'vector' models represent map-features (points, lines, polygons) by the precise co-ordinates of their defining points and segment-ends. This greatly increases accuracy but has the problem of requiring *two* sets of files for each map: one to store the position and shape of the map-features, and another to store the attributes associated with those features. Vector data can be stored by 'layers' (each containing one or several features), or by 'objects' (the latest approach now being developed) where the attention is on individual cartographic objects, their properties and their membership of different 'classes' and sub-classes, with the possibility of 'inheritance' of properties between them. Well known vector-based GIS include:

- ARC-INFO – expensive, still the market-leader worldwide, by ESRI (http://www.esri.com/);
- TIGRIS – also powerful, by INTEGRAPH (http://www.intergraph.com/).

'Integrated' GIS, capable of combining vector and raster data, as some of the more powerful systems (like ARC-INFO) are beginning to do today.

As an aside, other simple mapping programs with very limited GIS functionality are now available for use on PCs (or equivalent) with platforms such as Windows 95/98 or NT4. Examples include:

- AutoCad Map2000 (http://www.autocad.com)
- GeoMedia and GeoMedia Professional (http://www.intergraph.com)
- Map Maker Pro (http://www.mapmaker.com)
- Map Sheets (http://www.erdas.com *or* http://www.erdas.co.uk)
- PAMAP (http://www.pcigeomatics.com)

16.2.3 Data manipulation and analysis

Despite the cartographic sophistication of GIS, the tasks they can perform in terms of *spatial analysis* are quite limited and can be summarised as follows:

1. In two dimensions:

    - map 'overlay', superimposing maps to produce simple composite-maps, probably the single most frequent use of GIS functionality;
    - 'clipping' one map with the polygons of another to include (or exclude) parts of them, for instance to identify how much of the area of a proposed project overlaps with a sensitive area;
    - producing 'partial' maps containing only those features from another map that satisfy certain criteria;
    - combining several maps (weighted differently) into more sophisticated composite maps, using so-called "map-algebra", also referred to as "cartographic modelling" (Tomlin 1990, 1991), used, for instance, to do multi-criteria evaluation of possible locations for a particular activity, or calculating the composite effect of a set of factors on an area, e.g. as in Figure 16.2;
    - calculating the size (length, area) of the features of a map;
    - calculating descriptive statistics for the features of a map (frequency distributions, average size, maximum and minimum values, etc.);
    - doing some multivariate analysis (like standard correlation and regression) of the values of different attributes for different features in a map;
    - calculating minimum distances between features (some systems handle straight-line distances, others can also measure distances along 'networks');
    - using minimum distances to identify the features on one map nearest to particular features on another map;
    - using distances to construct 'buffer' zones around features, which can then be used to 'clip' other maps to include/exclude certain areas.

2. With a third dimension:

    - interpolating unknown attribute-values for new points (a 'third dimension' on a map) between the known values for existing points, using Triangulated Irregular Networks (TINs) to maximise the efficiency of interpolation;
    - drawing contour-lines using the interpolated values of an attribute (a 'third dimension');
    - constructing *Digital Terrain Models* (DTMs), also known as *Digital Elevation Models* (DEMs), using a third dimension (an attribute), which can then be displayed or manipulated;

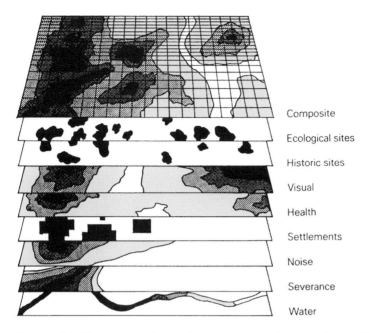

Composite

Ecological sites

Historic sites

Visual

Health

Settlements

Noise

Severance

Water

Figure 16.2 The use of overlays to show environmental impacts (*source*: Wathern 1988).

- calculating topographic characteristics of the terrain, like slope and orientation ('aspect') of different parts, their concavity and convexity, etc.;
- calculating volumes in a DTM, e.g. to calculate water-volumes in lakes or reservoirs;
- identifying 'areas of visibility' of certain features of one map from the features of another, for instance to define the area from which the tallest building in a proposed project would be visible;
- so-called 'modelling', identifying physical geographic objects from maps, like the existence of valleys, or streams forming a river-basin, river networks, etc.

The limited range of analytical tasks has been one of the main criticisms of GIS. Openshaw (1991) argued sometime ago, somewhat ironically, that of the 1000+ operations that a sophisticated GIS can perform, virtually none relate to true spatial analysis, and functions like those listed above really correspond to 'data description'. The GIS industry has reacted to these criticisms, and today an increasing number of the more sophisticated operations of spatial analysis can be found in the latest releases of some GIS, or in 'add-on' modules to them.

16.2.4 Presentation of results

The output of GIS is probably the best developed and most appealing aspect of these systems. Output can be produced for a variety of devices (the computer screen, plotters, printers) and can be classified by its dimensional level as:

Figure 16.3 Representation of a Digital Elevation Model with a visibility area draped on it.

- 2-D displays (maps), which are most common;
- so-called '2.5-D' representations of Digital Terrain Models (DTMs) which use a third (z) dimension over an x–y map. Other maps can be superimposed ('draped') on them so that they appear to be in 3D, or the slopes and aspects of the different facets in these models can be used to calculate sunlight-reflection and produce 'shaded' representations of the terrain (see Fig. 16.3);
- 3-D models, which are currently the object of considerable research, looking at the possibility of representing 3D objects as collections of 'sheets' using the standard functions of GIS (which are essentially two-dimensional), or maybe incorporating into GIS some of the features of Computer Aided Design (CAD) or Virtual Reality (VR).

A dominant current trend in GIS output when produced for the computer screen is towards interactive 'multimedia' output which combines maps, photographs, moving video-images, and even sound, as part of the emerging approach of 'hypermedia', in which the user can move between all these outputs by just 'zooming in and out' between them.

16.3 GIS and environmental impact assessment

16.3.1 Introduction

Bibliographical reviews show that by far the most common applications of GIS are concerned with environmental issues (Rodriguez-Bachiller 1998); applications of relevance to EIA are increasing and involve a variety of approaches. GIS should be well suited to EIA because it can answer questions that are central in the EIA process. As stated in ESRI (1995), these are:

1. **What is where**? – which is central in screening, scoping and baseline studies;
2. **What spatial patterns exist**? – which can help in understanding the baseline conditions, and in impact prediction and mitigation;
3. **What has changed since . . .** ? – which may be relevant to impact prediction, prediction of changes in the absence of the project, and monitoring;

4. **What if**? – which is the aim of impact prediction, and may be important in formulating mitigation measures.

16.3.2 Possible approaches

Possible relationships between GIS and EIA can be summarised as follows.

1. The GIS can be used to **provide data** for the impact assessment, which is done externally to the system. This may involve different levels of complexity in the role played by the GIS:

   - it can be used **just for mapping**, to provide a visual aid (for instance, maps of the project area and of sensitive areas nearby) with which EIA experts can perform the assessment manually;
   - it can **provide data to an external tool** (doing some kind of 'pre-processing' of the information about the project and its environment) programmed externally to the GIS and 'coupled' in some way to it. This in turn can vary depending on the nature and complexity of the tool, which can be:

     * a **simulation model** to predict certain impacts, like a noise-prediction model or a pollution-dispersion model;
     * an **expert system** or another type of decision-support tool to guide the experts in their impact assessment;
     * a **combination** of the two above, with an expert system being used to integrate GIS information and simulation models.

2. GIS can be used to **do the EIA** within the GIS. This approach can also involve different levels of complexity The assessment can be performed **by the user**, with the GIS functions discussed in §16.2.3, like (a) using 'buffering' and map-overlay to see if a project or its impact area overlap with sensitive spots, (b) working out areas of visibility. Also, the use of these functions can be programmed internally into subroutines (using some form of programming language internal to the GIS, like a macro-language) so that the whole process is done **automatically** for the user.

3. GIS can be used to **display the results** of the EIA, where the assessment is done by other means and the GIS is used only for the flexibility and quality it provides when displaying information in map form, and also sometimes to provide some degree of 'post-processing' of the results, e.g. drawing contour maps of predicted ground-pollution levels or noise levels, or showing visual impacts using 3-D representations of the type described above.

16.3.3 Resource implications

In EIA, the assessment and reporting of the likely environmental impacts of a development proposal is typically carried out by environmental consultancies working to tight deadlines within limited budgets. The resource implications of using GIS technology within an EIA (e.g. hardware and software costs, skilled staff, and the costs of acquiring or inputting data) indicate that its potential role must be carefully considered, preferably in the early stages of an EIA when the overall environmental assessment methodology is formulated. Project managers need to identify the ways in which GIS could be useful within a particular EIA and must then decide whether the outputs and analysis which GIS can facilitate will justify the resources required.

One of the crucial and most resource-intensive tasks – and one that may determine the feasibility of one or another of the approaches mentioned in §16.3.1 – is the setting up of the appropriate **map-base** (in digital form) for the GIS, including suitable maps of the area, maps of policy-areas (conservation areas, etc.) that may need to be taken into consideration, as well as maps of the project itself, in whatever detail is necessary.

In fact, GIS technology can provide a useful framework in which an **integrated spatial inventory** of environmental information can be developed, analysed and fed into EIA decisions. Typical layers of information may relate to biophysical, socio-economic, historical/cultural features and policy designations, and can include themes such as land use, habitats, soils, geology, hydrology, topography, pollution-monitoring data, census information, transport networks, archaeological resources, and conservation areas (including SSSIs and other designated sites). Although digitised forms of these maps or 'layers' of information are becoming increasingly available, most will typically be available in paper form. Consequently, the critical question remains: *what can GIS offer the EIA process beyond the provision of high-quality cartographic output* (which can also be produced using cheaper and less sophisticated software packages)?

The remainder of this chapter tries to help answer this question by focusing on the practical application of GIS technology at various stages in the EIA process. The intention is not only to highlight the different ways in which GIS *could* be used to good effect in EIA, but also to reflect pragmatically upon the limitations and restrictions to its application which can arise given the constraints facing EIA practitioners.

16.4 GIS in screening, scoping and baseline studies

16.4.1 Screening

Screening (deciding whether a project requires EIA) is usually based on: (a) *characteristics of the project* itself, e.g. the type of activity or construction it involves, the size or level of such activities, and whether they exceed certain thresholds; and (b) the *project's location* and the sensitivity of this and the area nearby. Examples of how GIS can facilitate screening include:

- Certain types of projects (e.g. industrial estates in the UK) will require an impact assessment if they *reach or exceed a certain area*, and a GIS will be able to calculate this automatically from a map of the project.
- Often it has to be established if a project lies *within an environmentally sensitive area* – in which case an EIS would be required. Although simple visual inspection of a map (GIS-produced or not) will often suffice, using GIS to overlay a map of the project and a map of the relevant sensitive areas will achieve the same result with increased accuracy, and with the additional advantage that the GIS may be programmed to do it *automatically* and report back.
- In some cases an EIA will be required if a project is within a given *distance from a certain type of feature* such as a road or a residential area. The 'buffering' capabilities of GIS can be used to good effect to answer such a query. A buffer-

zone at the critical distance around the project can be generated by the GIS, and then used to 'clip' a map containing all the roads or relevant features. If the result is an empty map it means the project is *not* within the critical distance, if the clipped area contains any roads or features it means they *are* within the critical distance.

The fact that GIS have technical capabilities to contribute to project-screening does not necessarily mean that they are the best way to do it. As mentioned before, whether it is cost-effective to use GIS for screening will depend largely on how central the GIS is to the whole information environment of the organisation doing this work, and how much preparatory work is needed (setting up the map-base, etc.) if the information is not already contained within the system.

16.4.2 Scoping

The logic here is similar to that of the last section because the considerations involved in scoping are quite similar to those affecting screening. Whilst it is the characteristics of the project that will determine many of the impacts to consider, the setting of the project will also determine impacts that need to be studied. For example:

- a project located on good-quality agricultural land will require a study of potential impacts on the soils and agricultural resources of the area;
- a project which involves the discharge of effluent to a nearby river will require a study of potential water pollution;
- a project located upwind from a nature reserve, and producing emissions to the atmosphere, will require an air-pollution study and an ecological study.

Placing a development proposal within its **geographical context** will help inform the scoping process through defining the project location, describing its environmental setting, and helping identify potential conflicts or impacts which will require detailed assessment in an EIA, and GIS can be used for this in ways not too different from those applicable to screening. For instance, in the EIA of a road scheme, GIS might be used to inform a scoping decision regarding the consideration of archaeology as follows:

1. The GIS could be used to create a 500 m buffer around the proposed route which could then be combined with a map of known archaeological sites using GIS overlay.
2. From this analysis, a map could be generated showing all the relevant features (i.e. road, 500 m buffer and archaeological sites), or alternatively the query could be structured such that only areas of archaeological interest falling *within* the buffer zone are identified and 'clipped'.

In this way GIS analysis can be used not only to scope the EIA in terms of identifying impact themes which require further investigation, but can also help to clarify the **spatial scope of the study**, i.e. the areas or receptor locations which will require detailed consideration in the assessment of a particular impact. To be effective, however, this requires that the criteria used (for instance, the distance used as a search radius for locating sensitive archaeological sites) be defined in an unequivocal way. This may present problems when such criteria have been defined in the law or in the practice-guidelines in 'fuzzy' terms, using expressions such as 'near', 'close', and the like.

It should also be recognised that GIS is only one of a number of methods which can be used for scoping an EIA (e.g. expert judgement, checklists, matrices, expert systems, public consultation) and that to be most effective it should be used to supplement and complement these techniques. It is probably fair to say that both screening and scoping can be done just as effectively *without* a GIS, and the potential for using a GIS really lies in the possibility of *programming* these activities so that they are carried out *automatically* in the system: finding the right maps, applying the right distances, identifying overlaps and buffers, etc., and then reporting back to the user.

16.4.3 Baseline studies

Building on the information generated as part of the scoping process, further data will be required in an EIA to describe and analyse the baseline environmental conditions for specific impact themes. In turn – reflecting the iterative nature of the EIA process – this information may influence and further refine the scope of the assessment as more data are gathered and the EIA progresses.

Once baseline data have been collected and input, GIS can be a powerful tool for displaying and visualising trends and patterns in spatial datasets.

Point-type data which relate to a specific sampling location (e.g. a pollution-monitoring station) can be displayed in the form of a proportional-symbol map or, where time series data are available, perhaps as a series of maps at various intervals to reflect the dynamic nature of the environmental baseline.

'Spatially continuous' data (e.g. noise, rainfall, topography, groundwater, air pollution) can be used (given a sufficient spatial sample) to produce contour (isoline) maps or, in the case of topography, as a DTM to describe the baseline terrain.

'Linear data' describing features such as rivers or roads can be represented using colour-coding, or perhaps with variation of line width in proportion to the data values, e.g. to illustrate traffic-flow data along roads.

Area data which relate to discrete spatial units (e.g. census data, designated sites and habitat patches), can be displayed as choropleth maps, where the intensity of shading is used to reflect the data values.

Whilst these types of graphical output can be produced using simpler software systems, GIS is ideally suited to organising and storing multi-disciplinary monitoring datasets into a framework which can be analysed, queried and displayed interactively in order to support and inform the EIA process. For instance, where comprehensive spatial datasets are available, the spatial query capabilities intrinsic to GIS can be used to highlight potential 'hotspots' (e.g. locations with pollution levels above specified thresholds) that may require particular attention in terms of impact prediction and assessment of significance, hence serving to refine the scope and focus of the EIA as more information becomes available. GIS are also ideal for determining the extent to which hotspots and sensitive locations are spatially concentrated across a variety of different environmental parameters.

Whilst GIS technology has some clear strengths which makes it appropriate for baseline studies, *its use is limited by the availability of data with a good spatial coverage*

(§16.2.1). It should be recognised that monitoring-data is costly to collect, and in many EIAs resources will be targeted towards a small number of receptor locations (which are likely to be most seriously affected by a project), rather than achieving a broad spatial sample which would satisfy the ideal requirements of GIS. Also, some of the information used for scoping and baseline studies is often presented in numeric form (e.g. socio-economic information about the area, levels of unemployment) without any need for a map to show it.

In recent years commercial services based upon spatial databases of environmental information which are of potential relevance to EIA baseline studies have appeared, e.g. "Envirocheck" from Landmark Information Group which, although not a fully fledged GIS, provides maps of environmental data centred on user-specified co-ordinates, in addition to a written report, all of which is dispatched to a client within two days.

16.5 GIS in impact prediction

16.5.1 Introduction

Impact prediction lies at the core of EIA and is intended to identify the magnitude and other dimensions of likely changes to the environment which can be attributed to a development proposal (Glasson *et al.* 1999). GIS are obviously most suited to dealing with the spatial dimension of impacts, and at the simplest level of analysis they can be used to make quantitative estimates of aspects such as:

- the 'land take' caused by development (e.g. the total area of agricultural land, grassland or wetland habitat which may be lost);
- the length of road or pipeline which passes through a designated landscape area such as an Area of Outstanding Natural Beauty (AONB);
- the number/importance of features, such as archaeological finds or ancient monuments, lost to the development.

More sophisticated predictions will require some form of **modelling** to represent or simulate the behaviour of the environment, and two broad ways in which GIS may be used for modelling in impact prediction can be identified.

1. The entire process of developing and implementing a model takes place *within* the GIS software, i.e. GIS is used for data input and preparation, modelling, and finally for the display and spatial analysis of model output;
2. Whilst GIS may be used in data preparation, the actual modelling is undertaken *outside* the GIS software using an independent computer model, the output from which is imported back into the GIS for purposes of display and further spatial analysis.

16.5.2 Modelling within the GIS

Modelling 'internal' to the GIS can vary in its level of sophistication. At its simplest, **GIS Mapping** is a form of modelling, essentially the same as conventional mapping but with the advantages provided by the overlay and buffering capabilities discussed in §16.2.3. To illustrate this level of GIS involvement in EIA, the case of ecological impacts can be used.

Because of their complexity, responses of ecosystems (§11.2.3) to impacts are notoriously difficult to predict (Chapter 11). Consequently, ecological assessment requires a high level of expertise and judgement. However, it can involve a substantial amount of mapping, and the facilities available in GIS can be very valuable. This can be illustrated in relation to the basic questions referred to in §16.3.1, plus an additional question – *Why is it there?* (Treweek 1999). To have a reasonable chance of understanding an ecosystem's current and likely behaviour it is important to have:

- a knowledge of the spatial relationships of its components (species populations, communities and environmental systems), i.e. to know *what is where* and *what spatial patterns exist*;
- an understanding of the factors that explain these relationships, i.e. *why is it there?* This will depend on a combination of present and past factors, and so may require –
- a knowledge of at least recent trends, i.e. *what has changed?*

Given adequate data, GIS overlay mapping can help to provide answers to these questions. For example:

- layers showing distributions and ranges of species, locations and extents of habitats and sites, and patterns of environmental parameters such as geology, soils, hydrology, or land use can clearly demonstrate spatial relationships;
- spatial relationships, e.g. between species and habitats or habitats and environmental patterns, often go a long way to explaining *why it is there*;
- layers created from past maps or records can illustrate *what has changed*, and help to explain the present patterns and relationships (Veitch *et al.* 1995).

Similarly, GIS mapping can be useful in attempting to answer some impact prediction (*what happens if?*) questions. For example, it can demonstrate locations and dimensions of:

- predicted impact areas, including 'buffer' zones along linear projects;
- direct 'land take' in relation to habitats and species, e.g.:
  * what parts and proportions of sites or habitat patches will be lost?
  * what will be the overall area loss of habitat types and what proportion of the current stock will this represent?
  * what parts and proportion of a species' habitat will be lost?

- habitat fragmentation, including sizes and isolation (as distances) of remaining habitat patches;
- new barriers to species dispersal, including the project itself (buildings, roads, etc.) and barriers created by habitat fragmentation;
- new pollution sources and likely dispersion patterns;
- environmental impacts such as changes in drainage patterns, soil moisture levels or sediment loads in aquatic ecosystems.

Moreover, GIS make it possible to answer 'what if' questions about alternative prediction scenarios, project characteristics, or locations, with relative speed and ease.

Of course, whilst such GIS mapping can provide information on the magnitude and spatial extent of impacts, it cannot provide precise predictions about their significance, assessment of which must rely on ecological interpretation.

At a higher level of sophistication, **Digital terrain models (DTMs)** can be used in various ways, a good example being **viewshed analysis**, i.e. the prediction of a project's 'viewshed' or Zone of Visual Influence (see for example, Howes & Gatrell 1993, Fels 1992, and Davidson *et al.* 1992). The main steps in this process are:

- Topographic data are digitised manually from a contour map or purchased in digital form (§16.2.1).
- These data are then used to create a DTM of the land surface within the GIS.
- Using the DTM and information describing the height of project structures and other elements in the landscape which could act as visual barriers, the 'viewshed' function commonly found within GIS software can be used to delimit the area over which the project will be visible.
- Finally the output from the visibility analysis can be mapped (or draped over the DTM) within the GIS, and further spatial analysis performed if required, e.g. the use of overlays to identify residential properties which lie within the viewshed.

A variety of refinements to the basic binary (yes–no) viewshed function have been developed in order to increase the information content of the output. At the simplest level these include: the use of options which serve to indicate *how much* of a development proposal is visible (e.g. how many turbines in a windfarm are visible from a given location); and weighting schemes to simulate the decline in visual impacts which occurs with increasing distance from the source. Other advances from the research domain include the use of fuzzy logic and probability to simulate project visibility under different atmospheric conditions (Fisher 1994).

Finally, **cartographic modelling** (§16.2.3) is a more generic approach to impact-prediction modelling within GIS. It involves the use of raster-based GIS overlay to combine individual layers of data in order to arrive at some form of composite. For example, as part of an EIA of a 1140 km electricity transmission line in the United States, Jensen & Gault (1992) developed a GIS model to assess the ground disturbance impacts associated with construction activities. The model used GIS overlay analysis to combine layers of information describing land cover (see Appendix F.6), slope and the transportation network in order to quantify the impacts into five levels of magnitude and create a map showing the spatial distribution of the disturbance.

Viewshed analysis and cartographic modelling represent relatively basic approaches to impact prediction modelling which are highly deterministic and can incorporate a strong degree of subjectivity, notably when determining weightings or classifications to be used to combine data layers. GIS currently lack the capabilities to undertake more powerful process-driven modelling, but have been used to good effect when combined with environmental models which operate external to the GIS.

16.5.3 Modelling external to the GIS

In this section, the example of air quality impact prediction is used to demonstrate how GIS may be used in combination with spatially distributed environmental models which operate outside the GIS software. The discussion then broadens to identify other impact themes where this approach is appropriate and briefly considers recent developments in environmental software which incorporate elements of GIS technology.

Air pollution impacts in EIA are typically predicted using Gaussian dispersion models (see Chapter 8) for which GIS have the potential to be used as a **pre-processor** or data preparation tool. For instance, many Gaussian models employ algorithms designed to simulate the dispersion of pollution in either urban or rural settings, and in most cases the criteria used to decide which option to adopt are based upon land-use data. With the Industrial Source Complex (ISC) model, urban dispersion coefficients should be used if more than 50% of the land use within a 3 km radius of a project is classified as either industrial, commercial or residential (Maitin & Klaber 1993). A GIS which holds land-use data (collected perhaps during the scoping stage of an EIA) is well placed to answer such a query accurately and efficiently. Air quality models also require terrain data for the receptor co-ordinates to be incorporated in the modelling and, again, GIS could be used to supply this information.

Once the calculation of air pollution impacts has been completed using the external model, GIS can be used effectively as a **post-processor**, particularly for purposes of presentation and display. Thus, output from the model could be fed into the GIS software where a contour map of impacts could be developed and perhaps combined with land-use data to assist in the interpretation of impacts. GIS can also facilitate further spatial analysis of the predicted impacts which might include overlaying contours on a proportional symbol map of baseline levels, or querying the GIS to identify residential properties which lie within a certain threshold level of pollution.

In recent years a number of research prototypes and commercial products have been developed which provide an **integrated approach** to combining air quality models and GIS. The software features 'user-friendly' interfaces which enable data to be transferred between the GIS and the model in a seamless fashion, e.g. ADMS3 from CERC Ltd. incorporates a link to the GIS package ArcView so that model output can be visualised and analysed spatially. Such approaches are most useful in that they facilitate the rapid simulation of alternative scenarios or mitigation strategies.

In the research domain, GIS have been used in combination with quantitative models for predicting ecological impacts (Hunsaker *et al.* 1993) although to date there is little evidence of such approaches being adopted in practice. Other impact themes in EIA for which GIS has been linked with environmental models include hydrology (to calculate runoff and flood risk), surface and groundwater quality, and noise. As with air quality, in recent years commercial products which combine environmental models and some elements of GIS (particularly in terms of mapping and overlaying data) within a single seamless software package have appeared on the market. Examples include RTA Acoustics 'Environmental Noise Model', Danish Hydraulics 'MIKE 11 GIS' for hydrology and 'Visual MODFLOW', a groundwater flow simulation package from Waterloo Hydrogeologic Software. The main limitation in the application of these models is likely to be the lack of adequate data specific to the study area.

16.5.4 Reflections on GIS in impact prediction

From the examples cited above it can be seen that the way in which GIS can be used for impact prediction varies according to the extent to which the analysis requires a comprehensive spatial database of information. In the case of ecological analysis, expert information about habitats and species will have to be put into GIS maps and, once created, these maps can be manipulated to provide much information relevant to impact prediction. In the case of viewshed analysis, GIS can be brought into use for a clearly defined 'one off' task within an EIA and, to be effective, the analysis does not require the development of a full GIS database, but can be conducted at the most basic level using only topographic data. Where GIS is used in combination with an external model, a limited number of data layers will be required, depending upon the requirements of a particular model and the degree of spatial analysis to be undertaken during post-processing of model outputs. In contrast, the cartographic modelling used for the Southwest Intertie Project emphasises the integrating capabilities of GIS and requires an extensive and comprehensive spatial database to be effective.

In terms of the use of models for prediction in EIA, the impacts suited to a spatial assessment using GIS appear to be those which exhibit continuous or semi-continuous variability over space and those which undergo diffusion or propagation through space, as opposed to through a functional structure such as the economy. However, the extent to which GIS is likely to be used for impact prediction in EIA will depend upon scoping decisions regarding the level of spatial detail required for decision-making and environmental management. As was the case with baseline studies, it may be that impact predictions are only deemed necessary for a limited number of receptor locations, and that a broad spatial assessment is surplus to requirements. To reiterate, the implication is that *there is a strong need for early planning and careful consideration over the extent to which GIS will be useful in EIA*. Finally, it must be stressed that however impressive the results from GIS may appear, they can only be as good as the data and models on which they are based.

16.6 GIS in mitigation

One of the most effective uses of GIS technology in terms of mitigation in the broadest sense relates to the identification and evaluation of **alternative locations** for a development project. Given a comprehensive spatial database and a series of clearly defined constraints or preferences, GIS overlay analysis can be used to good effect to identify and compare potential sites (or route alignments for linear developments). Two examples of practical EIA applications that have been documented in the literature include the work of Schaller (1990), who used GIS for the ecological assessment of alternative corridors for a Federal Motorway in Southern Bavaria, and Siegel & Moreno (1995), who applied GIS to identifying and assessing potential highway routes across the Tonto National Forest, near Phoenix, Arizona.

Beyond the application of basic overlay analysis, more sophisticated approaches to the identification and evaluation of siting alternatives from the research domain include the use of GIS technology in combination with multi-criteria analysis (e.g. Carver 1991), fuzzy logic (e.g. Bonham-Carter 1994) and genetic algorithms (Pereira & Antunes 1996).

As the focus of an EIA narrows to consider a specific site or route, the strengths of GIS in visualising and displaying the spatial distribution of impacts can be exploited to help identify and **target possible mitigation measures**. In particular, using criteria to define impact significance (determined by the EIA team or using published guidance), a GIS could be queried to identify locations which exceed thresholds and hence may require mitigation.

GIS can also be appropriate for simulating the effects of **alternative mitigation strategies** for individual impact themes. For instance, the effects on project visibility of planting screening-vegetation could be investigated or, in combination with environmental models, the implications of different project design characteristics or operational procedures could be looked at, e.g. the effects upon pollution dispersion of increasing the height of a stack or the velocity of exhaust gases.

In other situations, formulation of mitigation requirements may draw upon GIS analysis already conducted at an earlier stage in the EIA. For example, the maps produced for the baseline and impact assessment stages in an ecological assessment could be used to investigate:

- the potential for minimising impacts on nature conservation sites or habitat patches by project design modifications such as minor road realignments;
- the potential for species translocation (to suitable sites) or habitat creation, including the creation of stepping-stone or corridor habitats between fragmented habitat patches (§11.6.3);
- the suitability of options in particular localities, e.g. of new woodland planting in relation to existing woodland cover (Purdy & Ferris 1999);
- the optimum locations and dimensions of buffer zones to protect sensitive habitats.

16.7 GIS in monitoring

For large-scale development projects where a GIS system has been developed for use in EIA, it makes sense for the system to be used in the post-development phase as an integrative tool to store, analyse and display monitoring data. In this way the GIS becomes a tool for use in the actual operational environmental management, perhaps as part of an Environmental Management Plan (§1.6). Using GIS in this way also serves to recoup some of the costs of setting up a system for use in EIA.

Where monitoring datasets have a good spatial coverage GIS can be used productively in identifying patterns in the data and for examining change over time. It is worth mentioning here 'Monitor-Pro' as an example of a software product which, although not marketed as a GIS, does have some elements of this technology in terms of data mapping and visualisation (contouring, using proportional symbols, etc.) in addition to facilities for the automatic generation of reports.

Research by Wood (1999) has shown (using visibility, noise and air-quality impact assessment) how GIS can be used in a spatial approach to audit predictive techniques in EIA, where spatial patterns in the differences between predicted and actual impacts provided useful insights into the possible underlying causes of errors in impact assessment. Such 'prediction monitoring' is invaluable in terms of helping the EIA process to learn from experience.

16.8 Conclusions

It is probably fair to say that the development and diffusion of GIS has been *supply-led*, particularly in Europe, with developers and vendors of GIS and associated technologies (like remote sensing or digital cartography) 'sensing' a latent market for good-quality computerised mapping products, and investing in it long before the potential users were even aware of its existence. In this context, it is not surprising to find that the 'tone' of discussions about GIS applications (much of it in magazines which rely on advertising) tends to veer towards the positive side, often chanting the praise of this technology and its growing potential.

EIA as an area of application for GIS has been no exception, and articles like those quoted earlier in this chapter tend to present the use of GIS as a step forward, usually pointing in the direction of more – rather than less – involvement with GIS. In contrast, there is a severe shortage of literature which serves to critically discuss the limitations of applying GIS technology within the context of 'real world' EIA. This is why in this chapter we have attempted a more pragmatic evaluation of what GIS can realistically be used for, given the restrictions of time, money and data that face practitioners involved in carrying out an EIA.

It should be clear from the previous discussion that the technical potential of GIS for EIA is enormous. GIS is able to combine individual maps and databases and perform spatial analysis (such as overlay, buffering, viewshed analysis, etc.) which would be difficult and time consuming to achieve by hand, and which are not part of the armoury of standard mapping packages. In addition, all this can be achieved with maximum accuracy, and with the flexibility to combine data collected from a variety of sources and at a variety of scales.

However, the time and cost required to develop a full GIS database must be recognised, although suitable digital environmental information which can be imported directly into a GIS may be available commercially, at a price. Consequently, the use of GIS to do complex EIA is likely to be restricted to larger, well-funded projects for which the development of a full map-base is a viable option, and for which project managers have recognised the potential of the technology for use in *several* stages and for several aspects of the EIA.

On the other hand, the same argument can be turned around when applied to 'simple' EIA: when only a limited amount of (simple) impact analysis is needed (like viewshed analysis, or contouring ground pollution levels, etc.) the map-base required is very simple and easy to acquire, whilst the results can be quite impressive and considerably improve the overall quality of the final report. Also, an organisation, such as a local authority, may be (or is likely to be) engaged in a number of reports covering the same area. Whilst an individual project may not justify the expense of setting up a GIS, where a number of projects are proposed in an area, the use of GIS clearly becomes more viable. This also applies to cumulative impact assessment (e.g. Johnston *et al.* 1988).

Looking forward to the future, we can see current developments pointing in the direction of greater use of this technology via the Internet, as well as greater user-friendliness based on the notion of 'hypermedia' already mentioned, both opening the door to ever more *interactive* use of GIS, be it for EIA or for other areas of application. In addition to welcoming improvements like these in the GIS technology itself to suit the needs of EIA, what is crucial is that all these considerations about the potential role and limitations of GIS be an integral part of the planning of EIA work from its early stages. We need EIA managers who are aware of the potential of GIS, as much as we need the GIS industry to have a more thorough understanding of practical EIA issues.

References

Antenucci JC 1992. *Keynote address to the EGIS '92 Conference*, Munich (March 23–26).

Antenucci JC, K Brown, PL Croswell, MJ Kevany & H Archer 1991. *Geographic Information Systems: a guide to the technology*. New York: Van Nostrand Reinhold.

Bonham-Carter GF 1994. *Geographic Information Systems for geoscientists: modelling with GIS*. Oxford: Pergamon Press.

Carver SJ 1991. Integrating multi-criteria evaluation with Geographical Information Systems. *International Journal of Geographical Information Systems* **5**, 321–329.

Davidson DA, PH Selman, & AL Watson 1992. The evaluation of a GIS for rural environmental planning. *Proceedings of the EGIS '92 Conference*, Munich (March 23–26) **1**, 135–144.

DoE (Department of the Environment) 1993. *Countryside 1990 Series*: Vol. 1 *Ecological consequences of land use change*; Vol. 2 *Countryside survey 1990 Main report*; *Countryside survey 1990 Summary report*. London: DoE/HMSO.

ESRI (Environmental Research Systems Institute) 1995. Understanding GIS: the ARC/INFO method. Redlands, California: ESRI.

Fels JE 1992. Viewshed simulation and anlaysis: an interactive approach. *GIS World*, July, 54–59.

Fisher P 1994. Probable and fuzzy models of the viewshed operation. In *Innovations in GIS 1*, MF Worboys (ed.), 161–175. London: Taylor & Francis.

Glasson J, R Therivel, & A Chadwick 1999. *An introduction to environmental impact assessment*, 2nd edn. London: UCL Press.

Goodchild MF, BO Parks & LT Steyaert (eds) 1993. *Environmental modelling with GIS*. Oxford: Oxford University Press.

Goodchild MF, LT Steyaert, BO Parks, C Johnston, D Maidment, M Crane & S Glendinning (eds) 1996a. *GIS and environmental modelling*. Fort Collins, Colorado: GIS World Books.

Goodchild MF, BO Parks & LT Steyaert (eds) 1996b. *Third International Conference/Workshop on Integrating GIS and Environmental Modelling*, Santa Fe (New Mexico), January 21–25. Santa Barbara (California): the National Center for Geographical Information and Analysis (in compact-disk format).

Guariso G & B Page 1994. *Computer support for environmental impact assessment*. Amsterdam: North-Holland.

Howes D & T Gatrell 1993. Visibility analysis in GIS: issues in the environmental impact assessment of windfarm developments. *Proceedings of the EGIS '93 Conference*, 861–870.

Hunsaker CT, RA Nisbet, DCL Lam, JA Browder, WL Baker, MG Turner & DB Botkin 1993. Spatial models of ecological systems and processes: the role of GIS. In *GIS and environmental modelling*, MF Goodchild, BO Parks & LT Steyaert (eds), 248–264. Oxford: Oxford University Press.

Jensen J & G Gault 1992. Electrifying the impact assessment process. *The Environmental Professional* **14**, 50–59.

Johnston CA 1998. *Geographical Information Systems in ecology*. Oxford: Blackwell Science.

Johnston CA, NE Detenbeck, JP Bonde & GJ Neimi 1988. Geographical Information Systems for cumulative impact assessment. *Photogrammetric Engineering and Remote Sensing* **54**, 1609–1615.

Longley PA, MF Goodchild, DJ Maguire & DW Rhind 1999. *Geographical Information Systems*: Vol. 1 *Principles and technical issues*; Vol. 2 *Management issues and applications*, 2nd edn. Chichester: Wiley.

Mack EL, LG Firbank, PE Bellamy, SA Hinsley & N Veitch 1997. The comparison of remotely sensed and ground-based habitat area data using species-area models. *Journal of Applied Ecology* **34**, 1222–1229.

Maguire DJ 1991. An overview and definition of GIS. In *Geographical Information Systems: principles and applications*, DJ Maguire, MF Goodchild & DW Rhind, (eds), Vol. 1 (Principles), Ch. 1, 9–20, London: Longman.

Maguire DJ, MF Goodchild & DW Rhind (eds) 1991. *Geographical Information Systems: principles and applications* (2 vols). London: Longman.

Maitin IJ & KZ Klaber 1993. Geographic information systems as a tool for integrated air dispersion modeling. *Proceedings of the GIS/LIS Conference*, **2**, 466–474.

Melli P & P Zannetti 1992. *Environmental modelling*. London: Chapman & Hall.

O'Carroll P with A Rodriguez-Bachiller & J Glasson 1994. *Directory of digital data sources in the UK*. Working Paper No. 149, School of Planning, Oxford Brookes University.

Openshaw S 1991. Developing appropriate spatial analysis methods for GIS. In *Geographical Information Systems: principles and applications*, DJ Maguire, MF Goodchild & DW Rhind (eds), Vol. 1 (Principles), Ch. 25, 389–402. London: Longman.

Pereira A & P Antunes 1996. Extending the EIA process – generation and evaluation of alternative sites for facilities within a GIS. *Proceedings of the International Conference for Impact Assessment (IAIA'96)*, 479–484.

Purdy KM & R Ferris 1999. A pilot study to examine the potential linkage between and application of multiple woodland datasets: a GIS based analysis. *JNCC Report No. 298*. Peterborough: JNCC.

Rodriguez-Bachiller A 1998. *GIS and decision-support: a bibliography*, Working Paper No. 176, School of Planning, Oxford Brookes University.

Schaller J 1990. Geographical information system applications in environmental impact assessment. In *Geographical Information Systems for urban and regional planning*, HJ Scholten & JCH Stillwell (eds), 107–117. Dordrecht: Kluwer Academic.

Siegel MS & DD Moreno 1995. Geographical Information Systems: effective tools for siting and environmental impact assessment. In *Environmental analysis: the NEPA experience*, SG Hildebrand & JB Cannon (eds), 178–186. Florida: CRC Press.

Tomlin, CD 1990. *Geographic Information Systems and cartographic modeling*. London: Prentice-Hall.

Tomlin CD 1991. Cartographic modelling. In *Geographical Information Systems: principles and applications*, DJ Maguire, MF Goodchild & DW Rhind (eds), Vol. 1 (Principles), Ch. 23, 361–374. London: Longman.

Treweek J 1999. *Ecological impact assessment*. Oxford: Blackwell Science.

Veitch N, NR Webb & BK Wyatt 1995. The application of Geographic Information Systems and remotely sensed data to the conservation of heathland fragments. *Biological Conservation* **72**, 91–97.

Wadsworth R & J Treweek 1999. *Geographical Information Systems for ecology: an introduction*. Harlow: Addison Wesley Longman.

Wathern P 1988. An introductory guide to EIA. In *Environmental impact assessment – theory and practice*, P Wathern (ed.), Ch. 1, 3–30. London: Routledge.

Wood G 1999. Post-development auditing of EIA predictive techniques: a spatial analytical approach. *Journal of Environmental Planning & Management* **42**(5), 671–689.

17 Quality of Life Capital

Riki Therivel

17.1 Introduction

The concept of environmental 'capital' has been widely used by environmental managers and economists to describe the benefits that the environment accrues to humans, particularly in monetary terms. In theory, the idea that the environment consists of assets that can provide a stream of benefits or services so long as the capital is not damaged embodies the principle of sustainable development. However, in practice this approach has proven to be difficult and contentious to apply.

An integrated approach to identifying, analysing and managing all aspects of environmental capital was developed in 1997 by CAG Consultants and Land Use Consultants for the (then) Countryside Commission, English Heritage, English Nature and Environment Agency. This was published as *Environmental Capital: What Matters and Why?* (CC et al. 1997). Eighteen pilot studies were run in 1998/9 to test the application of this approach. The pilots not only showed that the technique could be useful in a wide range of circumstances, but also suggested that it can be used to consider social and economic as well as environmental capital. New guidance, on "quality of life capital", based on these pilots is available at http://qualityoflifecapital.org.uk. Parts of the approach are already incorporated in the Government's New Approach to Appraisal (§5.6.2).

This chapter summarises the Quality of Life Capital approach and gives an example of its use. It then discusses the benefits and limitations of the approach, and how it relates to EIA. It concentrates on those benefits for human quality of life that come from the environment, because the method has been most thoroughly piloted and tested on these, and because planners and practitioners often need a tool for this specific purpose. However the same method can also embrace social and economic capital.

17.2 The Quality of Life Capital approach

All applications of the approach involve the same six basic steps:

A. Purpose The first step is to be clear about the purpose of the study, since the details of what needs to be done vary greatly with the purpose. In the context of EIA, the purpose would normally be to compare the suitability of various sites for a given development proposal, to compare different proposals for the same site, and/or to optimally manage the development of a certain site.

B. Area/features This stage essentially involves collecting baseline information on the relevant area and/or features, as is done in the EIA baseline environmental description stage. The purpose will imply which area or features need to be studied. As in EIA, this area is likely to extend beyond the site boundary. For comparing potential development sites already identified, Quality of Life Capital could concentrate on the *differences* between them, whereas an exercise carried out to identify possible sites would need to consider the whole area. Where an area is diverse, this stage may involve classifying and describing areas of common character, as in landscape characterisation. In other cases – particularly for historical and cultural resources – it may involve determining why the resource matters, as a lead-in to Stage C.

C. Benefits and services This stage identifies what benefits and services the area or features provide. For instance a woodland (feature) could provide recreation, visual amenity, biodiversity and carbon fixing (benefits). Disbenefits are also identified at this stage.

D. Evaluation This stage examines the benefits/services systematically, using a series of questions (the last two of which do not apply to disbenefits):

- *whom do the benefits/services matter to, why, and at what spatial scale?* For example, habitat quality may matter for biodiversity at a regional or national scale, while recreational access may matter for quite specific groups of people from a small local area;
- *how important are the benefits/services?* A benefit that matters at national level is not necessarily more important than one that matters only locally;
- *are there enough of them?* It is more important to maintain benefits which are in short supply than ones that are plentiful. Where there are not enough, the aim should be to increase the level;
- *what (if anything) could make up for any loss or damage to the benefit?* Examples include other places where local people could go equally readily for the same type of recreation, or other areas that could be managed to support displaced bird populations.

This step needs to reflect the views of both experts (for internationally, nationally and regionally important benefits) and the local community (for locally important benefits). It thus draws on public consultation and involvement processes as well as technical appraisal methods such as characterisation studies.

E. Management implications This stage draws from the evaluation messages about the policies or 'rules' that would be needed to ensure that Quality of Life Capital is enhanced rather than damaged. In the EIA context, this step would aim to develop conditions (possibly couched as planning conditions) that any future development on the proposed site should fulfil. Where several sites for development are being compared, the number and complexity of the conditions for each site can give an indication of how appropriate the sites are for development, and can help to rank sites in terms of ease of development and the likely sustainability benefits that development could provide.

F. Monitoring The benefits and services identified as important in the process are the aspects of the environment which should be monitored. Quality of Life Capital thus provides its own performance indicators.

17.3 An example of the approach

Table 17.1 shows partial and simplified results from a Quality of Life Capital exercise that aimed to determine whether an existing stone quarry in Gloucestershire should be extended and, if so, how. The example focuses on those benefits for human quality of life that come from the environment, but similar principles apply to social and economic capital.

The proposed extension site was an agricultural field of about 4 ha bounded by hedges and crossed by a footpath. A road, Rock Road, ran between the existing quarry and the proposed extension. The proposal involved re-routing the footpath and road around the extension. Nearby residential properties had already been affected for many years by the noise and disruption of quarrying operations. The developer proposed to relinquish their existing planning permission for quarrying at a nearby site in return for permission to extract from the field.

Table 17.1 shows that the Quality of Life Capital approach suggested innovative ideas for pedestrian access which went beyond the 'replace like for like' approach initially proposed in the developer's EIA. It suggested enhancements regarding the footpath network, and acknowledged that some benefits – for instance footpath access between Rock Road and other footpaths – would not need to be replaced. It highlighted the importance of maintaining a small, rural scale for any realignment of Rock Road, in contrast to traditional 'engineering' design solutions that empha- sise safety and speed. It addressed the local residents' wish to gain certainty about the end of the quarrying operations. It also suggested importance rankings for the management implications. It showed that, in terms of virtually all benefits – recrea- tion, biodiversity, visual amenity, and badger habitat – the site with the existing planning permission was superior to the field, so that the developer's proposed 'trade' of planning permissions would be environmentally beneficial.

Overall the Quality of Life Capital approach seemed to formalise, and make more transparent and objective, the planning officers' existing good practice approaches to dealing with such sites. It also provided more flexible, less onerous 'rules' for the developers, and focused on enhancement as well as maintenance of environmental benefits.

17.4 Advantages of the approach

A key aspect of the approach is that it changes the focus of analysis from *things* to the *benefits* that they provide. By doing so, it can suggest more flexible, more cre- ative solutions that focus on compensatory action rather than on trying to prove that an area simply cannot accommodate one more development. The emphasis on benefits and services suggests management measures that may not normally be addressed, for instance replacing the benefits of a footpath with a lightly-used road.

By concentrating on the end-result, the benefits provided by an area, the ap- proach effectively considers secondary and indirect impacts, which may well be more significant than the primary impact. Essentially, it sets the primary impact into their social/quality of life context. For instance, it may identify that closing a small segment of footpath would preclude people from being able to complete a well- loved, longer circular walk. The switch from things to benefits also recognises the

Table 17.1 Example of Quality of Life Capital exercise applied to a stone quarry and its potential for extension

Feature	Benefit	Scale/importance	Trend/target	Substitutability	Management implications
footpath across the site, from Rock Road to (other) footpath X	recreation	of very local importance, possibly of greater than very local importance if part of walking loop or network of footpaths; *ask footpath officer to comment*	unlikely to be a problem, but ask footpath officer to comment	any reasonably direct footpath linking Rock Road (existing or diverted) to footpath X	The new road would provide an adequate replacement *if* it is not more built-up than the existing road. The new footpath currently proposed by the applicants is likely to be too indirect, and walkers would probably use the road anyway. Possible enhancement: changing any of the permissive footpaths in the area into formal footpaths, and provision of a continuous footpath from Rock Road to the A100.
Rock Road	access from A to B for drivers and horseback riders	local/ of limited importance, but would cause outcry if lost	?	road access from A to B that does not significantly increase journey times or accident risk for drivers or horseback riders	Diversion of Rock Road around the extension site, provided that the new road is not more built-up than the existing one.
field	agriculture/ grazing	local/ of very limited importance	going down, but not at all close to target	none needed	none needed
site as a whole	symbolic end to the quarry, lack of encroachment	local, very important	there have been about a dozen extensions over the years, well above 'target' for nearby residents	• definite end date to quarrying • shorten total *time* of future extraction • remove existing unused permissions (i.e. reduce *area* of future extraction)	The following are **very important** implications: Definite end date and shortening time of future extraction could be done through planning conditions and Section 106 agreements, but could be subject to future renegotiations, and thus do not offer complete certainty to local residents. Relinquishing planning permission for the site with existing permission would give such certainty.

interrelations between many impacts that are normally considered separately in EIA, for instance, air, water and ecology (LUC/CAG 2000).

Through its focus on trends and targets/'enoughness', the approach inherently also considers cumulative impacts. For instance, whilst individual development projects may have no significant impact on climate (and their EIAs would say so), cumulatively they would, especially when past and likely future development trends are taken into account. The approach would help to identify these changes, relevant targets (e.g. Government targets for reducing CO_2 emissions as well as the much tougher targets that need to be achieved to stop climate change), and necessary actions (e.g. replacing each unit of carbon fixing lost as a part of a development projects with several units, but this could be anywhere in the world) (LUC/CAG 2000).

The approach also provides a systematic and transparent framework for considering the views of experts and local residents in a complementary manner. Its focus on enhancement could bring forward development that people actively want, rather than proposals with pasted-on mitigation measures to minimise negative impacts. Similarly, the approach focuses on understanding what is important to a given area, rather than on designating and protecting a limited number of 'best' areas. It thus helps to promote uniqueness, representativeness and diversity, not just quality.

The approach suggests that there is no fixed capacity for development, but instead a rising 'sustainability tariff'. The technique gives an indication of the quality of life benefits that a development would have to provide before it was considered acceptable and, as a corollary, indicates where development may not be appropriate. The more benefits the site has (and thus the more attractive to developers it normally is), the more requirements – sometimes complex and expensive ones – the developer would need to fulfil under a Quality of Life Capital approach. Faced with the many demands that are likely to be linked to greenfield sites, developers might conclude that brownfield sites are rather good for development after all. The approach would thus help to most effectively protect those sites that provide the most benefits, possibly reversing the current perverse incentive on developers to develop out-of-town sites (LGA 1999).

The approach also suggests a more rational approach to betterment or planning gain by identifying desirable and relevant improvements, and guiding development to achieve them. Whereas the existing system charges developers based on their economic gain (with the private realm essentially penalising the public realm), the Quality of Life Capital approach highlights how developers could be charged for the removal of environmental benefits (i.e. the private realm 'refunds' the public realm) (LGA 1999).

17.5 Links between EIA and Quality of Life Capital

Table 17.2 shows that EIA and Quality of Life Capital are complementary approaches to identifying and managing the impacts of proposed development projects. Quality of Life Capital is a particularly useful input to the pre-application scoping stage of EIA, where the project context, alternatives and constraints are identified and analysed. It can be used to evaluate potential development sites, compare alternative sites, or establish whether or not there are opportunity sites within the area of search. As the project planning evolves, another Quality of Life Capital 'check' can be carried out to ensure that the final project really does maintain or enhance the quality of

Table 17.2 Complementary role of EIA and Quality of Life Capital approaches

Aspect	EIA	Quality of Life Capital
carried out by	the developer	the competent authority as part of development/design brief, or the developer
considers	impact of a project on the environment	impact of the environment on projects
deals well with	'things', technical issues, 'objective' impacts, primary impacts	benefits that things provide, perceived impacts, secondary/indirect/cumulative impacts
public participation	seen as a safeguard to ensure that EIA findings are comprehensive and accurate	seen as a key component in identifying and analysing locally important benefits, complementary to expert views on regionally, nationally and internationally important benefits
mitigation	minimisation and remediation of all significant impacts sought; protects designated areas	maintenance and enhancement of all important benefits sought; promotes the uniqueness and diversity of all areas
relation to decision-making	can stop environmentally harmful development, but is seen as restrictive by developers	encourages environmentally beneficial development, and may be viewed more positively by developers, but this has not yet been tested in practice

life benefits of the site. This can be used to set a management framework (e.g. Section 106 obligations, planning conditions, etc.) for any development on a given site.

Quality of Life Capital has the potential to merge into the EIA process so that it takes a minimum of additional time and effort. Stages A and B (purpose, area/features) of Quality of Life Capital are already virtually identical to the early stages of EIA. Stages C and D (benefits/services, evaluation) are different: however, by considering public views at this stage, it may be possible to minimise public opposition at the later phases of project planning. In a minimal form, the Quality of Life Capital approach could also easily be incorporated into the development of design briefs and/or the appraisal of different sites as part of an environmental appraisal of a development plan.

On the other hand, even using both processes in tandem can still have limitations. Neither technique effectively determines the area to be analysed: although both recommend analysing 'higher' or 'appropriate' scales, the focus is still clearly on the site under consideration. Although Quality of Life Capital in theory is based on (sustainability) targets, in practice few such targets are known and agreed, and EIA generally does not consider 'targets' beyond those enshrined in government policy or legislation (e.g. air and water quality criteria). Both techniques are perceived by local authorities and developers as being expensive and time-consuming, although this may change as Quality of Life Capital techniques become more widely used.

References

CC (Countryside Commission), EH (English Heritage), EN (English Nature) and EA (Environment Agency) 1997. *The new approach to environmental capital: what matters and why?* Cheltenham: CC, also at http://www.qualityoflifecapital.org.uk.

LUC/CAG (Land Use Consultants/CAG Consultants) 2000. Quality of life capital http://www.qualityoflifecapital.org.uk.

LGA (Local Government Association) 1999. *Environmental capital, sustainability and housing growth: a report to the LGA by CAG Consultants.* London: LGA.

Appendix A
Acronyms, addresses, chemical symbols, and quantitative units and symbols

A.1 Acronyms and addresses

Organisation websites normally provide postal, e-mail and telephone information; so where their Internet addresses are known, only these are given. Links to UK Government organisations are available at http://www.open.gov.uk/index/orgindex.htm. Internet addresses may not be included here when they are given in a chapter or another appendix, and this applies to many acronyms, e.g. full names of designated nature and countryside conservation sites/areas are given in Table D.3.

ALGAO – Association of Local Government Archaeological Officers, http://www.algao.org.uk/

Ancient Monuments Society, St Andrew-by-the-Wardrobe, Queen Victoria Street, London EC4 5DE

BBCS – British Butterfly Conservation Society, http://www.butterfly-conservation.org/

BBS – British Bryological Society, http://www.rbge.org.uk/bbs/

BCT – Bat Conservation Trust, http://www.bats.org.uk/

BENHS – British Entomological and Natural History Society, http://www.benhs.org.uk/

BGS – British Geological Survey, http://www.bgs.ac.uk/

BLS – British Lichen Society, http://www.argonet.co.uk/users/jmgray

BS/BSI – British Standard/British Standards Institution, http://www.bsi-global.com/

BSBI – Botanical Society of the British Isles, http://members.aol.com/bsbihgs/

BTO – British Trust for Ornithology, http://www.bto.org/

CA – Countryside Agency, http://www.countryside.gov.uk/

Cadw – Welsh Historic Monuments, http://www.cadw.wales.gov.uk/

CBA – Council for British Archaeology, http://www.britarch.ac.uk/

CBD – Convention on Biological Diversity (adopted at UNCED), http://www.biodiv.org/

CC – Countryside Commission (now replaced by CA)

CCMS – Centre for Coastal and Marine Sciences, http://www.ccms.ac.uk/. Includes the Proudman Oceanographic Laboratory (POL), and the Plymouth and Dunstaffnage Marine laboratories.

CCW – Countryside Council for Wales, http://www.ccw.gov.uk/

CEAA – Canadian Environment Assessment Agency, http://www.ceaa.gc.ca/

CEC – Commission of the European Communities – see European Commission (EC)

CEFAS – Centre for Environment, Fisheries and Aquaculture Science, http://www.cefas.co.uk/

CEH – Centre for Ecology and Hydrology, http://www.ceh.ac.uk/. Includes: CEH Merlewood which hosts the ECN (Environmental Change Network); CEH

Monkswood which hosts the EIC (Ecology Information Centre) and BRC (Biological Records Centre); CEH Wallingford; CEH Windermere.

CIPFA – Chartered Institute of Public Finance and Accountancy, http://www.cipfa.org.uk/

CIRIA – Construction Industry Research and Information Association, http://www.ciria.org.uk/

COE – Council of Europe, http://www.coe.int/

CORINE – Co-ordinated Environmental Information in the European Union

CPRE – Council for the Protection of Rural England, http://www.cpre.org.uk/

CPRW – Council for the Protection of Rural Wales, http://www.cprw.org.uk/

DARDNI – Department of Agriculture and Rural Development, Northern Ireland, http://www.dardni.gov.uk/

DCMS – Department for Culture, Media and Sport, http://www.culture.gov.uk/

DETR – Department of the Environment, Transport and the Regions, http://www.detr.gov.uk/

DfEE – Department for Education and Employment, http://www.dfee.gov.uk/

DoE – Department of the Environment (merged with DoT to form DETR, 1997)

DOENI – Department of the Environment for Northern Ireland, http://www.doeni.gov.uk/

DoT – Department of Transport (merged with DOE to form DETR, 1997)

EA – Environment Agency, http://www.environment-agency.gov.uk/

EA-W – Environment Agency Wales, http://www.environment-agency.wales.gov.uk/

EC – European Commission (formerly used for the European Community, now the EU)[1]

EC-EDG – EC Environment Directorate General, http://www.europa.eu.int/comm/environment/

ECNC – European Centre for Nature Conservation, http://www.ecnc.nl/

Ecology WWW, http://www.botany.net/Ecology/

EEA – European Environment Agency, http://www.eea.eu.int/. Includes designated European Topic Centres (ETCs): AE(Air Emission); AQ (Air Quality); CDS (Data Sources); IW (Inland Waters); LC (Land cover); MC (Marine & Coastal); NC (Nature Conservation); Soil; and Waste.

EH – English Heritage, http://www.english-heritage.org.uk/

EHS – Environment and Heritage Service, Northern Ireland, http://www.ehsni.gov.uk/

EN – English Nature, http://www.english-nature.org.uk/

ENTRUST – The Environmental Trust Scheme Regulatory Body Ltd, http://www.entrust.org.uk/

EPA – environmental protection agency (see Appendix B)

ERIN – Environment Australia Online, http://www.erin.gov.au/

ETB – English Tourist Board, http://www.etb.org.uk/

EU – European Union (formerly European Community (EC))[1]

FAO – Food and Agriculture Organisation of the UN, http://www.fao.org/

FBA – Freshwater Biological Association, http://www.fba.org.uk/

FC – Forestry Commission, http://www.forestry.gov.uk/. Includes the regional Forestry Authorities.

FOE – Friends of the Earth, http://www.foe.co.uk/

FRCA – Farming & Rural Conservation Agency, http://www.maff.gov.uk/

Georgian Group, http://www.heritage.co.uk/georgian/

GO – government organisation (e.g. department, agency)

Greenchannel, http://www.greenchannel.com/

GRO – General Register Office (Scotland), http://www.gro-scotland.gov.uk/

HMIP – Her Majesty's Inspectorate of Pollution (now incorporated in EA)

HMSO – Her Majesty's Stationary Office, http://www.hmso.gov/uk (see also TSO)

HS – Historic Scotland, http://www.historic-scotland.gov.uk/

HSE – Health and Safety Executive, http://www.hse.gov.uk/

IAIA – International Association for Impact Analysis, http://www.iaia.org/

IAU – Impacts Assessment Unit, Oxford Brookes University, http://www.brookes.ac.uk/iau/

IEA – Institute of Environmental Assessment (now incorporated in IEMA)

IEEM – Institute of Ecology and Environmental Management, http://www.ieem.org.uk/

IEM – Institute of Environmental Management (now incorporated in IEMA)

IEMA – Institute of Environmental Management and Assessment, http://www.iema.net

IGO – intergovernmental organisation (including the UN)

IHBC – Institute of Historic Building Conservation, http://www.ihbc.org.uk/

IHT – Institution of Highways and Transportation, http://www.iht.org.uk/

Institute of Field Archaeologists, http://www.archaeologists.net/

IUCN – International Union for the Conservation of Nature (or WCU), http://www.iucn.org/

JNCC – Joint Nature Conservation Committee, http://www.jncc.gov.uk/

LA/LPA – local authority/local planning authority (UK) – see directory at LGA

LAD – local authority district

Landscape Institute, http://www.l-i.org.uk

LGA – Local Government Association, http://www.lga.gov.uk/

LWTs – local Wildlife Trusts (affiliated to TWT)

MAFF – Ministry of Agriculture, Fisheries and Food, http://www.maff.gov.uk/

Marine Conservation Society, http://www.mcsuk.org/

MLURI – Macaulay Land Use Research Institute, http://www.mluri.sari.ac.uk/

MO – Meteorological Office, http://www.met-office.gov.uk/

MS – The Mammal Society, http://www.abdn.ac.uk/mammal/

Naturenet, http://www.naturenet.net/

NAW – National Assembly for Wales, http://www.wales.gov.uk/

NCCAs – nature and countryside conservation agencies (see Appendix B)

NGO – non-government organisation

NI – Northern Ireland

NOMIS – National On-Line Manpower Information System, http://www.dur.ac.uk/

NRA – National Rivers Authority (now incorporated in the EA)

NSCA – National Society for Clean Air and Environmental Protection, http://www.nsca.org.uk/

NT – National Trust, http://www.nationaltrust.org.uk/

NTS – National Trust for Scotland, http://www.nts.org.uk/

ONS – Office for National Statistics, http://www.statistics.gov.uk/

OPCS – Office of Population Censuses & Surveys (now incorporated in ONS)

OS – Ordnance Survey, http://www.ordsvy.gov.uk/

PCTPR – Ponds Conservation Trust Policy and Research Division, http://www.brookes.ac.uk/pondaction

Plantlife – The Wild Plant Conservation Charity, http://www.plantlife.org.uk/

RANI – Rivers Authority, Northern Ireland, http://www.darni.gov.uk/

RCAHMS – Royal Commission on the Ancient and Historical Monuments of Scotland, http://www.rcahms.gov.uk/

RCAHMW – Royal Commission on the Ancient and Historical Monuments of Wales, http://www.rcahmw.org.uk/

RCHME – Royal Commission on the Historical Monuments of England (now incorporated in EH)

Royal Commission on Environmental Pollution, http://www.rcep.org.uk/

RSPB – Royal Society for the Protection of Birds, http://www.rspb.org.uk/

SE – Scottish Executive, http://www.scotland.gov.uk/

SEA – strategic environmental assessment

SEPA – Scottish Environment Protection Agency, http://www.sepa.org.uk/

SERPLAN – London and South East Regional Planning Conference

SI – Statutory Instrument (of UK legislation)

SIA – socio-economic impact assessment

SNH – Scottish Natural Heritage, http://www.snh.org.uk/

SO – Scottish Office (now Scottish Executive (SE))

SPAB – Society for the Protection of Ancient Buildings, http://www.spab.org.uk/

SSLRC – Soil Survey and Land Research Centre, http://www.silsoe.cranfield.ac.uk/sslrc/

TSO – The Stationery Office, http://www.the-stationery-office.co.uk/

TWT – The Wildlife Trusts, http://www.wildlifetrust.org.uk/

UNEP – United Nations Environment Programme, http://www.unep.org/. Includes: Global Resource Information Database, http://www.grid.unep.ch/; Environmental Technology Centre, http://www.unep.or.jp/; Regional Office for Europe, http://www.unep.ch/

UNESCO – United Nations Educational Scientific and Cultural Organisation, http://www.unesco.org/. Includes: Man and Biosphere Programme, http://www.unesco.org/mab; World Heritage Centre, http://www.unesco.org/whc

USDA-NRCS – US Department of Agriculture, Natural Resources Conservation Service (formerly Soil Conservation Service (SCS), http://www.nrcs.usda.gov/. Includes various Sections/Centres, e.g. Soils, Water and climate, Watersheds and wetlands, Ecology, and Habitat Management.

USEPA – US Environmental Protection Agency, http://www.epa.gov/

USFWS – US Fish and Wildlife Service, Division of Ecological Services, http://www.fws.gov/

USGS – US Geological Survey, http://www.usgs.gov/. Includes various Divisions, e.g.: Biology, http://biology.usgs.gov/; Cartography, http://mapping.usgs.gov/, Earth Resources Observation Systems (EROS), http://edcwww.cr.usgs.gov/eros-home.html; Water, http://water.usgs.gov/.

USNTIS – US National Technical Information Service, http://www.ntis.gov/

Victorian Society, http://www.victorian-society.org.uk/

WCMC – World Conservation Monitoring Centre, http://www.wcmc.org.uk/

WCU – World Conservation Union – see IUCN

WHO – World Health Organisation, http://www.who.org/

WO – Welsh Office (now National Assembly for Wales (NAW))

World Bank http://www.worldbank.org/

WT – Woodland Trust, http://www.woodland-trust.org.uk/

WWF – Worldwide Fund for Nature, http://www.panda.org/; WWF-UK, http://www.wwf-uk.org/

WWT – Wildfowl and Wetlands Trust, http://www.wwt.org.uk/

Note

1 Throughout the book: **EU** refers to the European Union *and* the former European Community; and **EC** refers to the European Commission, which used to be known (and is still sometimes referred to) as the Commission of the European Communities (CEC).

A.2 Chemical symbols and acronyms

Al	aluminium	Na	sodium
Ca	calcium	N_2O	nitrous oxide
Cd	cadmium	NO_2	nitrogen dioxide
CFC	chlorofluorocarbon	NO_x	nitrogen oxides
CH_4	methane	O or O_2	oxygen
CO	carbon monoxide	O_3	ozone
CO_2	carbon dioxide	P	phosphorus
Cu	copper	Pb	lead
EDTA	ethylene diamine tetra-acetic acid	PAH	polycyclic aromatic hydrocarbon
HCFC	hydrochlorofluorocarbon	PCB	polychlorinated biphenyl
Hg	mercury	SO_2	sulphur dioxide
K	potassium	TOMP	toxic organic micro-pollutant
Mg	magnesium	VOC	volatile organic compound
N	nitrogen	Zn	zinc

A.3 Quantitative units and symbols

c.	*circa*/about/approximately	/ l	per litre
cm	centimetre	m	metre
cumec	cubic metres per second	mg	milligram ($g \times 10^{-3}$)
dB	decibel	min	minute
g	gram	mm	millimetre
ha	hectare (10,000 m^2 = 2.471 acres)	ng	nanogram ($g \times 10^{-9}$)
hectad	10 × 10 km square based on the UK national grid	MW	megawatt
		ppb	parts per billion
h	hour	ppm	parts per million
Hz	hertz	s	second
k	thousand, e.g. 25 k = 25,000	tetrad	2 × 2 km square
kg	kilogram	yr	year
km	kilometre	µg	microgram ($g \times 10^{-6}$)
kJ	kilojoule	µm	micrometre ($m \times 10^{-6}$)
J	Joule	>	below/less than
>	above/greater than	≤	equal to or less than
≥	equal to or greater than		

Appendix B
UK environment and heritage authorities and agencies

Internet addresses are given in Appendix A

Category	Organisation	Principal statutory environmental roles
Executive Authorities	Department of the Environment, Transport and the Regions (DETR)	Environmental policies for the UK and in England
	Department of Agriculture and Rural Development, N. Ireland (DARDNI)	Land drainage, flood defence and watercourse maintenance policies (in addition to agriculture)
	Department of the Environment for Northern Ireland (DOENI)	Environmental policies in Northern Ireland (NI)
	Ministry of Agriculture, Fisheries and Food (MAFF)	Land drainage and coastal and flood defence policies (in addition to agriculture and fisheries)
	National Assembly for Wales (NAW)	Environmental policies in Wales
	Scottish Executive (SE)	Environmental policies in Scotland
	Department for Culture, Media and Sport (DCMS)	Responsibilities include policies on the built heritage and tourism
Environment Protection Agencies (EPAs)	Environment Agency (EA)	Environmental protection[1] (under DETR & MAFF) in England and Wales
	Environment Agency Wales (EAW)	A branch of EA but now also responsible to NAW
	Environment & Heritage Service (EHS)	Environmental protection (under DOENI) in NI
	Farming and Rural Conservation Agency (FRCA)	Environmental protection, rural economy, etc., with particular reference to farming (under MAFF & NAW) in England and Wales

Category	Organisation	Principal statutory environmental roles
(EPAs) (continued)	Rivers Authority, Northern Ireland (RANI)	Drainage, flood defence, watercourse maintenance/ restoration, sustainable urban drainage systems
	Scottish Environmental Protection Agency (SEPA)	Environmental protection (under SE) in Scotland
Nature and Countryside Conservation Agencies (NCCAs)	English Nature (EN)	Nature conservation in England
	Countryside Agency (CA)[2] (formerly CC)	Countryside conservation[3] in England
	Countryside Council for Wales (CCW)	Nature and countryside conservation in Wales
	Scottish Natural Heritage (SNH)	Nature and countryside conservation in Scotland
	Environment & Heritage Service (EHS)	Nature and countryside conservation in NI
	Joint Nature Conservation Committee (JNCC)	Committee of the nature conservation agencies; advice and information on nature conservation
Heritage Agencies	Cadw (Welsh Historic Monuments)	Historic buildings/archaeological sites in Wales
	English Heritage (EH)	Historic buildings/archaeological sites in England
	Environment & Heritage Service (EHS)	Historic buildings/archaeological sites in NI.
	Historic Scotland (HS)	Historic buildings/archaeological sites in Scotland

1 Environmental protection includes soils, air quality, water quantity/quality, coastal and flood defences, and pollution control.
2 In England, EN is the statutory consultee for nature conservation in EIA, but CA should be consulted also.
3 Countryside conservation includes countryside character (including rural occupations/economy), landscape and amenity, especially in areas such as National Parks and Areas of Outstanding Natural Beauty.

Appendix C
Sources of historical information in the UK

Organisations' full names and Internet addresses are given in Appendix A. These websites usually provide links to related organisations that are often also useful sources of information.

Maps

Early OS maps (from 1820) are available for most areas, and digitised maps dating from 1843 are available from OS (e.g. 1:10,000, 1:2500 scales). Prior to this, tithe maps, estate maps and enclosure act maps may be available, but the information these provide is limited and often unreliable.

Aerial photographs

Early aerial photographs may be available from a variety of sources, including the government agencies listed in Table 15.1, p. 371. For example, EH holds RAF photographs from 1940 to 1945 (including the 1946 national survey) and OS photographs from 1952 to 1979.

General historical information

This is available from the Bodleian Library, local libraries and museums, county records offices, and local history societies.

Ancient monuments, cultural heritage and archaeology

Extensive information is held in the National Monuments Records (NMRs) held by EH, EHS, HS, RCAHMS and RCAHMW. EH holds (a) subject databases, e.g. NewHIS (on buildings and sites) and Listed Buildings System (details of listed buildings), and (b) an SMR address list (see §7.4.1). RCHMS holds an online database (CANMORE) and EH is developing similar facilities. Information is also available from ALGAO, CA, NT, NTS and local history and archaeology societies.

Trade directories

These are available for most urban areas, and can provide historical information about activities within commercial operations. In addition, local history societies often produce useful books and pamphlets on commercial activities of note.

The local authority planning record

This is also often a good source of information. Records frequently go back to the early 1940s, and liaison with the **county records office** will often identify previously archived planning records. It should be noted, however, that the change in authorities, and their boundaries, has not eased the collection of data. Even where records are available, it is often difficult to find them, and this situation is bound to be further exacerbated by the recent loss of several County Councils and the development of Unitary Authorities in England. A good example of this is the difficulty in accessing waste management records for sites that closed prior to 1976 (the date on which responsibility for waste management moved from District to County Councils in England).

Other local records

These can include parish records and newspaper articles (although the latter are often eye-witness accounts of catastrophic events such as storms, floods or landslips and must be treated with caution).

Scientific data

These can include the results of long-term studies or investigations using techniques to analyse past conditions from samples such as sediment cores (see §G.4). They can provide valuable information on changes that have occurred in a range of aspects, e.g. land cover, land use, climate, hydrology, geomorphology, and ecology. However, data for specific sites are rare.

Appendix D
Evaluating the conservation status of species, habitats and sites

Peter Morris

D.1 Introduction

Throughout the book, the term **high-status**, when applied to species, habitats or sites, means having high conservation status – or qualifying as VECs (see §11.4.1) – in terms of the criteria described in this Appendix. Some criteria can be considered alternatives, and many are complementary.

Interpretation of some criteria requires caution. For example, **rarity** is usually evaluated in relation to geographical range (e.g. local, regional, national, international) and in general its conservation importance increases accordingly. However, the perception of rarity varies in both space and time (Gaston 1994). For instance, a species may (a) have a wide geographical range, but exist only as small localised populations, or (b) have large populations but a small geographical range. Similarly, rarity within local areas varies in relation to different types of species distribution, e.g. restricted to, but abundant in, a few habitats, or widespread but infrequent; and for this reason, simple presence records in hectads, or even tetrads, may be of limited value for assessing impacts on local populations. Finally, rarity varies in time, e.g. by temporary population fluctuations or longer-term trends; and much attention is now paid to rates of decline, e.g. during the past 25 years. However, a further complication is that populations of a species may be declining in one area but increasing in another.

In general, there is normally little point in highlighting the local rarity of a species that is common elsewhere unless there is some other reason to justify its importance locally (see LBAP criteria in Table D.1). One reason for local or national rarity is that the area is near the normal limit of the species' range. Differences between national and international contexts explain some discrepancies between the rarity of some species and habitats in the UK and their status in the Habitats and Species Directive (HSD). This is because Annex I habitats and Annex II species are selected in the European context (§11.3.1) and consequently (a) include some examples that are locally common in the UK, but rare in the EU as a whole, and (b) exclude others that are rare in the UK but not threatened in the European context (e.g. see Palmer 1995).

D.2 Evaluation of species

The main criteria for evaluating the conservation status to species in the UK are listed in Table D.1. Given species are often included in more than one category.

Table D.1 Criteria used for identifying high-status species in the UK

Categories	Criteria
Protected status	Specified in Annexes/Schedules of International conventions and/or EU/UK legislation, or in specific legislation such as the Badgers Act (see Tables 11.2 and 11.3)
IUCN (1994) Red List CR = Critically endangered EN = Endangered (RDB1)[1] VU = Vulnerable (RDB2)[1] LR = Lower risk DD = Data deficient EW = Extinct in the wild	• population decline • limited extent of occurrence or area of occupancy in combination with fragmentation, decline or fluctuation • low numbers in combination with decline • very small or restricted population • analysis of the probability of extinction within a specific time
JNCC (http://www.jncc.gov.uk/): British Red Books (RDBs) R = Nationally rare (RDB3)[1] (Near threatened) NS = Nationally scarce (notable)	Include RDBI, 2 and 3 categories and criteria Terrestrial – present in ≤ 15 hectads Marine – present in ≤ 8 hectads within Britain's 3-mile territorial limit Terrestrial – present in 16–100 hectads Marine – present in 9–55 hectads within Britain's 3-mile territorial limit
UKBAP key (UKBG 1998): Priority Conservation concern	• globally threatened • rapidly declining in the UK, i.e. by > 50% in past 25 years • protected by legislation • threatened endemic and other globally threatened • the UK has > 25% of the world or relevant biogeographic population • numbers or range have declined by 25–49% in past 25 years • in some cases where found in < 15 hectads in the UK

LBAP other than species with UKBAP status (ULIAG 1997, Guidance note 4):	
Declining locally	25–49% decline in numbers or range in LBAP area in past 25 years
Rapidly declining locally	≥ 50% decline in numbers or range in LBAP area in past 25 years
Locally rare, scarce, common	Occurs in < 0.6%, 0.6–4.0%, > 4.0% of tetrads in LBAP area
Directly threatened	Habitat requirements threatened by lack of or inappropriate management
Indirectly threatened	Threatened indirectly by human activities, e.g. recreation, pollution
Historically 'endemic'	Believed to have always been 'endemic' to the LBAP area

Table D.1 (*continued*)

Categories	Criteria
Currently 'endemic'	Is now the only UK population, but previously occurred elsewhere
Highly localised	Comprises ≥ 20% of the UK population, or ≥ 10 times the proportion of the UK covered by the LBAP area (whichever is lower)
Localised	Comprises 10–19% of the UK population, or 5–9.9 times the proportion of the UK covered by the LBAP area (whichever is lower)
Isolated	Currently separated from other populations, and may enhance the species' genetic diversity
Outlying	Is at the edge of its range in the LBAP area
Flagship (high profile)	Has popular appeal that can influence other issues such as habitat protection
Keystone	Ecologically important (with a major/vital influence on the functioning of a community, e.g. a key role in a food web) and/or can be used: (a) as an ***indicator species*** of habitat health/quality (with fluctuations in abundance indicating habitat change); or (b) to identify genetic issues in the environment
Typical	Not necessarily identified as being of conservation concern, but is particularly associated with, or characteristic of, the area

1 RDB1, 2 and 3 are 'endangered', vulnerable' and 'rare' categories used in British RDBs.

Protected species lists are revised periodically. Current lists of species protected in the UK can be accessed at http://www.nbn.org.uk

The **IUCN Red List** criteria and categories were revised in 1994 (IUCN 1994). The criteria are applied quantitatively to assign species to the main categories (shown in the table). The Lower Risk category is subdivided into Conservation dependent, Near threatened, Nationally scarce and Least concern. A diagram showing the hierarchical relationships of the categories is available from http://www.jncc.gov.uk/. IUCN Red List publications are cited in Appendix E.

In addition to the **JNCC criteria** listed in Table D.1, account is also taken of criteria such as species' local importance and the status of UK populations in relation to global and regional contexts, e.g. if the UK population ≥ 1% of the global or European population. JNCC (and other) publications on the conservation status of British species are cited in Appendix E.

The **UKBAP key species** (§11.3.2) categories and criteria were proposed by UKBSG (1995), and subsequently modified by UKBG (1998). The UKBSG report included 116 SAPs, and UKBG have produced a further 275 (UKBG 1998, 1999). SAPs include an assessment of current status, actions to be taken, national targets and costings. The priority species list and SAPs are available from http://www.jncc.gov.uk/ukbg.

LBAP species automatically include species with UKBAP status, but may also be selected by criteria such as local importance or decline. This emphasis on local conservation status is particularly relevant to EIA. The keystone species criterion includes indicator species of habitat health/quality, which focus on habitat (rather than species) evaluation. For example, lichens can be used to indicate whether or not a habitat is subject to air pollution (Richardson 1992).

D.3 Evaluation of habitats and sites

The main criteria currently used for evaluating UK habitats and sites are listed in Table D.2.

D.3.1 Sites with protected status

Types of protected site in the UK are listed in Table D.3. Sites may have more than one designation, e.g. AOSPs, Biosphere and Biogenetic Reserves, NNRs, SACs, SPAs and Ramsar sites are also notified as SSSIs. The relevant legislation is outlined in Tables 11.2 and 11.3, and further information on designation criteria, protection, and locations is available from the DETR and JNCC websites. Sites with international designations have the highest conservation status, followed by sites designated under UK statute. However, the ecological and conservation value of non-designated sites should not be discounted, and these should be awarded high status if they host species or habitats that qualify in terms of other criteria (outlined below).

D.3.2 Nature Conservation Review criteria

The NCR criteria (Ratcliffe 1977) have been widely used for evaluating sites in the UK, and are employed, together with the NVC (Appendix F.4), in the selection of SSSIs (JNCC 1998).

Large size generally enhances habitat/site value (see §11.5.4). However,

- Minimum viable size varies for different species and communities, e.g. (a) in a farmland area, a 10 ha fen may dry out while a 1 ha drier meadow may retain its floristic composition if well managed (JNCC 1998), (b) small habitats can support some high-status species.
- The edge habitats of sites with small area : edge ratios can host species-rich communities and high-status species.
- Small sites may be valuable as *stepping-stone habitats*.

Diversity (biodiversity) can be assessed in terms of: (a) **habitat diversity**, which is the variety of habitats/communities in an area, and/or (b) **species richness/ diversity** (see Table 11.1). High habitat diversity and/or species diversity is generally considered valuable. However, caution is needed in interpreting species diversity values because:

- they are area dependent, i.e. normally increase with increasing area, so data are strictly compatible only when obtained from sampling areas of similar size;
- strictly, they should be derived from all the species of a community, which is normally only possible for plant communities (see Appendix G.3.1);

Table D.2 Criteria used for identifying high-status habitats and sites in the UK

Sites with protected status (especially national or international designations)

Sites hosting protected habitats or species, e.g. HSD Annex I habitats and Annex II species, and species listed under the Bern Convention, Bonn Convention and Wildlife & Countryside Act (see Tables 11.2 and 11.3)

Nature Conservation Review (NCR) habitats/sites (Ratcliffe 1977, JNCC 1998)
Primary criteria: size; diversity; rarity; naturalness; typicalness; fragility/sensitivity; non-recreatability
Secondary criteria: recorded history; position in an ecological/geographic unit; potential value; intrinsic appeal

UKBAP priority habitats (UKBG 1998):
- for which the **UK has international obligations;**
- are **at risk,** e.g. have declined rapidly in the past 20 years (see also Rodwell *et al.* (1997);
- may be **functionally critical,** i.e. are part of a wider ecosystem but provide reproductive or feeding areas for particular species (e.g. seagrass communities that are spawning grounds for fish);
- are **important for UKBAP priority species** (species–habitat associations are listed in EN 1999).

LBAP key habitats (ULIAG 1997, Guidance note 4) – which have UKBAP priority status and/or are:
Declining or rapidly declining – 25–49% or ≥ 50% decline in extent in LBAP area in past 25 years;
Endemic – Comprises 100% of total UK resource of the same habitat;
Significant or highly significant – Comprises 10–19% or 20–99% of total UK resource of the same habitat;
Rare or scarce – Covers < 0.6% or 0.6–4.0% of total LBAP area;
Threatened – Direct or indirect threats as for species (Table D.1);
Continuous or fragmented – with or without potential for increase in area (and linking fragments);
Viability in terms of size – Viable, potentially viable (has potential for increase in area), or not viable;
Local distinctiveness – e.g. characteristic of local area, of special historical/cultural importance;
Key species' habitat – Important for UKBAP species including keystone or 'flagship' species.

Additional criteria
If a habitat:
- is in an area that contains a **nationally** or **internationally important number** (≥ 1% of the UK, European or world population) **of a bird species,** as a resident or regular visitor, e.g. seasonal or during migration;
- is evidently **suitable for a high-status species even if this is not now present,** e.g. 1) it includes a suitable breeding habitat near to a known population of a protected bird species; 2) it is suitable for a threatened amphibian, reptile or invertebrate species, and (a) has previously hosted this or (b) if no record exists, is within its geographical range – especially if it occurs in similar habitats nearby;
- is on the **Invertebrate Site Register** (Table 11.4) or is of high value for invertebrates (Kirby 1992);
- is **ancient woodland.**

Sites and countryside areas designated under international conventions and EU Directives

Areas of Special Conservation Interest (ASCIs) – to protect habitats in a pan-European (not just EU) EMERALD network, equivalent to (and encompassing) the EU *Natura 2000* network of SACs and SPAs.

Biogenetic Reserves – to protect representative examples of European flora, fauna and natural areas.

Biosphere Reserves – to conserve globally significant examples of *biomes*.

Environmentally Sensitive Areas (ESAs) – areas with landscape, wildlife or historic features of national importance. Protection by incentives for landowners to adopt environmentally sensitive management.

European Diploma Sites (EDSs) – awarded by COE for natural heritage importance and current protection.

Ramsar Sites or **Wetlands of International Importance (WIIs)** – to protect wetlands of international importance, especially as waterfowl habitats; damage to any part requires an equivalent compensatory designation.

Special Areas of Conservation (SACs) – to protect **Annex I habitats/priority habitats** and **Annex II species/priority species**. Requirements for (a) "conservation measures involving, if need be, management plans", (b) measures to avoid habitat deterioration or disturbance of species, (c) EIA (see Table 11.2).

Special Protection Areas (SPAs) – to protect important habitats for naturally occurring wild bird species.

World Heritage Sites (WHSs) – mainly selected for heritage, but usually have nature conservation interest.

Sites and countryside areas designated under UK national statute

Areas of Outstanding Natural Beauty (AONBs) – to conserve natural scenic beauty.

Areas of Special Protection (AOSPs) (formerly Bird Sanctuaries) – for special protection of birds.

Country Parks – declared and managed by LAs. Primarily intended for recreation and leisure.

Limestone pavement order sites (LPOs) – created by LAs; limestone removal or damage prohibited.

Local Nature Reserves (LNRs) or **Local Authority Nature Reserves (LANRs)** in N. Ireland – declared by LAs on grounds of local rather than national importance.

Marine Nature Reserves (MNRs) – to protect marine wildlife, geological or geomorphological features.

National Nature Reserves (NNRs) – for particularly important ecosystems. Most are managed by NCCAs.

National Parks (NPs) – for outstanding areas of countryside, and their amenity and socio-economic value.

Natural Heritage Areas (NHAs) and **National Scenic Areas (NSAs)** – for wildlife/landscape (Scotland).

Sites of Special Scientific Interest (SSSIs) or **Areas of Special Scientific Interest (ASSIs)** in N. Ireland – The main UK site protection category. Notified by NCCAs using a set of quality and rarity criteria. Protected by agreements with landowners/occupiers (c.40% are owned or managed by public bodies or NGOs).

Table D.3 (*continued*)

Non-statutory sites and countryside areas

Ancient Woodland Sites – Inventories (AWIs) kept by EN and LAs.

UKBAP sites – containing UKBAP priority habitats or species, but lacking statutory status.

Forest Nature Reserves (FNRs) – designated by the Forestry Authority/Forest Enterprise on its land.

Heritage Coasts – about 33% of the coastline in England and Wales; designated for scenic value by CA/CCW.

National Trust/NT for Scotland properties – include areas of scenic and nature conservation value.

NCR sites – identified as valuable by NCR criteria, but lacking statutory status.

NGO nature reserves – established and usually owned (or leased) and managed by NGOs.

Preferred Conservation Zones (PCZs) (Scotland) – coastal areas unsuited to industrial development.

Regional Landscape Designations (Scotland) – sites afforded 'strong presumption against development'.

Sensitive Marine Areas (SMAs) (England) and **Marine Consultation Areas (MCAs)** (Scotland) – to promote awareness of their importance and sensitivity or because they are adjacent to statutory sites.

Sites of importance for nature conservation (SINCs) – recognised as having local or regional importance; usually adopted by LAs and shown on local planning maps to protect them from development.

- species diversity varies intrinsically in different communities, and it should only be used to compare like with like, e.g. plant species diversity is normally high in meadows and ancient woodlands, but low in heaths and bogs;
- animal species diversity is not necessarily correlated with plant species diversity, e.g. invertebrate species diversity is usually high in lowland heaths.

Rarity, naturalness and **typicalness** can be assessed by comparison with types defined in habitat or vegetation classifications such as the NVC (Appendix F.4). Naturalness is the degree to which a habitat or community approximates to a natural state, and typicalness is the degree to which it is a good example of those that are, or have been, characteristic of an area.

Sensitivity/fragility is the susceptibility of a habitat/community to environmental changes including project impacts. Assessment, which can include consideration of **resilience/recoverability** (see Table 11.1) requires an understanding of the ecology of the ecosystems in question, and will be related to other criteria such as size. In general, semi-natural communities, such as calcareous grasslands, are more sensitive than modified communities, such as improved grasslands.

Non-recreatability is usually related to naturalness because "the more natural an ecosystem, the greater the difficulty of re-creating it in original richness and complexity once it has been destroyed" (JNCC 1998). It applies particularly to long-established habitats with a complex community structure (e.g. *ancient woodlands*) and is important in EIA because it rules out habitat creation as a mitigation method for such habitats.

Recorded history can enhance a site's potential for education and research, and as a model for management.

Position in an ecological unit is when a site may be judged to belong to (and be an important component of) a larger ecological unit, e.g. part of a network. This can also apply to *linear habitats* and *stepping-stone habitats* which may increase the *connectivity* between larger sites/habitats or provide 'green networks' in urban areas.

Potential value acknowledges that a site's current ecological value may increase or decrease by natural change (e.g. ecological succession), management, degree of protection, or external influences (including climate change).

Intrinsic appeal takes account of public perception. It can include criteria such as visual/landscape, social/amenity, education, accessibility to residents, and presence in an area of deficiency. These are often highly valued, especially in urban environments (e.g. see Collis & Tyldesley 1993).

D.3.3 UKBAP and LBAP criteria

The **UKBAP priority habitat criteria** are designed primarily for the selection of priority habitats within the framework of a **broad habitat classification** (Appendix F.2). The system was originally devised by UKBSG (1995) and has subsequently been revised by UKGB (1998). UKBSG produced 14 **habitat action plans** (HAPs) for priority habitats (each including ecological information, an assessment of current status, actions to be taken, national targets, and costings) and UKBG have produced a further 20 (UKBG 1998, 1999). HAPs are available from http://www.jncc.gov.uk/ukbg.

UKBAP criteria are bound to be increasingly applied, and LBAP criteria (ULIAG 1997, Guidance note 4) should be useful in EIA because of their local context. The 'functionally critical' criterion (Table D.2) can be interpreted more widely to incorporate the need for sensitive management outside protected areas, e.g. by consideration of small and linear habitats that may be important in their own right, may function as refuges, or may increase *connectivity*. Two linear habitats, cereal field margins and old/species-rich hedges, are UKBAP priority habitats.

There is no specific evaluation method for cereal field margins, but **hedgerows** can be evaluated using the system devised by Clements & Tofts (1992), which divides them into four grades based on:

- **structure**, e.g. height, width, presence of standard trees, length and connectivity (including number and size of gaps and end-connections);
- **species richness** of woody species, and degree of dominance by native species;
- **associated features**, e.g. the presence of a bank, ditch, grass verge, and notable ground flora – especially relict woodland species (woodland *indicator species*).

Other criteria can be added, e.g.: (a) the nature of adjacent farm land, which has been found to affect the value of hedges for wildlife (Green *et al.* 1994); (b) the presence of animals such as birds, especially breeding pairs, and of high-status species. EN (1999) provides a list of UKBAP priority species associated with hedgerows. Similar criteria are used in the *Hedgerows Regulations* 1997, but these are under review because they are considered to be over-complex, and to neglect some attributes.

There is a link between a hedge's ecological and historical interest because the number of woody species present is usually related to its age (Pollard *et al.* 1974). A simple 'rule of thumb' is that the average number of woody species per 30 m length of hedgerow indicates its approximate age in 100 yr increments, e.g. 5 species per 30 m ≈ 500 yr. The relationship does not always hold, e.g. hedges that are relics of woodland tend to have more woody species than planted hedges, regardless of age (Wolton 1999). Consequently, estimates should be checked against historical evidence where possible (Appendix C). It is also worth remembering that straight field boundaries are often associated with the enclosure acts of the 1700s–1800s, and are unlikely to pre-date this period. Further information on the history and ecology of hedges can be found in Rackham (1986), Dowdeswell (1987) and Muir & Muir (1987).

D.3.4 Additional criteria

Most of the additional criteria listed in Table D.2 are factors suggested by IEA (1995) for triggering Phase 2 surveys. They can be interpreted as evaluation criteria, although in most cases the relevant habitats will already have high conservation status in terms of other criteria.

Ancient woodland is included here because it has no statutory protection and is not specifically designated in habitat or vegetation classifications (although similarity to a UKBAP, HSD or NVC woodland type is highly indicative). EN holds an inventory of ancient woodlands of > 2 ha (see Table 11.4) but many remaining fragments are smaller and may not have been identified. Ancient woodland can usually be identified by the features outlined in Table D.4. Further guidance is given in EN (1996), Marren (1990, 1992) and Rackham (1980, 1986, 1990).

Table D.4 Ancient woodland indicators

Documentary evidence

- It is shown on old maps, e.g. OS 1820.
- Its name includes: the name of a nearby settlement; an old name for 'wood' (e.g. grove, hanger, lea); reference to an old industry (e.g. kiln, tanner); or tree names (e.g. oak, ash, beech, hazel).

Location, form and historical features

- It has sinuous or irregular external boundaries, often with ditches and banks (may not apply if an original wood has been fragmented), lacks straight internal boundaries, and does not fit a field enclosure pattern.
- It is sited along parish boundaries, adjacent to common land or heath, on a steep slope, or in a stream valley.
- It shows evidence of coppicing, pollarding or other traditional uses, e.g. charcoal hearths, kilns.

Vegetation structure and composition

- It has a well-developed vertical structure (canopy, shrub, field and ground layers).
- It has a rich flora of mainly native species including **indicator species**.
- It contains rare species or species that are local to the area.
- The trees vary in age, and are not evenly distributed (as in plantations). A simple method for estimating the approximate age of trees is given in Mitchell (1974).

The most reliable indicator species are considered to be Ancient Woodland Vascular Plants (AWVPs), which are often taken to be species having ≥ 55% of their locations in ancient woods (Peterken 2000). However:

- Although they are generally indicative of woodland age (Peterken & Game 1984, Peterken 1993): (a) many can occur in more recent woodlands, especially when these are or have been located near ancient woods; (b) because of variations in climate, soils and past management, few are consistently associated with ancient woodland throughout the UK. Marren (1992) provides guidance on regional variations, and local lists are given in some county floras.
- Rose (1999) and Peterken (2000) stress that: (a) the presence of one or a few AWVPs may have little or no significance; (b) the number of AWVPs also tends to increase with woodland size; (c) taken alone, AWVPs cannot be taken as proof of a wood's antiquity, and reference should always made to historical data where these are available.

D.4 Numerical and semi-numerical methods

A number of evaluation methods have been developed which aim to increase objectivity by the use of numerical values. However, they often involve rather arbitrary procedures or have limited application, and only two examples are given here.

Scorecards provide a simple method for comparing and ranking 'subjects'. Scores for several criteria can be assigned to each subject; and a variety of scales can be used, although summation is only possible if the same scale applies to all criteria (Table D.5). The method can be used for objectives such as (a) selecting VECs (§11.4.1) to include in a survey, or (b) assessing the relative importance of different impacts. The method has the advantages of being simple and transparent, but limitations are:

- unless based on quantitative data, the scores are inevitably subjective, and experts with differing viewpoints may assign different scores for a given criterion;
- scorecards provide no assistance in determining if criteria overlap/interact or should be given different weightings;
- most people cannot accurately compare more than about seven issues, so large scorecards can lead to erroneous conclusions.

Table D.5 Hypothetical scorecard to compare subjects in relation to several criteria

	Criterion 1 (% scale)	Criterion 2 (1–10 scale)	Criterion 3 (0–5 scale)	Criterion 4 (+/– scale)	Sum (if applicable)	Rank (if possible)
Subject 1	15	5	5	+		2
Subject 2	40	3	2	0		3
Subject 3	60	6	4	++		1
Subject 4	10	4	1	–		4

Table D.6 Construction of a descriptive HSI (as ratings) and conversion of this to a numerical HSI (as scores) using ranking and the ratio of study-habitat rank to optimal rank. (Based on USFWS 1981.)

Descriptive HSI (ratings)	Rank	Ratio	Numerical HSI (scores)
Optimal	4	4:4	1.0
Good	3	3:4	0.75
Fair	2	2:4	0.5
Poor	1	1:4	0.25
Unsuitable	0	0:4	0

The **habitat suitability index** (HSI) was developed by the US Fish and Wildlife Service for use in their *Habitat Evaluation Procedure* (HEP) (USFWS 1980, 1981). HEP evaluates a habitat in terms of its **carrying capacity** for selected animal species (**evaluation species**) when compared with optimum habitats for these species – using measurable habitat 'quality' criteria such as vegetation composition. For each evaluation species, this is expressed as the HSI, which is derived from a habitat quality ratio (study-habitat conditions:optimum-habitat conditions) and ranges from zero (totally unsuitable) to one. Given adequate information on a species' habitat requirements, a simple model can be constructed using descriptive terms and ranking these to derive numerical values, e.g. as in Table D.6.

Once this has been achieved the HSI score is multiplied by the area of available habitat to obtain habitat units (HUs). For EIA, HUs are calculated for the habitat with and without the proposed development and can thus be used (a) to predict the potential loss of suitable habitat for the evaluation species, and (b) to formulate mitigation measures to avoid or minimise this loss. They are increasingly being employed with a GIS (Chapter 16) which facilitates comparison of the various scenarios.

Limitations of the method include:

- HEP only evaluates habitats in relation to evaluation species, selection of which is necessarily limited, and does not necessarily imply suitability for other species;
- HSIs assume a linear relationship between HSI values and carrying capacities, which may not always apply (Treweek 1999);
- as with all models, the output (a) can only be as good as the input information (on the species' habitat requirements and the relationship between these and the habitat variables measured) and (b) should be validated, e.g. against measured populations, before being widely applied;
- HSI models have been produced for many US species, but are largely restricted to these.

D.5 Evaluation in terms of 'capital resources'

A growing number of methods aim to evaluate natural resources, including biodiversity, as 'capital resources' analogous to economic capital (finance, goods, utilities, etc.). Like economic capital, these may be depleted, enhanced and in some cases replenished or 'traded'. Criteria used include notional monetary value, socio-economic benefits, and replacement value.

An example of the 'monetary value' approach is the *Habitat Replacement Cost Method* advocated by MAFF (1999, 2000). This suggests that SACs, SPAs, Ramsar sites and SSSIs could be considered to have a 'national economic value' based on the cost of protecting them *in situ*, or (if lower) the cost of replacing them. Other designated sites would have a lower 'local value'. Given the difficulties of satisfactory habitat creation (§11.6.3) this approach may be viewed with concern. Indeed, MAFF accepts that it:

- tends to favour habitats that are the most expensive to create, regardless of their ecological value;
- should not be taken to imply that habitat replacement is the most appropriate option and, particularly for *European* sites, there should normally be a presumption in favour of *in situ* habitat protection (with habitat replacement undertaken only as a last resort);
- cannot strictly apply to technically irreplaceable habitats such as ancient woodland.

Irreplaceability is included in the qualitative concept of **Natural Capital**, which was adopted by EN (1994) for application in ***Natural Areas***. There are two main categories:

1. *Critical Natural Capital* which refers to valuable 'aspects of biodiversity' that are virtually irreplaceable;
2. *Constant Natural Assets* which should not be allowed to fall below minimum levels within a Natural Area, but which in individual cases are replaceable or 'tradable'.

This concept is applied in NATA, in which some assessment scores refer to 'compensation' in terms of net gain, or no net loss, in a Natural Area (§11.5.8). However, 'natural capital' may be superseded by the 'Quality of Life Capital' approach (Chapter 17).

References

Clements DK & RJ Tofts 1992. *Hedgerow evaluation and grading system (HEGS) – a methodology for the ecological survey, evaluation and grading of hedgerows (test draft)*. Cirencester: Countryside Planning and Management.

Collis I & D Tyldesley 1993. *Natural assets: non-statutory sites of importance for nature conservation*. Local Government Nature Conservation Initiative, Hampshire County Council.

Dowdeswell WH 1987. *Hedgerows and verges*. London: Allen & Unwin.

EN (English Nature) 1994. *Sustainability in practice*. Peterborough: EN.

EN 1996. *Guidelines for identifying ancient woodland* (pamphlet). Peterborough: EN.

EN 1999. *Biodiversity: making the links*. Peterborough: EN.

Gaston KJ 1994. *Rarity*. London: Chapman & Hall.

Green RE, PE Osborne & EJ Sears 1994. The distribution of passerine birds in hedgerows during the breeding season in relation to characteristics of the hedgerow and adjacent farmland. *Journal of Applied Ecology* **31**, 667–692.

IEA (Institute of Environmental Assessment) 1995. *Guidelines for baseline ecological assessment*. London: E & F N Spon.

IUCN 1994. *IUCN Red List Categories.* Prepared by the IUCN Species Survival Commission. As approved by the 40th meeting of the IUCN Council. Gland, Switzerland: IUCN.

JNCC (Joint Nature Conservation Committee) 1998. *Guidelines for the selection of biological SSSIs* – revised edition of NCC (1989), with new chapters (available separately): Hodgetts NG 1992. *Non-vascular plants*; JNCC 1994a. *Bogs*; JNCC 1994b. *Intertidal marine habitats and saline lagoons.* Peterborough: JNCC.

Kirby P 1992. *Habitat management for invertebrates: a practical manual.* Sandy: RSPB.

MAFF 1999. *Flood and coastal defence project appraisal guidance: economic appraisal, FCDPAG3.* London: MAFF.

MAFF 2000. *Flood and coastal defence project appraisal guidance: Environmental appraisal, FCDPAG5.* London: MAFF.

Marren P 1990. *Britain's ancient woodland: woodland heritage.* Newton Abbot, Devon: David & Charles.

Marren P 1992. *The wild woods: a regional guide to Britain's ancient woodland.* Newton Abbot, Devon: David & Charles.

Mitchell A 1974. *A field guide to the trees of Britain and Northern Europe.* London: Collins.

Muir R & N Muir 1987. *Hedgerows: their history and wildlife.* London: Michael Joseph.

Palmer M 1995. *A UK plant conservation strategy: a strategic framework for the conservation of the native flora of Great Britain and Northern Ireland*, 2nd edn. Peterborough: JNCC.

Peterken GF 1993. *Woodland conservation and management*, 2nd edn. London: Chapman & Hall.

Peterken GF 2000. Identifying ancient woodland using vascular plant indicators. *British Wildlife* **11**(3), 153–158.

Peterken GF & M Game 1984. Historical factors affecting the number and distribution of vascular plant species in the woodlands of central Lincolnshire. *Journal of Ecology* **72**, 155–182.

Pollard E, MD Hooper & NW Moore 1974 *Hedges.* New Naturalist No. 58. London: Collins.

Rackham O 1980. *Ancient woodland: its history, vegetation and uses in England.* London: Arnold.

Rackham O 1986. *The history of the countryside.* London: Dent.

Rackham O 1990. *Trees and woodland in the British landscape.* London: Dent.

Ratcliffe DA (ed.) 1977. *A nature conservation review.* (2 vols). Cambridge: Cambridge University Press.

Richardson DHS 1992. *Pollution monitoring with lichens.* Slough: Richmond Publishing.

Rodwell JS et al. 1997. *Red data book of British plant communities.* Lancaster: Lancaster University Press.

Rose F 1999. Indicators of ancient woodland – the use of vascular plants in evaluating ancient woods for nature conservation. *British Wildlife* **10**(4): 241–251.

Treweek J 1999. *Ecological impact assessment.* Oxford: Blackwell Science.

UKBG (UK Biodiversity Group) 1998. *Tranche 2 Action Plans:* Vol. I *Vertebrates and vascular plants*; Vol. II *Terrestrial and freshwater habitats.* Peterborough: EN.

UKBG 1999. *Tranche 2 Action Plans:* Vol. III *Plants and fungi*; Vol. IV *Invertebrates*; Vol. V *Maritime species and habitats*; Vol. VI *Terrestrial species and habitats.* Peterborough: EN.

UKBSG (UK Biodiversity Steering Group) 1995. *Biodiversity: The UK Steering Group Report:* Vol. I: *Meeting the Rio challenge*; Vol. II: *Action Plans.* London: HMSO.

ULIAG (UK Local Issues Advisory Group) 1997. *Guidance for local biodiversity action plans: Guidance notes 1–5.* Bristol: UKBG.

USFWS (US Fish and Wildlife Service) 1980. *Habitat evaluation procedures (HEP).* Washington, DC: Division of Ecological Services, Department of the Interior.

USFWS 1981. *Standards for the development of habitat sustainability index models for use in the habitat evaluation procedure (HEP).* Washington, DC: Division of Ecological Services, Department of the Interior.

Wolton R 1999. Do we need hedges anymore? *Biologist* **46**(3), 118–122.

Appendix E
Publications on species conservation status, distribution, habitats and identification

The publications are listed in two sections: **E.1 Conservation status, distribution and habitats**; and **E.2 Identification**. However, some 'E.1' titles include identification aids, and identification books and keys normally include information on distributions and habitats. When an organisation publishes many relevant texts, space is saved here by grouping titles under the organisation's name; and (when details are available at the website) only citing the first author and abridging some titles.

E.1 Conservation status, distribution and habitats

Academic Press (T & AD Poyser), London (http://www.academicpress.com/)
Batten C et al. 1991. *Red Data Birds in Britain*; Gibbons D et al. 1993. *The New Atlas of Breeding Birds in Britain and Ireland*; Hagemeijer W 1997. *The EBCC Atlas of European Breeding Birds*; Mitchell-Jones A et al. 1999. *The Atlas of European Mammals.*

BES (British Ecological Society) 1992. Biological Flora listing. *Journal of Ecology* **80**, 879–882. (Biological Flora papers discuss the ecology and distributions of plant species.)

Birdlife International, Cambridge (http://www.wing-wbsj.or.jp/birdlfe)
Tucker 1994. *Birds in Europe: their conservation status*; Heath 2000. *Important Bird Areas in Europe: priority sites for conservation*. Vol. 1 *Northern Europe*, Vol. 2 *Southern Europe.*

Birks JDS & AC Kitchener (eds) 1999. *The distribution and status of the polecat Mustela putorius in Britain in the 1990s*. London: Vincent Wildlife Trust.

Blockeel TL & DG Long 1999. *A check-list and census catalogue of British and Irish bryophytes*. Cardiff: BBS, National Museum and Gallery of Wales.

BSBI (Botanical Society of the British Isles) (http://www.members.aol.com/bsbihgs)
Perring FH & SM Walters (eds) 1982. *Atlas of the British Flora*, 3rd edn; Jermy AC, HR Arnold, L Farrell & FH Perring 1978. *Atlas of ferns of the British Isles.*

COE (Council of Europe), Strasbourg (http://www.coe.fr/)
COE 1999. *Red Data Book of European Butterflies*; Heredia 1996. *Globally threatened birds in Europe: Action plans*; Maitland 1994. *Conservation of freshwater fish in Europe*; Wells 1992. *Threatened non-marine molluscs of Europe.*

Corbet GB & S Harris 1991. *The handbook of British mammals*, 3rd edn. Blackwell Scientific.

Cramp S & KEL Simmons 1977 et seq. *The birds of the Western Palaearctic*. Oxford: Oxford University Press.

Curtis TGF & HN McGough 1988. *The Irish Red Data Book 1: Vascular plants.* Dublin: The Stationery Service.

EIC-BRC (Ecology Information Centre-Biological Records Centre) (http:// www.nmw.ac.uk/ite/eicbrc)

Atlases and **Provisional atlases** (of Britain, British Isles or Britain and Ireland): Arnold 1993. *Mammals;* Arnold 1995. *Amphibians and reptiles;* Barber 1998. *Centipedes;* Drake 1991. *Larger Brachycera (Diptera);* Edwards 1997, 1998. *Aculeate Hymenoptera, Parts 1 & 2;* Haes 1997. *Grasshoppers, crickets and allied insects;* Harding 1985. *Woodlice – distribution and habitat;* Johnson 1993. *Cryptophagidea-Atomaritimae (Coleoptera);* Mendel 1990. *Click beetles;* Merritt 1996. *Dragonflies;* Norton 1985. *Marine algae;* Plant 1994. *Lacewings and allied insects;* Pont 1986. *Sepsidae (Diptera);* Stubbs 1992. *Long-palped craneflies;* Stubbs 1993. *Ptychopterid craneflies.*

English Nature Research Reports (RR) (complete list available from EN):

Drake 1998. *Invertebrates and their habitats in Natural Areas,* Vols 1 & 2, RR 298; Swan & Oldham 1993. *Herptile sites. Vol. 1: National amphibian survey,* RR 38; Vol. 2: *National common reptile survey,* RR 39; Grice 1994. *Birds in England: a Natural Areas approach,* RR 114; Kirby P. 1994. *Habitat fragmentation; species at risk. Invertebrate group identification,* RR 89; Mitchell-Jones 1996. *Status and woodland requirements of the dormouse in England,* RR 166; Mitchell-Jones 1997. *Priority Natural Areas for mammals, reptiles and amphibians,* RR 242; Porley 1997. *Rare and scarce vascular plants and bryophytes in Natural Areas,* RR 267.

Fox R & J Asher (in prep.) *Butterflies for the new millennium atlas.* Oxford: Oxford University Press.

Gibbons DW *et al.* 1996. Bird species of conservation concern in the United Kingdom, Channel Islands and Isle of Man: revising the Red Data List. *RSPB Conservation Review* **10**, 7–18.

Green R & J Green 1997. *Otter survey of Scotland 1991–94.* London: Vincent Wildlife Trust.

Grime JP & S Lloyd 1973. *An ecological atlas of grassland plants.* London: Arnold.

Grime JP, JG Hodgson & R Hunt 1988. *Comparative plant ecology.* London: Unwin Hyman.

Harley Books, Colchester

Hill MO, CD Preston & AJE Smith (eds). *Atlas of the bryophytes of Great Britain and Ireland:* 1991. Vol. 1. *Liverworts (Hepaticae and Anthocerotae);* 1992. Vol. 2. *Mosses (except Diplolepidideae;* 1994. Vol. 3. *Mosses (Diplolepideae).* Kerney MP 1999. *Atlas of the land and freshwater molluscs of Britain and Ireland.*

Heath J, E Pollard & J Thomas 1984. *Atlas of butterflies in Britain and Ireland.* London: Viking.

Hutson AM 1993. *Action plan for the conservation of bats in the UK.* London: Bat Conservation Trust.

Ing B 1992. A provisional Red Data List of British fungi. *The Mycologist* **6**, 124–128.

IUCN (The World Conservation Union), Geneva (http://www.iucn.org)

Bailie *1996 IUCN Red List of threatened animals.* Walter *1997 IUCN Red List of threatened plants,* Oldfield 1998. *The world list of threatened trees.* (All have online searchable databases.)

Herman JS & JF Haddow 1997. Recorded distribution of bats in Scotland. *Scottish Bats* **4**, 16–21.

Jalas J. *et al.* (eds) 1979–99. *Atlas Florae Europaeae,* Vols 1–12. (http://www.helsinki.fi/ kmus/)

JNCC (Joint Nature Conservation Committee) (http://www.jncc.gov.uk)

>   **Red Data Books (British or Britain & Ireland)**: Bratton 1991. *Invertebrates other than insects*; Church 1996. *Lichens*, Vol. 1 *Britain*; Shirt 1987. *Insects*; Stewart 1992. *Stoneworts*; Stewart (in press) *Mosses and liverworts*; Wigginton 1999. *Vascular plants*, 3rd edn.

>   **Reviews of the scarcer/scarce and threatened . . . of Great Britain**: Bratton 1990. *Ephemeroptera and Plecoptera*; Hyman 1992, 1994 *Coleoptera – Part 1, Part 2*; Falk 1991. *Bees, wasps and ants*; Falk 1991. *Flies, Part 1*; Kirby 1991. *Neuroptera*; Kirby 1992. *Hemiptera*; Parsons 1993. *Pyralid moths*; Parsons 1995. *Ethmiine, stathmopodine and gelechiid moths*.

>   **Other publications:** Fowles 1994. *Invertebrates of Wales: a review of important sites and species*; Harris 1995. *Review of British mammals: population estimates and conservation status. Guidelines for selection of biological SSSIs: non-vascular plants*; Hodgetts 1997. *The pink book of plants in Great Britain*; Hutchings 1996. *The current status of the brown hare*; Seaward: 1990. *Distribution of the marine molluscs of north-west Europe*; 1993. *Additions and amendments*; Stewart 1994. *Scarce plants in Britain*; Stone 1995. *Atlas of seabird distribution in north-west European waters*; Strachan 1996. *Pine marten survey of England and Wales 1987–1988*; Wallace 1991. *Review of the Trichoptera of Great Britain*.

Maitland PS & AA Lyle 1991. Conservation of freshwater fish in the British Isles: the current status and biology of threatened species. *Aquatic Conservation* **1**, 25–54.

Morris PA 1993. A Red Data Book for British mammals. London: Mammal Society.

NCC (Nature Conservancy Council), Peterborough

>   Ball SG 1986. *Terrestrial and freshwater invertebrates with Red Data Book, notable or habitat indicator status. Site register report*; Cresswell P, S Harris, DJ Jefferies 1990. *The history, distribution, status and habitat requirements of the badger in Britain*; Merrett P 1990. *A review of the nationally notable species of spiders of Great Britain*; Palmer MA & C Newbold 1983. *Wetland and riparian plants in Great Britain. An assessment of their status and distribution*; Smith A 1988. *European status of rare British vascular plants*.

Seaward MRD (ed.) 1995. *The lichen atlas of the British Isles*. London: British Lichen Society.

SNH (Scottish Natural heritage), Edinburgh (http://www.snh.org.uk)

>   SNH Reviews (R): *Local Biodiversity Action Plans – Technical information on species*: Ward 1999. *I. Cryptogamic plants and fungi*, R 70; McKinnell 1999. *II. Vascular plants*, R 125; Sivell 1999. *III. Invertebrate animals*, R 5; Ward 2000. *IV. Vertebrate animals*, R 10.

>   Research, Monitoring & Survey Reports (RSM): Balharry 1996. *Distribution of pine martens in Scotland as determined by field survey and questionnaire*, RSM 48; Balharry 1998. *Wild living cats in Scotland*, RSM 23.

Stewart NF 1995. *Red Data Book of European bryophytes, Part 1 (Parts 2 & 3 are in prep.)*. Dragvoll, Norway: Euopean Committee for the Conservation of Bryophytes (ECCB).

Strachan R & DJ Jefferies 1993. *The water vole Arvicola terrestris in Britain 1989–1990: its distribution and changing status*. London: The Vincent Wildlife Trust.

UKBG (UK Biodiversity Group), Bristol (http://www.jncc.gov.uk/ukbg)

>   *Tranche 2 Action Plans*: Vol. I *Vertebrates and vascular plants* (1998); *Tranche 2 Action Plans*: Vol. III *Plants and fungi* (1999); Vol. IV *Invertebrates* (1999); Vol. V *Maritime species and habitats* (1999); Vol. VI *Terrestrial species and habitats* (1999).

Ward D, N Holmes & P José (eds) 1994. *The new rivers and wildlife handbook*. Sandy: RSPB.

Warren *et al.* 1997. Assessing conservation priorities: an improved red list of British butterflies. *Biological Conservation* **82**, 317–328.

Whilde A 1993. *The Irish Red Data Book 2: Vertebrates*. Dublin: The Stationery Service.

Willing M 1993. Land molluscs and their conservation – an introduction. *British Wildlife* **4**, 145–153.

E.2 Identification

BSBI (Botanical Society of the British Isles) (http://www.members.aol.com/bsbihgs) BSBI Handbooks: Dudman 1997. *Dandelions*; Graham 1993. *Roses*; Jermy 1982. *Sedges*; Kent 1992. *List of Vascular Plants of the British Isles*; Lousley 1981. *Docks and knotweeds*; Meikle 1984. *Poplars and willlows*; Moore 1986. *Charophytes*; Rich 1991. *Crucifers*; Preston 1995. *Pondweeds*; Tutin 1980. *Umbellifers*.

Cambridge University Press, Cambridge (http://www.cup.cam.ac.uk) Clapham AR *et al.* 1987. *Flora of the British Isles*, 3rd edn; Miller PL 1987. *Dragonflies. Naturalists' Handbook 7*; Smith A 1978. *The moss flora of Britain and Ireland*; Smith A 1990. *The liverworts of Britain and Ireland*; Stace C 1997. *New flora of the British Isles*, 2nd edn; Stace C 1999. *Field flora of the British Isles*; Watson E 1995. *British Mosses and Liverworts*, 4th edn.

Collins (HarperCollins), London (http://www.harpercollins.co.uk) **Collins Field Guide** or **Pocket Guide . . . of Britain and Europe/N. Europe/ W. Europe** including: Arnold 1978. *Reptiles and amphibians*; Barrett & Yonge 1996. *Sea shore*; Chinery 1993. *Insects*; Courtecuisse 1995. *Mushrooms and toadstools*; Fitter 1984. *Grasses, sedges, rushes and ferns*; Fitter 1996. *Wild flowers*; MacDonald 1993. *Mammals*; Maitland & Campbell 1992. *Freshwater fishes*; Miller 1997. *Fish*; Mitchell 1998. *Trees*; Peterson 1993. *Birds*; Roberts 1995. *Spiders*; Sample 1998. *Bird call identification*; Tolman 1997. *Butterflies*; Svensson 1999. *Collins Bird guide*.

Daniels RE & A Eddy 1990. *Handbook of European Sphagna*, 2nd edn. London: HMSO.

Dobson FS 1992. *Lichens: an illustrated guide to the British and Irish species*. Slough: Richmond.

Drake CM 1991. Ephemoptera and Plecoptera in freshwater and brackish ditch systems on British grazing marshes. *Entomologist's Gazette* **41**, 45–59.

Elliott JM, JP O'Connor & MA O'Connor 1979. A key to the larvae of Sialidae (Insecta: Megaloptera) occurring in the British Isles. *Freshwater Biology* **9**, 511–514.

Ellis AE 1978. *British freshwater bivalve mollusca. Synopses of the British Fauna. N.S. No. 11 Linnean Society*. London: Academic Press.

FBA (Freshwater Biological Association), http://www.fba.org.uk – about 60 keys (usually with ecological notes) for fresh water taxa.

FSC (Field Studies Council) (web.ukonline.co.uk/fsc.dalefort) Keys of **British/British Isles** taxa, including: Friday 1988. *Adults of water beetles*; Haslam *et al.* 1982. *Water Plants*; Hiscock 1979. *Brown Seaweeds*; Hiscock 1986. *Red Seaweeds*; Hopkin 1991. *Woodlice*; Jones-Walters 1989. *Families of spiders*;

Plant 1997. *Adults of lacewings and their allies*; Wheeler 1994. *Shore fishes*; Wheeler 1997. *Freshwater fishes*; Willmer 1985. *Bees, Ants, and Wasps*; Wright 1990. *Sawflies*.

Hamlyn, London (http://www.hamlyn.co.uk)
Hamlyn guide to . . . of Britain and Europe including: Bruun B 1992. *Birds*; Campbell A 1994. *Seashores and shallow seas*; Humphries C 2000. *Trees*; Maitland PS 2000. *Freshwater fish*.

Harley Books, Colchester
Askew RR 1988. *The dragonflies of Europe*; Goater B 1986. *British pyralid moths. A guide to their identification*; Hammond CO 1983. *The dragonflies of Great Britain and Ireland*; Heath J et al. 1983 et seq. *The moths and butterflies of Great Britain and Ireland*, Vols. 1–10; Marshall JA & ECM Haes 1988. *Grasshoppers and allied insects of Great Britain and Ireland*; Paton JA 1999. *The liverwort flora of the British Isles*; Preston CD & JM Croft 1997. *Aquatic plants in Britain and Ireland*; Roberts MJ 1993. *The spiders of Great Britain and Ireland*.

Hayward PJ & JS Ryland (eds) 1995. *Handbook for the marine fauna of north-west Europe*. Oxford: Oxford University Press.

Hill MO (revised by N Hodgetts & A Payne) 1992. *Sphagnum: a field guide*. Peterborough: JNCC.

Hodgetts NG 1992. *Cladonia: a field guide*. Peterborough: JNCC.

Hubbard C 1984. *Grasses*, 3rd edn. London: Penguin Books.

Natural History Museum (NHM), London (http://www.nhm.ac.uk/science/)
Seaweeds of the British Isles: Irvine LM & YM Chamberlain 1994. Vol. 1 *Rhodophyta*, Part 2b *Corallinales, Hildenbrandiales*; Burrows EM 1991. Vol. 2: *Chlorophyta*. Jermy C & J Camus 1991. *The illustrated field guide to ferns and allied plants of the British Isles*. Purvis et al. 1992. *The lichen flora of Great Britain and Ireland*.

Pond Action 1994. *A guide to the identification of British freshwater snails*. Oxford: Pond Action, Oxford Brookes University.

Prys-Jones OE & SA Corbet 1991. *Bumblebees*. Slough: Richmond Publishing.

Rose F 1981. *The wild flower key for the British Isles and North-west Europe*. London: Warne.

Rose F 1989. *Colour identification guide to the grasses, sedges rushes and ferns of the British Isles and North-west Europe*. London: Viking.

Royal Entomological Society, London – *Handbooks for the identification of British Insects*, e.g. Chandler PJ 1998. *Checklists of insects of the British Isles (New Series)* Part 1: *Diptera*.

Sargent G & P Morris 1999. *How to find and identify mammals*. London: The Mammal Society.

Tutin TG et al. 1968–96. *Flora Europea*, Vols 1–5. Cambridge: Cambridge University Press.

Wolff W 1987. Identification. In *Biological surveys of estuaries and coastal habitats*, JM Baker & WJ Wolff (eds). Cambridge: Cambridge University Press.

Prys-Jones OE & SA Corbet 1991. *Bumblebees*. Slough: Richmond Publishing.

BioImages Virtual Field Guide (UK) (http://www.bioimages.org.uk/) (images of species).

Yalden DW 1985 (revised 1993). *Identification of British Bats*. London: The Mammal Society.

Appendix F
Habitat, vegetation and land classifications (and their limitations)

Classifications specific to freshwater ecosystems are referred to in Chapter 12, and the BioMar marine biotopes classification is outlined in Chapter 13.

F.1 The JNCC Phase 1 habitat classification

This is an integral part of the JNCC (1993) Phase 1 survey method (§11.4.4). It is a hierarchical system, with major (top level) habitats subdivided into sub-types (Table F.1). The habitats are defined in the broad sense, i.e. to include both the abiotic environment and associated communities (§11.2.4). The main criteria are (a) vegetation physiognomy (e.g. woodland, grassland), (b) environmental features of vegetated habitats (e.g. saltmarsh, sand dune, calcareous grassland) *or* substratum of non-vegetated habitats (e.g. rock, mud), (c) characteristic plant species, and (d) land use (e.g. improved grasslands and most category J types).

Surveyed habitats may not precisely match any designated type, or may be variants within a type such as broadleaved woodland, which includes a range of variants dominated by different tree species. The problem can be alleviated by using target notes, mapping codes and labels (e.g. for dominant species). It is also permissible to assign a name under 'J5 other habitats'. Addition of such categories should be normally kept to a minimum, but an exception in EIA may be to increase the number of some J-class types (e.g. urban, commercial and industrial buildings), thus extending the land use component of the classification (see F.6).

Table F.1 Outline of the JNCC Phase 1 habitat classification

A Woodland and scrub[1]

A1 Woodland – Dominated by trees > 5 m tall when mature, forming a definite (but sometimes open) canopy.

A1.1 Broadleaved – Dominated by broadleaved deciduous trees with ≤ 10% conifers in the canopy. Can be mono-dominated, e.g. oak or beech woodland; co-dominated, e.g. oak-ash woodland, or mixed (or at least have stands of different species). Varies in relation to climate, altitude, soils and management (past and present).

A1.1.1 semi-natural – < 30% obviously planted, e.g. *ancient woodlands* and plantations > c.120-yr old. Trees mainly native (but can include self-sown exotics, e.g. sweet chestnut, sycamore). Shrub and ground floras mainly native and often species-rich. Includes tall (> 5 m) alder or willow (except *Salix cineria*) carr.

Table F.1 (*continued*)

A1.1.2 plantation – > 30% of the canopy obviously planted (regardless of age). Often even-aged stands, with poorly developed and species-poor sub-canopy layers.

A1.2 Coniferous – Dominated by conifers with ≤ 10% broadleaved species in the canopy.

A1.2.1 semi-natural – Equivalent to A.1.1.1. The only native coniferous trees are *Pinus sylvestris* (Scots pine) (native in Scotland but reintroduced elsewhere) and *Taxus baccata* (yew).

A1.2.2 plantation – Equivalent to A.1.1.2. Usually commercial plantations (e.g. of non-native larches, firs, pines and spruces) with little or no sub-canopy vegetation.

A1.3 Mixed – 10–90% of either broadleaved or conifer species in the canopy.

A1.3.1 semi-natural – as above; **A1.3.2 plantation** – as above.

A2 Scrub – Dominated by native shrubs < 5 m tall. Includes montane willow scrub, willow carr < 5 m and all *Salix cinerea* carr (even if > 5 m). Lowland scrub is a seral community that will be replaced by woodland, but some upland scrub is climax vegetation. Can be: **A.2.1 continuous** or **A.2.2 scattered**.

A3 Parkland/scattered trees – Tree cover < 30%. Includes historically managed wood-pasture and parkland on grassland or heath. Can be: **A3.1 broadleaved**, **A3.2 coniferous** or **A3.3 mixed**.

A4 Felled woodland – Only used when future land use is uncertain, e.g. may be replanted or used for agriculture. Can be: **A4.1 broadleaved**, **A4.2 coniferous** or **A4.3 mixed**.

B Grassland and marsh[2]

Dominated by grasses and/or, in marshy areas, by sedges, rushes or marsh *forbs* (see B5). **Unimproved (semi-natural) grasslands** are usually maintained by traditional methods, mainly grazing and mowing. **Improved grasslands** may be sown and/or treated with fertilisers and herbicides.

B1 Acid grassland – On *oligotrophic*, acid soils (pH < 5.5) in the uplands or lowlands. Often unenclosed. Relatively species-poor. Can be: **B1.1 unimproved** or **B1.2 semi-improved**.

B2 Neutral grassland – On *mesotrophic*, circumneutral soils (pH 5.5–7.0). Usually lowland and enclosed or roadside verges, etc. May be moist and/or periodically water-logged or inundated. Often species-rich in grasses and forbs, e.g. **meadows** which can contain > 25 grass species, about twice as many forbs, and several sedges (*Carex* spp.) and rushes (*Juncus* spp.). Can be: **B2.1 unimproved** or **B2.2 semi-improved**.

B3 Calcareous grassland – On calcareous soils (pH > 7.0) over chalk or limestone. Sward short and usually species-rich when close-grazed, but taller (dominated by coarse grasses) and less species-rich when under-grazed. Can be: **B3.1 unimproved** or **B3.2 semi-improved**.

B4 Improved grassland – Usually species-poor with > 50% of species sown (e.g. rye-grass and clovers).

B5 Marsh/marshy grassland – On mineral soils or peat < 0.5 m deep. Dominated by grasses, e.g. *Molinia caerulea* (purple moor grass) or by sedges, rushes or marsh forbs.

C Tall herb and fern

C1 Bracken – Bracken dominant (C1.1) or in scattered patches (C1.2) (often invasive on grassland and heathland).

C2 Upland species-rich ledges – Mainly dominated by forbs and ferns.

C3 Other tall herb and fern – Stands of tall forbs and ferns.

Table F.1 (*continued*)

D Heathland[3]

Usually dominated (≥ 25% cover) by dwarf shrubs (but see D3 and D4); on oligotrophic, acid soils and thin peats (< 0.5 m deep). **Upland heaths** usually overly siliceous rock. **Lowland heaths** usually overly sands or gravels.

D1 Dry dwarf shrub heath – On well-drained soils, usually over sand or gravel. Dominated by ericoids, e.g. *Calluna vulgaris* (heather) and dwarf gorses, with a ground flora of mosses and lichens.

D2 Wet dwarf shrub heath – On wetter, peatier substrates than D1, with more hyrdophilous species, e.g. *Erica tetralix* (cross-leaved heath), *Molinia caerulea*, sedges and bryophytes – especially **Sphagna** (species of the genus *Sphagum*, commonly called 'bog mosses').

D3 Lichen/ bryophyte heath – Largely montane type, but with variants on sandy soils in some lowland areas, e.g. the Brecklands. Usually dominated by bryophytes and lichens, with < 30% *vascular plant* cover.

D4 Montane heath/dwarf forb – Montane and snow-bed vegetation dominated by sedges and rushes or dwarf forb communities (not dwarf shrubs).

D5 Dry heath/acid grassland mosaic – Mixture of D1 and B1; common in upland areas.

D6 Wet heath/acid grassland mosaic – Similar to D5, but a mixture of D2 and B1.

E Mires[4]

Peatlands with peat normally > 0.5 m deep (see §12.2.5)

E1 Bog – Ombrotrophic, oligotrophic and acid. Dominated by Sphagna, ericoids and cotton sedges.

E1.6.1 Blanket bog – Confined to cool, wet climates in the north and west. Covers the surface except on steep slopes. Often has a hummock-hollow complex with *Sphagnum*-rich pools and heath vegetation on hummocks.

E1.6.2 Raised bog – In estuarine and lowland floodplains and to moderate altitudes, where it may intergade with blanket bog. Typically has a central dome (with vegetation like blanket bog) and a marginal *lagg* stream or fen.

E1.7 Wet modified – Mainly on degraded (e.g. drained or cut) blanket or raised bog. Sphagna replaced by *Molinia* (purple moor grass), *Tricophorum* (deer grass), or ericoids, with frequent bare patches.

E1.8 Dry modified – Areas subject to heavy draining, burning or grazing. Sphagna replaced by *Eriophorum vaginatum* (hare's tail cotton sedge) or ericoids (e.g. *Calluna*) with mosses & lichens.

E2 Flush and Spring – Minerotrophic and soligenous (where groundwater seeps to the surface). Peat depth often < 0.5 m. Usually rich in bryophytes, sedges and rushes.

E2.1 Acid/neutral flush – Typically species-poor, with Sphagna, rushes and/or cotton sedges.

E2.2 Basic flush – Typically have a carpet of mosses with sedges.

E2.3 Bryophyte-dominated spring – Only at up-welling points. Vegetation usually mainly mats of mosses.

E3 Fen – Minerotrophic mires. Can be: (a) **Rich-fen** – fed by mineral-rich, calcareous waters (pH ≥ 5), vegetation species-rich with sedges, rushes, forbs and bryophytes; (b) **Poor-fen** – fed by mineral-poor, oligotrophic waters (pH < 5), vegetation usually species-poor with a high proportion of Sphagna.

Table F.1 (*continued*)

E3.1 Valley mire – On the lower slopes and floor of small valleys (e.g. in heathlands). Soligenous (with lateral water flow), so the vegetation can be rich-fen or poor-fen (depending on catchment geology).

E3.2 Basin mire – In basins with little through-flow of water (topogenous) or hence nutrients. Vegetation usually poor-fen with stands of swamp or woodland, often on a floating raft over a lens of water.

E3.3 Flood-plain mire – On mineral and/or peat substrate, usually inundated periodically, e.g. in winter. Generally topogenous, with vegetation similar to E.3.2.

F Swamp, marginal and inundation[4]

Have standing water permanently or for most of the year.

F1 Swamp – Dominated by tall, emergent *graminoids* (reeds, etc.) in standing water.

F2 Marginal and Inundation

F2.1 Marginal – Narrow strips (< 5 m wide) of emergent vegetation at margins of lowland watercourses. May include swamp species, but also large 'aquatic' forbs.

F2.2 Inundation – Open, unstable communities, periodically submerged, e.g. on river gravels and lake margins.

G Open water[4]

Beyond the limit of swamp or other emergent vegetation.

G1 Standing waters – Ponds, lakes, etc. (see §12.2.2)

G1.1 Eutrophic – Nutrient rich; pH > 7; water often turbid/green (due to algae); substrate often organic mud.

G1.2 Mesotrophic – Intermediate nutrient levels; pH *c*.7; water sometimes turbid due to phytoplankton.

G1.3 Oligotrophic – Nutrient-poor; pH 5.5–7; water clear (plankton sparse); substrate rocky, sandy or peaty.

G1.4 Dystrophic – Very nutrient-poor; pH 3.5–5.5; water often peat-stained; plankton and macrophytes sparse.

G1.5 Marl – Usually meso-eutrophic; pH > 7.4; water clear, calcium rich/*alkaline*; calcareous (tufa) deposits.

G1.6 Brackish – Usually coastal, e.g. **lagoons** (see §13.2.3) which are usually classed as littoral habitats (H1). Often host unusual communities that include algae, vascular plants, and invertebrates that rarely occur elsewhere.

G2 Running waters – Rivers, streams, etc. Divided (as standing waters) into G2.1–G2.6.

H Coastlands[5]

Includes littoral (H1, H2) and supralittoral (H3–H8) but not sublittoral habitats.

H1 Intertidal (littoral) – Habitats located between the extreme high-water spring-tide (EHWS) and extreme low-water spring-tide (ELWS) levels. Strictly includes H2 saltmarsh.

H1.1 Sand and mud – Host animals in the substratum (infauna) but generally lack surface dwelling organisms (epibiota) (see §13.2.3). However, *Zostera* (seagrass) beds (H1.(1–2).1) occur on some muddy sands.

H1.2 Shingle/cobbles – Are an unstable, hostile environment in the littoral zone.

Table F.1 (*continued*)

H1.3 boulders/rocks – Rocky shores (see §13.2.3): H1.(1–3).2 green algal beds; H1.(1–3).3 brown algal beds

H2 Saltmarsh – Develops where terrestrial vegetation can colonise sheltered mudflats.

H2.3 Saltmarsh/dune interface – Vegetation usually shrubby.

H2.4 Scattered plants – Usually lower marsh dominated by *Salicornia* (glasswort) spp.

H2.6 Dense/continuous – Dense stands of *Spartina anglica* (cord grass), or more species-rich swards with *Puccinellia maritima* (sea poa) and forbs.

H3 Shingle above high tide – **vegetated shingle banks** sometimes support scrubby vegetation or a grass sward, but more exposed areas have open vegetation with scattered vascular plants and lichens.

H4 Rock above high tide – Mainly lichen-dominated platforms in the 'splash zone'.

H5 Strandline vegetation – Open community at high tide level on shores.

H6 Sand dunes – Include a range of habitats. There are usually several dunes (aligned approximately parallel to the coastline and increasing in age along the sea–land axis) interspersed with depressions (dune slacks).

H6.4 Dune slacks – Depressions between dunes; usually wet with swamp, marsh or carr vegetation.

H6.5–H.6.7 Consolidated and flattened dunes:

H6.5 Dune grassland – Dominated by grasses such as *Festuca rubra* (red fescue). Includes machairs (§13.2.4).

H6.6 Dune heath – Similar to inland dry heaths (D1) with *Calluna* usually dominant.

H6.7 Dune scrub – Dominated by inland and/or coastal scrub species, e.g. *Hippophae rhamnoides*.

H6.8 Open dune – Semi-consolidated, including: embryo dunes; mobile dunes (dominated by *Ammophila arenaria* (marram grass); and grey dunes (older, more stabilised, often dominated by mosses and lichens).

H8 Maritime cliffs and slopes – Vary in relation to their geology and local landforms.

H8.1 Hard cliff (rock including chalk) and **H8.2 Soft cliff** (e.g. clay) – With < 10% vascular plant cover.

H8.3 Crevice and ledge vegetation – With ≥ 10% vegetation cover; on cliffs or in the 'splash zone' (H4).

H8.4 Coastal grassland – Often on cliff tops. Contains maritime species, e.g. *Scilla verna, Armeria maritima*.

H8.5 Coastal heathland – Like inland dry heath (D1) but with maritime species.

I Rock exposure and waste

Exposed inland surfaces with < 10% vegetation cover.

I1 Natural exposures – **I1.1 Inland cliff, I1.2 Scree/boulder scree, I1.3 Limestone pavement; I1.4 Other exposure** (I1.4.1 acid/neutral, I1.4.2 basic); **I1.5 Caves.**

I2 Artificial exposures – **I2.1 Quarry** (gravel, sand and chalk pits and stone quarries); **I2.2 Spoil** (abandoned industrial areas, coal spoil/slag), **I2.3 Mine; I2.4 Refuse-tip.**

Table F.1 (*continued*)

J Miscellaneous

J1 Cultivated/disturbed land: J1.1 Arable – croplands, leys, and horticultural land;

J1.2 Amenity – intensively managed grassland, e.g. lawn, golf course fairway;

J1.3 Ephemeral – short patchy vegetation on freely drained, usually thin, soils of derelict land.

J1.4 Introduced shrub – dominated by non-native shrubs.

J2 Boundaries: J2.1 Intact hedge (species-rich, species-poor);

J2.2 Defunct hedge (with gaps);

J2.4 Fence; J2.5 Wall; J2.6 Dry ditch; J2.8 Earth bank.

J3 Built-up areas: J3.4 Caravan site; J3.5 Sea wall (artificial material); **J3.6 Buildings**.

J4 Bare ground – Any bare soil or other substrate not included elsewhere in the classification.

J5 Other habitat – Any habitat not covered by the classification, and justifies mapping as a unit.

1 Information on woodlands can be found in Peterken (1993, 1996).
2 Information on grasslands can be found in Crofts & Jefferson (1999), Duffey *et al.* (1974), Hillier *et al.* (1990).
3 Information on heathlands can be found in Webb (1986).
4, 5 For further information and references, see Chapters 12 and 13, respectively.

F.2 UKBAP Broad habitats and priority habitats

The UKBSG report (§11.3.2) defined **broad habitat types** as a framework for selecting **priority habitats** in need of HAPs. Some of the habitats have since been redefined by UKBG (see Appendix D.3.3). The system is effectively a two-level habitat classification, except that priority habitats are only a selection of the types that may occur within broad habitats. Table F.2 lists the revised broad and priority habitats, and indicates their approximate correspondences with JNCC habitats.

F.3 The Habitats and Species Directive (HSD) Annex I habitat classification

HSD Annex I habitats are only a small selection of European habitat types. Those in the original Directive 92/43/EEC were selected from draft versions of the CORINE Habitat Classification (EC 1991). This is similar to the JNCC classification in that it is hierarchical and uses similar criteria; but it is more complex, with some major habitats progressively subdivided into five subsidiary levels.

Problems arose because the final CORINE classification (a) contained numerous revisions, which caused ambiguities in the interpretation of Annex I, and (b) was subsequently revised and extended to include the whole of the Palaearctic region, and was therefore superseded by the Palaearctic Habitat Classification (Devilliers & Devilliers-Terschuren 1996).

To rectify the problems, the *Interpretation manual of European Union habitats* was developed. The final version of this, EUR15 (EC 1996) includes the HSD habitats,

Table F.2 UKBAP broad habitats, priority habitats and related JNCC habitats

---

Broadleaved, mixed and yew woodland – >20% of the cover composed of broadleaved trees or these and yew trees. Includes recently felled stands, **carr**, and patches of scrub of > 0.25 ha with continuous canopy

Upland oakwood, Upland mixed ash wood, Wet woodland, Lowland beech and yew woodland, Lowland wood pasture and parkland (may be mainly grassland)

A1.1, A1.2 (if yew), A1.3, A2, A3, A4?

Coniferous woodland – all conifer stands (except yew) with < 20% cover composed of broadleaved trees

Native pine wood   A1.2.1 (if Scots pine), A1.2.2, A1.3, A3, A4?

Boundary and linear features – e.g. hedges, walls, field margins, road and railway verges, dry ditches.

Ancient and/or species-rich hedgerows, Cereal field margins   J2

Arable and horticulture – e.g. croplands, orchards, rotational set aside and fallow, horticultural land   J1

Improved grassland – species-poor, on any soil, sown or modified (e.g. by drainage, fertilisers or herbicides)

Coastal and floodplain grazing marsh (may be only semi-improved)   B4

Neutral grassland – un- or semi-improved; on circumneutral soils; managed or unmanaged; dry or wet.

Lowland meadows, Upland hay meadows   B2.1, B2.2

Calcareous grassland – un- or semi-improved, on shallow lime-rich soils, normally over chalk or limestone

Lowland calcareous grassland, Upland calcareous grassland   B3.1, B3.2

Acid grassland – unimproved or semi-improved, on acid soils

Lowland dry acid grassland   B1.1, B1.2

Bracken – areas of ≥ 0.25 ha dominated by continuous bracken.   C1.1

Dwarf shrub heath – vegetation dominated by heaths, dwarf gorses or mosses and lichens

Lowland heathland, Upland heathland   D1–D3, D5, D6

Fen, marsh and swamp – **minerotrophic** wetlands not dominated by grasses other than purple moor grass, reeds or sweet-grass.

Purple moor grass and rush pastures, Fens, Reedbeds   B5, E2, E3, F

Bogs – ombrotrophic mires   Blanket bog, Lowland raised bog   E1

Standing open water and canals – natural systems and man-made waters. Includes the open water zone (with submerged, floating or floating-leaved vegetation) and fringe vegetation

Mesotrophic standing waters, Eutrophic standing waters, Aquifer-fed naturally fluctuating water bodies   G1.1–G1.5

Table F.2 (*continued*)

Rivers and streams – includes the open-water zone, exposed shingle banks, etc., and fringe vegetation (to bank top or mean annual flood level)

| Chalk rivers | G2.1–G2.5 |

Montane habitats – heath, snow-bed, dwarf forb and lichen/bryophyte habitats

| | D3, D4 |

Inland rock – natural or man-made exposures with little vegetation

| Limestone pavements | I1, I2 |

Built-up areas and gardens – urban and rural developments, including parks but not amenity grassland

| | J3 |

Supralittoral rock – soft or hard rock, above the EHWS mark, but influenced by wave splash and sea spray

| Maritime cliff and slopes | H4, H8 |

Supralittoral sediment – mainly sand or shingle, above the EHWS mark but influenced by sea spray

| Coastal sand dunes, Machair, Coastal vegetated shingle | H3, H5, H6 |

Littoral rock – intertidal rock (rocky shores)

| Littoral chalk, *Sabellaria alveolata* reefs | H1.3 |

Littoral sediment – intertidal sands and muds forming features such as beaches, sandbanks and mudflats

| Coastal saltmarsh, Mudflats, Seagrass beds (littoral), Sheltered muddy gravels | H1.1 |

Inshore sublittoral rock – mainly reefs and near-shore rock, e.g. subtidal zones of rocky shores

| Sublittoral chalk, *Sabellaria spinulosa* reefs, *Modiolus modiolus* beds, Tidal rapids |

Inshore sublittoral sediment – sediment types as in the BioMar classification

| Sublittoral sands and gravels, Mud in deep water, Seagrass beds (sublittoral), Maerl beds, Serpulid reefs, Saline lagoons | G1.6 |

Offshore shelf rock – isolated reefs, usually of hard rock and swept by strong tidal currents

Offshore shelf sediment – unconsolidated benthic material and the overlying water column

| Sublittoral sands and gravels (offshore shelf) |

Continental shelf slope – the seabed and water column *Lophelia pertusa* reefs

Oceanic seas – the seabed and water column within UK waters beyond the continental slope.

and labels each with three codes: the original Annex I code; the Palaearctic Classification code, and a 4-digit 'Natura 2000' code. It was adopted by EU Directive 97/62/EC (EC 1997) in which Annex I is revised and the habitats are given the Natura 2000 codes. Some ambiguities remain, however, because a few EUR15 codes and/or habitat names are amended in the new Annex I.

Table F.3 lists the 1997 Annex I habitats and priority habitats that occur in the UK, and indicates their approximate relationships with UKBAP priority habitats, NVC communities, and BioMar categories. Partly because Annex 1 habitat types are selected on the basis of their pan-European importance, correspondences are often imprecise. For instance, some HSD habitats are not UKBAP priority habitats, and some UKBAP priority habitats (lowland wood pasture and parkland, ancient and/or species-rich hedgerows, cereal field margins, coastal and floodplain grazing marsh, upland calcareous grasslands, lowland dry acid grassland, poor-fens, reedbeds, and tidal rapids) are poorly represented in Annex I.

Table F.3 The codes[1] and names of HSD Annex I (1997) habitats and priority habitats (*) occurring in the UK, and their approximate relationships with:

UKBAP Priority habitats (see Table F.2)	NVC communities (see Table F.4)

BioMar habitat complexes (also > biotope complexes > biotopes related to UKBAP priority habitats)[2]	
ELR/MLR/SLR	Exposed/Moderately exposed/Sheltered **Littoral Rock**
EIR/MIR/SIR	Exposed/Moderately exposed/Sheltered **Infralittoral Rock**
ECR/MCR/SCR	Exposed/Moderately exposed/Sheltered **Circalittoral Rock**
COR	**Circalittoral Offshore Rock**
LGS/LMS/LMU/LMX	**Littoral** Gravels and Sands/Muddy Sands/Muds/Mixed **sediment**
IGS/IMS/IMU/IMX	**Infralittoral** Gravels and Sands/Muddy Sands/Muds/ Mixed **sediment**
CGS/CMS/CMU/CMX	**Circalittoral** Gravels and Sands/Muddy Sands/Muds/ Mixed **sediment**

1 4-digit numbers (bold) are Natura 2000 codes; bracketed numbers are CORINE codes as in the 1992 Annex I.
2 The BioMar biotope classification is explained in §13.4.5.

1. COASTAL AND HALOPHYTIC HABITATS

11. Open sea and tidal areas

1110 (11.25) Sandbanks which are slightly covered by sea water all the time

Sublittoral sands and gravels, Maerl beds[1], Seagrass beds	IGS > Maerl beds; IMU > Seagrass beds	SM1

1130 (13.2) Estuaries – comprise a range of habitats, e.g. can include **1140**, **1310–40** and **1410–20**.

1140 (14) Mudflats and sandflats not covered by sea water at low tide

Seagrass beds (littoral), **Mudflats**, Sheltered muddy gravels	LMS > seagrass beds; LMU, LMX	SM1

1150 (21) *Coastal lagoons

Saline lagoons	IMU > Angiosperm communities (lagoons)	A12, A21, S4, S20, S21, SM1, SM2

Table F.3 (*continued*)

1160 (12) Large shallow inlets and bays

Maerl beds, Mud in deep water	IMU; IMX > Maerl beds; CMU > Seapens and burrowing megafauna

1170 (11.24) **Reefs** – rock or biogenic concretions; sublittoral or extending into the littoral, e.g. as rocky shores

Sabellaria alveolata reefs, *Sabellaria spinulosa* reefs, Littoral and sublittoral chalk, *Modiolus modiolus* beds, Serpulid (*Serpula vermicularis*) reefs, *Lophelia pertusa* reefs
ELR, MLR > *Sabellaria alveolata* reefs, SLR; MCR > *Sabellaria spinulosa* reefs, Soft rock communities; MCR, SCR or CMX > *Modiolus modiolus* beds, CMS > Serpulid reefs; COR > *Lophelia* reefs.

12. Sea cliffs and shingle or stony beaches

1210 (17.2) Annual vegetation of drift lines	SD2, SD3

1220 (17.3) Perennial vegetation of stony banks

Coastal vegetated shingle	SD1 (others may be present)

1230 (18.21) Vegetated sea cliffs of the Atlantic and Baltic coasts

Maritime cliff and slopes	MC1–MC12; H6–H8

13. Atlantic and continental saltmarshes and salt meadows

1310 (15.11) *Salicornia* and other annuals colonising mud and sand

Coastal saltmarsh	LMU > Saltmarsh (pioneer)	SM8–SM10, SM27

1320 (15.12) *Spartina* swards (*Spartinion maritimae*)

Coastal saltmarsh	LMU > Saltmarsh (pioneer)	SM4–SM6

1330 (15.13) Atlantic salt meadows (*Glauco-Puccinellietalia maritimae*)

Coastal saltmarsh	LMU > Saltmarsh (low–mid, mid–upper)	SM10 – SM23

1340 (15.14) * Inland salt meadows	SM23

14. Mediterranean and thermo-Atlantic saltmarshes and salt meadows

1410 (15.15) Mediterranean salt meadows (*Juncetalia maritimi*)

Coastal saltmarsh	LMU > Saltmarsh (mid–upper)	SM15, SM18

1420 (15.16) Mediterranean and thermo-Atlantic halophilous scrubs (*Sarcocornetea fruticosi*)

Coastal saltmarsh	LMU > Saltmarsh (drift-line)	SM7, SM21, SM25

2. COASTAL SAND DUNES AND INLAND DUNES

21. Sea dunes of the Atlantic, North Sea and Baltic coasts

2110 (16.211) Embryonic shifting dunes	Coastal sand dunes	SD4, SD5

2120 (16.212) Shifting dunes along the shoreline with *Ammophila arenaria* ('white dunes')

Coastal sand dunes	SD6, SD7

Table F.3 (*continued*)

2130 (16.221 to 16.227) * Fixed coastal dunes with herbaceous vegetation ('grey dunes')

Coastal sand dunes	SD7–SD12

2140 (16.23) * Decalcified fixed dunes with *Empetrum nigrum*

Coastal sand dunes	H11b

2150 (16.24) * Atlantic decalcified fixed dunes (*Calluno-Ulicetea*)

Coastal sand dunes	H1d, H11a,c

2160 (16.25) Dunes with *Hippophaë rhamnoides*

Coastal sand dunes	SD18

2170 (16.26) Dunes with *Salix repens* ssp. *argentea* (*Salicion arenariae*)

Coastal sand dunes	SD14–SD16

2190 (16.31 to 16.35) Humid dune slacks

Coastal sand dunes	SD13–SD17

21A0 (1.A) Machairs (* in Ireland)

Machair	SD6–SD8, MG8, MG10, MG11

2250 * Coastal dunes with *Juniperus* spp.

Coastal sand dunes	no precise correspondences

23. Inland dunes, old and decalcified

2330 (64.1 × 35.2) Inland dunes with open *Corynephorus* and *Agrostis* grasslands

Lowland dry acid grassland	some forms of SD10–SD12

3. FRESHWATER HABITATS

31. Standing water

3110 (22.11 × 22.31) Oligotrophic waters containing very few minerals of sandy plains (*Littorelletalia uniflorae*)

No precise correspondences, but A22–A24 are similar

3130 (22.11 × (22.31 and 22.32)) Oligotrophic to mesotrophic standing waters with vegetation of the *Littorelletea uniflorae* and/or of the *Isoëto-Nanojuncetea*

Included in Mesotrophic waters, but this is a much broader category	A22, A23 and (less strictly) A24

3140 (22.12 × 22.24) Hard oligo-mesotrophic waters with benthic vegetation of *Chara* spp.

Included in Mesotrophic waters, but this is a much broader category	A11 and A13a are similar

3150 (22.13) Natural eutrophic lakes with *Magnopotamion* or *Hydrocharition*-type vegetation

Included in Eutrophic waters, but this is a much broader category	A1, A3, A4, A9, A13

3160 (22.14) Natural dystrophic lakes and ponds

	A24

3170 (22.34) * Mediterranean temporary ponds (and 3180 * Turloughs in Ireland)

Aquifer-fed naturally fluctuating water bodies	No strictly equivalent communities

Table F.3 (*continued*)

32. Running water

3260 (24.4) Water courses of plain to montane levels with *Ranunculion fluitantis* and *Callitricho-Batrachion* vegetation

Chalk rivers	A8–A20 (A17 is characteristic of chalk rivers)

4. TEMPERATE HEATH AND SCRUB

4010 (31.11) Northern Atlantic wet heaths with *Erica tetralix*

Lowland heathland, Upland heathland	M14–M16, H5 (see also under 4040)

4020 (31.12) * Temperate Atlantic wet heaths with *Erica ciliaris* and *Erica tetralix*

Lowland heathland	M16, H3, H4

4030 (31.2) European dry heaths

Lowland heathland, Upland heathland	H1–H4, H7–H10, H12, H16, H18, H21

4040 (31.234) * Dry Atlantic coastal heaths with *Erica vagans*

Lowland heathland	H6 (H5 is relatively wet and hence not strictly equivalent)

4060 (31.4) Alpine and Boreal heaths   H13–H15, H17, H19, H20, H22

4080 (31.622) Sub-Arctic *Salix* spp. scrub   W20

5. SCLEROPHYLLOUS SCRUB (MATORRAL)

5110 (31.82) Stable xerothermophilous formations with *Buxus sempervirens* on rock slopes (*Berberidion*)

W12c, W13 (some stands)

5130 (31.88) *Juniperus communis* formations on heaths or calcareous grasslands

some W19 & W21

6. NATURAL AND SEMI-NATURAL GRASSLAND FORMATIONS

61. Natural grasslands

6130 (34.2) Calaminarian grasslands[2] of the *Violetalia calaminariae*

No priority habitat, but included in broad habitat Inland rock	OV37

6150 (36.32) Siliceous alpine and boreal grasslands   U7–U12, U14

6170 (36.41 to 36.45) Alpine and subalpine calcareous grasslands

Upland calcareous grassland	CG12–CG14

62. Semi-natural dry grasslands and scrubland facies

6210 (34.31 to 34.34) Semi-natural dry grasslands and scrubland facies on calcareous substrates (*Festuco-Brometalia*) (* if important orchid sites)

Lowland calcareous grassland	CG1–CG9

6230 (35.1) * Species-rich *Nardus* grasslands, on siliceous substrates in mountain areas (and submontane areas in Continental Europe)

some stands of CG10, CG11 and U5c

Table F.3 (*continued*)

---

64. Semi-natural tall-herb humid meadows

6410 (37.31) *Molinia* meadows on calcareous, peaty or clayey-silt-laden soils (*Molinion caeruleae*)

Purple moor grass and rush pastures	M24, M26

6430 (37.7 & 37.8) Hydrophilous tall herb fringe communities of plains and of the montane to alpine levels

	U17

65. Mesophile grasslands

6510 (38.2) Lowland hay meadows (*Alopecurus pratensis, Sanguisorba officinalis*)

Lowland meadows (includes unimproved neutral pastures)	MG4, MG5 and MG8

6520 (38.3) Mountain hay meadows

Upland hay meadows	MG3

7. RAISED BOGS AND MIRES AND FENS

71. Sphagnum acid bogs

7110 (51.1) * Active raised bogs

Lowland raised bog	M1, M2, M3, M18

7120 (51.2) Degraded raised bogs still capable of natural regeneration

Lowland raised bog	Often largely M15, M20, M25, W4, but some 7110 communities usually present

7130 (52.1, 52.2) Blanket bogs (* if active)

Blanket bog	M1–M3, M15, M17, M18, M19, M20, M25

7140 (54.5) Transition mires and quaking bogs

Various communities associated with both fens and bogs, but mainly M4, M5, M8, M9, S27

7150 (54.6) Depressions on peat substrates of the *Rhynchosporion*

No strict correspondence, but there are affinities with M1, M2, M16 and M21

72. Calcareous fens

7210 (53.3) * Calcareous fens with *Cladium mariscus* and species of the *Caricion davallianae*

Fens	S2, S24, S25; SD14, SD15 (dune slacks); some stands of M9 and M13

7220 (54.12) * Petrifying springs with tufa formation (*Cratoneurion*) | M37, M38 |

7230 (54.2) Alkaline fens

Fens	M9–M11, M13, M14 (see also fen-meadows under 6410)

7240 (54.3) * Alpine pioneer formations of the *Caricion bicoloris-atrofuscae* | M12 |

8. ROCKY HABITATS AND CAVES

8110 (61.1) Siliceous scree of montane to snow levels (*Androsacetalia alpinae* & *Galeopsietalia ladani*)

	U18, U21

8120 (61.2) Calcareous and calcshist screes of the montane to alpine levels (*Thlaspietea rotundifolii*)

	OV38

Table F.3 *(continued)*

8210 (62.1 and 62.1A) Calcareous rocky slopes with chasmophytic vegetation[3]

| | CG14, U15 |

8220 (62.2) Siliceous rocky slopes with chasmophytic vegetation | U21 |

8240 (62.4) * Limestone pavements

| Limestone pavements, some Upland mixed ash woodland | various, e.g. CG9 or W8, W9 on some areas |

8330 (none) Submerged and partly submerged sea caves

9. FORESTS

91. Forests of Temperate Europe

9120 (41.12) Atlantic acidophilous beech forests with *Ilex* and sometimes also *Taxus* in the shrublayer

| Lowland beech and yew woodland, some Lowland wood pasture and parkland | W14, W15 |

9130 (41.13) *Asperulo-Fagetum* beech forests

| Lowland beech and yew woodland | W12, W14 |

9160 (41.24) Sub-Atlantic and medio-European oak or oak-hornbeam forests of the *Carpinion betuli*

| Some stands of: W8 (especially W8a & W8b); and W10 (especially W10a & W10b) |

9180 (41.4) * *Tilio-Acerion* forests of slopes, screes and ravines

| Upland mixed ash woodland | W8, W9 |

9190 (41.51) Old acidophilous oak woods with *Quercus robur* on sandy plains

| Some Lowland wood pasture and parkland | W10, W16 |

91A0 (41.53) Old sessile oak woods with *Ilex* and *Blechnum* in the British Isles

(a) mainly lowland | W10e, W16b |

(b) mainly upland | Upland oakwood | W11, W17 |

91C0 (42.51) * Caledonian forest | Native pine wood | mainly W18, some W19 |

91D0 (44.A1 to 44.A4) * Bog woodland | Wet woodland, Native pine wood | W4 |

91E0 (44.3) * Alluvial forests with *Alnus glutinosa* and *Fraxinus excelsior*

| Wet woodland | W5–W7 |

91J0 (42.A71 to 42.A73) * *Taxus baccata* woods of the British Isles

| Lowland beech and yew woodland, some Upland mixed ash woodland | W13 |

---

1 **Maerl beds** are calcareous encrustations on the seabed formed by calcium-fixing algae (maerls).

2 **Calaminarian grasslands** are open grasslands on natural rock outcrops rich in heavy metals (e.g. lead, zinc); serpentine soils (rich in nickel, chromium and magnesium, but with low nitrogen, phosphorus, calcium and molybdenum); or man-made sites such as spoil heaps around old mines.

3 **Chasmophytic vegetation** is vegetation of rocks, cliffs, screes and exposed summits, and is dominated by plants adapted to these habitats (chasmophytes).

Table F.4 Outline of the NVC as published in *British Plant Communities*

Volume number and title	Major categories (volume sections)	Community codes[1]
1 Woodlands and scrub	Woodlands and scrub	W1–W25
2 Mires and Heaths	Mires (including wet heaths)	M1–M38
	Heaths (dry)	H1–H22
3 Grasslands and montane communities	Mesotrophic grasslands	MG1–MG13
	Calcicolous grasslands	CG1–CG14
	Calcifugous grasslands and montane communities	U1–U21
4 Aquatic communities, swamps, and tall-herb fens	Aquatic communities	A1–A24
	Swamps and tall herb fens	S1–S28
5 Maritime communities and vegetation of open habitats	Saltmarsh communities	SM1–SM28
	Shingle, strandline and sand-dune communities	SD1–SD19
	Maritime cliff communities	MC1–MC12
	Vegetation of open habitats	OV1–OV42

1 for full community names and descriptions see the relevant volumes of Rodwell (1991–2000).

F.4 The National Vegetation Classification (NVC)

The NVC is published in the five volumes of *British plant communities* (Rodwell 1991–2000). The structure of the classification is outlined in Table F.4. The major categories are characterised by vegetation physiognomy and environmental criteria. Each category contains a number of communities, most of which are further subdivided into two or more sub-communities. Definition of communities and sub-communities is phytosociological, i.e. they are characterised by full plant species composition. Each community, and its sub-communities, is described in a chapter of the relevant volume, which includes information on associated aspects such as climate, soils, succession, and distribution. Application of the system involves four key aspects which are outlined below.

Field sampling can be restricted to vascular plants and/or simple presence records in individual **quadrats**, but wherever possible it should include quantitative observations of all taxa, preferably in batches of ≥ 10 quadrats. The recommended procedures include the use of:

- **selective sampling** (Fig. G.1) within apparently homogeneous vegetation, with the aim of ensuring that observations are representative of a specific community type;
- large quadrats (aimed at including most of the community's species within each) with specific sizes for different vegetation types (Rodwell 1991–2000);
- **Domin values** of species' cover-abundance within quadrats (Table F.5).

Table F.5 Domin cover-abundance values and Braun-Blanquet constancies as used in the NVC

Domin scale of cover-abundance			Braun-Blanquet constancy classes and equivalent bands of % presence in samples	
Domin value	% cover	Category when cover is less than 4%	Constancy class	% of samples in which a species is present
1	< 4%	few individuals	I	1–20%
2	< 4%	several individuals	II	21–40%
3	< 4%	many individuals	III	41–60%
4	4–10%		IV	61–80%
5	11–25%		V	81–100%
6	26–33%			
7	34–50%			
8	51–75%			
9	76–90%			
10	91–100%			

Data analysis can be achieved by reference to the keys and **floristic/diagnostic tables** in Rodwell (1991–2000). Each table characterises the relevant community and sub-communities on the basis of:

- **species composition**, with an expected Domin-value range for each species;
- a **constancy profile**, which lists the constancy (frequency) with which each species is expected to occur in samples, expressed as **Braun-Blanquet constancy classes** (Table F.5). Species with high constancies (IV and V) contribute most to the diagnosis.

Analysis is greatly facilitated by the use of a computer program such as MATCH (Malloch 2000). This uses a coefficient to calculate the similarity of survey data to NVC communities on a scale of 0–100, from which approximate similarity ratings can be assigned as follows: 0–49 = very poor; 50–59 = poor; 60–69 = fair; 70–79 = good; and 80–100 = very good.

The findings should be interpreted with care. For example:

- Apparently good floristic matches are not always supported by the habitat requirements and distributions of the relevant NVC communities. Consequently, it is essential to refer to the information given in Rodwell (1991–2000);
- Similarities < 60 should rarely be considered significant, although they may result from inaccurate sampling (which can also sometimes produce erroneous 'good matches'). Genuine similarities > 80 are rare, and samples often show (usually poor or fair) similarity to more than one NVC type. A major reason is the limitations of any classification to accommodate community variability (§F.7). Indeed, the NVC does not pretend to be a fully comprehensive and precise classification of all British plant communities;
- It follows that (a) very close similarities between survey data and NVC communities should not be generally expected; (b) whilst similarity to an NVC type is a good measure of naturalness, **a poor match should not necessarily be**

taken to mean that a community has low ecological or conservation value. In addition, the NVC is not really appropriate for evaluating vegetation that may be of value for 'non-ecological' reasons, especially in urban environments (Appendix D.3).

Approximate correspondences of NVC communities with HSD Annex I habitats and UKBAP priority habitats are shown in Table F.3. Correspondences with JNCC Phase 1 habitats are given in JNCC (1993); and Hall & Kirby (1998) provide guidance on relationships of NVC woodland communities with Peterken (1993) and Forestry Authority (1994) woodland classifications. 'Woodland' rather than 'forest' is generally used in the UK because 'forest' originally meant an area where deer were kept for hunting, whether or not it was wooded (Rackham 1986).

F.5 The Countryside Vegetation System (CVS)

The CVS (Bunce *et al.* 1999) is based on the data collected for the *Countryside Survey 1990* (CS90) and is used in the analysis of data collected for CS2000 (Table 11.4). It aims to provide a national classification that focuses on the wider countryside and is suitable for monitoring changes.

It consists of (a) 100 vegetation classes, produced by a TWINSPAN analysis of the data, and (b) eight larger aggregate classes, created by analysis using DECORANA (Appendix G.2). Some classes are rather general (e.g. grassy roadsides); most represent more disturbed vegetation types than NVC communities; and correspondence between the CVS and NVC is poor. A computer program MAVIS (available from http://www.ceh.ac.uk/) allows a user to compare survey data with CVS types, but those from semi-natural vegetation can produce strange results, e.g. data from a calcareous fen were interpreted as 'acid streamsides/flushes'.

As intended, the main application of the CVS is likely to be at the strategic level. It may be useful in some EIA surveys, e.g. to check if hedgerow data conform to CVS type 21 'species-rich lowland hedges'. In general, however, the current dataset seems unlikely to provide more useful information than manual application of the JNCC Phase 1 habitat classification.

F.6 Land classifications

Land classifications are usually intended mainly for application on a regional scale, and hence generally consist of relatively few categories. The three main attributes employed are:

- **land capability**, which is mainly concerned with the suitability of land for agriculture or forestry (see §9.3.6);
- **land cover**, which is the observed physical cover (seen from the ground or by remote sensing) including water, ice, bare ground (rock, etc.), vegetation (natural or planted) and human constructions (buildings, roads, etc.). Land cover classifications are primarily of value in landscape and ecological assessments;
- **land use**, which is the purpose for which land is being used, i.e. the activities currently taking place to produce goods or services. Land use is partially

incorporated in land cover, but (a) it includes more human-activity categories (e.g. commerce, heavy industry) and (b) several land uses may occur on the same piece of land. This provides a basis for environmental *and* socio-economic impact analysis (Di Gregorio & Jansen 1996).

UK and EU land cover classifications include those used for the LCMGB, LC2000 and CORINE land cover maps (Table 16.1). For EIAs of large-scale or linear projects, they have the advantage that digitised data are available for mapping and GIS applications. For smaller-scale EIAs, however, disadvantages are (a) classes often correspond poorly with habitat classifications, and (b) mapping resolutions are generally low, although methods of integrating the data with ground survey data have been demonstrated (§16.2.1).

The FAO **Land Cover Classification System** (LCCS) is intended as a flexible system for worldwide use. It consists of two phases: a dichotomous phase in which eight major land cover classes are distinguished; and a modular-hierarchical phase in which each major class is subdivided using a pre-defined set of attributes (classifiers) that are specific to that class.

The FAO **Land Use DataBase** (LUDB) is not a formal classification. A user defines land use classes by applying a set of provided classifiers, and can enter additional user-defined classifiers. Information on this, and on the LCCS, is available from http://www.fao.org/. The UNECE **Statistical Classification of Land Use** (SCLU) consists of seven major categories with subdivisions into one or two sub-levels, especially under the urban/commercial/industrial category.

F.7 Limitations of classifications

A major purpose of classifications is to provide a mechanism by which records from different investigations can be compared in terms of accepted categories that are meaningful to all users, and they are an essential tool in environmental assessment and management. As evident above, however, a common problem is that different classifications are not fully compatible, so it is frequently difficult to 'translate' between them (Gibson 1998). An expert can interpret most discrepancies; but these can lead to serious misinterpretation by non-experts, and also hinder the development of a computerised matching program that does not generate false matches.

To rectify these problems, a EUNIS classification is being developed by EEA-ETC-NC. This will be based largely on the Palaearctic classification, and the aim is to integrate other European classifications. However, a fundamental problem will always remain. This is that no classification can fully accommodate the variability of ecological systems, and their tendency to intergrade (§11.2.5). Consequently:

- no two examples of a designated habitat or community type will be precisely the same;
- communities found in given locations will effectively represent points on gradients of variation within or between designated types;
- data from a habitat/community in a particular location may not match exactly (or even closely) any designated type, and this does not necessarily diminish its ecological or conservation value.

Failure to appreciate the limitations of classifications, especially those used in legislation such as the HSD, can lead to errors such as under-valuation of habitats that do not closely match designated types; and it is essential that this is made clear in EIAs, so avoiding misinterpretation by developers and decision-makers.

References

Bunce RGH, HM van de Poll, JW Watkins, WA Scott, SM Smart & DCH Howard 1999. *ECOFACT research report series*. Vol. 1: *Vegetation of the British countryside*. London: HMSO. (http://mwnta.nmw.ac.uk/ite/ecofact/volume1.htm).

Crofts A & RG Jefferson (eds) 1999. *The lowland grassland management handbook*, 2nd edn. Peterborough: EN/The Wildlife Trusts.

Devilliers P & J Devilliers-Terschuren 1996. *A classification of Palaearctic habitats*. (Nature and Environment No. 78). Strasbourg: Council of Europe.

Di Gregorio A & LJM Jansen 1996. *FAO Land Cover Classification: A dichotomous, modular-hierarchical approach*. Rome: FAO Land and Water Development Division. (http://www.fao.org/).

Duffey E, MG Morris, J Sheail, K Lena, DA Wells & TCE Wells 1974. *Grassland ecology and wildlife management*. London: Chapman & Hall.

EC (European Commission) 1991. *CORINE biotopes manual*. Vol. 3: *Habitats of the European Community*. Luxembourg: Office for Official Publications of the European Communities.

EC 1996. *Interpretation manual of European Union habitats, Version EUR 15*. Brussels: European Commission, DG XI. (http://europa.eu.int/comm/environment/).

EC 1997. Council Directive 97/62/EC adapting to technical and scientific progress Directive 92/43/EEC on the conservation of natural habitats and of wild fauna and flora. Brussels: *Official Journal of the European Commission* L305/42-65. (http://europa.eu.int/comm/environment/).

Forestry Authority 1994. *Forestry Practice Guides – The management of semi-natural woodlands*. Edinburgh: Forestry Commission.

Gibson CWD 1998. *Harmonisation of habitat classifications*. JNCC Reports No. 279. Peterborough: JNCC.

Hall JE & KJ Kirby 1998. *The relationship between Biodiversity Action Plan Priority and Broad habitat types, and other woodland classifications*. JNCC Report No. 288. Peterborough: JNCC. (http://www.jncc.gov.uk/).

Hillier SH, DWH Walton & DA Wells (eds) 1990. *Calcareous grasslands – ecology and management*. Huntingdon: Bluntisham Books.

JNCC 1993. *Handbook for phase 1 habitat survey – a technique for environmental audit*. (Separate *Field manual* also available). Peterborough: JNCC.

Malloch AJC 2000. *MATCH II: A computer program to aid the assignment of vegetation data to the communities and subcommunities of the National Vegetation Classification*. Version 2.15 for Windows NT/95/98. Unit of Vegetation Science, University of Lancaster.

Peterken GF 1993. *Woodland conservation and management*, 2nd edn. London: Chapman & Hall.

Peterken GF 1996. *Natural woodland: ecology and conservation in northern temperate regions*. Cambridge: Cambridge University Press.

Rackham O 1986. *The history of the countryside*. London: Dent.

Rodwell JS (ed.) *British plant communities*: 1991a. Vol. 1 *Woodlands and scrub*; 1991b. Vol. 2 *Mires and heaths*; 1992. Vol. 3 *Grasslands and montane communities*; 1995. Vol. 4 *Aquatic communities, swamps and tall-herb fens*; 2000. Vol. 5 *Maritime communities and vegetation of open habitats*. Cambridge: Cambridge University Press.

Webb N 1986. *Heathlands: a natural history of Britain's lowland heaths*. London: Collins.

Appendix G
Phase 2–3 ecological sampling methods

Peter Morris and David Thurling

G.1 Introduction and sampling options

Phase 2 ecological surveys can be valuable or essential in EIAs, but they are generally time-consuming, expensive and may be hindered by unavailability of specialists. In addition, the development cycle may preclude the time required for undertaking appropriate surveys.

It is vital that the work on different aspects is co-ordinated, and that the findings are clearly presented. These should include concise descriptions of the methods employed (including sampling times), and clear presentations of the results and of their interpretation/evaluation, including limitations and uncertainties.

This appendix gives guidance on general principles and methods, but it must be emphasised that surveys should be carried out by, or under the supervision of, experienced ecologists who already have the necessary skills. The principles described here apply to all ecological survey work; the methods focus on terrestrial species, vegetation and environmental factors. Specific methods for freshwater and marine systems are described in Chapters 12 and 13, respectively.

Whether surveying individual species or whole communities, it is important to check species conservation status, distributions and habitat requirements. Relevant publications are listed in Appendix E.1. It is also important to remember that many species are legally protected (see §11.3.1 and Appendix D.2) and that legal protection imposes restrictions. In particular, any activity likely to involve handling or disturbance may require a licence from the relevant NCCA.

Frequent **sources of error** in field survey data are misidentification and inadequate or inappropriate sampling. Publications that can provide valuable assistance in identification are listed in Appendix E.2. However, identification should always be carried out by competent personnel. Specialists are likely to be needed for many taxa (especially of invertebrates) and all doubtful identifications should be confirmed by relevant experts. Careful consideration should be given to the **selection of sampling methods** including data collection methods, species abundance measures, sampling pattern (in space and time) and sample size. In selecting options, it is also important to consider what data analysis will be needed, and whether the data collected will be suitable for the purpose.

Data collection methods include:

- **Plot sampling**, which involves taking observations (e.g. of species presence or abundance) within defined plots, usually *quadrats*;

- **Plotless sampling**, which is any method in which sampling is not conducted within defined areas. Simple methods include observations taken along **transects**, e.g. for bird surveys, or using the *line intercept method*, which can be applied (a) as a habitat-measurement method (§11.4.4) or (b) for estimation of plant species cover in sparse vegetation (see Kent & Coker 1992). Most other plotless methods involve **distance measurements** from sampling points. Those employed in vegetation studies, e.g. for estimating tree densities in woodlands, are reviewed in Mueller-Dombois & Ellenberg (1974) and Goldsmith *et al.* (1986); methods used for animals are discussed in Greenwood (1996), Buckland *et al.* (1993), Krebs (1998);
- **Specialised collecting equipment** is often needed in faunal sampling (see G.3).

The main options for estimating **species abundance** are outlined in Table G.1. In theory, an additional option is *biomass*, but measurement of this in terrestrial ecosystems generally involves destructive sampling and is too time-consuming for most EIA surveys.

The design of **spatial sampling patterns** involves the questions *where to sample?* and *what pattern of sampling locations is appropriate?* The main options are outlined in Figure G.1.

The design of **temporal sampling patterns** can be related to aims such as **monitoring**, e.g. in selecting intervals between sampling times, but the commonest reason for considering timing in EIA is **seasonal constraints**. In Britain, some taxa can be sampled throughout the year (although there are usually optimal sampling periods for these), but appropriate and reliable data on other taxa can only be obtained during short sampling seasons (Fig. G.2). In most cases, this means late spring–summer, but there are exceptions, e.g. winter migrant birds. Community surveys pose additional problems because most communities contain species that are inconspicuous or absent during part of the normal sampling season, and some have components with distinctly different seasonalities – so failure to carry out repeat surveys on at least two occasions can often lead to error.

Sample size can be critical because data obtained from small samples are generally unreliable, and cannot be 'improved' by the application of sophisticated analytical procedures. For instance, there is little chance that a few randomly or subjectively placed quadrats will provide representative data for a site. There is no completely objective way of determining the minimum requirement, and the number of observations taken is usually a compromise between the need for precision and the cost in terms of labour and time (Krebs 1998). A percentage-of-area target is sometimes applied in vegetation surveys, e.g. to sample 5% of a study area (Mueller-Dombois & Ellenberg 1974). However, this is rarely achieved (especially on large sites) and Greig-Smith (1983) emphasises that sample accuracy is more dependent on the **number of observations**. For example, it is generally preferable to use a large number of small quadrats than a small number of large quadrats of equivalent total area.

G.2 Plant species and vegetation

In general, **vascular plant** species are relatively easy to sample, but the most appropriate methods will depend on their **life forms** and population distributions. For example:

Table G.1 Abundance measures

Semi-quantitative abundance ratings are visually estimated using systems such as DAFOR in which:
D = dominant; A = abundant; F = frequent; O = occasional; R = rare (with the prefix l = locally added to any category if required). They are quick to record, but are subjective, approximate, and have limited potential for analysis and presentation. Consequently, they are generally more suited to Phase 1 rather than Phase 2 studies.

Number of individuals is a suitable measure for species which have readily discernible individuals that can be counted. It is not usually applicable in community studies because it has little meaning when comparing species of widely differing size. When measured in defined areas, numbers can be expressed as **density** (number per unit area) and/or as **population size** (in the study area). There are two counting methods:

- **Direct counting**, which is only generally valid for plants, near-sedentary animals or small populations of animals within defined areas. Occasionally, whole populations (e.g. of trees or nesting birds) can be counted in small areas. More usually, population estimates are derived from samples, e.g. in quadrats or by plotless sampling;
- **Indirect counting methods**, which can provide estimates of fairly small populations (e.g. of small mammals in a study area), although certain assumptions must apply, at least approximately. **Mark-recapture methods** (see Begon 1979, Blower *et al.* 1981, Greenwood 1996, Krebs 1998) involve capturing and marking a number of individuals, releasing them, and re-sampling after a suitable time interval. Formulae are used to derive the population estimate from the proportion of marked individuals in the recapture sample.

Cover (%) is the percentage of ground occupied by the aerial parts of a species. It is usually measured by visual estimation in quadrats, although there are other methods such as the line intercept method or 'point quadratting' (see Bullock 1996, Kent & Coker 1992). It is suitable for studies of communities which include species of differing size. Visual estimates are prone to observer error (accuracy > the nearest 5% should not be attempted) and species present as small scattered individuals tend to be under-estimated.

Cover-abundance scales, such as the **Domin scale** (Table F.5) aim to: (a) accommodate the cover-estimation error by designating bands of % cover; (b) avoid the under-estimation of small, scattered species, by using cover for most species, but abundance (in the strict sense of numbers) for species with low cover, e.g. < 4%.

Frequency (%) is the percentage of observations in a sample that contains the species, and is derived from presence/absence observations, e.g. in quadrats. Limitations are: (a) it is strictly a measure of distribution rather than abundance and does not discriminate between high density and density that is just sufficient for a species to be present in a large proportion of quadrats; (b) it tends to over-represent small species; (c) it increases in value with increasing quadrat size, so results using different-sized quadrats are not strictly comparable (it should generally be calculated from observations in ≥ 20 small quadrats). However, frequency can be a cost-effective method for obtaining large representative samples of communities because it is relatively rapid and free from observer error. It is also a measure of **constancy**, which is the criterion employed in the similarity coefficient in MATCH for comparing vegetation samples with NVC communities (Appendix F.4).

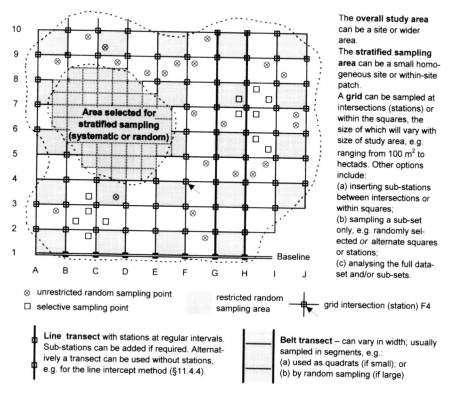

The **overall study area** can be a site or wider area.

The **stratified sampling area** can be a small homogeneous site or within-site patch.

A **grid** can be sampled at intersections (stations) or within the squares, the size of which will vary with size of study area, e.g. ranging from 100 m² to hectads. Other options include:
(a) inserting sub-stations between intersections or within squares;
(b) sampling a sub-set only, e.g. randomly selected *or* alternate squares or stations;
(c) analysing the full dataset and/or sub-sets.

⊗ unrestricted random sampling point
☐ selective sampling point

restricted random sampling area

grid intersection (station) F4

Line transect with stations at regular intervals. Sub-stations can be added if required. Alternatively a transect can be used without stations, e.g. for the line intercept method (§11.4.4).

Belt transect – can vary in width; usually sampled in segments, e.g.:
(a) used as quadrats (if small); or
(b) by random sampling (if large)

Systematic sampling is conducted at regular intervals, e.g. along transects or in relation to a grid system. Transects are useful for walkover surveys (transect walking) or studying gradients. Their placement can be random, systematic (at regular intervals) or selective, e.g. along linear habitats. **Grid sampling** is the best way to obtain representative data showing the patterns of variation within a study area.

Random sampling points are randomised throughout a study area. This is the most statistically acceptable method, but (a) spatial distributions of variables are rarely random; (b) the random location of points inhibits the detection of gradients and requires very large samples to ensure that the whole of a large area is represented. Consequently, **restricted random sampling** is usually employed in fairly small areas, e.g. within selected grid squares *or* habitat/vegetation patches (when it is called **stratified random sampling**).

Stratified sampling involves the selection of fairly small areas of similar character (the 'strata') such as patches of a habitat type on a site, or on different sites. The areas can be sampled systematically or by stratified random sampling; and can be of equal or varying size (in which case the number of samples is adjusted accordingly). It is widely used where whole-area sampling is unnecessary or impracticable.

Selective sampling points are selected because: (a) access to other points is difficult; (b) a variable only occurs in scattered locations; or (c) they are judged to be within homogeneous patches of vegetation that will provide representative samples for comparison with designated types in classifications such as the NVC (Appendix F.4). The method is generally considered to be too subjective for other purposes.

Figure G.1 Spatial sampling pattern options in relation to hypothetical study areas.

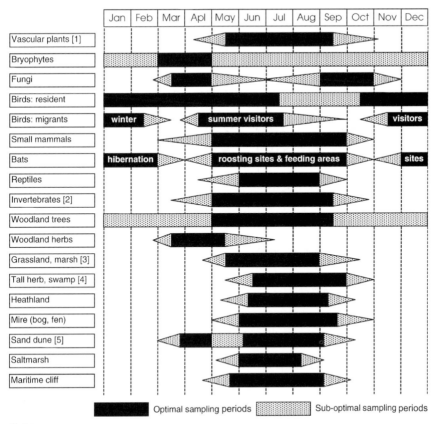

	Jan	Feb	Mar	Apl	May	Jun	Jul	Aug	Sep	Oct	Nov	Dec
Vascular plants [1]												
Bryophytes												
Fungi												
Birds: resident												
Birds: migrants		winter			summer visitors						visitors	
Small mammals												
Bats		hibernation			roosting sites & feeding areas						sites	
Reptiles												
Invertebrates [2]												
Woodland trees												
Woodland herbs												
Grassland, marsh [3]												
Tall herb, swamp [4]												
Heathland												
Mire (bog, fen)												
Sand dune [5]												
Saltmarsh												
Maritime cliff												

■ Optimal sampling periods   ▨ Sub-optimal sampling periods

Notes:
[1] There are some exceptions, e.g. see woodland herb layer.
[2] Preferably with three samples taken during early, mid and late parts of the optimal sampling period.
[3] Hay meadows should not be sampled after cutting (usually in June).
[4] Short vegetation in tall herb and swamp should be sampled in March-April (as woodland herb layer).
[5] Sand dune animals and Spring annual plants should be sampled in March-April.

> The periods shown are approximate only, and vary somewhat in different locations. For example, growing seasons in Britain generally start later in the north than in the south.

Figure G.2 Possible sampling periods for terrestrial taxa and communities. Suitable sampling periods for freshwater and marine species and communities are given in Chapters 12 and 13, respectively

- Herbaceous species can normally be sampled by recording % cover, frequency or numbers (if individuals can be readily counted) in small quadrats;
- Tree species can be sampled by (a) counting in large quadrats or habitat patches, (b) estimating % cover or frequency by looking up at the canopy, or (c) using plotless sampling methods.
- **Stratified sampling** will usually be appropriate, but **selective sampling** may be necessary in some cases (see Fig. G.1).

Sampling **bryophytes** and lichens can be more problematical because many species are inconspicuous and/or difficult to identify. However, they should not be ignored because they are often important components of communities, and can be better environmental indicators than higher plants (e.g. see Gilbert 2000, Hodgetts 1992).

Phase 2 vegetation surveys utilise vegetation classifications, such as the *National Vegetation Classification* (NVC) which is described in Appendix F.4. This is widely used in the UK, and is recommended by IEA (1995) as the main Phase 2 vegetation survey method in EIA. Surveys may also include studies on particular aspects such as **species diversity** (Table 11.1), which is usually calculated from species abundance data using numerical indices (see Kent & Coker 1992, Hawksworth 1995, Krebs 1998). However, species diversity values must be interpreted with caution (see §D.3.2).

Occasionally, it may be desirable to carry out a **Phase 3 vegetation survey**. For example, if a community may be affected by changes in groundwater level or quality, it may be important to quantify the relationships in order to assess the threats and formulate mitigation measures. Phase 2 NVC data are sufficiently detailed, and can be analysed using the multivariate methods outlined below (e.g. see Yeo *et al.* 1998). However, NVC selective sampling (Appendix F.4) is likely to be inadequate. Instead, it is probably necessary to employ a systematic sampling pattern for collection of both the floristic and environmental data.

This type of data is best subjected to **multivariate analysis**, which is the simultaneous analysis of a number of variables that relate to the same set of observations. There are two main approaches, classification and ordination, which can utilise the same data and provide complementary information. **Classificatory methods** seek to identify groups of similar units (e.g. samples or species with similar distributions) that represent communities. These communities can be (a) represented on maps, and (b) compared with NVC communities, although they may not correspond closely with the latter because they are unique to the ecosystem under study (see §11.2.5 & Appendix F.7). **Ordination methods** seek to identify gradients by plotting units (e.g. samples) along one or more dimensions. Ordination can also be used to analyse community–environment relations. Further information on these methods can be found in texts such as Digby & Kempton (1987), Gauch (1982), Jongman *et al.* (1995), Kent & Coker (1992) and Krebs (1998).

Multivariate analysis requires computer assistance. Programs for many of the methods are available in general statistical packages, but some are designed specifically for ecological use. The latter include TWINSPAN for classification, and DECORANA for ordination. These are available separately (Hill 1994) or in VESPAN (Malloch 2000).

G.3 Animals

G.3.1 Introduction

Thorough faunal surveys are generally difficult and time-consuming. The main problems are:

1.   Whilst the number of vertebrate species may be small, invertebrate species are usually numerous (there are c.22,500 insect species in Britain).

2.   Different taxa require very different sampling methods and associated expertise, including identification, especially of invertebrates (experts are often needed for specific groups).

3.   Many animals are inconspicuous, fugitive or nocturnal, and many hibernate or have inaccessible life-cycle stages during parts of the year (e.g. most invertebrates).

4.   Many animals are very mobile, and periodically move between habitat patches, sites or wider areas. For example:

   •   whilst some site-resident species may only need a particular habitat patch, others may utilise different parts of a site, or wider area, for different purposes such as breeding and feeding;

   •   a species present at a given time may be a casual or regular (e.g. seasonal) visitor, utilising a site for shelter, feeding or breeding and then moving on, perhaps during dispersal (e.g. of adolescents) or migration. Even transitory migrants are dependent on the site, especially if it is on a regular migration route.

5.   As a consequence of points 3 and 4:

   •   a simple site-presence record may not adequately reflect a species' spatial requirements or use of the site;

   •   it may be possible to study some residents at any time, but effective sampling seasons are restricted for many species (see Figure G.2);

   •   repeat sampling may be needed to provide reliable data.

6.   Determination of distribution and abundance can be difficult, time-consuming and imprecise; and most species vary in abundance from year to year at any site.

The problems inevitably impose limitations on what can be achieved in Phase 2 field surveys, which must be carefully targeted on key and feasible objectives, perhaps using a scorecard to assist in selection (Appendix D.4). They are bound to involve focusing on high-status species and high-profile groups, but other aspects should also be considered. For example, a development may have an indirect impact on a plant species or community, through effects on the fauna, especially **keystone species**. In any case, surveys will usually be restricted to **partial species lists** (certainly of invertebrates) and limited quantitative data. Distribution and abundance data may be vital in assessing a species' dependence on a site and the likely viability of the population in the face of impacts; but fully quantitative studies are effectively Phase 3 surveys. Similarly, animal **species richness** or **species diversity** estimates are nearly always limited to partial community data, and often to high-profile and/or easily identified taxa such as butterflies.

G.3.2 *Assessing habitat suitability*

Most animals depend directly or indirectly on particular types of vegetation. In general, semi-natural vegetation (e.g. similar to NVC communities) can be assumed to provide a variety of habitats, although other vegetation and bare ground can also be important (Key 2000).

   Consequently, it can be argued that the value of a site for animals can be assessed largely from a habitat/vegetation study. However, while this can indicate where animal sampling is likely to be profitable, the presence of animal species cannot be assumed from apparent habitat suitability because their distributions depend on other factors such as local climate, past site conditions, and degree of isolation from other suitable habitat patches.

On the other hand, if a species is absent because of factors such as past management, the habitat conditions and vegetation should give a good indication of the potential for re-colonisation. This is why IEA's (1995) evaluation criteria include sites that are evidently suitable for some species even if these are not now present (see Table D.2).

G.3.3 Field survey methods for vertebrates

Appropriate methods vary widely for the various vertebrate groups. It is important to remember that any activity likely to involve handling or disturbance of protected species requires a licence.

Amphibians and reptiles

Methods for amphibians are given in §12.4.6. The sampling season for reptiles is restricted (Fig. G.2), and field observations tend to be fleeting glimpses in good weather. Guidance on sampling is provided in Blomberg & Shine (1996), EN 1996, and Gent & Gibson (1998). Reading (1996) suggests a quantitative method using arrays of artificial refuges, coupled with transect walking between them, but this is time-consuming and requires repeat sampling.

Birds

The main aims of a bird survey in EIA will be to evaluate habitat patches and sites (including small and linear habitats) for birds in general, and for high-status species in particular, and to assess the vulnerability of the populations to potential impacts. It must be remembered that sites may be utilised for various purposes (e.g. roosting/shelter, feeding, breeding) and that most species need to move between sites.

Regional, national and supra-national population estimates are available for many bird species (e.g. Gibbons *et al.* 1993); so if the local populations can be quantified, both they and the sites that support them can be evaluated in terms of their representation (see Table D.2).

Reviews of bird census techniques, including specific methods for a range of species, are provided in Bibby *et al.* (2000), Gibbons *et al.* (1996) and Gilbert *et al.* (1998). Specific survey methods for wetland birds and seabirds are outlined in §12.4.6 and §13.4.6, respectively. There are sampling problems associated with all methods. For example:

- they require expertise, e.g. in both visual identification and the recognition of bird calls/song;
- they are time-consuming, require repeat sampling, and involve extensive site walking;
- they are affected by seasonal variations (e.g. in relation to breeding and migration) and by weather conditions (birds may be less active and conspicuous in wet and windy weather);

- it is usually not possible to count all species equally well. Some can be relatively easy, especially during the breeding season when rival males are often conspicuous; others are difficult to detect, let alone count;
- some habitats (e.g. dense scrub and woodland) are more difficult to sample than others.

In most cases, the most suitable method for EIA is likely to be **transect walking** along line transects (located randomly, systematically or along linear habitats). This method can be used to estimate densities, e.g. by counting in relation to a prescribed band each side of the transect. However, its value may be limited in small and/or heterogeneous sites, when the **point count method** (using randomly located observation points) may be more appropriate. Other options include the selection of observation points in relation to specific aims, e.g. at central points in the range of selected species. This can be useful in the study of scarce species that tend to be neglected by the other methods.

Mammals

Survey methods for marine mammals and species associated with fresh waters (e.g. otter and water vole) are outlined in §13.4.7 and §12.4.6, respectively. A number of the British mammals are specifically protected by legislation, and for some species a surveyor must be licensed. For survey purposes, terrestrial mammals can be divided into three groups – bats, small mammals and larger mammals – each requiring different survey techniques. General guidance is provided by Corbet & Harris (1991) and Sutherland (1996b). Where specific methods have been devised for particular species or groups, relevant publications are cited.

All British **bats** and their roosts are protected, and a licence is needed for any survey method that involves catching bats, or that may disturb them in their roosts or hibernation sites. Bat survey techniques require expertise (see Hutson 1993, Mitchell-Jones 1999), and the Bat Conservation Trust (BCT) must be contacted to advise on methods and personnel. Methods include the detection of roosts, foraging bats, and flight pathways. Roosts may be found in places such as buildings, trees, caves, mines and tunnels. They may be detected by the presence of droppings and insect remains, although it may be necessary to confirm their presence by sound or a visual search. Flying bats may be observed (visually and by ultrasonic detectors) from fixed points or along transects. Suitable sampling periods are indicated in Figure G.2.

Small mammals include the shrews, voles and mice. The presence of shrews may be detected by high-pitched squeaks, and of species such as field voles by turning over objects to expose runs. Taxa can sometimes be identified using **hair tubes** (sections of plastic pipe containing sticky pads to which hairs adhere); but identification and enumeration is best achieved by **live trapping** and mark-recapture (Table G.1) using Longworth traps (see Gurnell & Flowerdew 1995). A licence is required for trapping shrews. Somewhat different methods are needed for certain species which spend much of the time above ground, e.g. the dormouse (Bright & Morris 1990), fat dormouse (Hoodless & Morris 1993) and harvest mouse.

Larger mammals include badger (Harris *et al.* 1989), brown hare (Langbein *et al.* 1999), deer (Buckland 1992), fox, hedgehog, mole, mountain hare (Angerbjörn 1983), pine marten (Balharry *et al.* 1996, Strachan *et al.* 1996), polecat, rabbit (Trout *et al.* 1986), squirrels (Bryce *et al.* 1997, Gurnell & Pepper 1994), stoat, weasel (King 1974) and wildcat. Although these can be identified by direct observation, many are fugitive or nocturnal, and survey methods often utilise: (a) hair tubes and live traps (for smaller species); (b) identification of tracks, droppings, excavations, feeding damage, and habitations such as setts, holts or dreys. Within the scope of an EIA survey it is rarely possible to determine population sizes, but communal groups such as badgers can be counted when emerging from a sett. It is important to remember that, under the *Protection of Badgers Act 1992*, it is an offence to disturb a badger sett. A major problem with larger mammals is that individuals are often wide-ranging and may use a site on a seasonal basis; so limited periods of recording may miss important species or misrepresent the importance of a site to a species.

G.3.4 Field survey methods for invertebrates

Sampling terrestrial invertebrates can be difficult and time-consuming. Even a limited survey will produce a large number of individuals and species; and specimens from a day's sampling may require at least two days for sorting and identification, usually involving specialists if identifying to species level. Surveys are seasonally restricted, and should ideally involve repeat sampling (Fig. G.2). Species can be easily missed if they are in a concealed phase when the survey is conducted, e.g. soil dwelling and stem boring larvae, and the egg phase of many species. In addition, the activity of many species is restricted to particular times of day or weather conditions.

Consequently, surveys must be carefully targeted, e.g. on high-status species, target groups and **indicator species** (which can sometimes attest the general suitability of habitats for invertebrates). Target groups suggested by IEA (1995) include Carabidae (ground beetles), Lepidoptera (butterflies and moths), Orthoptera (crickets and grasshoppers) and Syrphidae (hoverflies).

The question of **where to sample** is critical. Habitats likely to be important for invertebrates are fairly easy to recognise (§G.3.2), but target species and groups will vary with habitat type (see Brooks 1993). Moreover, different species, and even different life stages of the same species, may utilise different **microhabitats**, e.g. ranging from ground level to the vegetation canopy, or on different plant species or even different parts of the same plant. Fry & Lonsdale (1991) and Kirby (1992) provide information on invertebrate habitat requirements.

Brooks (1993) provides guidelines for invertebrate site surveys, and sampling techniques are discussed by Ausden (1996), New (1998) and Southwood & Henderson (2000). They can be divided into **observer-dependent methods**, which are carried out by the investigator in the field, and **observer-independent methods**, which employ traps of various types (Table G.2).

G.4 Environmental variables and site history

Information on factors such as habitat management, soil conditions, local climatic conditions and pollution/contamination levels may be important for several reasons, e.g. to investigate environmental relationships of species or communities; to assist in impact prediction; or to evaluate a site's potential for habitat creation (§11.6.3). In addition, an understanding of current ecological systems often requires some knowledge of past conditions, i.e. of site history.

Table G.2 Methods for sampling invertebrates

Observer-dependent methods

Direct searching and recording in selected habitat/vegetation patches. It is not normally quantitative, can lead to misidentification, only records species that are active at the time, and tends to be limited to species that are conspicuous and/or common in the study area.

Transect walking involves the observation, identification and enumeration of species along a set route, within prescribed time and weather conditions. It is usually restricted to butterflies and day-flying moths (see Brooks 1993, Pollard 1977, Thomas 1983).

Sweep netting involves a hand-held net swept through vegetation up to 1 m in height. It collects most species from the vegetation (except those occupying the basal parts), but active flying insects often escape. It can be quantitative if a standard number of sweeps is taken, but sweeps in different vegetation types are not directly comparable because of differing resistance to the net. It is not suitable for woody, thorny or wet vegetation.

Swish netting is like sweep netting but is restricted to the air boundary immediately above vegetation. It is especially good at collecting Diptera (flies) and Hymenoptera (bees and wasps).

Suction sampling uses a portable vacuum to collect invertebrates from the ground layer and/ or basal parts of vegetation. It can be efficient in dry conditions and where there is little vegetation litter, and can provide quantitative data if a set number of samples are obtained.

Soil samples can be taken for identification and enumeration of soil invertebrates. A variety of physical or chemical extraction methods are used to extract the organisms from the soil samples.

Beating uses a stout stick to knock invertebrates off vegetation onto a sheet, from which they are collected. It is usually used to sample the fauna of individual tree species. With care, it can be used to obtain quantitative data, but it is not practical in wet conditions.

Subsidiary methods are used by many experts for particular invertebrate groups. They include observing flower visitors, hand searching vegetation for plant grazers (especially molluscs), stone turning especially for beetles, molluscs and millipedes, and investigating litter and dead wood for decomposers.

Observer-independent methods

Pitfall traps are placed on a regular grid within selected areas, and provide quantitative data, mainly for ground-dwelling beetles, which fall into the traps. They usually contain a killing and preserving fluid.

Malaise traps intercept flying insects by a net, and funnel them into a collection vessel. They can collect large numbers of insects (especially Diptera and Hymenoptera) and obtain quantitative and comparative data, but do not discriminate between insects resident in or simply flying through the area.

Sticky traps usually consist of a mesh screen on which a viscous oil is applied. They can be used like malaise traps or placed within vegetation. Fragile species may become damaged in trying to escape from the trap, and samples have to be removed by a solvent.

Water traps rely on the fact that a variety of flying insects (especially flower visitors) are attracted to coloured surfaces. They are simple to use but selective (e.g. see Usher 1990).

Light traps attract night-flying insects, especially if they emit ultra-violet wavelengths. They are useful but require a power source, are not easily transported, and may sample species that are flying over a site rather associated with it (see Waring 1994).

Emergence traps usually consist of a closed mesh canopy (placed over vegetation) and a collecting vessel. They are designed to collect most adult flying insects which were in a developmental stage on the vegetation or in the soil when the trap was erected. They can be used quantitatively, but must be in place for long periods.

Existing data on physico-chemical variables may be available, and new data may be collected for other EIA components such as climate, soils and water. If necessary, additional new data can be obtained using the methods described for these components (see Chapters 8, 9 and 10).

Sources of documentary historical information are given in Appendix C. If required, new evidence of past conditions can be provided by investigations (on or in the vicinity of the study site) of:

- current ecological features such as floristic richness and the presence of indicator species, e.g. of ancient hedgerows or woodlands (see §D.3.3 & §D.3.4);
- the nature of stratified sediments, and of their fossil (e.g. pollen) content (see Moore 1986);
- archaeological features (see Chapter 7).

The collection and analysis of new environmental data is generally time-consuming and expensive; and if a study seeks to establish relationships between past or present environmental conditions and biotic variables such as species or community distribution patterns, it may be important to employ a sampling pattern from which one environmental measurement can be associated with the mean of several biological observations.

References

Angerbjorn A 1983. The reliability of pellet counts as density estimates of mountain hares. *Finnish Game Research* **41**, 433–448.

Ausden M 1996. Invertebrates. See Sutherland (1996a), 139–177.

Balharry EA, GM McGowan, H Kruuk & E Halliwell 1996. *Distribution of pine martens in Scotland as determined by field survey and questionnaire.* SNH Research, Survey and Monitoring Report N. 48. Perth: SNH.

Begon M 1979. *Investigating animal abundance: capture–recapture for biologists.* London: Arnold.

Bibby CJ, ND Burgess, DA Hill & S Mustoe 2000. *Bird census techniques,* 2nd edn. London: Academic Press.

Blomberg S & R Shine 1996. Reptiles. See Sutherland (1996a), 218–226.

Blower JG, LM Cook & JA Bishop 1981. *Estimating the size of animal populations.* London: Allen & Unwin.

Bright PW & PA Morris 1990. *A practical guide to dormouse conservation. Occasional Publication No.11.* London: The Mammal Society.

Brooks SJ 1993. Guidelines for invertebrate site surveys. *British Wildlife* **4**(5), 283–286.

Bryce J, JS Pritchard, NK Waran & RJ Young 1997. Comparison of methods for obtaining population estimates for red squirrels in relation to damage due to bark stripping. *Mammal Review* **27**(4), 165–170.

Buckland ST 1992. *Review of deer count methodology.* Edinburgh: Scottish Office.

Buckland ST, DR Anderson, KP Burnham & JL Laake 1993. *Distance sampling: estimating abundance of biological populations.* London: Chapman & Hall.

Bullock J 1996. Plants. See Sutherland (1996a), 111–138.

Corbet GB & S Harris 1991. *The handbook of British mammals,* 3rd edn. Oxford: Blackwell Scientific.

Digby PGN & RA Kempton 1987. *Multivariate analysis of ecological communities.* London: Chapman & Hall.

EN (English Nature) 1996. *Reptile survey methods.* English Nature Science No. 27. Peterborough: EN.

Fry R & D Lonsdale (eds) 1991. *Habitat conservation for insects – a neglected green issue.* Middlesex: The Amateur Entomologists' Society.

Gauch Jr HG 1982. *Multivariate analysis in community ecology.* Cambridge: Cambridge University Press.

Gent T & S Gibson (eds) 1998. *Herpetofauna worker's manual.* Peterborough: JNCC.

Gibbons DW, JB Reid & RA Chapman (eds) 1993. *The new atlas of breeding birds in Britain and Ireland.* Calton: T & A Poyser.

Gibbons DW, DA Hill & WJ Sutherland 1996. Birds. See Sutherland (1996a), 227–259.

Gilbert G, DW Gibbons & J Evans 1998. *Bird monitoring methods: a manual of techniques for key UK species.* Sandy: RSPB.

Gilbert O 2000. *Lichens* (New Naturalist Series). London. HarperCollins.

Goldsmith FB, CM Harrison & AJ Morton 1986. Description and analysis of vegetation. In *Methods in plant ecology*, 2nd edn, PD Moore & SB Chapman (eds), 437–524. Oxford: Blackwell Scientific.

Greenwood JJD 1996. Basic techniques. See Sutherland (1996a), 11–110.

Greig-Smith P 1983. *Quantitative plant ecology*, 3rd edn. Oxford: Blackwell.

Gurnell J & JR Flowerdew 1995. *Live trapping small mammals, a practical guide*, 3rd edn. Occasional Publication No. 3. London: The Mammal Society.

Gurnell J & H Pepper 1994. *Red squirrel conservation: field study methods.* Research Information Note 255. Edinburgh: Forestry Commission.

Harris S, P Cresswell & DJ Jefferies 1989. *Surveying badgers.* London: The Mammal Society.

Hawksworth DL 1995. *Biodiversity: measurement and estimation.* London: Chapman & Hall.

Hill MO 1994. *DECORANA and TWINSPAN, for ordination and classification of multivariate species data; a new edition together with supporting programs, in FORTRAN 77 (Version 1.0).* Huntington: CEH Monkswood.

Hodgetts NG 1992. *Guidelines for selection of biological SSSIs: non-vascular plants.* Peterborough: JNCC.

Hoodless A & PA Morris 1993. An estimation of the population density of the fat doormouse (*Glis glis*). *Journal of Zoology* **230**, 337–340.

Hutson AM 1993. *Action Plan for the conservation of bats in the United Kingdom.* London: The Bat Conservation Trust (BCT).

IEA (Institute of Environmental Assessment) 1995. *Guidelines for baseline ecological assessment.* London: E & F N Spon.

Jongman RHG, CJF Ter Braak & OFRVan Tongeren 1995. *Data analysis in community and landscape ecology.* Cambridge: Cambridge University Press.

Kent M & P Coker 1992. *Vegetation description and analysis: a practical approach.* London: Belhaven.

Key R 2000. Bare ground and the conservation of invertebrates. *British Wildlife* **11**(3), 183–191.

King CM 1974. A system of trapping and handling live weasels in the field. *Journal of Zoology* **171**, 255–264.

Kirby P 1992. *Habitat management for invertebrates: a practical manual.* Sandy: RSPB.

Krebs CJ 1998. *Ecological methodology*, 2nd edn. New York: Addison Wesley Longman.

Langbein J, MR Hutchings, S Harris, C Stoate, SC Tapper & S Wray 1999. Techniques for assessing the abundance of brown hares, *Lepus europaeus. Mammal Review* **29**(2), 93–116.

Malloch AJC 2000. *VESPAN III: A computer package to handle and analyse multivariate species data and handle and display species distribution data.* Version 3.31 for Windows NT/95/98. Unit of Vegetation Science, University of Lancaster.

Mitchell-Jones AJ (ed.) 1999. *The bat worker's manual*, 2nd edn. Peterborough: JNCC.

Moore PD 1986. Site history. In *Methods in plant ecology*, 2nd edn, PD Moore and SB Chapman (eds). Oxford: Blackwell Scientific, 525–556.

Mueller-Dombois D & H Ellenberg 1974. *Aims and methods of vegetation ecology.* Chichester: Wiley.

New TR 1998. *Invertebrate surveys for conservation.* Oxford: Oxford University Press.

Pollard E 1977. A method for assessing changes in the abundance of butterflies. *Biological Conservation* **12**, 115–134.

Reading CJ 1996. *Evaluation of reptile survey methodologies: Final report, English Nature Research Report No. 200.* Peterborough: English Nature.

Southwood TRE & PA Henderson 2000. *Ecological methods,* 3rd edn. Oxford: Blackwell Science.

Strachan R, DJ Jefferies & PFR Chanin 1996. *Pine marten survey of England and Wales 1987– 1988.* Peterborough: JNCC.

Sutherland WJ (ed.) 1996a. *Ecological census techniques: a handbook,* Cambridge: Cambridge University Press.

Sutherland WJ 1996b. Mammals. See Sutherland (1996a), 260–280.

Sykes JM & AMJ Lane (eds) 1996. *Protocols for standard measurements at terrestrial sites.* London: TSO.

Thomas JA 1983. A quick method of estimating butterfly numbers during surveys. *Biological Conservation* **27**, 195–211.

Trout RC, SC Tapper & J Herradine 1986. Recent trends in the rabbit population in Britain. *Mammal Review* **16**, 117–123.

Usher MB 1990. Assessment of conservation values: the use of water traps to assess the arthropod communities of heather moorland. *Biological Conservation* **53**, 191–198.

Waring P 1994. Moth traps and their use. *British Wildlife* **5**, 137–148.

Yeo MJM, TH Blackstock & DP Stevens 1998. The use of phytosociological data in conservation assessment: a case study of lowland grasslands in mid Wales. *Biological Conservation* **86**, 125–138.

Glossary

The terms defined below are highlighted in **bold italic** the first time they appear in a chapter. Terms highlighted within definitions are defined elsewhere in the glossary.

abundance   See **species abundance**.

acid deposition   Dry deposition (gravitational settling, impact with vegetation) and wet deposition (scavenging by precipitation) of acidic substances such as sulphates and nitrates. It is often called **acid precipitation** or 'acid rain', but these terms strictly refer to wet deposition only.

Agenda 21   Encompasses the principles adopted at Earth Summit '92 (§1.4) for *sustainable development*. *Local Agenda 21* is a UK initiative for implementing this at the local level by local authorities.

air pollutants   Substances or energy (e.g. waste heat) in the atmosphere in such quantities and of such duration likely to cause harm to people, plants or animals, or damage to materials (e.g. fabrics) and structures (e.g. buildings), or changes in the weather and climate, or interference with the comfortable enjoyment of life or property (e.g. due to the effects of odours or noise).

air quality standard   The concentration of a pollutant, over a specified period, above which adverse effects on health (or the environment) may occur and which should not be exceeded.

algae   Primitive, mainly aquatic, unicelled or multicelled plants that lack true stems, roots or leaves. They include *phytoplankton*, filamentous 'pond scum' species and seaweeds.

algal bloom   Rapid growth of *algae* in water bodies, facilitated by high nutrient levels and/or other physical and chemical conditions. They may increase water turbidity and reduce dissolved oxygen levels at night and when the algae decay. Blooms of some algae and **cyanobacteria** may also produce toxins that may affect fish and other wildlife, and present a hazard to human health.

alkalinity   (a) the state when the *pH* of a solution is > 7; (b) more strictly the concentration of carbonates in water (its carbonate hardness) and hence its ability to resist (buffer) changes in pH, in which terms it is possible for water with pH < 7.0 to have high alkalinity and for water with pH > 7 to have low alkalinity. Values are often quoted in mg/l calcium carbonate but are better quoted in milliequivalents of acid per litre, i.e. the amount of acid needed to change the pH.

alluvial soil   A soil that has accumulated by deposition of water-borne *sediments*, e.g. from successive floods in a *floodplain*.

ancient woodland   Woodland that has existed continuously since at least AD 1600 (often much longer). It has normally been managed for centuries and, in addition to having a rich native fauna and flora, can provide a record of early settlements and of traditional practices such as coppicing, pollarding and charcoal burning.

anoxia   Complete lack, or a pathological deficiency, of oxygen.

anthropogenic   Generated and maintained by human activities.

aquifer   A stratum of porous or fractured rock that contains groundwater and allows this to flow through.

audit trail   A record of all analyses, decisions, etc. during a process such as EIA, to assist in (a) explaining how options were considered and why decisions were made, and (b) reviewing the study, e.g. if conditions change.

bioaccumulation   The process by which some pollutants accumulate in the tissues of living organisms.

bioamplification (biomagnification)   The increase in concentration of bioaccumulating pollutants along food chains, culminating in high concentrations in top carnivores. It is associated with the trend of decreasing *biomass* along food chains (see Fig. 11.3, p. 250).

bioassay   A method using the biological response of a species to test the toxicity of a pollutant.

biochemical oxygen demand (BOD)   The quantity of dissolved oxygen in water (mg/l) consumed (under test conditions) by microbial degradation of organic matter during a given period (5 days). It is one of the standard tests used to characterise *effluent* quality and measure organic pollution in surface waters, e.g. in the Environment Agency's General Quality Assessment (GQA).

biodiversity   The variety of life, globally or within any area – defined in the UN Convention on Biological Diversity (1992) as "The variability among living organisms from all sources, including terrestrial, marine and other aquatic ecosystems and the ecological complexes of which they are part; this includes diversity within species, between species and of ecosystems."

biomass   The amount of organic matter in a community's living organisms at a given time, usually measured as dry weight per unit area (e.g. $g\,m^{-2}$) or (in aquatic systems) volume (e.g. $g\,m^{-3}$).

biomes   The major climatic climax communities (Fig. 11.2) on a given continent, characterised largely by the vegetation and the governing climate. Similar biomes on different continents belong to global **biome types**, e.g. tropical rainforest, tundra. The principal biome types in the British Isles are: temperate deciduous forest (now represented by semi-natural broadleaved woodland); boreal (conifer and birch) forest (now represented mainly by the Caledonian pine forest); blanket bog; and alpine communities (on mountains above c.650 m).

biotransformation   The conversion by organisms (usually bacteria) of chemical pollutants to more toxic forms/compounds, e.g. of inorganic mercury to methyl mercury.

bryophytes   Mosses and liverworts.

buffer zones/strips   Permanently vegetated strips of land designed to manage various environmental concerns, e.g. (a) to intercept water-borne pollutants and hence protect groundwaters and surface waters; (b) to slow *runoff* and enhance infiltration (within the buffer), so stabilising streamflows; (c) to reduce soil and streambank erosion; (d) to provide visual/noise/odour screens and landscape features; (e) to protect wildlife habitats/sites from pollution and disturbance; and (e) to provide *wildlife corridors* and habitats/refuges. Buffer types include: wellhead protection zones, *riparian* buffers, grassed waterways, shelterbelts/

windbreaks/snowbreaks, contour strips, roadside verges and field borders. Further information, and an interactive program for selecting and sizing buffers, can be found at http://www.nhq.nrcs.usda.gov/CCS/Buffers.html.

carrying capacity Can have various meanings, e.g.: (a) the population size of a species (including human) which a given environment can support; (b) the ability of a habitat to support one or more given species; (b) the capacity of an ecosystem to tolerate a given stress such as pollution level.

catchment A drainage basin/area within which precipitation drains into a river system (and possibly into lakes and wetlands) and eventually to the sea. Catchment boundaries are generally formed by ridges, on different sides of which rainfall drains into different catchments. In the UK, these are usually called watersheds; but in the US, the term **watershed** is used in place of catchment.

climate The totality of the weather experienced at a given place. This is not simply 'average weather' since climate includes the extremes or deviations from the mean state of the atmosphere (e.g. the occurrence of fogs, frosts and storms). The behaviour of the atmosphere at a given place over periods of weeks, months, seasons, years and decades is its climate, i.e. the integration of its *weather* over long periods. The climate of a location is usually characterised using long-term records of, say, 30 yr.

competent authority General term for a decision-making body in the UK.

connectivity The degree to which habitat patches in an urban or agricultural matrix are interconnected by *linear habitats* and/or *stepping-stone habitats* between the main patches.

controlled waters Surface waters, groundwaters, and coastal waters (to three nautical miles out to sea) to which UK pollution legislation applies. They include virtually all fresh waters except small ponds and reservoirs (not used for public supply) that do not supply other waters. It is an offence, with certain exceptions, to cause or knowingly permit trade or sewage effluent, toxic pollutants, or solid matter to enter controlled waters without a *discharge consent* (see *designated waters*).

critical load An amount of one or more pollutants below which significant harmful effects on specified ecosystem components (e.g. sensitive species or vegetation) evidently do not occur. It is most commonly used for deposited air pollutants such as *acid deposition* (including that containing nitrogen). 'Exceedance' of critical loads in soils and waters may affect organisms directly, or indirectly, e.g. increased dissolved aluminium concentrations associated with acidity.

culvert A pipe or box-type conduit through which water is carried under a structure such as a road.

cyanobacteria Photosynthetic bacteria that have features similar to free-floating algae and are often called **blue-green algae**. They can be a problem in *algal blooms*.

denitrification The process of nitrogen removal from waterlogged soils by the action of denitrifying bacteria which utilise nitrate and release nitrogen gas (to the atmosphere).

design event An event such as a rainstorm or flood of given magnitude and probability, derived from past records. A **design rainfall/storm** can be formulated from depth–duration–frequency (DDF) data of past rainstorms, e.g. can be the maximum rainfall (mm) likely to occur at a location during a given period (e.g. 1 h or 24 h) within a specified *return period* – so a 50-yr, 1-h design storm is the maximum rainfall probable in a 1 h period within any 50 yr interval. Similar models can be constructed for maritime storms and storm surges.

designated waters   Water bodies or sections of river that are designated under one or more EU Directives, and must comply with the relevant water quality objectives (WQOs).

Development plans   Statutory documents produced by LPAs (under the *Town and Country Planning Act 1990*) outlining their strategies for development over a 10–15 yr period. They include County Structure Plans, Unitary Development Plans (UDPs) and Local Plans.

dewatering   Pumping of water to reduce the flow of groundwater into an excavation, or to reduce its pressure, e.g. to allow dry working for mineral extraction or deep foundations.

discharge consent   Statutory document issued by the EA (under schedule 10 of the *Water Resources Act 1991*) setting limits and conditions on the discharge of an **effluent** into a **controlled water**.

dominant species   The species of highest **abundance** or **biomass** in a community. It usually has a major influence, but is not necessarily a **keystone species**, e.g. may be replaceable by a similar species without significantly affecting the community organisation. A community may have two or more co-dominant species.

drawdown   Lowering of the water table or piezometric surface (Fig. 10.2) usually caused by **dewatering**, e.g. adjacent to mineral workings.

dystrophic   Very nutrient-poor soils, waters, or ecosystems; they also have low **pH** (< 5.5).

ecotoxicology   (strictly, The study of) the effects on living organisms of chemicals released into the environment.

effluent   Treated or untreated liquid waste material that is discharged into the environment from a point source such as a wastewater treatment plant or an industrial facility.

electrical conductivity (of an aqueous solution)   The 'ease' with which an electrical current passes through the solution. Conductivity increases with total ion concentration, and provides a measure of overall amount of solutes present – but gives no indication of the relative amounts of different solutes.

emission standard   The maximum amount or concentration of a pollutant allowed to be emitted from a specified source.

emissions inventory   An organised collection of data relating to the characteristics of processes or activities which release pollutants to the atmosphere across a study area.

environmental components (of an EIA)   The aspects of the natural or man-made environment (e.g. people, landscape, heritage, air, soils, water, ecosystems) that may be significantly affected by a proposed project, and are individually assessed in an EIA. They are **receptors**, but can also include many of these, e.g. individual species or buildings.

environmental impact statement (EIS)   The document that presents the findings of an EIA, including proposed mitigation measures, and is submitted (with the planning application) to the **competent authority** responsible for deciding if the proposal may proceed, which in the UK is normally the LPA.

ericoids   Shrubby plants mainly belonging to the heath family (*Ericaceae*), e.g. heather, heaths, bilberry, cranberry – typically the main constituents of heathland vegetation.

erosion   The wearing away of rock or soil by water, ice, wind, or chemical processes such as solution. Natural (geologic) erosion refers to natural processes occurring over long periods. Accelerated erosion refers to erosion that exceeds estimated naturally occurring rates as a result of human activities.

eutrophic Nutrient-rich soils, waters, or ecosystems, which also usually have high *pHs* (> 7).

eutrophicated Refers to an ecosystem that contains excessive nutrient levels.

eutrophication The process or trend of soil or water enrichment by plant nutrients – especially nitrogen and phosphorus. It can occur naturally, but usually refers to *anthropogenic* enrichment (sometimes called enhanced eutrophication) which can lead to excessive nutrient loading and consequent ecosystem degradation (see Table 12.2).

evapotranspiration Total evaporative loss from a land area, including evaporation from soils and surface waters, and *transpiration* (which is the major component in well-vegetated ecosystems).

field capacity The moisture content of a soil when water percolating downwards under gravity has drained out; usually expressed as cm^3 water per cm^3 soil.

floodplain (a) A **river floodplain** is the land adjacent to a watercourse over which water naturally spills and flows (unless prevented by flood defences) when floodwaters exceed the capacity of the channel, (b) a **coastal floodplain** is land adjacent to a coastline or estuary that is naturally inundated (unless prevented by coastal defences) by very high tides or (in the case of estuaries) by a combination of high tides and river flows.

food chain A major route of energy flow (in food) through a community, e.g. from green plants to herbivores and then carnivores (see Fig. 11.3). In reality, most communities have a complex network of feeding relationships between species, i.e. a **food web**.

forbs Generally broadleaved (non-grass-like) herbaceous flowering plants, usually having conspicuous flowers; often called 'herbs' or 'flowers' (see *graminoids*).

french drain A trench over a drainage line, backfilled with layers of material (coarse at the bottom and grading to fine-grained at the top) to act as a *sediment* filter; usually with a vegetated surface.

graminoids Grasses and grass-like plants, i.e. rushes and sedges (see *forbs*).

heath/heathland A habitat/vegetation type usually dominated by dwarf shrubs such as *ericoids*. European heathland is an *anthropogenic* community that was created by forest clearance (often in the bronze age) and maintained by grazing, fire, and the use of materials for fuel, etc.

heavy metals Metals with atomic weight > 63.5 and specific gravity > 4.0. Some (e.g. cobalt, copper, iron, manganese, molybdenum, zinc) are essential nutrients, although more than trace amounts of most are toxic, especially to some taxa. Others (e.g. silver, cadmium, mercury, lead) are highly toxic, and the term heavy metals is often restricted to these – which have an atomic weight > 100.

hydraulic conductivity The permeability of soil or rock, and hence the ease, and potential rate, of water flow through it.

hydraulics Processes and regimes of water flow (velocities, volumes, duration, frequency, etc.) in hydrological systems such as surface waters and groundwaters.

hydrophillous 'Water loving'; tolerant of wet conditions.

indicator species (or groups) Species, or groups of species, that can be used as biological indicators: (a) to define and identify community or habitat types, e.g. *ancient woodland* vascular plants (§D3.4) and high-constancy species of NVC communities (§F.4); (b) to assess the conservation value of habitats, e.g. threatened and protected species; or (c) to assess environmental/habitat quality and monitor change in this, e.g. lichens in relation to atmospheric pollution (§D.2) and invertebrate families in relation to river pollution (§10.7.3).

keystone species   A species having an important (perhaps vital) influence on the structure and functioning of a community, e.g. with a key role in a food web. Sometimes also used to include *indicator species* of habitat health/quality (Table D.1).

leachates   Solutes, including pollutants, in water (or a non-aqueous liquid) that has leached from a 'solid' matrix such as a soil or landfill (see *leaching*).

leaching   The removal of soluble nutrients and other chemicals from a 'solid' matrix – such as a soil horizon (see §9.3.2), whole soil or landfill – by water percolating through it.

leakages (economic)   The flows of money out of a national, regional or local economy, following from an initial injection of money into that economy. The most significant leakages are for taxation (direct and indirect), savings and improved goods and services.

life forms   Types of animals or plants characterised by their morphology (body form) rather than taxonomy, e.g. herbaceous plants (subdivided into *graminoids* and *forbs*) *or* woody plants (subdivided into trees, shrubs and climbers).

linear habitats   Linear (much longer than wide) features that support biological communities. They can be valuable habitats in their own right, and may also act as *buffer zones* and *wildlife corridors*. Examples include hedgerows, field margins, road and railway verges, habitat edges, woodland rides and fire breaks, transmission line routes, urban green belts, avenues of trees, ditches, streams, river corridors, and lake and coastal shorelines.

macroinvertebrates   Invertebrate animals that are large enough to be seen by eye or can be captured using a sieve of mesh 0.5–1.0 mm.

macronutrients   Nutrient elements needed by organisms in relatively large amounts, i.e.: (a) carbon, hydrogen, and oxygen – which plants obtain from air and water by photosynthesis; (b) calcium, iron, magnesium, nitrogen, phosphorus, potassium and sulphur – which plants obtain from soil (although some have root nodules in which nitrogen-fixing bacteria assimilate gaseous nitrogen).

macrophytes   Plants large enough to be seen by eye (as opposed to microscopic).

meadow   Grassland maintained primarily for hay, often on poorly-drained land. Meadows are usually species-rich, partly as a result of traditional management which involves taking one late hay crop and then introducing grazing stock until winter or early spring, with no use of artificial fertilisers or *pesticides* (see *pasture*).

mesotrophic   Refers to soils/peats, waters, or ecosystems with nutrient levels intermediate between *eutrophic* and *oligotrophic*, and usually near-neutral *pH*.

microclimate   The climate associated with very localised factors such as topography, aspect, soils, waterbodies, vegetation and buildings. Microclimates may differ quite markedly from meso- (small-area/region) and macro- (large-region) climates.

microhabitat   A small habitat, with localised environmental conditions and resources, within a larger habitat, e.g. in small patches or vegetation layers (canopy, etc.).

micronutrients (trace elements) Nutrient elements needed by organisms in small quantities, e.g. boron, chlorine, copper, manganese, molybdenum, and zinc. Some (e.g. copper) are toxic if present in more than small amounts (see *macronutrients*).

multiplier A measure of the scale of the increase in income or employment in a local, regional or national economy resulting from an initial injection of an amount of money into that economy.

Natural Areas Areas (97 terrestrial and 23 maritime) of England defined by EN as "biogeographic zones which reflect the geological foundation, the natural systems and processes, and the wildlife in different parts of England, and provide a framework for setting objectives for nature conservation".

niche separation The mechanism by which competition between cohabiting species in a community is minimised by the divergence of **ecological niches**. Each species has a niche that determines (a) how it utilises the habitat resources, and (b) its role in the community. The niches of all species evidently differ at least slightly in one or more ways, e.g. trophic (eating different foods), spatial (e.g. in different *microhabitats*) and temporal (e.g. active during the day or night).

non-labile organics Organic compounds that are resistant to decay, as opposed to labile organics that are easily degraded in the aquatic environment.

non-point source (diffuse) pollution Pollution that cannot be attributed to discharges at specific locations. Typical causes are *runoff* to surface waters, or percolation of *leachates* to groundwater, from farmland, roads, urban and industrial areas, or many minor point sources (e.g. land drains, leakages from sewers, etc.). It is generally more difficult to control than *point source pollution*.

oligotrophic (nutrient-poor) Refers to soils/peats, waters, or ecosystems with low nutrient levels, and usually low *pH* (5.5–7).

pasture Grassland maintained primarily for and by grazing, and on which grazing stock is kept for a large part of they year (see *meadow*).

pesticide Defined under the *Food and Environment Protection Act 1985* as "any substance, preparation or organism used for destroying any pest" – including herbicides, fungicides, insecticides, molluscicides, rodenticides, growth regulators and masonry and timber preservatives.

pH Scale of 0–14 defining the acidity/*alkalinity* of solutions including those in soils and water bodies; 0 = extremely acid, 14 = extremely alkaline, and 7 = neutral (although soils and waters with pHs between *c*.6.5 and *c*.7.5 are often referred to as neutral).

phytoplankton The 'plant' component of *plankton*. They are the primary producers (Fig. 11.3) of open-water communities.

plankton The usually small (often microscopic) freshwater or marine 'plants' (*phytoplankton*) and animals (*zooplankton*) that are suspended in, and drift with, a waterbody.

point source pollution Pollution from specific locations such as (a) sewage outfalls and industrial *effluent* discharge points into surface water, and (b) *wells* or the bases of quarries and disposal sites into groundwater. It is generally easier to control than *non-point source pollution*.

pollutant see *pollution*.

pollution Any increase of matter or energy to a level that is harmful to living organisms or their environment (when it becomes a **pollutant**). It thus includes

physical pollution (e.g. thermal, noise and visual) and **biological pollution** (e.g. microbial or by non-native plants and animals), but most commonly refers to **chemical pollution**. Chemical pollutants can be: (a) man-made compounds such as *pesticides*; (b) toxic chemicals, such as *heavy metals*, harmful levels of which are not normally present in ecosystems; or (c) normally benign or even essential substances such as nutrients, either because these are *micronutrients* that are toxic in more than trace amounts, or because of excessive nutrient loading (*eutrophication*).

population equivalent   Unit used to quantify populations served by sewage treatment works (STWs). A single population equivalent (pe) is the organic biodegradable load having a five-day *biochemical oxygen demand* of 60 g per day (approximately the load from a single person's domestic waste).

precautionary principal   An approach that takes avoiding action based on the possibility of a significant environmental impact before there is conclusive evidence that it will occur.

project alternatives   Options that should be considered in a project proposal, including: location/siting; alignment of linear projects; design (scales, layouts, etc.); processes; procedures employed during the construction, operational and decommissioning phases; and the 'no action' option that the project should not go ahead. Assessment may result in the selection of **preferred options**.

quadrat   Strictly a four-sided (usually square) sampling plot, but can include shapes such as circles. Quadrats can be any size, e.g. from portable frame quadrats (usually $\leq m^2$) to national grid squares.

receptor   Any component of the natural or man-made environment that is potentially affected by an impact from a development.

return period/interval   A period within which there is a given probability/risk of a *design event* occurring. For instance, a 1-in-100 year event is likely to occur once in any 100-year period. Return periods are based on long-term average time intervals between past (recorded) events, and it is statistically possible for a 1-in-100 year event to occur more than once within a year (or shorter period) or not for several hundred years – so they are often expressed as '% chance', e.g. a 1-in-50 year event has a 2% chance of occurring in any one year, a 45% chance of occurring within any 30-year period, and a 76% chance of occurring within any 70-year period.

riffle   Area of a stream/river with a rocky or gravel substrate and shallow, turbulent, fast-moving water.

riparian   Relating to the banks of streams/rivers (sometimes also used to refer to the shorelines/fringes of standing waterbodies such as lakes).

river corridor   A river and the adjacent land (that has physical, ecological and visual links to the watercourse) considered together as a linear feature of conservation importance. It can be taken to include the river *floodplain*, but is often restricted to the channel, banks and narrow strips of adjacent floodplain land (see *riparian*).

runoff   The part of precipitation that flows as surface water from a site, catchment or region and eventually reaches the sea. It is effectively the excess of rainfall over *evapotranspiration*, making allowance for storage in surface, soil and groundwaters, and excluding groundwater seepage. Most runoff occurs in streams/rivers, and the term is often restricted to this.

screening (in the EIA process) Examination of a development proposal to determine if, under the EIA regulations, it: (a) is a Schedule 1 project requiring mandatory EIA; (b) is a Schedule 2 project and hence qualifies for a discretionary EIA; or (c) does not require a formal EIA but should be subject to an informal environmental assessment.

secondary treatment see *sewage treatment levels*.

sediments Organic or inorganic material that has precipitated from water to accumulate on the floor of a waterbody, watercourse or trap – or as *alluvial* deposits on a *floodplain*. It commonly consists of *silt*, but can include coarser particulates and material such as calcium carbonate that has precipitated through chemical reaction. Suspended particulates that have not yet undergone sedimentation are usually called **suspended solids** or (incorrectly) **suspended 'sediments'**.

seed bank The accumulation of viable seeds in a soil (mainly the top 40 cm) which may germinate if conditions become suitable – often when the soil is disturbed.

semi-natural (ecological system or landscape) A habitat, ecosystem, community, vegetation type or landscape that has been modified by human activity – but largely consists of, or supports, native species (and/or has relatively undisturbed soils, waters and geomorphological features) – and appears to have a similar structure and functioning to a natural type. Very few completely natural systems now exist, so conservation is largely concerned with protecting semi-natural systems.

sewage treatment levels (in the UK) **Primary**: usually physical treatment to (a) remove gross solids and (b) reduce suspended solids by c.50% and **Biochemical Oxygen Demand** by c.20%. **Secondary**: biological treatment to significantly reduce suspended solids, BOD and ammonia. **Tertiary**: additional treatment, e.g. nutrient removal/stripping or ultra-violet treatment to kill pathogenic bacteria.

silt Fine particulate organic and inorganic material; strictly with an average particle size intermediate between sands and clays (see §9.3.1) but often taken to include all material finer than sands.

siltation trap A hard-lined stilling well/basin with inflow and outflow pipes for drainage water; designed to slow the flow sufficiently for collection of fine suspended solids by sedimentation.

soil moisture deficit (SMD) State when the soil moisture content is below *field capacity*, usually expressed in mm (rainfall equivalent) to indicate the amount of rain needed to cancel the deficit. SMDs develop during periods when *evapotranspiration* (ET) exceeds precipitation (Pn) and can be estimated by simple accounting based on ET – Pn values (mm) for weekly periods. As SMDs increase, the availability of soil water to plants decreases and they tend to wilt.

species abundance The 'amount' of a species in an area or community, expressed by a quantitative measure such as number, density, cover or *biomass* (see Table G.1).

stepping-stone habitats Small sites/habitats that may be scattered and apparently isolated in a landscape, but which may assist in the migration or dispersal of species by providing **staging posts** between larger sites/habitats. Larger staging post sites are also needed by long-distance migrants such as migratory birds, especially at points along their regular migration routes.

sustainable development Defined in the 1987 Report of the World Commission on Environment and Development (the Brundtland Report) as "Development that meets the needs of the present without compromising the ability of future generations to meet their own needs".

synergism   The mechanism by which the combined effect of two or more pollutants is greater than the sum of their separate effects, i.e. the effect of one is exacerbated by another.

transpiration   Evaporative loss of water from plants. When the plants are 'in leaf', it is normally the largest component of *evapotranspiration* from well-vegetated ecosystems, and can return > 50% of precipitation water to the atmosphere.

turbidity   The opacity of (and hence the degree of light attenuation in) water, due to the presence of suspended matter and *plankton*. High turbidities are harmful to aquatic life.

vascular plants   'Higher' plants which transport water and nutrients in a specialised structural system that is not present in simple (non-vascular) plants such as *bryophytes*, algae and lichens. They include angiosperms (flowering plants), gymnosperms (mainly conifers) and pteridophytes (ferns, horsetails and clubmosses).

vice-counties   System of county-like areas, covering the British Isles, which are often used for biological recording. Many have boundaries similar to those of the administrative counties.

visual amenity   The popularity of an area, site or view in terms of visual perception.

weather (in a given place)   The condition of the atmosphere at a given time with respect to the various elements, e.g. temperature, sunshine, wind, precipitation. Refers to the behaviour of the atmosphere over a few hours or at most over a few days (see *climate*).

weathering   The physical and chemical breakdown of geological materials which contributes to soil formation.

well   Strictly a hand-dug shaft to a groundwater body, but used in the text to include **boreholes**, which are constructed by machinery, and are usually deeper but smaller in diameter than traditional wells. Both are used for abstraction and observation of groundwaters including water table levels and water quality.

wildlife corridors   Linear habitats/landscape features, such as *river corridors*, hedgerows, field margins and roadside verges, that may increase *connectivity* by acting as routes between habitat patches, and hence: (a) increasing the overall extent of habitat for animals with large range requirements; (b) facilitating migration or dispersal of species between habitats; (c) facilitating access to, and hence colonisation of, new habitats. Together with *stepping-stone habitats* they may be (a) particularly important in areas in which there is severe habitat fragmentation, and (b) the only remaining wildlife habitats in urban or intensively cultivated areas.

zooplankton   The animal component of *plankton*, many of which graze on *phytoplankton* and are thus equivalent to the herbivores of terrestrial communities.

Index

Bold page numbers indicate where terms and concepts are defined/explained.
* indicates that the entry is also indexed under individual environmental components.
and: indicates relationships between the entry and the topics subsequently listed